Introduction to
Live Sound Reinforcement

The Science, the Art, and the Practice

Teddy Boyce

Copyright

FriesenPress

Suite 300 - 990 Fort St
Victoria, BC, V8V 3K2
Canada

www.friesenpress.com

ISBN
978-1-5255-6508-3 (Hardcover)
978-1-5255-6509-0 (Paperback)
978-1-5255-6510-6 (eBook)

1. TECHNOLOGY & ENGINEERING, ACOUSTICS & SOUND

Distributed to the trade by The Ingram Book Company

Table of Contents

Preface

Imagine a crowd of thousands, stretching perhaps over a hundred meters, gathered to hear a popular band of acoustic musicians live on stage. Without some form of sound amplification, those people in the audience further than a few meters from the stage would hear no intelligible and useful sound, if any sound at all, from the band. Live sound reinforcement, the science and art of taking soft sounds and amplifying them hundreds of times in order to cover a large space which the unamplified sources could never cover, is the most effective way to solve this problem. Many people may be involved in delivering the live sound experience, but there is no doubt that, to a great degree, the success of the performance and the level of audience satisfaction lie in the hands of the live sound engineer.

The first time that I really paid attention to the sound at a live performance, I found that it was both inspiring and enlightening. There was that certain power in the sound—a power to evoke so many reactions from the audience. But I was looking beyond the performers and looking to the engineer behind the mixer. And I wondered—what would it be like to be that guy? With interest so piqued, I put together a modestly sophisticated sound system and started volunteering to mix sound for one of the up and coming youth groups in my hometown. At the same time, I realized that I knew very little about the technical and practical aspects of engineering live sound. I read everything I could get my hands on, but always wanted more. While I was able to gather what to do, and to a great extent, how to do it, I found that too many times I was left asking myself why. Over the years, as I gained experience in mixing and managing the delivery of live performances, I also started to create courses to teach others what I had learned. After nearly two decades, I had amassed enough knowledge and material to write the book I wish I had found when I first started out in this field: a book with enough detail but yet not overwhelming, and that presented a well-structured approach to the study of live sound engineering. I believe I have succeeded in this objective, as this book continues to be adopted by many educators as the basis to teach the subject matter to aspiring engineers.

The book can be divided into three parts. The first twelve chapters present a detailed study of many audio concepts, as well as operational descriptions of many sound reinforcement tools. The book starts with a review of fundamental audio concepts of sound and interaction with the environment, before moving on to examine the design and operation of microphones, loudspeakers, amplifiers, effects processors, and mixers. Additionally, a thorough examination of the workings of digital audio, digital processors, digital mixers, and digital networks is presented. Review questions at the end of every chapter help to reinforce the newly encountered concepts.

The second part of the book goes on to present practical applications of the live sound reinforcement tools. The expectation is that by that stage, the reader will be able to see why things are done in a certain way, having acquired knowledge of fundamental audio theory and a behind-the-scenes look at the functioning of various components in the sound reinforcement domain. The reader will learn such things as optimal house and monitor loudspeaker placement, microphone-loudspeaker interaction, determination of amplifier requirements, and much more while designing an entire live sound system. Many examples of the proper use and application of various tools are given.

Finally, the last three chapters are devoted to the planning and execution of the pre-performance, performance, and post-performance phases of a live show from the audio engineer's perspective.

New in the Second Edition

The second edition of this book has numerous minor improvements as well as a number of major updates and new sections. In Chapter 1, the discussion around phase and polarity has been completely rewritten and expanded, while in Chapter 13, I have added new discussion on loudspeaker and PA management systems. Two new chapters have been added to cover digital networks and their implementation, as well as look at the different approaches to digital audio distribution and a comparison of popular distribution protocols. The glossary has also been greatly expanded.

Acknowledgements

I would like to thank Monique Boyce, Regius Brown, Helena Dang, Augusto Lima, Ilya Panfilov, and Donovan Wallace for reviewing various sections of the manuscript, providing valuable feedback on its content, for asking challenging questions, and running simulations and tests of various scenarios. Their help over the course of this project is much appreciated, and allowed me to create a book that is more than what I could have achieved on my own.

Teddy Boyce

Chapter 1

Fundamentals

1.1 What is Sound Reinforcement?

Have you ever been to a presentation or meeting where you were so far from the presenter that you could only hear and make out every few words or so? "The ... and ... men ... fire." Very frustrating, and you are left guessing what the speaker is saying most of the time. This is a good example of where sound reinforcement could be used to solve this very basic problem. Sound reinforcement is the process of manipulating and amplifying sound for the purpose of allowing a low-level source to be heard and understood in and over an area that the source, by itself, would be incapable of covering adequately. Every sound reinforcement system is made up of many individual components, connected together, so that working as a whole they can accomplish the purpose of sound reinforcement.

A live sound reinforcement system is used primarily to amplify sound in a live environment, i.e., one where a large percentage of the sound source is not prerecorded and the environment itself is subject to changes during the process of sound reinforcement. The sound system must be able to do its job and overcome the problems of this environment. To be useful, all live sound reinforcement systems must satisfy at least the following primary requirements:

- **Intelligibility** – Can the audience hear and understand? If the audience cannot understand what is being reproduced, then the system is basically useless.
- **Fidelity** – Does it faithfully reproduce sound from input to output?
- **Power** – Is the level of output sound high enough to cover the intended audience? If there is too little power, intelligibility in the distance will suffer as ambient noise overcomes the source material.
- **Monitoring and Controllability** – Can the operator adequately listen to and adjust what the system is doing? A live sound system accomplishes much of its intelligibility requirements by being highly controllable. This means ability to perhaps move speakers around, move microphones to the proper position to sample the sound, adjust levels of all sources independently, and apply equalization selectively.

If the live sound system is intended to be moved from event to event, then other secondary requirements would be portability, longevity, and flexibility. The system must be rugged; and equipment should not be easily damaged during transportation.

1.2 Sound System Model

Models are used to simplify the analysis and presentation of a complex subject. Our sound system model as shown in Figure 1-1 is very basic. We start with such a basic model because it allows us to focus on the essentials that are required to understand how we create and interact with our sound system. The model shows us that we have five blocks. Devices that capture the sound and get it into a form that the system can process are called input transducers. Once we have the sound in the system, we use signal processors to manipulate the signals that represent the sound. When we are done transforming these signals to meet our needs, we use devices called output transducers to turn those signals back into actual sound again. We refer to these three blocks as the core blocks of the system. The arrows connecting the core blocks show the flow of sound signals. This process all happens in the environment in which these conversions take place, i.e., the space in which the system operates, and so we must somehow take that into account. To complicate matters further, we need to factor in how we as humans perceive the sound (the psychoacoustic interface) because we do not always fully hear what the sound waves are actually carrying. We notice that this model has three parts that are fully under our control (the core blocks) and two parts that are not. Nonetheless, any sound system that we actually put together will follow this model and should attempt to account for all the parts. All of the sound reinforcement issues we talk about in this book will basically be in the context of this model.

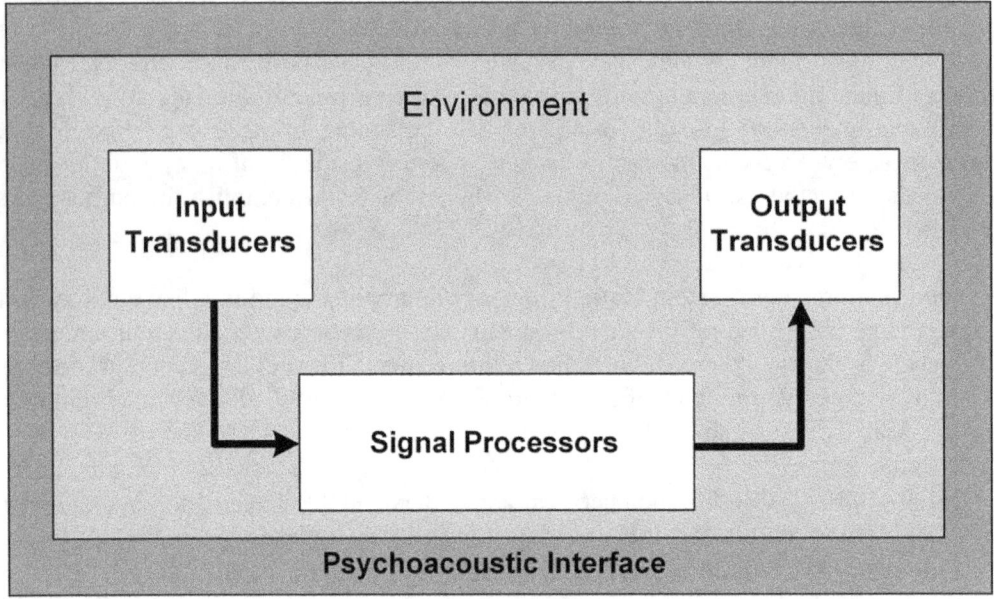

Figure 1-1   Basic Sound System Model

For example, consider Figure 1-2 which shows a high-level schematic of the core blocks of a sound system. It basically fleshes out these three parts of the model and shows the type of equipment that may be involved in each block. Input transducers usually consist of microphones. Microphones convert sound energy to electrical energy. This is done so that we can process the sound more effectively. Output transducers are loudspeakers. Loudspeakers convert electrical energy into sound energy after it has been manipulated by the signal processors. The major processors in a live sound system are the mixer, front of house (FOH)[a] amplifier, and the stage monitor amplifier. Other processors consist of effects that help us shape the sound—equalizers, reverb, compressors, and others. Again, the arrows in the figure show us the direction of signal flow and represent audio cables in the real world. As Figure 1-2 only deals with the core blocks of the system, the environment is not represented there. We will deal extensively with the environment in Chapter 2, and we will highlight important psychoacoustic issues as we encounter them as we proceed through the book.

---

[a] The term *front of house* (FOH) refers to equipment or resources whose primary function is to provide audio coverage to the main audience area.

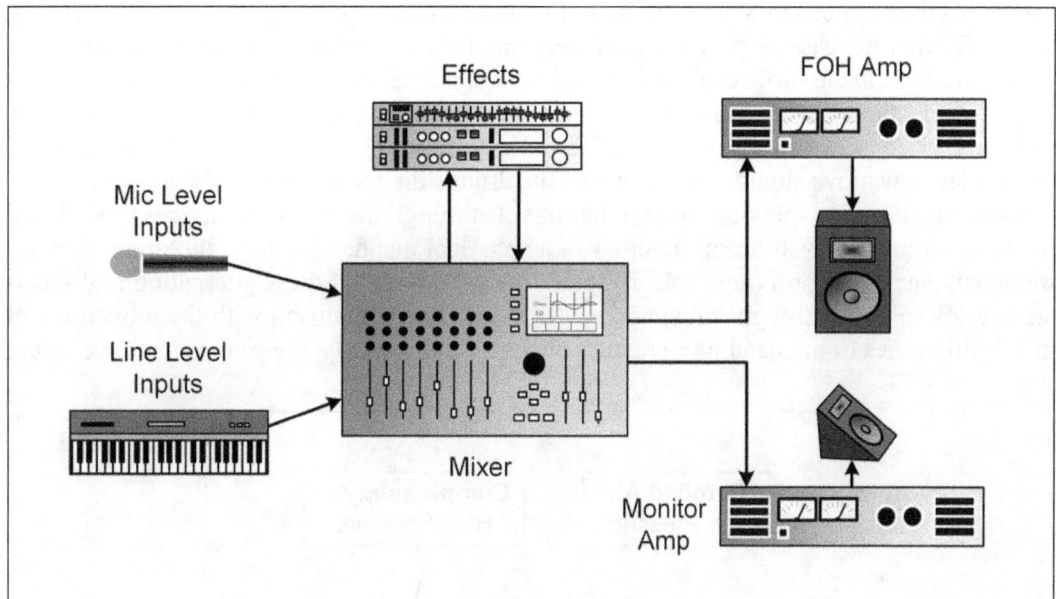

Figure 1-2 Basic Sound System Schematic

Practically all basic sound reinforcement systems use a layout similar to that shown in Figure 1-2. We can summarize this layout and signal flow as follows:

- Inputs (microphones, keyboards, etc.) feed into the mixer.
- The mixer sends and receives signals to and from effects processors.
- The mixer sends signals to the front of house amplifiers.
- The mixer sends signals to the stage monitor amplifiers.
- The amplifiers send signals to the loudspeakers.

1.3 What is Sound?

When we think of sound, we think of something we can hear. We hear because vibrating air particles impinging on our eardrums transfer energy from the air to our eardrums, causing it to move. This motion is transformed further into electrical impulses that our brain can understand and interpret as sound. Sound is also produced by vibration of matter other than air. For example, sound can be heard when under water, and sound travels through solid objects. However, in this book, we will restrict ourselves to considering vibrations in air.

While air particles always exhibit small random vibrations, they just don't start vibrating wildly in sync on their own. Something typically disturbs the air and causes its particles in turn to vibrate. For example, when a drum is struck and its skin vibrates, it disturbs the air in contact with it. This disturbance travels through the air for long distances. Note that the air itself is not traveling for long distances (that would be wind), but rather the disturbance is traveling through the air. This happens because an air molecule that is disturbed will in turn disturb others that are in close proximity to it. This can be compared to a person in a crowd moving around, jostling other persons as they move.

Let us take a closer look at this process as illustrated in Figure 1-3. If an object moves rapidly in air, it will compress the air just in front of it and cause the air just behind it to be rarefied. When the drum skin moves outwards, it pushes on the air molecules in front of it. Because air is not rigid, this squeezes the air molecules in that vicinity together, thus compressing the air and increasing the air pressure. This increase in pressure (a disturbance) creates a high-pressure crest that starts to travel outwards from the surface of the drum skin in a spherical manner. When the drum skin moves rapidly inwards, the opposite situation obtains, with a slight void created between the skin and the air in front of it. This causes the pressure of the air next to it to be reduced—the air is rarefied. The air molecules now move slightly back towards the drum skin, trying to fill the void. This reduction in pressure creates a low-pressure trough that travels through the air in the same way as the high-pressure disturbance—away from the drum in a spherical manner. Because the drum skin moves continuously back and forth (until it is damped), we get a series of these alternating high-pressure crests and low-pressure troughs moving away from the drum in unison with the movement of its skin. It is this series of high and low-pressure changes of the air that create the sound we hear.

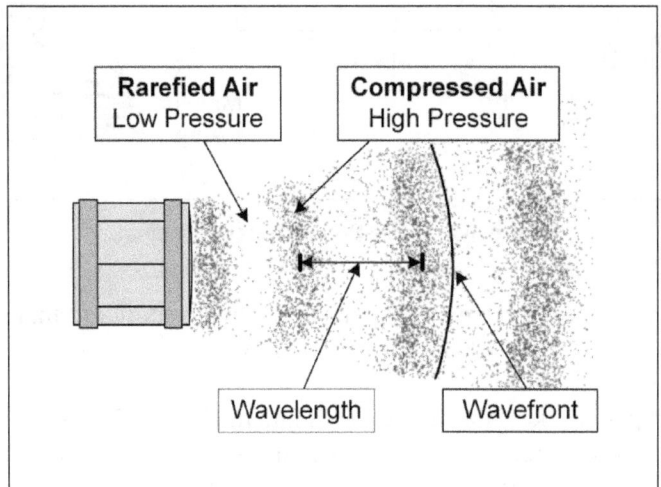

Figure 1-3 Vibrating Drum Skin Creates Sound Waves in Air

1.4 Sound Waves

So now we know that sound is caused by a disturbance that is propagating through the air. This disturbance travels in the form of a wave. As shown in Figure 1-3, the advancing surface of this wave is called the *wavefront*. Because the wavefront can be thought of as an expanding spherical surface, the wave is sometimes referred as a plane wave. It is the sound wave that transports sound energy through the air and allows us to hear.

Sound waves repeat themselves and are called *periodic*. In a periodic wave, the smallest part of the wave that can be identified as repeating in its entirety (i.e., when we get to the end of the repeating section, it starts over at the beginning of the section) is called a *cycle*. The length of time that it takes for a cycle to complete is called the *period* of the wave.

The *wavelength* is the physical length of the cycle. It is measured in meters (m). The *frequency* of the wave is the number of cycles occurring per second. It is measured in hertz (Hz).

The speed at which sound waves travel in air is about 340 m/s at room temperature, but it varies quite a bit with changes in temperature and humidity. We will discuss that in more detail later. There is a relationship between frequency (f), wavelength (l), and speed of sound (c). It is given by:

$$l = \frac{c}{f}$$

If we take the speed of sound to be fixed at any instant, then we can see that as the frequency of the sound increases, the wavelength decreases. Table 1-1 shows the relationship between frequency and wavelength of sound waves for a selected set of frequencies. Notice the large difference in wavelength at the opposite ends of the spectrum. A 20 Hz sound wave is over 17 m long, while a 20 kHz sound wave is less than 2 cm.

Frequency	Wavelength
20 Hz	17.2 m
50 Hz	2.3 m
400 Hz	85.8 cm
1 kHz	34.3 cm
2 kHz	17.2 cm
4 kHz	8.6 cm
10 kHz	3.4 cm
20 kHz	1.7 cm

Table 1-1   Relationship between Frequency and Wavelength

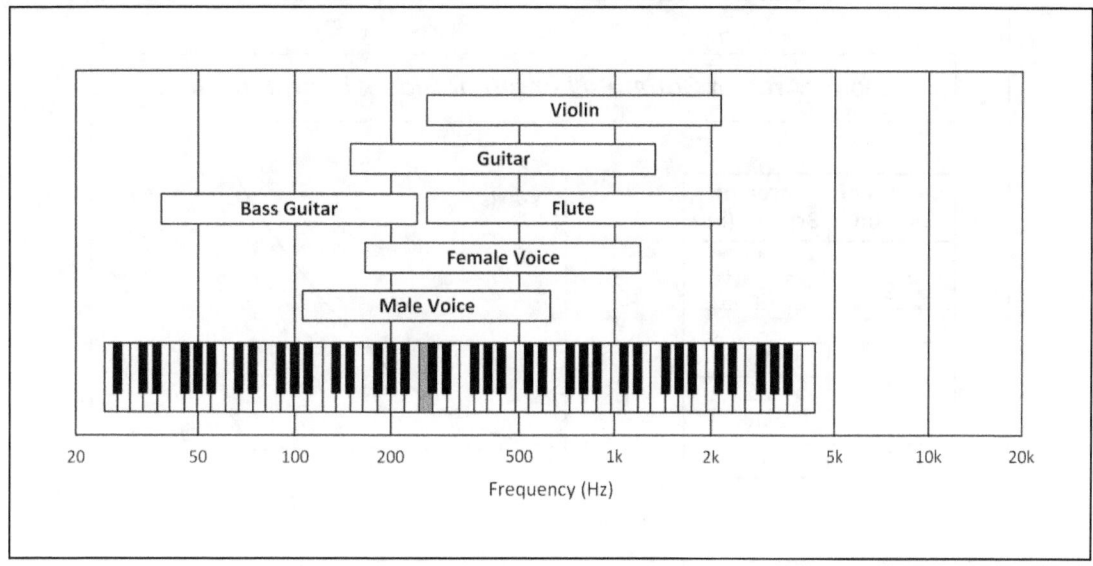

Figure 1-4   Fundamental Frequency Ranges of Some Common Sources

Figure 1-4 shows the fundamental frequency ranges of the typical human voice, as well as a few common instruments. On the piano keyboard, the gray key is middle C, with a frequency of 261.63 Hz. As can be seen, the fundamental frequencies do not get down extremely low, nor do they go very high. The harmonics (i.e., additional sonic energy generated when a note is sounded, with frequencies at integer multiples of the fundamental frequency) of even the lowest notes of these instruments and voices go well into the kilohertz range, however. The range of human hearing is commonly taken to be 20 Hz to 20 kHz. However, as humans get older, the range of hearing is reduced, especially at the high end of the spectrum. Many older people cannot hear sound above 16 kHz or below 25 Hz.

1.5   Graphical Representation of Sound Waves

It is common to view sound waves in a graphical form. This makes visualizing and analyzing them easier and aids our understanding of their behavior. Before looking at such representations of sound waves, let us review the fundamentals of graphs in general.

A two-dimensional (2-D) graph is used to pictorially display the relationship between two quantities—say X and Y. When one quantity Y depends on, varies with, or is derived from the second quantity X, we say that Y is a function of X. X and Y are called variables because their actual values can vary from instant to instant. For example, the area of a square depends on the length of its sides, and is equal to its length multiplied by its width (which for a square are equal). If Y represents the area of the square and X represents the length of its side, then we can write the area function as:

$$Y = X \times X$$

Once we write down the function, we can always figure out the value of Y for any given value of X. We could even make a table of values of Y that correspond to some particular values of X, as shown on the left in Figure 1-5. These numbers may contain an interesting pattern in the relationship, but it may not be obvious because all we see is a bunch of numbers when we look at the table. Consequently, we use graphs to visually show us the behavior of a function and make obvious these patterns in the relationship.

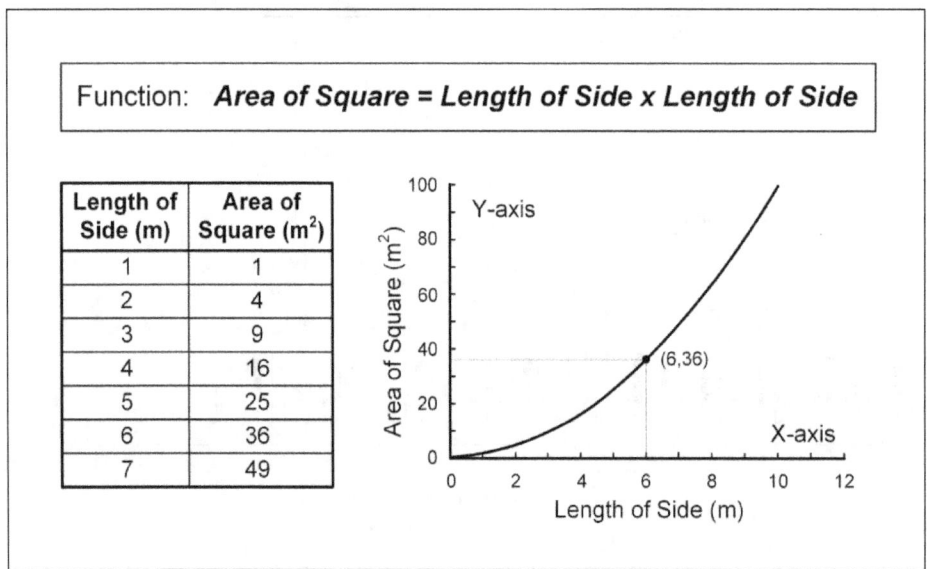

Figure 1-5   Function – Tabular and Graphical Display

The right side of Figure 1-5 shows that a 2-D graph is drawn as having two axes on a plane (like a sheet of paper). In a rectangular system, one axis is drawn in the horizontal direction and the other is drawn in the vertical direction. It is customary to have values of the independent variable X represented on the horizontal axis (the x-axis) and values of the dependent variable Y represented on the vertical axis (the y-axis). In the rectangular space above the x-axis and to the right of the y-axis, we can represent the position of any point by the coordinates (x-value, y-value). The point where the two axes intersect is called the origin of the rectangular coordinate system and has coordinates of (0, 0). To represent a function pictorially on the graph, we draw a line in this rectangular space so that for all points (x, y) on the line, when we choose any value of X, the correct value of

Y is determined as satisfying the function. Thus, in Figure 1-5 we highlight the example where the length of the side of the square is 6 m, and we see that the graph shows us that the corresponding area is 36 m^2.

In Figure 1-5, all of the derived values of *Y* are positive for all values of *X*, but this may not be the case for other functions. Sometimes the derived *Y* values can be negative or below the normal value represented by the *Y*-origin. In such cases, we extend the y-axis below the origin so that these negative values can be represented. In Figure 1-6, we see an example of this.

Figure 1-6 shows the graphical representation of a sound wave and how it relates to the pressure changes in air. This graphical model is used to simplify the presentation of information pertinent to the sound wave. The graphical wave model is drawn as a transverse wave and bears no resemblance to the actual sound wave in air (which is a longitudinal wave), but since this is only a model, that is perfectly ok.

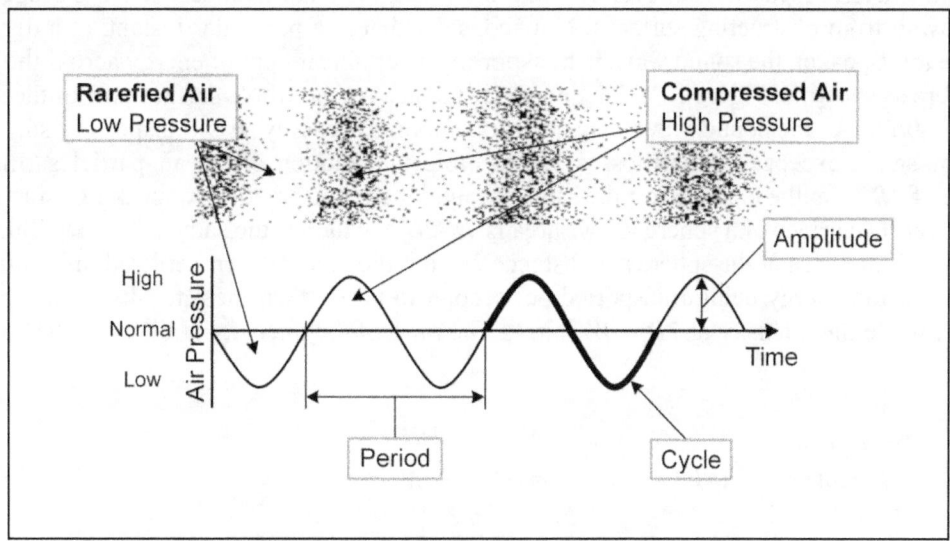

Figure 1-6   Graphical Representation of Sound Wave

Let us take a closer look at this model. The y-axis in our graph is used to represent air pressure, but in other models can be used to represent sound pressure level, intensity, loudness, or some other attribute of the sound wave. The high-pressure (relative to the normal undisturbed air) portions of the wave are shown as points above the origin, while the low-pressure portions of the wave are shown as points below the origin. The *amplitude* of the wave is the maximum displacement (positive or negative) from the normal undisturbed position. Amplitude is related to the intensity and loudness of the sound, which we will discuss below. On the x-axis of the graph, we show time, increasing to the right. Thus, the graph shows a picture of how the pressure component of the sound wave behaves with the passing of time. One cycle of the wave is highlighted in bold on the graph, and other elements, such as period, are also readily apparent.

1.6   Intensity of Sound

In Sections 1.3 and 1.4, we suggested that sound was a form of energy and that energy is transported through the air via sound waves. We also noted above that the amplitude of the wave was related to its intensity. In fact, the intensity of sound is proportional to the amplitude of the sound wave.

That means that waves with bigger amplitude transport more energy and possess the ability to deliver louder sound. Intensity is a physical quality and is actually a measure of the amount of energy that the wave is transporting through a given cross-sectional area of space in a specific time. This is the same as sound power. To better understand the concept of intensity, imagine that we have a window in a wall, with a sound source on one side. Let's say the window measures 1 m by 1 m; i.e., it has an area of 1 square meter. Now we open the window and measure how much energy the sound wave is moving through that open window in one second. That measurement gives us the intensity, or sound power, of the sound wave at that point. Intensity is measured in watts per square meter (W/m^2).

1.7 Inverse Square Law for Sound Wave Propagation

When we were discussing how sound waves develop from the vibrations of a drum skin, we stated that they expand in a spherical fashion. This is the case for any point source, e.g., a loudspeaker. The sound waves expand like an ever-growing ball. Imagine that this loudspeaker is suspended in midair away from obstructing surfaces. Now consider that at a particular instant, at a distance **D** from the loudspeaker, the sound wave is transporting a certain amount of energy across the surface of a sphere as shown in Figure 1-7. That energy is spread over the whole surface area of the sphere,[b] which is **4πD²**. As the sound wave expands, the amount of energy it is transporting stays pretty much the same, except for small losses due to molecular motion of the air particles. Thus, at a distance of **2D** from the loudspeaker the same amount of energy, practically speaking, must now be spread over the surface of a sphere whose area is **4π(2D)²**, which is the same as **16πD²**. This is four times the surface area of the sphere at a distance **D** from the loudspeaker, depicted in Figure 1-7. If we consider the energy being transported per second to be **W**, then the intensity at distance **D** is **W/4πD²**, while the intensity at **2D** is **W/16πD²**. The ratio of the intensity at distance **2D** compared to **D** is:

$$\frac{W}{16\pi D^2} \div \frac{W}{4\pi D^2} = \frac{W}{16\pi D^2} \times \frac{4\pi D^2}{W}$$

$$= \frac{4}{16}$$

$$= \frac{1}{4}$$

This means that at distance **2D**, at any given location on the surface of the sphere, the intensity of the sound is four times less than at distance **D**. This four-times reduction in intensity with doubling of distance from the source is called the inverse square law for sound propagation. More generally, we see that the intensity varies inversely with the square of the distance from the sound source, leading to very rapid falloff in intensity as we move away from the source. This has serious implications for sound systems using only loudspeakers at the front of house position and trying to cover a large volume venue. If the level is set so that it is correct for the people at the front of the audience, then those at the back may have trouble hearing. If it is set so that those at the back can hear well, then it may be uncomfortably loud for those at the front.

---

[b] The surface area of a sphere is equal to **4πr²**, where **r** is the radius of the sphere, and π is a constant equal to 3.1416.

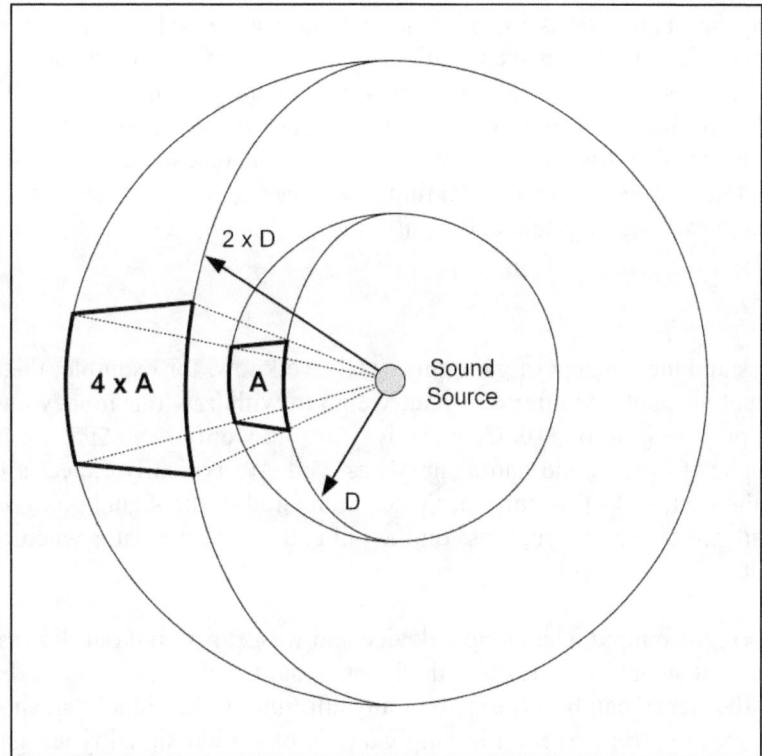

Figure 1-7   Inverse Square Law for Sound Propagation

Do not confuse the reduction of intensity due to the inverse square law with the reduction of intensity due to air particle motion. As the sound wave travels through the air, some of its energy is robbed by making the air particles move back and forth, and this causes the sound to become weaker and weaker. Eventually, the energy left in the sound wave will become so small that no sound will be heard.

If the loudspeaker is not suspended in mid-air, but rather sitting on the floor away from any walls, most of the sound waves that would have propagated downward are instead reflected upward to join those propagating in those directions in free space. Thus, we get a hemispherical propagation pattern with the waves now transporting practically double the energy when compared to the fully suspended loudspeaker.[c] Since the intensity of the sound is directly proportional to the energy being transported, the intensity of the sound is also doubled. Similarly, if the loudspeaker is placed at the juncture of the floor and a wall, waves that would have propagated towards the wall are now reflected forward. The propagation pattern becomes a quarter of a sphere, and double the amount of energy again is transported by the waves as compared to the loudspeaker just sitting on the floor. If we place the loudspeaker in a corner, thus creating three reflecting surfaces, the amount of energy transported doubles yet again. In each case, the sound intensity is also doubled as the energy transported doubles.

---

[c] Since the floor is not a perfect reflector, not all of the energy is reflected, and there will also be some loss due to phase cancellation, but we will ignore these for now.

1.8 The Decibel

We won't get very far in any discussion of sound without sooner or later running into the decibel. Our measurement equipment will give us readings in it; you will find it on the labels of the controls on your gear; the meters on our amplifiers, effects processors, and mixers are calibrated in some form of it; and the specifications of all our gear are littered with references to it. It is everywhere. So, let us take a look at what this thing is, how it came about, and how we use it in describing audio. Before we get to that, though, there are two things we need to look at first that play a big role in understanding decibels. These are gain and logarithms.

1.8.1   Gain

Most of us understand the concept of gain quite well. We know, for example, that if we put $100 into our investment account and after one year, we go to withdraw our money and we then have $110, we have a positive gain of $10. Conversely, if we now only have $95, we have a loss. The same concept applies to sound and audio signals as well. We typically expect a positive gain in signal level out the back end of an amplifier when compared to the signal going in the front end. On the other hand, we expect a signal loss on the tail end of an attenuator when compared to the signal at the input.

In sound engineering, we input a signal to a device and we get a signal out. Formally, the ratio of the level of the output signal to the level of the input signal is called the *gain* of the device. That's all there is to it. The 'level' can be a measure of any attribute of the signal that shows its strength, e.g., power, voltage, or current. Since it is simply a ratio of similar signal types, gain has no units. We can state the following obvious attributes about gain:

- If Output *is bigger than* Input *then* Gain *is bigger than* 1.
- If Output *is less than* Input *then* Gain *is less than* 1.
- If Output *is equal to* Input *then* Gain *is equal to* 1.

1.8.2   Logarithms

The logarithm is the second fundamental aspect to our understanding of the decibel. Wherever there are decibels, there are logarithms; so, let us do a short review of this subject before going further.

Suppose *a* represents some given number. We can use the following notation to show certain multiples of *a*:

$$a \times a = a^2$$
$$a \times a \times a = a^3$$
$$\text{and } a \times a \times a \times a = a^4$$

The emerging pattern should be visible; the exponent (the numbers *2*, *3*, and *4* in our example) is equal to the number of times that *a* is multiplied by itself. Thus, we could write:

$$a^n = a \times a \times \cdots \times a \quad (n \text{ times})$$
where a^n is called '*a* to the power *n*.'

For the purposes of our discussion, we will refer to *a* as the *base*, while *n* is called the *exponent*. For example, suppose we take a base of 10, which is the common logarithm base, then:

$$10^2 = 10 \times 10 = 100$$
and $10^3 = 10 \times 10 \times 10 = 1000$
etc.

We can now define the logarithm of a number y, to be another number n, such that

if $a^n = y$
then $\log_a(y) = n$

The expression **log$_a$(y)** is called the logarithm of y, and the logarithm is said to have a base of a. Thus, for example:

$$\log_{10}(100) = 2$$

because we know that $10^2 = 100$.

Logarithms have rules that allow easy manipulation. These are listed here for reference.

$$\log_a(y_1 \times y_2) = \log_a(y_1) + \log_a(y_2)$$ *Rule 1*
$$\log_a(y_1/y_2) = \log_a(y_1) - \log_a(y_2)$$ *Rule 2*
$$\log_a(y^r) = r \times \log_a(y)$$ *Rule 3*
$$\log_a(1) = 0$$ *Rule 4*

Note *Rule 4* which says that the logarithm of 1 has a value of zero. Some further things to note are:

- The logarithm of any number bigger than 1 is positive.
- The logarithm of any number less than 1 is negative.

We will use these rules many times throughout the book, e.g., when we calculate power requirements for venues. Additionally, we will no longer explicitly show the base of the logarithm when we write equations, as all references to a logarithm are understood to be to base 10. Thus, when you see *log(x)*, this is understood to be the same as *log$_{10}$(x)*.

To highlight the use of logarithms, let us look at two sequences of numbers. First, a simple sequence consisting of increasing multiples of the number 10:

$$(10 \times 1, 10 \times 2, 10 \times 3, 10 \times 4, 10 \times 5) = (10, 20, 30, 40, 50)$$

Here we can see that each number in the sequence increases in value by the same amount each time. We call this a *linear* progression. By comparison, let us look at the following sequence consisting of powers of the number 10:

$$(10^1, 10^2, 10^3, 10^4, 10^5) = (10, 100, 1000, 10000, 100000)$$

Here we see that each number increases by a different amount each time. Not only that, but we also see that each subsequent increase is getting bigger and bigger. The difference between the first and last numbers in such a sequence can be very large, and the numbers can get unwieldy to manipulate. The kind of increase shown above is said to be *exponential*. We can in fact represent the progression of this sequence by using the logarithm of the numbers in the sequence, which in our case are the exponents, giving the following:

$$(1, 2, 3, 4, 5)$$

We observe two things as a result of doing this. First, the numbers we must deal with are now much more manageable. Secondly, this new sequence is now linear as opposed to exponential (even though it still *represents* the original exponential sequence). Both of these resulting changes help us better deal with situations where the values of a sequence of numbers change exponentially and the range of values is large.

As it turns out, the way the human ear perceives changes in sound level is logarithmic. Therefore, it also makes sense to use a logarithmic scale when dealing with sound level changes as it applies to human hearing.

1.8.3   The Decibel is Born

We know that for a particular device or system, the gain is simply the ratio of the output level of its signal to its input level. If some device has a large loss or large amplification, the gain will be correspondingly quite tiny or very large, e.g., 0.000001 or 100000. As you can see, gain numbers can acquire lots of zeros, and the range between the smallest and the largest value can be huge. The straight gain numbers can become unwieldy, making it difficult to work with them directly. In 1929, early telephone engineers wanted a way to express large losses and gains without having to deal with large numbers, so they invented the bel (after Alexander Graham Bell). The bel is simply the logarithm of the *power* gain. Expressing the power gain as a logarithm reduces the number of digits that has to be manipulated considerably. Let us write this as an equation so that it is clearer:

$$bel = \log\left(\frac{output\ power}{input\ power}\right)$$

The decibel is one-tenth (1/10) of a bel. Therefore, to convert bels to decibels, we have to multiply the number of bels by 10. Thus, we can write:

$$decibel = 10\log\left(\frac{output\ power}{input\ power}\right)$$

The symbol for the decibel is **dB**. From our discussion on logarithms we can now also say that:

- If Gain *is bigger than* 1 *then* dB gain *is positive.*
- If Gain *is less than* 1 *then* dB gain *is negative.*
- If Gain *is equal to* 1 *then* dB gain *is zero.*

Let us look at a few examples. First, positive gain. Say that the input power to an amplifier is 0.2 W, while its output power is 800 W. Then:

$$Gain\ of\ amplifier\ =\ \frac{800}{0.2}$$

$$=\ 4000$$

$$\textit{Gain of amplifier in dB} \quad = \quad 10 \log \left(\frac{800}{0.2} \right)$$

$$= \quad 10 \log(4000)$$

$$= \quad 36 \text{ dB}$$

So, a power gain of 4000 is equivalent to a gain of +36 dB.

Now let us look at a situation where there is a loss. Say that the input power to an attenuator is 2 W and the output power is 0.03 W. Then:

$$\textit{Gain of attenuator} \quad = \quad \frac{0.03}{2}$$

$$= \quad 0.015$$

$$\textit{Gain of attenuator in dB} \quad = \quad 10 \log \left(\frac{0.03}{2} \right)$$

$$= \quad 10 \log(0.015)$$

$$= \quad -18.2 \text{ dB}$$

Thus, we see that a gain of 0.015 (which, because it is less than 1, represents a loss), when expressed as a decibel gain, is equivalent to −18.2 dB. As you can see from the above examples, the dB gain numbers tend to have fewer digits than the straight gain numbers.

As originally conceived, the decibel gives the logarithmic gain of a device in terms of power ratios. However, we can express dB power gain in terms of other source units if we take care to convert properly between them and power units. Let us look at the power gain of an amplifier again, but now in terms of voltage. To express dB power gain in terms of voltage, we have to express power (W) in terms of voltage (V). You may know already that electrical power is related to voltage and electrical impedance (Z) by the formula:

$$W = \frac{V^2}{Z}$$

If we consider, for the purposes of establishing a dB reference, that Z is the same for input and output, then we can express power gain in terms of voltage, as follows:

$$\textit{Gain in dB} \quad = \quad 10 \log \left(\frac{V_{out}^2}{Z} \bigg/ \frac{V_{in}^2}{Z} \right)$$

$$= \quad 10 \log \left(\frac{V_{out}^2}{V_{in}^2} \right)$$

$$= \quad 10 \log \left(\frac{V_{out}}{V_{in}} \right)^2$$

But according to logarithm Rule 3, the last line can be written as:

$$Gain\ in\ dB \quad = \quad 2 \times 10 \log\left(\frac{V_{out}}{V_{in}}\right)$$

$$= \quad 20 \log\left(\frac{V_{out}}{V_{in}}\right)$$

The use of the decibel evolved rapidly. While it was originally conceived to express the ratio of two power levels, it can be used to express the ratio of any two measurements of the same attribute. It is widely used to express voltage gain, current gain, sound pressure levels, bit rates, to name but a few.

1.8.4 Decibel Reference Values

Note that the decibel expresses a ratio of two values—this is always the case. If someone simply said that a signal had a level of 30 dB, you would have to ask "with reference to what?" To avoid this confusion, reference values have been defined for various signal and measurement types. These reference values are taken as 0 dB. Once the dB reference is known, technicians can simply say that a particular signal is a given number of dB.

Some reference values for various measurement domains are given below:

- 0 dBm – Reference power level equal to 1 mW
- 0 dBW – Reference power level equal to 1 W
- 0 dBu – Reference voltage level equal to 0.775 Vrms (derived from a power of 1 mW into 600 Ω impedance)
- 0 dBV – Reference voltage level equal to 1 Vrms
- 0 dB SIL – Reference sound intensity level equal to 10^{-12} W/m^2
- 0 dB SPL – Reference sound pressure level equal to 20 μPa (same as 20 N/m^2)

1.9 Signal Levels

Many times, reference will be made to a particular signal level when discussing connections between devices. These levels are sometimes stated as voltages, but will often be given in decibels. The most common of these that you will encounter is line level. However, you will also see references to instrument level, microphone level, and speaker level.

Line level is the signal level expected on the line output or line input of typical signal processor (such as the mixer or effects processor). There are two common references for this level, *viz.* pro-audio and consumer. The pro-audio line level is taken to be +4 dBu. Because we are interested in the voltage ratio, we know from Section 1.8.3 that we would define the decibel dBu level as:

$$dBu = 20 \log\left(\frac{V_u}{V_{ref}}\right)$$

where V_u is the voltage being measured, and V_{ref} is the reference voltage for this measurement domain. From above, we know that 0 dBu is referenced to 0.775 Vrms. Thus, we can calculate the corresponding nominal voltage, V_{pro}, of the pro-audio line level as follows:

$$+4\ dBu = 20\log\left(\frac{V_{pro}}{0.775}\right)$$

Therefore:

$$\frac{4}{20} = \log\left(\frac{V_{pro}}{0.775}\right)$$

$$0.2 = \log\left(\frac{V_{pro}}{0.775}\right)$$

$$10^{0.2} = \frac{V_{pro}}{0.775}$$

$$1.6 = \frac{V_{pro}}{0.775}$$

And thus:

$$V_{pro} = 1.6 \times 0.775 = 1.23\ \text{Vrms}$$

We see then, that the pro-audio live level of +4 dBu corresponds to a nominal line voltage of 1.23 Vrms. Note that the peak voltage of the line will be higher than the nominal voltage (the peak voltage will be around 1.7 V).

The consumer line level is set at −10 dBV. Going through a similar calculation to above, we can see that the nominal consumer line level is equivalent to 0.316 Vrms, which is quite a bit lower than the pro-audio level. Many signal processors will therefore allow users to select between consumer and pro-audio input levels to better match the signal to gain settings of the preamplifier stage of the processor.

Microphone level is also a common term used to describe the nominal level of the signal output by a pro-audio low-impedance microphone. This level is in the millivolt range (1 millivolt = 0.001 V). Microphone level is not referenced to any set level like line level, but you will see the output level for microphones published in dBV. There is considerable variation in the output levels of microphones, from close to −65 dBV (approximately 0.6 mV) for a large diaphragm dynamic microphone, to −32 dBV (25 mV) for a large diaphragm condenser.

Instrument level is another term we will encounter many times. This term is used to describe the output signal level from instruments like electric guitars and basses, and the level associated with Hi-Z (high-impedance) instrument inputs that you will see on some preamps. There is no defined standard reference for instrument level. The level is higher than microphone level, but less than line level. Again, there is wide variation in level from one instrument to the next (depending on things like passive vs. active pickups). Typically, instrument level is in the range of −20 dBV to 0 dBV, but can be even higher for instruments with a built-in preamplifier.

Speaker level is what we expect to find at the output terminals of a power amplifier driving a loudspeaker system. This voltage level is thus very high, ranging from tens of volts for a system delivering a few watts of power to well over 100 V peak for a system delivering over 1500 W. If calculated using the dBV reference, a 100 V amplifier output would be represented as +40 dBV peak.

1.10 Decibel Sound Intensity Level – dB SIL

Recall that in Section 1.6 sound power was defined in terms of intensity, measured in W/m^2. Since this is a measurement of power, we could directly define a logarithmic relationship for sound intensity level, or SIL, as:

$$dB\,SIL \;=\; 10\log\left(\frac{I_a}{I_{ref}}\right)$$

where I_a is the sound intensity and I_{ref} is the reference intensity, taken to be 10^{-12} W/m^2.

Suppose now we revisit the inverse square law discussed in Section 1.7 and see what is the reduction in SIL as we move away from the sound source in free space. If we arbitrarily take the intensity I_D at distance D to be 1 W/m^2, then we know that the intensity I_{2D} at $2D$ will be 1/4 W/m^2. Thus, we can write the dB SIL change as follows:

$$
\begin{aligned}
dB\,SIL\;change \;&=\; 10\log\left(\frac{I_{2D}}{I_D}\right)\\
&=\; 10\log\left(\frac{1}{4}/1\right)\\
&=\; 10\log\left(\frac{1}{4}\right)\\
&=\; 10\,(\log(1)-\log(4)) \qquad (Rule\ 2)\\
&=\; 10\,(0-0.602)\\
&=\; -6.02
\end{aligned}
$$

We see that as we double the distance from the sound source, the SIL is reduced by approximately 6 dB. We can work out the change in SIL for the case where we place the sound source on the floor and in corners too. For example, we said that the intensity is doubled when we place the loudspeaker on the floor (I_{floor}) away from the walls as compared to when the loudspeaker is suspended in midair (I_{air}). Thus, the dB SIL change would be:

$$
\begin{aligned}
dB\,SIL\;change \;&=\; 10\log\left(\frac{I_{floor}}{I_{air}}\right)\\
&=\; 10\log\left(\frac{2}{1}\right)\\
&=\; 10\log(2)\\
&=\; 10\times0.301\\
dB\,SIL\;change \;&=\; 3.01
\end{aligned}
$$

So, every time we move the loudspeaker against a reflecting surface—floor, wall, corner—the SIL increases by approximately 3 dB.

1.11 Decibel Sound Pressure Level – dB SPL

We just defined a logarithmic relationship for sound intensity. However, since it is easier to measure or deduce sound pressure than intensity, it is more usual to express sound level using sound pressure rather than intensity. Similar to how electrical power is related to voltage and electrical impedance, sound intensity (I_a) is related to sound pressure (p) and acoustic impedance (Z_a). The equation is very similar:

$$ I_a = \frac{p^2}{Z_a} $$

We see from this equation that sound intensity—and hence sound power—is proportional to the square of sound pressure. Just as we expressed the electrical dB power ratio in terms of voltage in Section 1.8.3, we can express the dB ratio for sound power in terms of sound pressure. We can write this as:

$$
\begin{aligned}
dB\,SPL &= 10\log\left(\frac{p^2}{p_{ref}^2}\right) \\[2mm]
&= 10\log\left(\frac{p}{p_{ref}}\right)^2 \\[2mm]
&= 20\log\left(\frac{p}{p_{ref}}\right)
\end{aligned}
$$

where p is the sound pressure and p_{ref} is the reference pressure. The logarithmic ratio expressed in this way is called the *sound pressure level* and is usually written as dB SPL. The reference pressure, p_{ref}, is 20 micropascals (µPa) and is commonly taken to be the threshold of hearing at 1 kHz.

One interesting thing to note is that for sound propagation in free space, i.e., when neither the sound source nor the measurement position is close to reflecting surfaces, SPL is roughly equal to SIL, and changes in level of SPL are equivalent to changes in SIL.[1] This means that when intensity increases or decreases at a given location, sound pressure will also increase or decrease by the same proportion. Thus, for example, in free space when we double our distance from a given sound source, the SPL will be reduced by approximately 6 dB, just as we derived for the SIL.

Sound pressure level can be measured with an SPL meter, like the one shown in Figure 1-8. Meters used for professional work should conform to the relevant standard, e.g., IEC 61672 or ANSI S1.4. When measuring SPL, the proper weighting, or equalization profile, should be selected on the meter. A number of weighting curves have been defined. These equalization profiles are shown in Figure 1-9. Z-weighting is a flat equalization profile, while profiles A and C tailor the SPL meter to more closely conform to how humans perceive the loudness of sounds. The original intention was to use a different weighting curve depending on the loudness of sound, with A-weighting used for soft sounds up to 55 dB and C-weighting used for loud sounds above 85 dB.[2] However, for noise related hearing loss risk assessment, A-weighting is currently used for all sound levels.

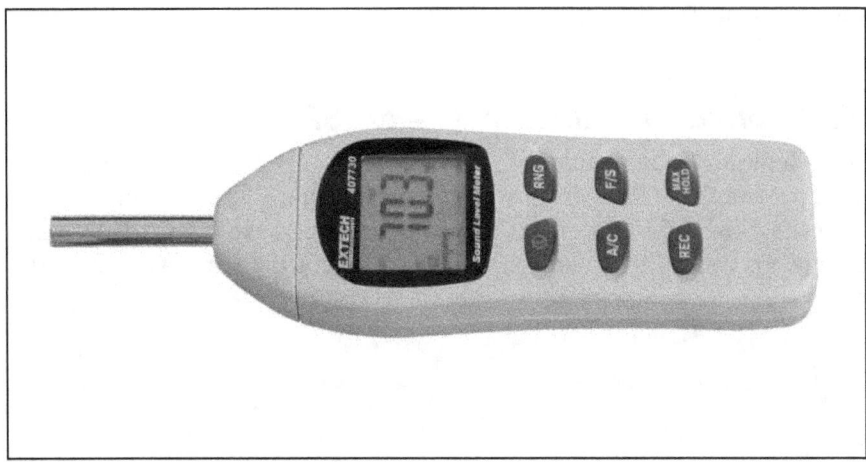

Figure 1-8   Sound Pressure Level Meter

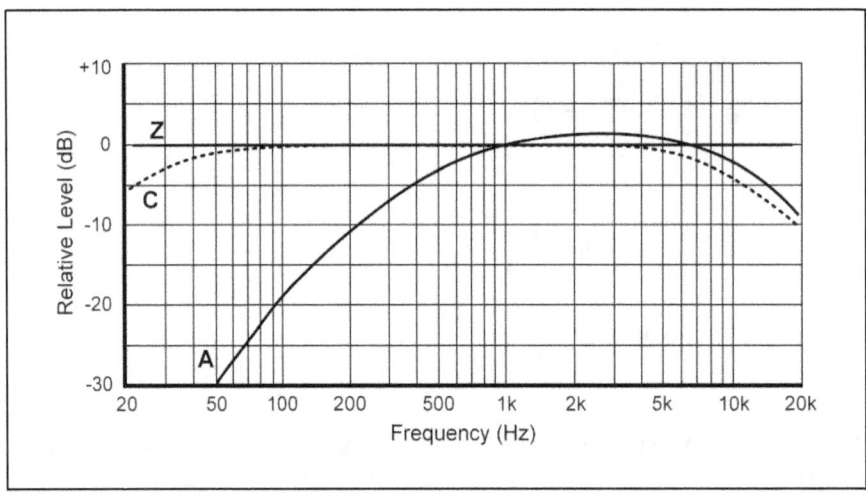

Figure 1-9   Weighting Curves for SPL Meter

For a comparison to the A-weighting and C-weighting schemes of the SPL meter, look at the frequency response of an average human ear shown in Figure 1-10. We can see that there are certain similarities between these curves. We also see that humans do not perceive sounds of different sound pressure levels the same way. Note that at lower levels the ear is less sensitive, especially in the lower frequency range, but also somewhat at the higher frequencies. Thus, some manufacturers of consumer audio equipment have in the past incorporated a 'loudness' switch that boosts the low frequencies below a certain volume setting. Note too, that the ear does not have a flat response, and there is a bump in sensitivity from 2 kHz to 6 kHz. We will see later that this range is where most consonants, predominant for speech intelligibility, are articulated.

To get an idea of what different SPLs sound like, take a look at Table 1-2. These are some typical sound levels for common sounds. The threshold of hearing, 20 μPa, is taken to be 0 dB SPL. With the threshold of pain generally around 120 dB, it can be seen that a loud rock concert is, literally, a painful experience! And at 160 dB, one can expect some damage to the ear, possibly a broken eardrum.

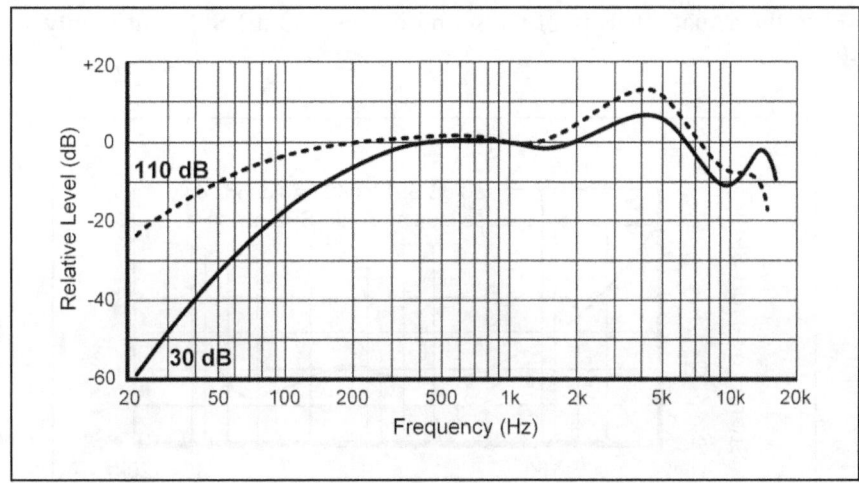

Figure 1-10   Comparative Relative Response of Human Ear

Sound Source	dB SPL
Loud rock concert	120
Siren at 30 m	100
Busy street traffic	75
Ordinary conversation at 1 m	60
Quiet radio	40
Whisper	20
Threshold of hearing	0

Table 1-2   SPL of Common Sounds

1.12 Loudness of Sound

Loudness is one of those psychoacoustic phenomena that we have to consider in sound reinforcement. It is something that is perceived by humans and cannot be measured directly by physical instruments. The loudness of a sound is related to its intensity. The higher the intensity of the sound wave, the greater the loudness of the sound. However, this relationship is not linear, but logarithmic. In fact, to get a sound that is perceived as being twice as loud requires a sound that has about ten times the intensity. Thus, if we have a single tenor singing his part in a choir, we must add nine more tenors in order for us to perceive that the tenor part is twice as loud. Interestingly, this perception only holds if the sounds are close in frequency. If a single tenor is singing high notes and a bass singer starts singing his low notes, the combined parts may indeed be perceived as being twice as loud.

The unit used to represent loudness is called the phon and is referenced to an SPL at 1 kHz. In other words, if we say that a particular sound has a loudness of 60 phons, that means that it sounds as loud as a 60 dB SPL sound at 1 kHz. If we consider that the ear is more sensitive to some frequencies than others, then we can see that sounds of the same SPL will sound louder at some frequencies than others. Stated another way, for different frequency sounds to sound just a loud as some particular loudness level, the actual SPL will be different for different frequencies. Figure 1-11 shows just such a curve for 60 phons. This graph tells us that a sound of any SPL that can be represented on this curve will sound roughly as loud as a 60 dB SPL sound at 1 kHz. Thus, we can see that at 40 Hz, the actual intensity of the sound is about 83 dB SPL, but it only sounds as if it was 60 dB SPL.

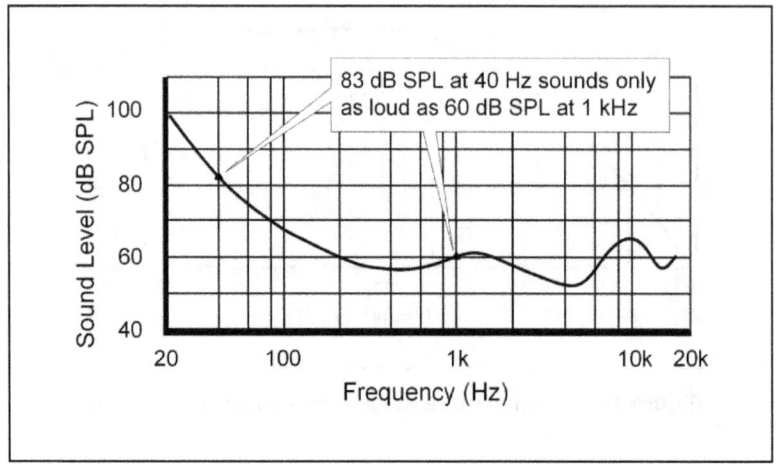

Figure 1-11   Equal Loudness Curve for 60 Phons

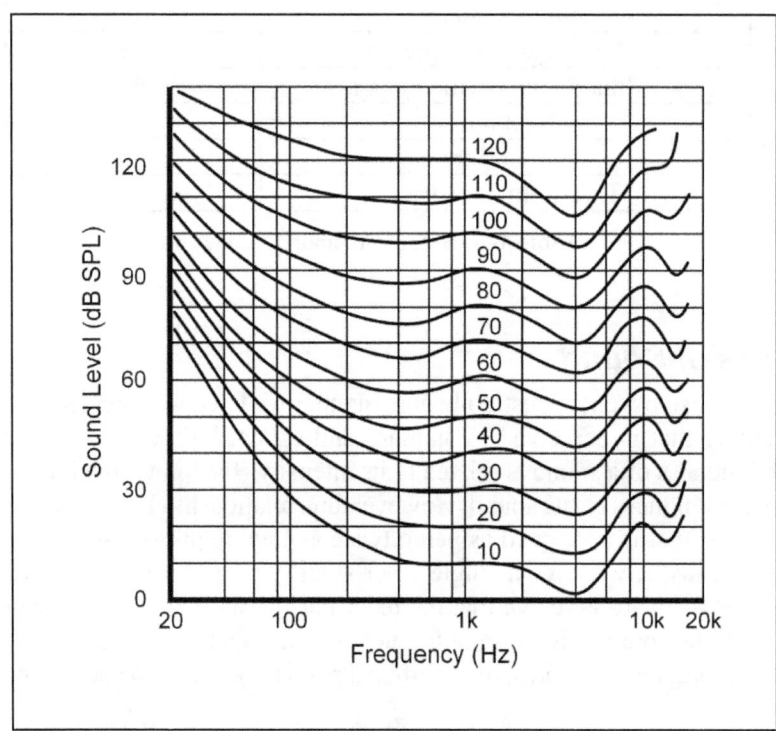

Figure 1-12   Equal Loudness Curves

We can construct a whole series of curves for different sound pressure levels, as shown in Figure 1-12, which shows similar information as Figure 1-11. When presented like this, they are known as equal loudness curves. There are different versions of these curves—Fletcher-Munson, Robinson-Dadson, IS0-226 to name a few—and all are somewhat different from each other. The important thing for us to take away from these curves is that for softer sounds, the change in perception is very dramatic in the lower frequencies when compared to higher level sounds. In other words, the ear is fairly insensitive to soft low-frequency sounds. Again, we see that the ear is most sensitive to sounds in the region 2 to 6 kHz which, as we hinted before, is the range of frequencies that are predominant for speech intelligibility.

1.13 Phase Relationships and Interference

If a point moves around the center of a circle, we can use its angular displacement from some arbitrary reference point on the circle to describe its position as it moves. As we know, there are 360 degrees in a circle. Thus, the point at 0° and the point at 360° actually represent the same position. Just like we can use angular displacement of a point on a circle to describe its position as it moves around the center of the circle, we can use angular displacement to describe any point in a cycle relative to some reference point in the cycle if we translate positions on the circle to positions in the cycle. To do this, we fit one cycle of the wave into a 360° cycle, no matter what the actual frequency of the wave. Figure 1-13 shows the circle-to-cycle translation.

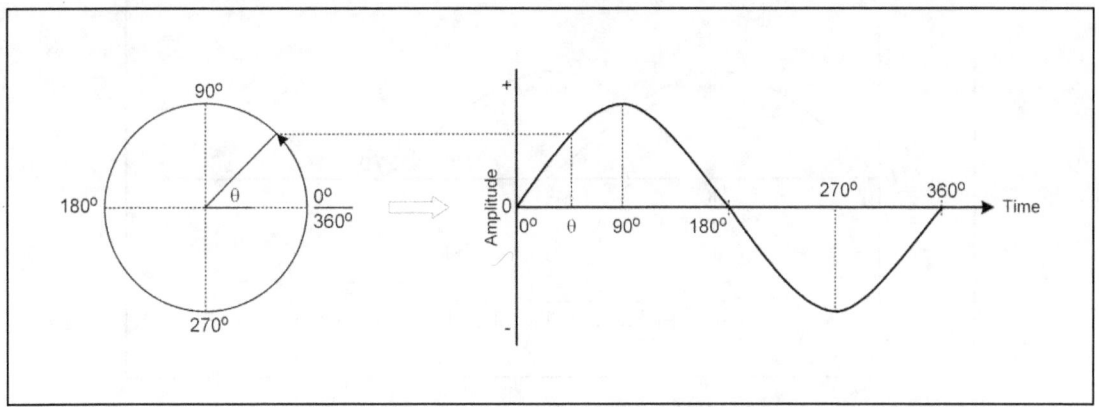

Figure 1-13   Translating 360 Degrees of Angular Displacement to One Cycle

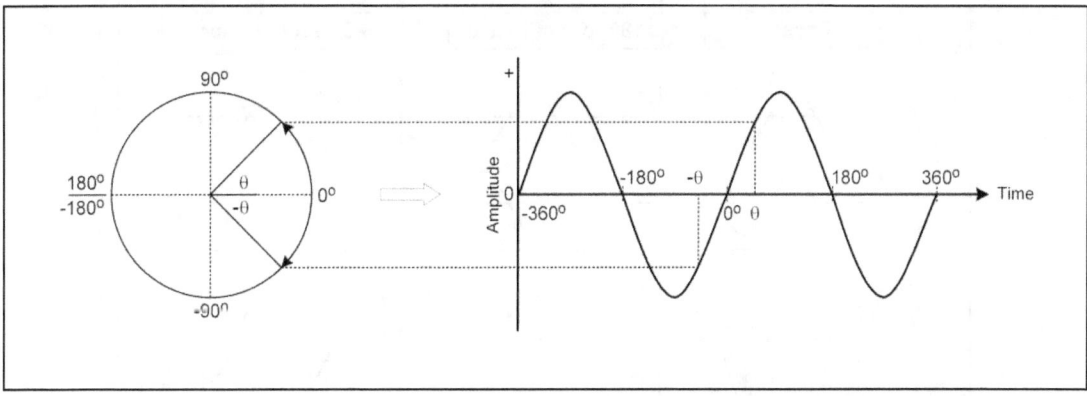

Figure 1-14   Positive and Negative Displacement Angle Translation from Zero Reference

There is nothing that says we must use only positive angular displacement to describe the position of the point on the circle, and hence any corresponding translated point in the cycle. For example, Figure 1-14 shows us that we can traverse the circle in either direction starting from our 0° reference position, where one direction uses positive angular displacement to describe the position of the point, and the other uses negative displacement. In this case, we see that the point at +180° and the point at -180° actually represent the same relative position in a cycle.

Phase difference is a way of describing the relative time positions of two waves that have the same wavelength, and hence, frequency. Two waves are in phase if identical points in the cycle occur at the same time, i.e., they reach their maximum, zero, and minimum amplitudes at the same time. This is a rather informal definition.

Instead of using time to describe how the waves align with each other, the number of degrees of relative displacement from a fixed reference position in the cycle can be used. By using degrees of displacement rather than literal time, it becomes much easier to visualize the relative positions of two waves. All relative positions of the two waves of the same frequency fit within 360°. Take a look at Figure 1-15 where we have added a second wave on top of the wave shown in Figure 1-13. The second wave occurs later in time, and you can see that it seems to lag behind the first wave by 90°, i.e., its zero-amplitude point is 90° after the corresponding zero-amplitude point of the first waveform. Thus, we say that the second wave is 90° out of phase with the first wave.

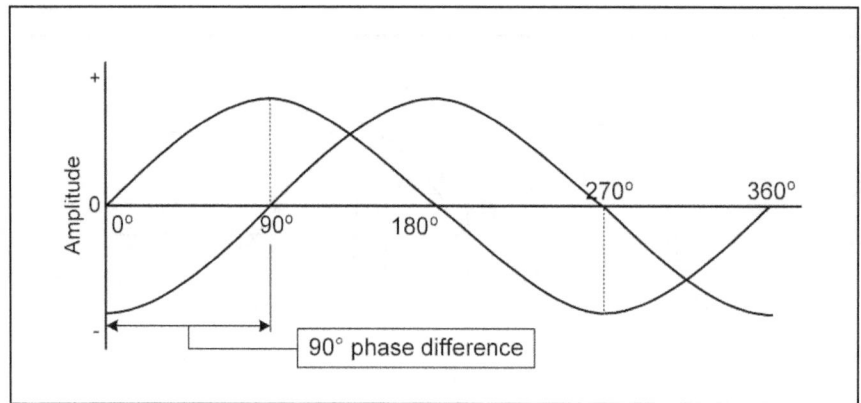

Figure 1-15 Waves 90 Degrees Out of Phase

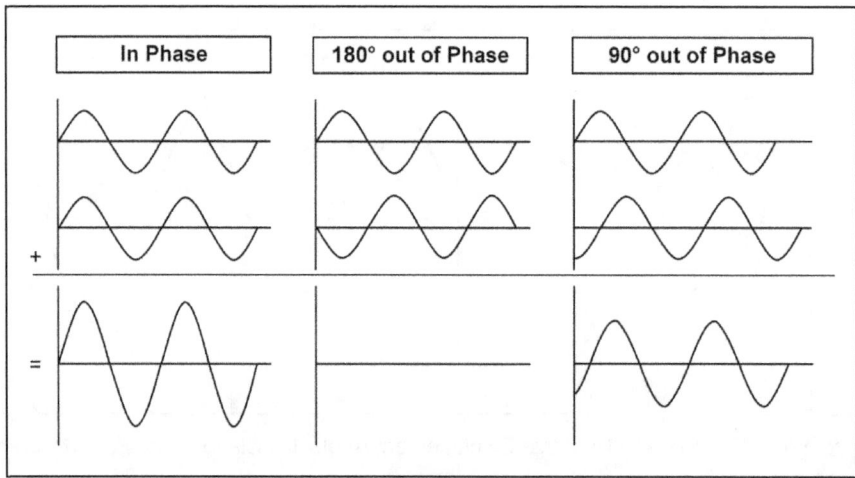

Figure 1-16 Results of Interfering Waves

1.13.1 Interference Due to Phase Difference

Sound waves that have the same frequency and some clearly discernible phase relationship are said to be correlated. When two correlated sound waves interact, they interfere with each other—whether constructively or destructively depends on the relative phase of the waves. When the phase difference at a particular location is such that the amplitudes of the waves are both positive or both negative at a given instant, they interfere constructively and the resulting amplitude is bigger than either of the two. When one wave is positive and the other is negative, they interfere destructively. The resulting amplitude is smaller than the larger of the two, and sometimes the waves cancel each other completely. Figure 1-16 shows the results of interfering waves for the cases where the waves are in phase, 180° out of phase, and 90° out of phase. In the diagram, the waveform at the bottom is the sum of the two waves above it. Note that in the case where the waves are 180° out of phase, the end result is total cancellation.

1.13.2 Phase Shift vs. Polarity Inversion

It bears repeating that when we talk about phase shift, we are comparing two waves of the same frequency at some reference point or reference plane, and one of those waves has some time delay (which may be zero) with respect to the other. Note too, that due to the simple relationship between frequency and wavelength given in Section 1.4, we can always transform a phase difference of time into some fraction of a cycle or a number of degrees.

The discussion in the previous section used simple sine waves to illustrate the various outcomes of combining two waves with various phase relationships. It is worth noting that real-world sound waves consist of complex combinations of several sinusoidal waves and have subtle variations from cycle to cycle. As a result, real-world sound waves are never fully canceled due to phase shifting, though individual sinusoidal components of these sounds may be canceled. For example, Figure 1-17 shows 1.5 cycles of such a wave, W_1, composed of multiple sinusoids. It somewhat resembles a reversed sawtooth waveform. In this figure, W_2 represents a copy of the same wave, phase shifted by 180°, completely out of phase with W_1. When these two waves are added, the result is W_3, not total cancellation.

Figure 1-18 shows the most common scenarios in which we will encounter phase shifting in the real world. We only mention them briefly here, but will look at these in more detail in later chapters. In (A), we have two loudspeakers some distance apart playing the same signal, with a listener closer to one loudspeaker than the other. The sound wave, represented by W_1, from the closer loudspeaker will arrive at his ear before the sound wave, W_2, from the farther loudspeaker. Thus, W_2 will lag W_1 by some amount of time.

In Figure 1-18(B), we have a single loudspeaker playing a signal where some of the sound waves, W_1, reach the listener directly and some, represented by W_2, reach the listener after reflecting off a wall. Since the waves represented by W_2 travel farther than W_1 to reach the listener, they will arrive later, and thus be phase shifted by some degree.

Figure 1-18(C) represents another common scenario where we have a person speaking into two microphones. The microphone closer to the speaker generates signal W_1, while the one further away generates signal W_2. Due to the longer time it takes for the sound waves from the speaker's mouth to reach the farther microphone, W_2 will be phase delayed relative to W_1 when combined at the summing amplifier (the triangle symbol with the + sign).

Finally, Figure 1-18(D) shows the use of an electronic delay, such as may be found in an effects processor, to deliberately introduce some phase shift to wave W_2. When combined at the summing amplifier, the two waves will add to produce a final result that depends on their phase difference.

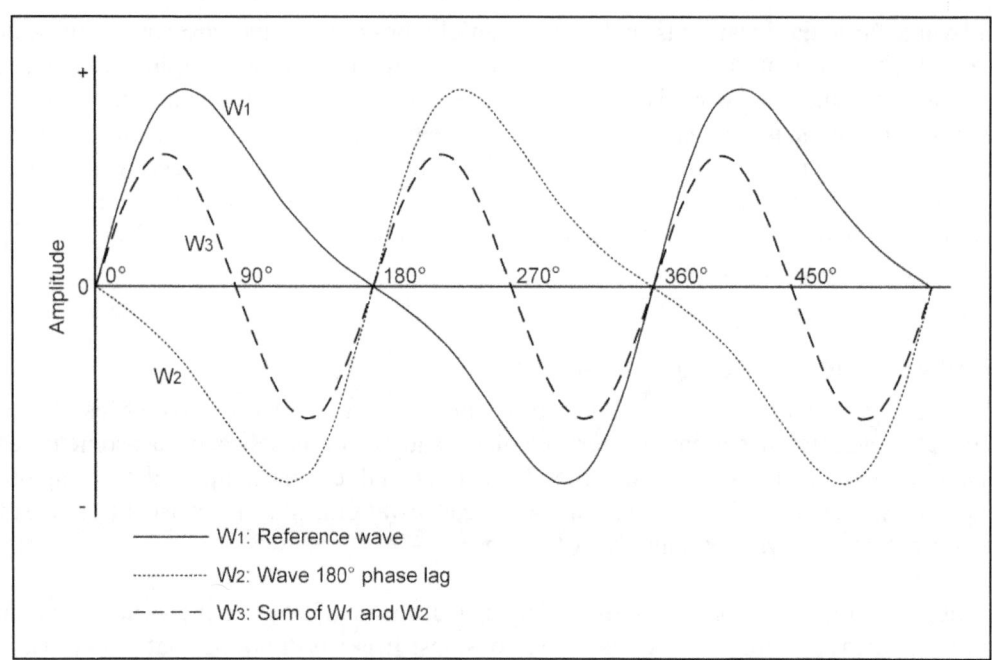

Figure 1-17 Non-Canceling Complex Wave Addition at 180° Phase Shift

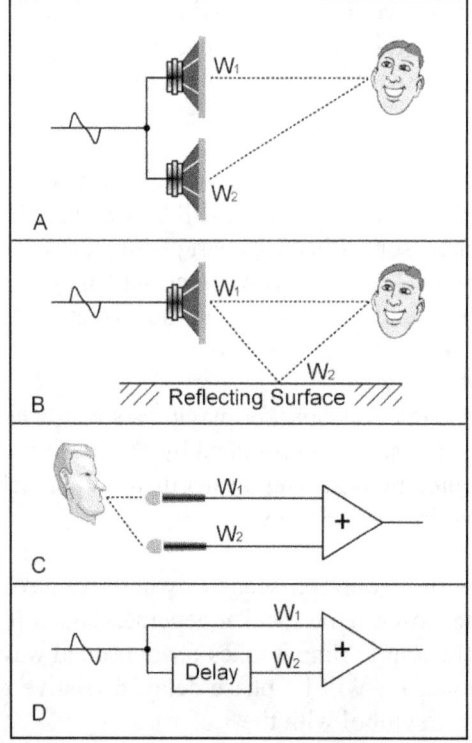

Figure 1-18 Common Phase Shift Scenarios

In all the above cases, the combination of the waves at the reference point (the listener's ear or the summing amplifier in our case) will look something like Figure 1-17, with W_2 lagging W_1 by some amount. However, the exact result of their addition will vary based on just how much W_2 lags W_1.

If we refer back to Figure 1-17, there is no degree of phase shift that we can apply to W_2 that will produce total cancellation when the two waves are added. However, there is a scenario that will produce cancellation, and that is by addition of the waves where W_2 is a polarity inverted version of W_1. By itself, polarity inversion does not involve any phase shift, and is produced by a simple electronic circuit called an inverter. Shown in Figure 1-19, it is depicted by the triangle with the minus sign on its input. The inverter simply takes the input signal and outputs a signal with the opposite voltage level as the input, but without any delay, and thus without any phase shift. For example, if the input signal had a voltage of +1.2 V, the output of the inverter would be −1.2 V. If the input signal had a level of −0.7 V, the output signal would be +0.7 V. Because we used simple sine waves in the previous section when discussing phase shift, it is impossible to visually distinguish a 180° phase shift from an inverted signal. However, the difference should be clearly visible when comparing Figure 1-17 with Figure 1-20, which shows the result of inverting the same wave as used in Figure 1-17 and adding it to the original. We can see that the inverted wave is a mirror image of the original, and addition of W_1 and W_2 results in total cancellation.

Signal polarity inversion can be a useful tool in some miking situations, and you will often find a switch to do just that on the input channels of some mixers.

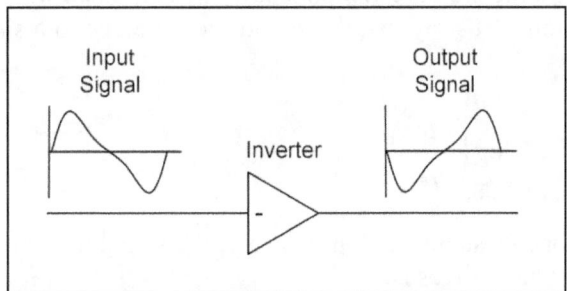

Figure 1-19   Inverter Changes the Polarity of the Input Signal

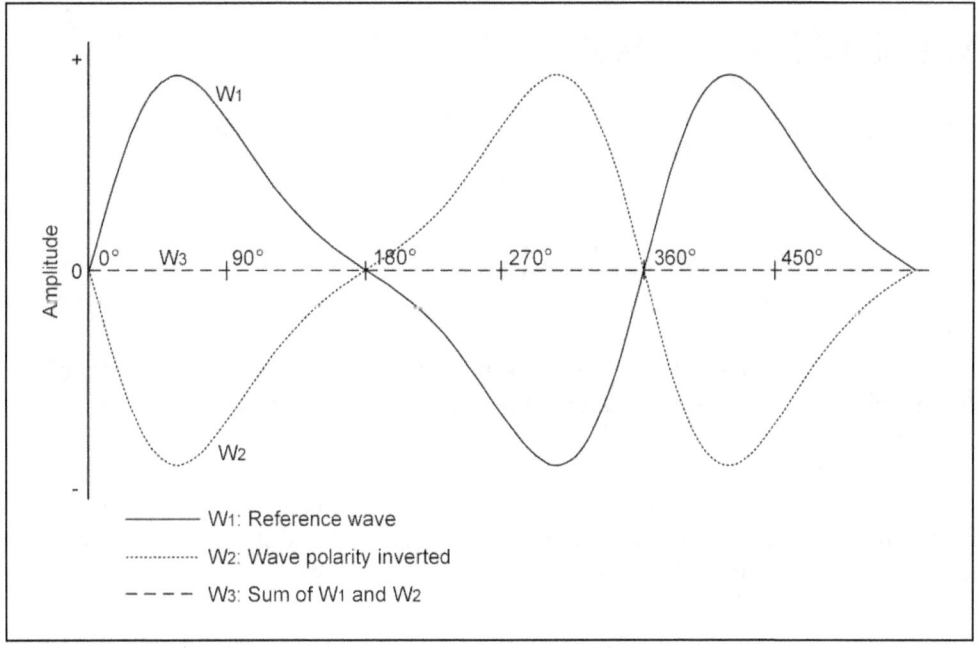

Figure 1-20   Polarity Reversed Wave Cancels Original when Added

1.14 Decibel SPL Levels of Added Sound Sources

Suppose we have our sound system operating with one loudspeaker and we want to add another loudspeaker to the setup. If the newly added loudspeaker is radiating at the same intensity as the first one, what will be the increase in SPL at a given location in the sound field? The first thing to recognize is that the sounds from these two sources will be correlated. These sounds will be correlated because the two loudspeakers are driven from the same source. Another way that we can get correlated sounds is if we have a single source and sounds are reflected only a few times from the walls of a room.

We saw in the previous section that the end result of combining two correlated sounds depends on the relative phase of their waves. While we will look at the role of relative phase in combining sound sources more closely in Section 8.9, let us say now that the relative phase of the two waves at any given location depends on where that point is relative to the two loudspeakers. In other words, there are an infinite number of phase relationships between the two waves, from completely in phase to completely out of phase.

If we are at a position where the waves are in phase, then, as we saw above, the amplitude of the waves will simply add, and the pressure component of the combined wave will be double that of a single wave at that position. Let's say that the sound pressure due to a single wave is p. Then the SPL of one wave is:

$$dB\ SPL \quad = \quad 20 \log\left(\frac{p}{p_{ref}}\right)$$

Since the pressure components simply add in this case, the sound pressure of the two waves is $2p$. Thus, the SPL of the combined waves is:

$$dB\ SPL \quad = \quad 20 \log\left(\frac{2p}{p_{ref}}\right)$$

Using Rule 1 for logarithms, we can rewrite this as:

$$dB\ SPL \quad = \quad 20 \left(\log(2) + \log\left(\frac{p}{p_{ref}}\right)\right)$$

$$= \quad 20 \log(2) + 20 \log\left(\frac{p}{p_{ref}}\right)$$

We see that compared to the single wave, the combined waves have an increase in SPL of *20 log(2)*, or approximately 6 dB. This is thus the greatest increase in SPL that we can expect when combining two correlated sound waves.

Another common scenario is the result of combining two uncorrelated sounds. Uncorrelated sounds come from sources that are not strictly related, e.g., two tenors in the choir. Sounds can also become uncorrelated after numerous reflections in a room, e.g., the reverberation tail of a sound. When uncorrelated sounds combine, there is no simple summation of the pressure components of the waves like we did for correlated sounds because there is no strict phase relationship between the two waves. In fact, to get the combined intensity of two uncorrelated waves, we must add their sound powers together. As we saw in Section 1.11, the sound power is proportional to the square of sound pressure. Thus, we would have to add the square of the sound pressures when combining uncorrelated sound waves:

$$dB\ SPL = 10 \log \left(\frac{p_1^2 + p_2^2}{p_{ref}^2} \right)$$

If the uncorrelated sounds were equally strong at the location where they are combined, each with sound pressure of *p* say, we could write:

$$
\begin{aligned}
dB\ SPL &= 10 \log \left(\frac{p^2 + p^2}{p_{ref}^2} \right) \\
&= 10 \log \left(\frac{2p^2}{p_{ref}^2} \right) \\
&= 10 \log(2) + 10 \log \left(\frac{p^2}{p_{ref}^2} \right)
\end{aligned}
$$

So, we can see that the greatest increase in SPL when two uncorrelated sounds combine is *10 log(2)*, or approximately 3 dB.

If we take this a step further and generalize the process, we can determine the final SPL of any number of added sound sources. Suppose we have a number of individual sound sources, s_i, $i = 1$ *to n*, with individual levels SPL_i. Extending the above equation, we can write the expression for all sources combined:

$$
\begin{aligned}
SPL_{sum} &= 10 \log \left(\frac{p_1^2 + p_2^2 + \cdots + p_n^2}{p_{ref}^2} \right) \\
&= 10 \log \left(\frac{p_1^2}{p_{ref}^2} + \frac{p_2^2}{p_{ref}^2} + \cdots + \frac{p_n^2}{p_{ref}^2} \right)
\end{aligned}
$$

From the definition of the logarithm given in Section 1.8.2, we know that since

$$SPL_i = 10 \log \left(\frac{p_i^2}{p_{ref}^2} \right)$$

then

$$\frac{p_i^2}{p_{ref}^2} = 10^{SPL_i/10}$$

Substituting for p_i^2 / p_{ref}^2 in the summation, it becomes:

$$SPL_{sum} = 10 \log \left(10^{SPL_1/10} + 10^{SPL_2/10} + \cdots + 10^{SPL_n/10} \right)$$

For example, suppose we have two uncorrelated sounds with levels of 100 and 105 dB SPL at some location. The combined SPL of these sounds would then be:

$$
\begin{aligned}
SPL_{sum} &= 10 \log\left(10^{100/10} + 10^{105/10}\right) \\
&= 10 \log(10^{10.0} + 10^{10.5}) \\
&= 10 \log(41622776601.7) \\
&= 106.2 \text{ dB}
\end{aligned}
$$

We see then, that adding these two sources produces a mere 1.2 dB increase in SPL over the louder sound at the location in question.

Review Questions

1. Discuss the primary and secondary requirements of a live sound reinforcement system and why they are important to the successful implementation of live sound.
2. What is a transducer? What types of transducers do we have in a live sound system and why do we need them?
3. Describe the basic signal flow in a live sound reinforcement system.
4. What is meant by the term *front of house*?
5. What is the generally accepted frequency range of human hearing? What happens to this range as a person ages?
6. What is the relationship between the period and cycle of a sound wave?
7. What is the relationship between the period and frequency of a sound wave?
8. If we assume the speed of sound is 340 m/s, what is the wavelength of a 100 Hz wave?
9. Discuss how the air pressure at a particular point in space varies over time when a sound wave moves past that point.
10. Explain how the amplitude of a sound wave is related to the intensity of the sound and the amount of energy that the sound wave is transporting.
11. What is meant by sound power of a sound wave?
12. By how much will the SPL increase if we halve our distance to a sound source?
13. What is weighting curve as it pertains to measurements of SPL? Describe the main characteristics of the A-weighting curve and why it may be used when taking SPL measurements.
14. If an amplifier has a power gain of 3000, what is its gain in dB?
15. If a pad (an attenuator) has a voltage gain of 0.5, what is its gain in dB?

Chapter 2

The Environment

2.1 Why is the Environment Important?

When we refer to the environment, we are not talking about forests, swamps, rivers, or pollution. Rather, the environment is the collection of objects, conditions, and circumstances in which the sound system operates and with which the sound engineer must interact. The environment affects the final sound produced by the system, and thus what the audience hears. There are numerous aspects to the environment that affect the sound system, but we will limit our discussion to basic interactions with air temperature, humidity, and objects in the path of the sound.

2.2 Environmental Effects of Air on the Speed of Sound

The speed of sound changes when the physical attributes of the air through which it is traveling change. This change in speed can be quite dramatic. When using multiple loudspeakers that are widely distributed around the audience area, it is common to use delay devices to align the sound from these loudspeakers so that echo-like effects are eliminated in the audience area. The change in the speed of sound could have a bearing on delay times when using multiple loudspeakers.

The speed of sound is 331 m/s at 0 °C. This speed will increase 0.6 m/s per °C with increasing temperature. To a much lesser extent, the speed of sound will also increase with increasing humidity. Table 2-1 shows how the speed of sound varies with temperature and humidity. It is based on research presented by Cramer in 1993.[1] Interestingly, the speed of sound in air is not affected by changes in the air pressure.

Temp °C	Relative Humidity (%)										
	0	10	20	30	40	50	60	70	80	90	100
0	331.5	331.5	331.5	331.6	331.6	331.6	331.6	331.7	331.7	331.7	331.8
10	337.5	337.5	337.6	337.7	337.7	337.8	337.9	337.9	338.0	338.1	338.1
20	343.4	343.5	343.6	343.7	343.9	344.0	344.1	344.3	344.4	344.5	344.6
30	349.2	349.4	349.6	349.9	350.1	350.3	350.6	350.8	351.0	351.3	351.5

Table 2-1   Changes in Speed of Sound (m/s) with Temperature and Humidity

When people enter an enclosed space, both the temperature and the humidity of the air will rise. The degree to which these factors increase depends on the size of the audience and what they are doing, and obviously on the ability of the building's HVAC system to control the air quality of the space. A large, active crowd will generate more heat and humidity than a small, quiet group.

2.3 Reflection of Sound

Unless your loudspeakers are floating outdoors in midair, there will always be some interaction of the sound waves with solid objects in the environment. As a technician, you must be concerned not only with the direct sound from the loudspeakers, but also with reflected, absorbed, diffracted, and, to a lesser extent, transmitted sound that has interacted with objects in the environment.

When considering reflection of sound waves, the object of interest is usually large and flat, such as a wall, floor, or ceiling. When the sound source is close (less than 3 m) to the reflective surface, we consider that the wavefront is spherical. When the sound source is far (greater than 15 m) from the wall, the wavefront has expanded so much that it can be considered to be flat over small areas. In either case, the laws of reflection are the same. The waves will be reflected at an angle that is equal to the angle of incidence, even for spherical wavefronts. Therefore, for a spherical wavefront, the reflections themselves form a spherical wavefront. When the sound source is far from the reflecting surface, the reflected wavefront is also flat. This is shown in Figure 2-1.

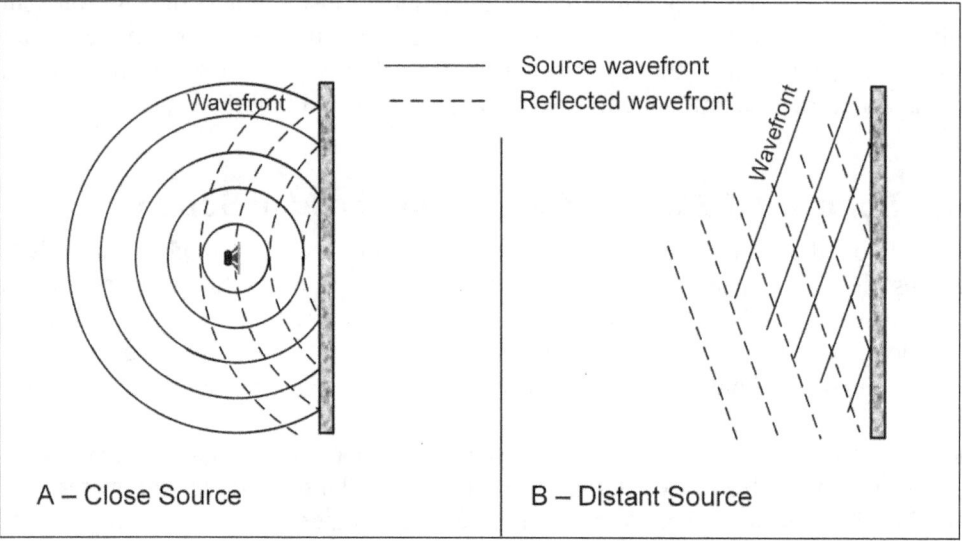

Figure 2-1   Sound Wave Reflection

Clearly, a surface that reflects a lot of sound is one that does not absorb a lot of sound. These surfaces tend to be hard, rigid, and not porous. Examples are glass panes, tiled floors, painted wood panels and painted walls

In a live venue, strong reflections can cause problems with feedback and echo. Rooms that have very reflective surfaces are said to be 'live' and may pose problems with reverberation. Sound engineers therefore need to be aware of these issues and take the appropriate steps when aiming loudspeakers and microphones to best use the reflective properties of the venue to their advantage. We will discuss approaches to such problems in Chapter 13.

2.4 Diffusion of Sound

When reflection occurs from a hard, flat surface such that the sound energy in a small area is all propagated in the same direction (as in a ray of light reflecting off of a mirror), this is called specular reflection. This is shown in the left side of Figure 2-2. Specular reflection may exacerbate certain problems, such as feedback, in certain situations.

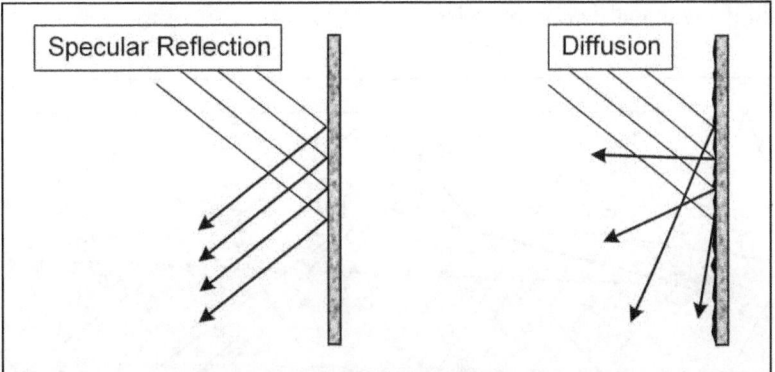

Figure 2-2   Reflection and Diffusion

When reflection occurs from a rough or irregular surface such that the sound energy in a small area is propagated in many directions, as shown in the right side of Figure 2-2, this is called diffuse reflection or simply diffusion. Because the sound energy is spread over a larger area, the harsh directional qualities associated with specular reflection are reduced. Diffusion therefore has the effect of 'taking the edge off' the sound in a room and adds to the sense of spaciousness.

2.5 Reverberation

Whenever there is reflection of sound waves in an enclosed area (even partially enclosed), there will be reverberation. Consider a listener in a room with a loudspeaker emitting sounds. The first sound heard by the listener is the direct sound, which takes the shortest path from the source. This sound is heard at a higher intensity than any single reflection. The listener will next hear sounds reflected from various surfaces. Figure 2-3 graphically shows this situation. The more times the sound is reflected, the less the intensity will be when it arrives at the listener. The intensity decays exponentially.

Sounds that arrive at the listener within the first 80 ms of the direct sound add to the perceived loudness of the sound and are called early reflections. These sounds are directionally well defined and correlated to the room geometry.

Sounds that arrive at the listener after 100 ms of the direct sound add to the persistence of the sound. That is, they cause the sound to be heard even after the source has stopped transmitting the sound. The rate of arrival of these reflections is much higher than that of the early reflections and it increases very rapidly. Additionally, the direction of these reflections appears more random. These reflections are called late reflections, diffuse reverberation, or simply reverberation. Reverberation gives the listener a sense of size and spaciousness of the enclosure.

Rooms with too little reverberation sound dead and sterile. Rooms with too much reverberation cause the sound to become muddy and difficult to understand, as new sounds blend into the reflections of the old sounds. The reverberation time of a room is a well-defined concept, being equal to the time it takes for the sound level to decay to 60 dB below the initial SPL of the direct sound. The reverberation time of a room depends on the type of walls, ceiling, and floor treatment, as well as the volume of the space. There is no hard and fast rule for optimal reverberation times, but generally, speech requires shorter reverberation time than music. Concert halls have a reverberation time of about 1.5 to 2 seconds. Traditional liturgical music can tolerate longer times than pop music which has short attacks and quick rhythms. A reverberation time of about 0.5 to 1.5 seconds seems like a reasonable compromise.

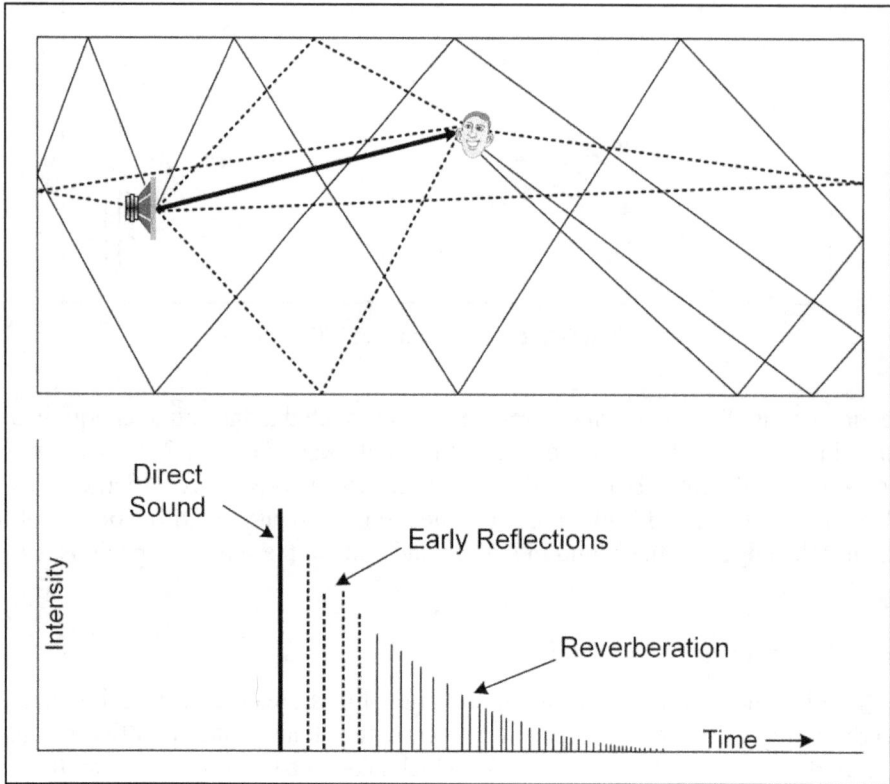

Figure 2-3   Reverberation

2.6   Absorption of Sound by Objects

If a sound wave can penetrate into an object, then some or all of the sound will be absorbed by that object. The best absorbers are ones that are composed of solid material with some air spaces interspersed. Absorption occurs because the energy carried by the wave, which was previously used to displace only air molecules, is now being used to displace the molecules of the object. It takes more energy to displace the molecules of the object than of air. That is why denser materials absorb more sound than less dense ones. High-frequency sounds, which must displace molecules more times per second, are absorbed more than low-frequency sounds. The energy that is absorbed from the sound wave is transformed into a very small amount of heat in the object. This is shown in Figure 2-4. Whatever sound is not absorbed (or reflected) is transmitted through the object.

Every material has an absorption coefficient which tells how effective the material is in absorbing sounds. The absorption coefficient is the percentage of incident sound energy that is absorbed. For example, if 20% of the sound falling on the material is absorbed, then the material has an absorption coefficient of 0.2. If the material absorbed all of the incident sound energy, it would have an absorption coefficient of 1. For a given material, absorption coefficients are usually published for the frequencies 125 Hz, 250 Hz, 500 Hz, 1 kHz, 2 kHz, and 4 kHz. Occasionally, you may see the absorption of a specific absorber published in sabins, the unit of sound absorption (after Wallace C. Sabine, a Harvard University professor c1900). For example, one square meter of a material having an absorption coefficient of 0.2 gives 0.2 metric sabins of absorption. To get the total absorption of an object in sabins, simply multiply the surface area of the object by its absorption coefficient.

One of the things that the sound engineer needs to be aware of is sound absorption by people. Most times, when the sound check is performed in a venue, the space is mostly empty—the audience has not yet arrived. After the show starts, though, the venue will be filled with efficient high-frequency absorbers, i.e., people. The amount of sound absorbed by the audience depends on how they are dressed, whether they are sitting or standing, and the spacing between them. What this means to the engineer is that things will sound different after the audience arrives than during the sound check, and small EQ adjustments will have to be made to restore the overall tonal balance of the system.

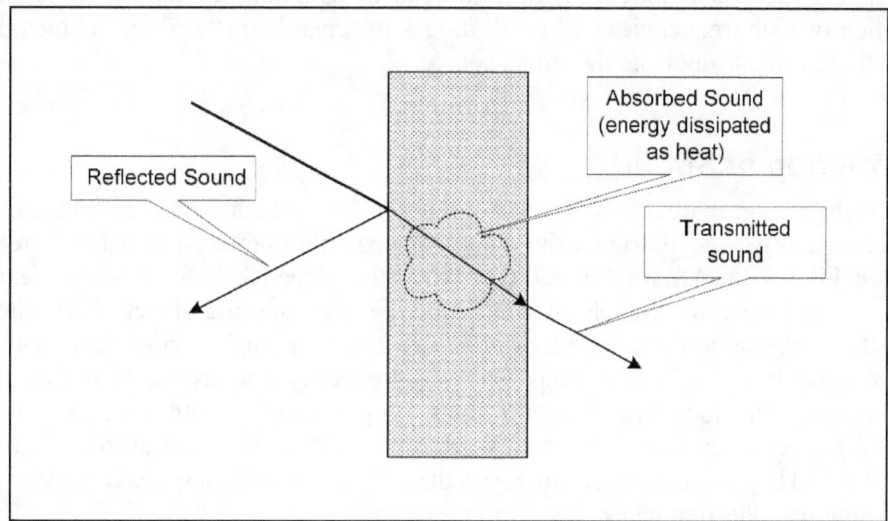

Figure 2-4   Reflected, Absorbed, and Transmitted Sound

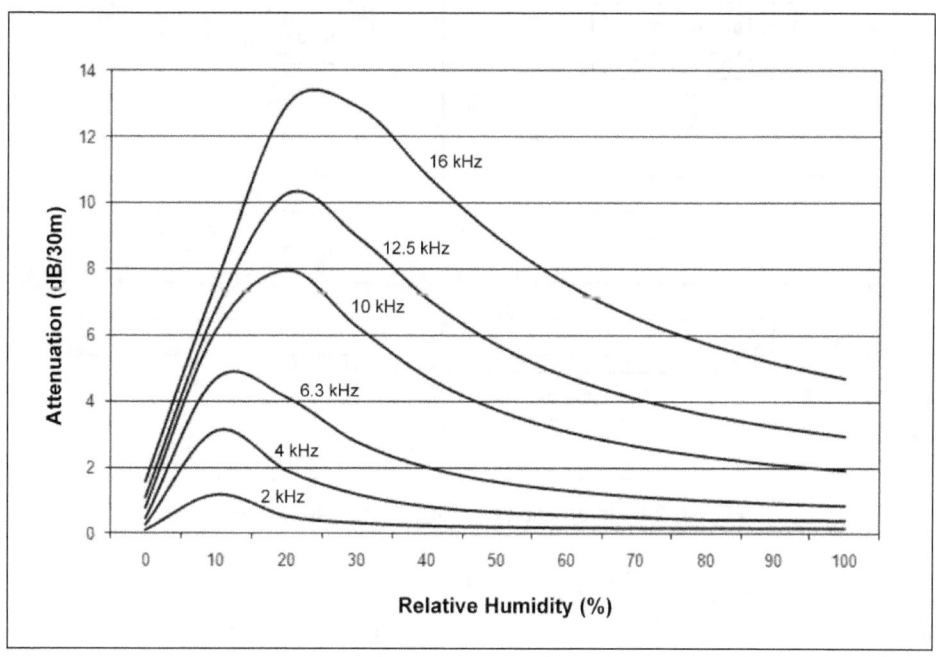

Figure 2-5   Sound Absorption in Air at 20°C

2.7 Absorption of Sound by Air

Within a certain range of relative humidity, from about 5% to 50%, damp air will absorb sound better than dry air. The critical range is between 15% and 40%. This is also the most common range encountered.

Figure 2-5 is derived from data published by Bohn in 1988.[2] As can be seen from this figure, at 12.5 kHz, a change in relative humidity from 10% to 20% results in an increase in absorption of almost 4 dB per 30 m. This could significantly alter the sonic quality of direct sounds in large auditoria and the quality of reflected sounds in smaller spaces. Note that in poorly climate controlled enclosed spaces, the relative humidity will rise as people fill the space.

Look again at Figure 2-5 and notice that the high frequencies are absorbed much more that the low. The absorption of high frequencies will result in lack of 'sparkle' or 'zip' for the members of the audience sitting farther away from the sound source.

2.8 Diffraction of Sound

Diffraction is the change in direction of travel of the sound waves caused by the interaction with an object, where the waves bend around the object. This should not be confused with reflection or refraction. In Figure 2-6(A), we can see that when the object is smaller than or equal to the wavelength of the sound, the wavefront will simply bend around the object. When the object is larger than the wavelength of the sound, as in Figure 2-6(B), a shadow zone develops behind the object. This again is one of the reasons why low-frequency sounds travel farther than high-frequency sounds. The long wavelength sounds simply bend around obstacles, while short wavelength high-frequency sounds get dissipated in the shadow zone. Recall from Table 1-1 that the length of a 4 kHz wave is already much less than 10 cm, so it does not take very large objects to start affecting the high-frequency waves of the sound.

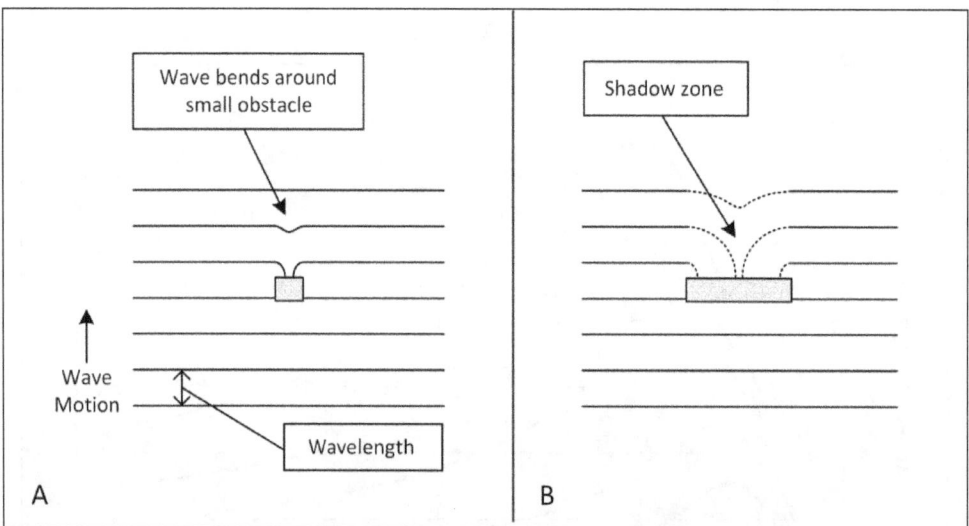

Figure 2-6 Interaction with Small and Large Objects

When the object is larger than the wavelength of the sound, its edge becomes a new point source for the sound. This is shown in Figure 2-7. That is why a small (relative to the wavelength) opening in a wall acts a point source of the sound striking the wall from the opposite side.

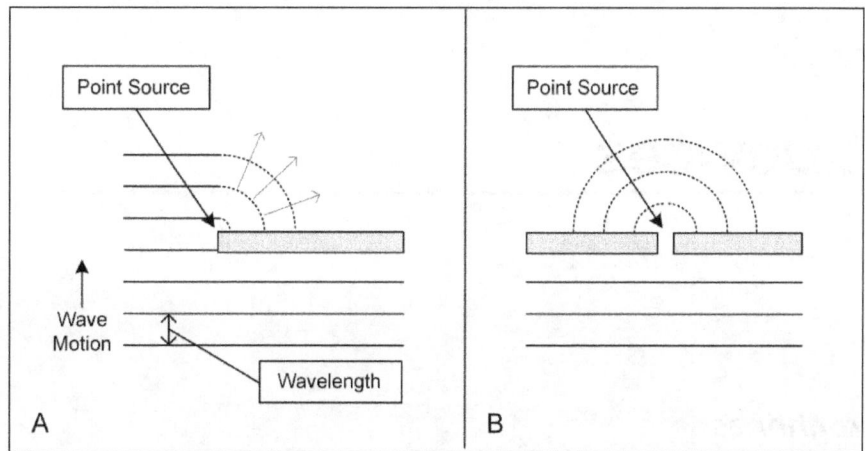

Figure 2-7   Diffraction Point Sources

2.9 Working with the Environment

As we have seen, there are many things that can dramatically alter the sound as it travels from our loudspeakers to the members of the audience. Even the bodies in the audience itself will affect the way the sound is perceived. Understanding how the various elements of the environment affect the sound allows the engineer to apply reasoned and considered adjustments to the sound system, working with the environment rather than blindly fighting it, and removes much of the guesswork when trying to shape the quality of the audio to achieve the end goal of intelligible, pleasing sound.

Review Questions

1. How do changes to the temperature of the air affect the speed of sound traveling through it?
2. How do changes to the humidity of the air affect the speed of sound traveling through it?
3. Consider two rooms: a hall with hardwood floor and bare paneled walls, and a hall with carpeted floor and draped walls. Which room is likely to cause problems with feedback? Why?
4. Explain what is meant by diffusion of sound.
5. What are early reflections?
6. What is sound reverberation?
7. How does reverberation affect the intelligibility of sound in a room?
8. What effect does the audience have on the quality of the perceived sound in the listening space and why?
9. Discuss how the level of relative humidity of the air affects the absorption of sound waves traveling through the air.
10. What are the minimum dimensions of an object that will just create a shadow zone for a 2 kHz sound wave?
11. What happens to the sound energy that is absorbed by an object?
12. An absorption panel is 2 m tall and 1.5 m wide, and has an absorption coefficient of 0.25. What is its total absorption in metric sabins?
13. What kind of surface would you expect to be good at diffusing sound waves?
14. What is the difference between diffusion and diffraction of sound?

Chapter 3

Input Devices

3.1 Microphones

In this chapter we will talk about input transducers and input devices. Let us begin with a look at microphones.

Recall that a transducer is a device that converts one form of energy into another form. A microphone is a transducer that converts sound energy to electrical energy. We convert sound energy to electrical energy so that we can process it in a more effective manner. A microphone captures some of the sound energy falling on it and uses that energy to do the work of the conversion process. There are many types of microphones, including dynamic, condenser, ribbon, and electret. In live sound reinforcement, the dynamic and condenser types are most common.

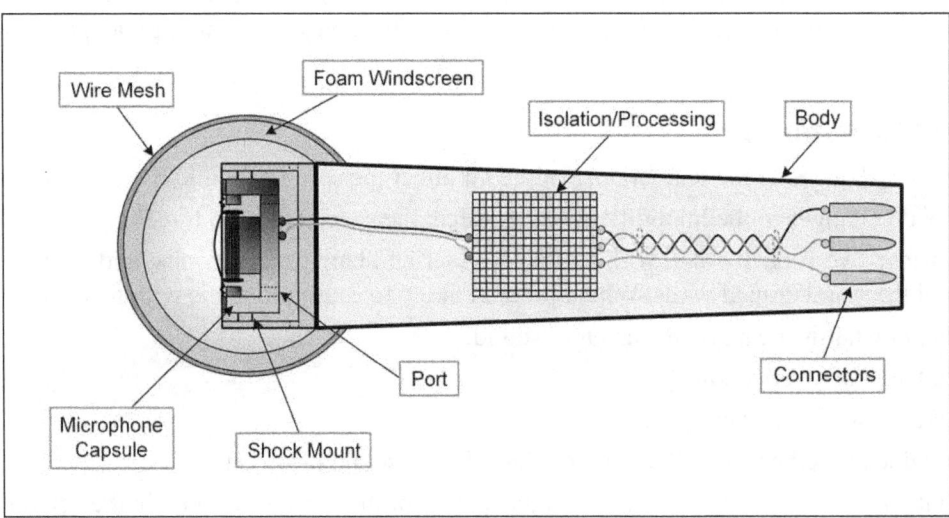

Figure 3-1 Microphone Cross-Section

From a high level, most microphones have the same basic design. Figure 3-1 shows the design of a typical handheld microphone. The differences in the types of microphones will be in the details of the components, such as the capsule and isolation/processing device. As can be seen in Figure 3-1, the actual microphone capsule is held in place by a shock mounting system to reduce handling noise and damage from hard bumps. Ports situated behind the capsule help establish the directional characteristics of the microphone. The capsule is further protected by a foam windscreen and wire mesh at the front of the microphone. The windscreen helps to reduce wind noise and prevents moisture from the user's breath from getting into the capsule. Located in the body of the microphone is some sort of isolation and processing device which serves to isolate the capsule from, and match the electrical characteristics of the capsule to, the external processing devices. Depending on the microphone type, the isolation and processing device may either be a simple transformer or a more complex amplifier. The capsule is connected to this device by shielded wire, and the isolation and processing device in turn is connected to the pins on the end of the microphone by a shielded twisted pair of wires. In professional microphones, these pins are typically in the configuration of an XLR plug. We will now take a more detailed look at the more common types of microphones in use.

3.2 Dynamic Microphones

Dynamic microphones are the most common and least expensive microphone type used in live sound reinforcement. An example of this type of microphone is the popular Shure SM58, shown in Figure 3-2. The principle of the dynamic microphone is simply that of a wire moving perpendicularly through a magnetic field, so as to cut the lines of flux. When a wire moves in a magnetic field, a current is induced in the wire, and a voltage difference—sometimes called electromotive force, or EMF—appears at the ends of the wire. This is shown in Figure 3-3. The direction in which the induced current flows depends on the direction of travel of the wire. If the direction of travel is reversed, so is the direction of current flow. Thus, the dynamic microphone works just like a tiny electric generator.

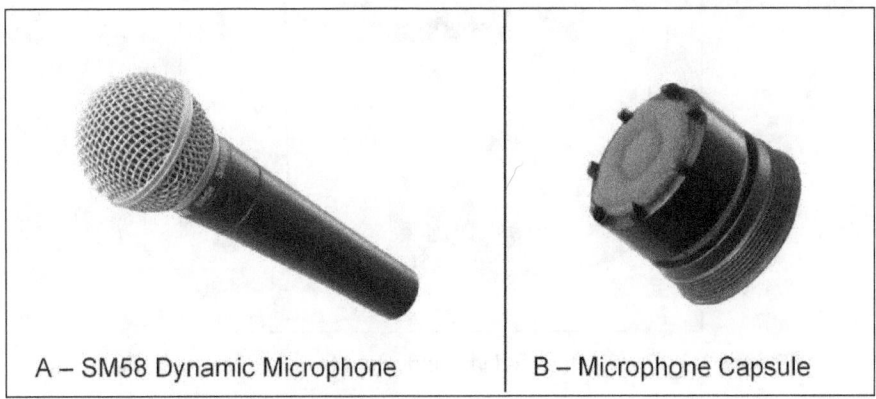

| A – SM58 Dynamic Microphone | B – Microphone Capsule |

Figure 3-2   Ubiquitous Shure SM58 Vocal Dynamic Microphone and its Capsule

For use in a dynamic microphone, we want to have the induced current and voltage as high as practical. The voltage induced across the wire can be increased primarily in two ways. The first is by making the magnetic flux greater. This is accomplished by using a more powerful magnet over a smaller area. The second is by increasing the length of the wire in the magnetic field. This is accomplished by making it into a tightly wound coil.

The heart of the dynamic microphone is the capsule, like that of the SM58 shown in Figure 3-2(B). This is where the principles of the wire moving through a magnetic field are practically realized. A schematic diagram of a dynamic capsule is shown in Figure 3-4. The capsule consists of a diaphragm which intercepts the sound wave. The sound wave causes the diaphragm to vibrate back and forth, in sync with the changing pressure of the air. Thus, the diaphragm must be as light as possible to be able to capture as much of the intricacies of the sound as possible. The diaphragm is connected to a cylindrically formed voice coil, which is suspended between the poles of the magnet.

Because dynamic microphones generate their output by having a changing magnetic field induce a current in the voice coil, they are also susceptible to stray magnetic fields that can induce unwanted signals in their coils. For example, a strong power line field may induce hum in the signal output of the capsule. Manufacturers therefore sometimes incorporate an additional coil, called a hum-bucking or hum compensating coil, wound in the opposite direction to the voice coil. It is placed in close proximity and same orientation to the voice coil (but is not attached to the diaphragm), and wired in series with it. The idea is that any external varying magnetic field will induce the same current in both coils, but since they are wound in opposite directions and wired in series, the externally induced current will cancel, rendering the microphone insensitive to external electro-magnetic interference.

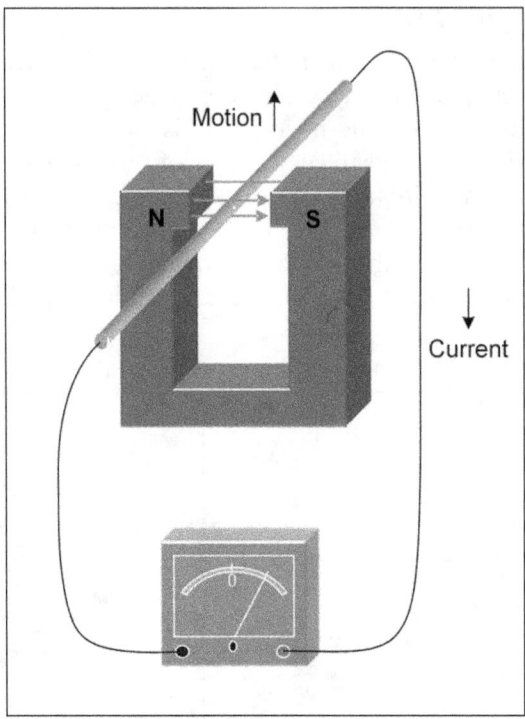

Figure 3-3 Current Induced in Conductor Moving in Magnetic Field

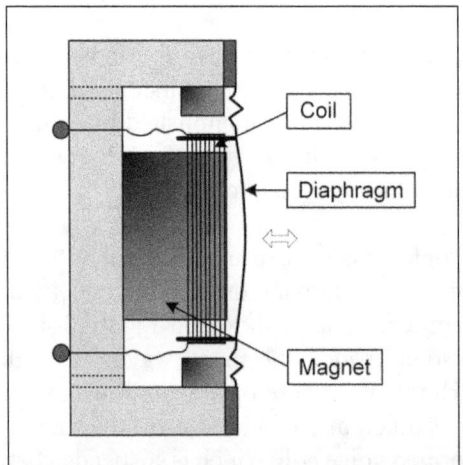

Figure 3-4 Dynamic Microphone Capsule

In Figure 3-5, we can see the circuitry of the dynamic microphone. The circuitry is conceptually simple. The output of the capsule is connected to the primary windings of an isolation transformer. The transformer converts the two-wire output of the capsule to a three-wire configuration that can be used in a balanced connection. We will discuss balanced connections in detail in Chapter 5. The secondary windings of the transformer are connected to the XLR output plug built into the microphone body. Note that pin 1 is ground (0 V signal reference), pin 2 is hot (positive signal), and pin 3 is cold (negative signal).

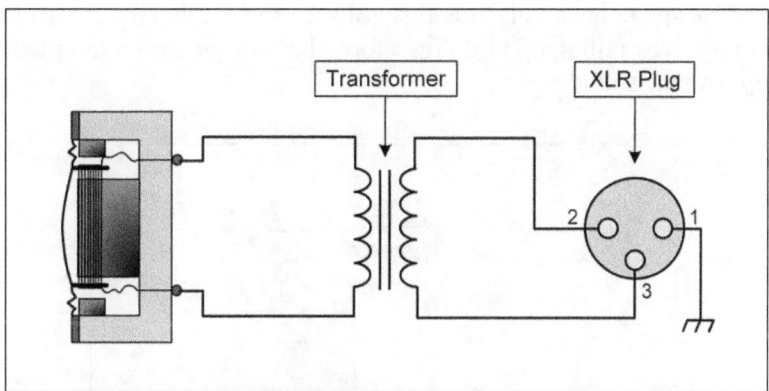

Figure 3-5   Internal Circuitry of a Dynamic Microphone

3.3 Condenser Microphones

Condenser is the old name for capacitor. A capacitor consists of two conductive plates separated by an insulator, called a dielectric. When a voltage is applied to the plates of a capacitor, a negative charge will be built up on the plate connected to the negative voltage source and a positive charge built up on the plate connected to the positive voltage source. While the capacitor is charging, a current will appear to flow through the capacitor, but in reality, the current flows between the source and the plates. Except for leakage, no current flows through the insulator. When the capacitor is fully charged, its plates will have a voltage equal, but opposite to, the charging source and no current will be seen to flow. Thus, capacitors store electrical energy in the form of opposite electrical charges on its two plates. *Capacitance* is a measure of the ability of the capacitor to store this electrical energy and depends on such things as the surface area of the plates and the distance between them.

If the charging source is removed from the plates of a charged capacitor and a voltmeter is connected in its place, it will read a voltage that is equal to the charging voltage, except for leakage losses. This is shown in Figure 3-6. Suppose now that we vary the distance between the plates of this charged capacitor and observe the voltmeter. We will see that the voltage that is read by the voltmeter varies in step with how we move the plates. Since the capacitance is inversely proportional to the distance between the plates, the voltage will be lower when the plates are closer together and higher when the plate are further apart. (This is true over a small working distance, where the plate separation distance is much smaller than the plate dimensions. If the plates are moved far enough apart, the voltage will drop to zero.) If the plates of a charged capacitor are connected with a conductor, the excess charge on one plate will flow to the other until they are both neutralized and no more current flows. In this state, the capacitor is said to be discharged and the voltage across the plates will now be zero.

An example of a condenser microphone is the AKG C535 shown in Figure 3-7, whose capsule is shown on the right of the figure. Figure 3-8 shows the cross-section of the capsule of a condenser microphone. If you look closely, you will see that it is nothing but a capacitor. One plate is the diaphragm, and the other is a backplate a short distance from the diaphragm. The insulator is air. The backplate may be perforated to help tune the stiffness of the diaphragm, or to provide a path for sound waves to reach the back of the diaphragm. The working principle is simple. A charge is maintained on the capacitor by a source voltage. When the capacitor is charged, there is a voltage across the plates of the capacitor, i.e., the diaphragm and the backplate. If the distance between the diaphragm and the backplate is varied, so is this voltage. The diaphragm moves back and forth in sync with the sound waves falling upon it. Therefore, the voltage across the plates of the capsule also varies in step with the sound.

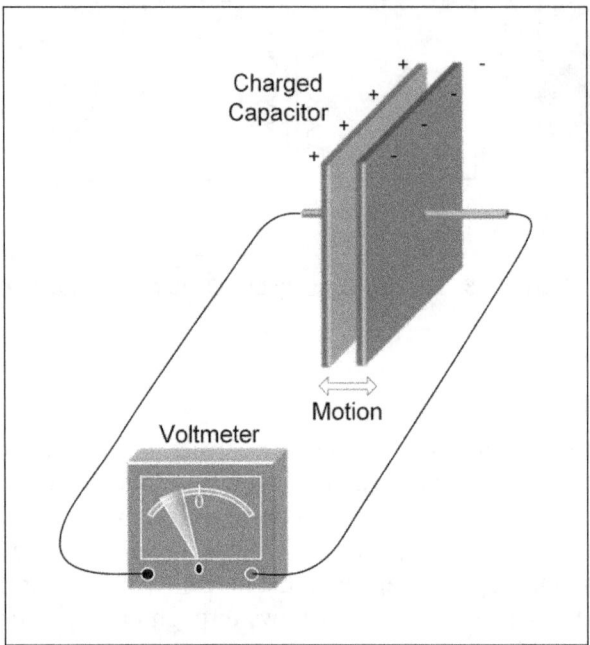

Figure 3-6   Charged Capacitor – Voltage Varies with Plate Separation

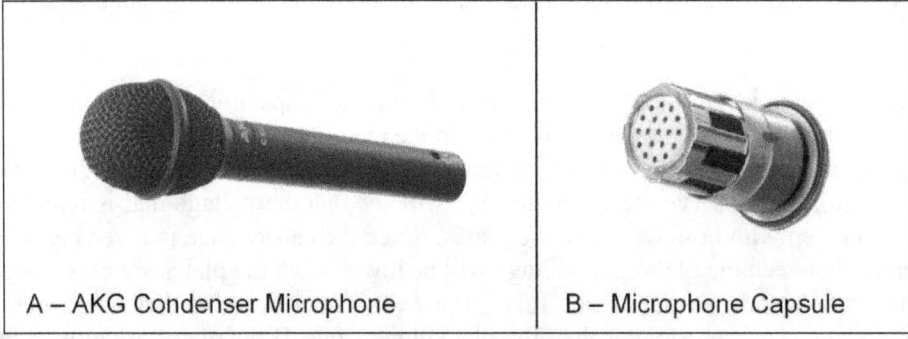

A – AKG Condenser Microphone   B – Microphone Capsule

Figure 3-7   AKG C535 Condenser Microphone and its Capsule

In comparison to a dynamic microphone capsule, the diaphragm of a condenser microphone can be very light, since there is no coil attached to it. Because of this low mass, the sound waves can more easily move the diaphragm. That is why condenser microphones are so sensitive and can also capture higher frequency sounds better than a dynamic microphone.

The voltage fluctuations produced by a condenser microphone capsule when the diaphragm vibrates are very small. These voltage fluctuations must be amplified before being sent to an external device, such as a mixer. Handheld condenser microphones therefore contain a small amplifier built right into the microphone's body. Thus, unlike the dynamic microphone, the condenser microphone must be powered. This is shown in Figure 3-9. In addition to powering the amplifier, voltage must also be applied to the capsule to charge it.

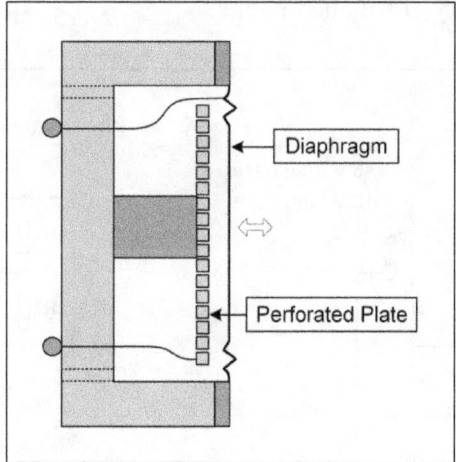

Figure 3-8   Condenser Microphone Capsule

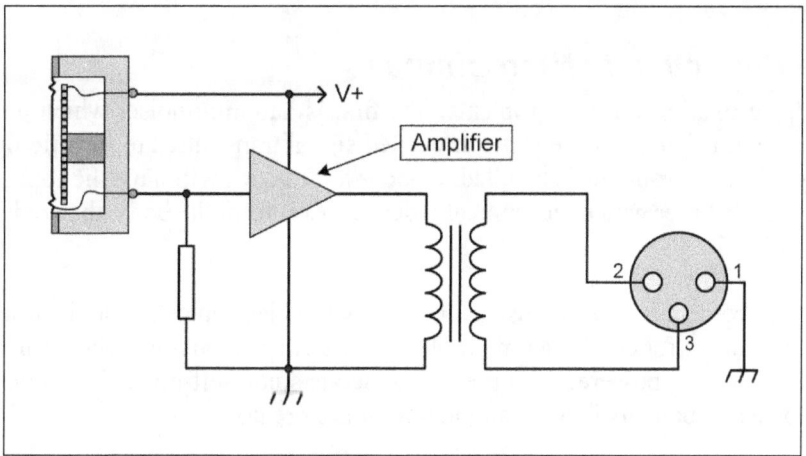

Figure 3-9   Internal Circuitry of a Condenser Microphone

3.4 Phantom Power

As we saw above, condenser microphones require power to operate. This can be supplied by a battery that is housed in the body of the microphone. Most times, however, power is supplied by the mixer, in the form of phantom power. Look at Figure 3-10 to see how this works.

When phantom power is turned on at the mixer, +48 V[a] is supplied to pins 2 and 3 of the XLR plug of the microphone. These are the same pins on which the audio signal travels. The addition of this voltage does not interfere with the signals because the line uses differential signaling (discussed in Chapter 6), and +48 V is added to both the positive and negative signals. Inside the microphone, the circuitry is supplied with the voltage taken off the center tap of the output transformer, or tapped via a pair of precisely matched resistors, shown as R1 and R2 in Figure 3-10(B), for the case where there is no output transformer. The transient voltage suppression diodes, D1 and D2, protect the circuit from damage in case the supplied voltage is too high. Except for that used in powering the amplifier, no current flows through the microphone when phantom power is turned on because no potential difference (i.e., voltage) is developed across the secondary winding of the output transformer or other output circuitry of the microphone.

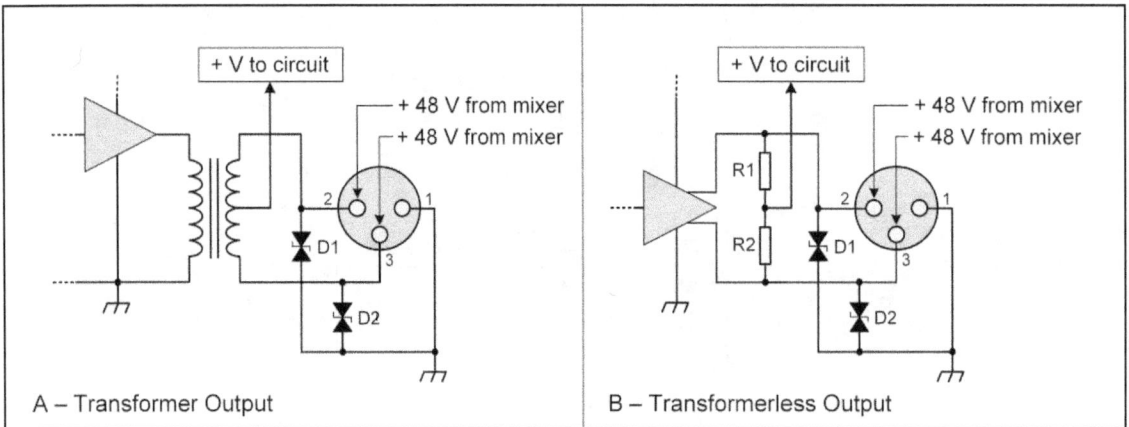

Figure 3-10   Phantom Power Extraction in a Condenser Microphone

3.5 *Noise Reduction in Microphones*

Microphones have two basic noise problems. The first is handling noise, which all microphones have to some extent. Handling noise is the unwanted signal transmitted to the microphone capsule when the body of the microphone is handled. Good microphones will have the capsule mounted in an effective shock absorber which mechanically decouples it from the body, thus reducing handling noise considerably.

The second noise problem is that of noise generated by the electronics inside the microphone. This is usually only a concern for condenser microphones or some ribbon microphones that have a built-in preamp. All active (i.e., powered) components generate some self-noise. The better microphones will use low-noise components in the manufacture of the preamp.

3.6 *The Polar Coordinate System*

In Section 1.5, we discussed how data can be displayed in a two-dimensional (2-D) coordinate system called a rectangular coordinate system. In this section, we will examine another two-

---

[a] IEC 61938, the standard that defines the characteristics of phantom power, actually specifies a number of different variants for powering equipment via this scheme. Currently, the most common version implemented in the pro-audio industry is the +48 V variant, referred to as P48. P12 and P24 (12 volts and 24 volts, respectively) variants are also specified in the standard.

dimensional coordinate system—the polar coordinate system—commonly used to display data about a microphone's spatial sensitivity.

The polar coordinate system can be considered to be based on a set of concentric circles whose center is the origin (or pole) of the coordinate system. The polar coordinate system uses an angular displacement from an axis (the polar axis) that starts at the pole, and a linear displacement from the pole to define a location in 2-D space. Take a look at Figure 3-11 to see how this all fits together.

Recall that a circle is divided into 360 degrees. For our purposes, we will consider the circle to be divided into two halves of 180 degrees each by a vertical line which forms the polar axis. With reference to the pole, any point can then be described as being a certain number of degrees to the left or right of the axis.

We mentioned above that to completely specify the location of a point in polar space, we need to also give its distance from the pole. In Figure 3-11, we have arbitrarily labeled the linear displacement as 5, 10, 15, and 20; thus, the farther away a point is from the pole, the bigger is its value. When applied to microphones, this distance will represent the relative sensitivity of the microphone to sound, and shown in dB.

Combining the two notions of angular displacement and linear displacement, we can see that the coordinates of point **A** in Figure 3-11 is (60,15), while the coordinates of point **B** is (–30,18). Can you determine the coordinates of point **C**?

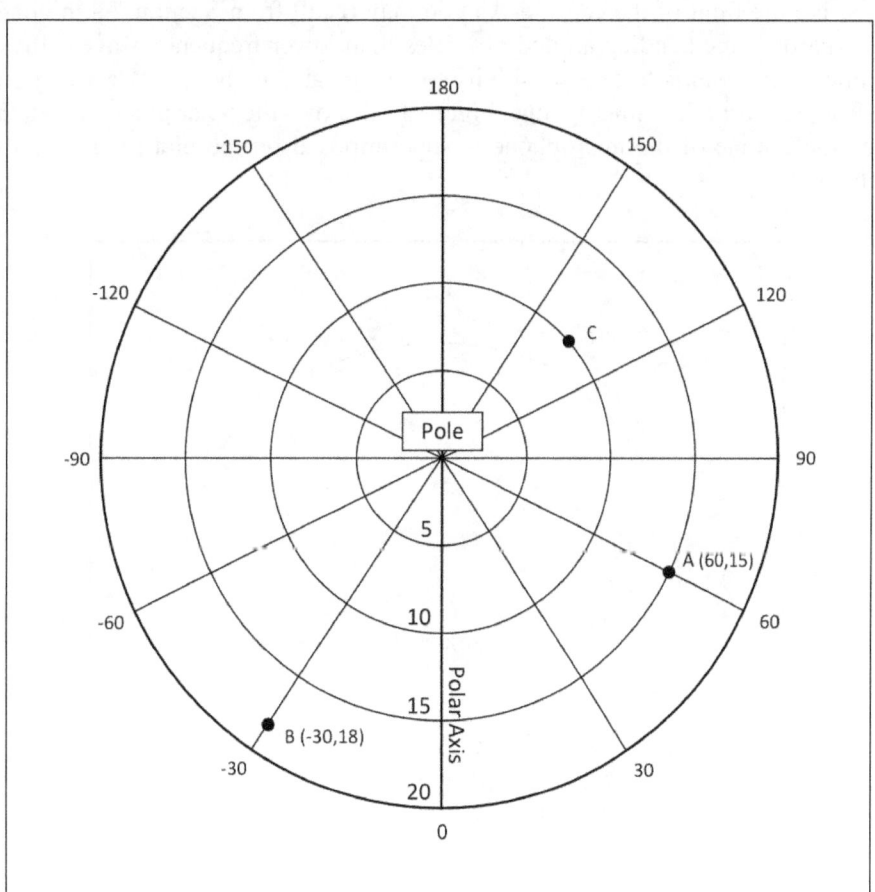

Figure 3-11   Polar Coordinate System

3.7 Microphone Pickup Patterns

A microphone picks up sound from all around it. However, the microphone may not be equally sensitive in every direction. The sensitivity of the microphone to sound coming from a particular direction is shown on a polar plot. This graph depicts what is called the pickup pattern of the microphone.

Remember that in our application, we will use the linear displacement of a point to represent the relative sensitivity, in dB, of the microphone to sound coming from that direction. When we place several such points on the diagram and join these points with a line, we get the polar plot of the pickup pattern.

3.7.1 Omnidirectional Pickup Pattern

With the exception of very high frequencies, an omnidirectional microphone picks up sound equally well from all directions around the microphone. The smaller the diaphragm of the capsule, the higher the frequency to which the microphone remains omnidirectional. An omnidirectional microphone capsule based on the design shown in Figure 3-4 is sealed, except for a very small capillary tube that allows the gradual equalization of air pressure inside the capsule with ambient atmospheric pressure. Thus, only the front face of the diaphragm is impacted by the sound wave. This is referred to as a *pressure* microphone capsule, as the diaphragm responds only to the pressure changes at the front face of the capsule. Since an omnidirectional microphone capsule does not have ports, the sound waves from behind the microphone capsule must therefore bend around the microphone structure to strike the front of the capsule. As you may recall from Section 2.8, higher frequency waves have a harder time bending around obstacles than lower frequency waves. Thus, an omnidirectional microphone starts to lose sensitivity to sounds above about 8 kHz when these sounds are coming from the rear of the microphone. Figure 3-12 shows the polar plot of an omnidirectional microphone. The outline of the microphone is superimposed on the plot to show its orientation relative to the pole.

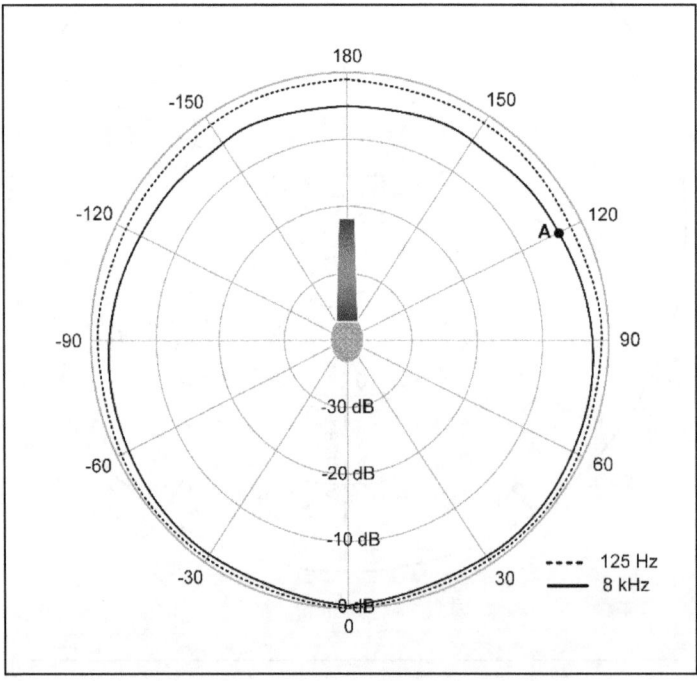

Figure 3-12 Omnidirectional Pickup Pattern

As an example of how to interpret the information in polar plots, look again at Figure 3-12 and notice the point **A**. It is located on the 8 kHz line, at 120° off the axis of the microphone, and its displacement value is about −4 dB. This tells us that at 120° off the axis, the microphone is about 4 dB less sensitive to 8 kHz sounds compared to the on-axis orientation.

Because an omnidirectional microphone does not greatly discriminate in its directional pickup, it is useful as an ambience microphone. It will pick up audience sounds as well as stage sounds. It is also good for groups of singers who are not clustered tightly together. However, using too many omnidirectional microphones on stage may cause feedback problems, since they will easily pick up the sound from the stage and FOH loudspeakers and send it back to the mixer.

3.7.2 Cardioid Pickup Pattern

The cardioid pattern is one of many unidirectional pickup patterns. It is called cardioid because the polar plot resembles a heart shape. This can clearly be seen in Figure 3-13. A cardioid microphone is most sensitive to sound arriving directly in front of the microphone. Sounds from the rear are rejected quite well. Sensitivity falls off fairly rapidly once the source moves off to the side of the microphone.

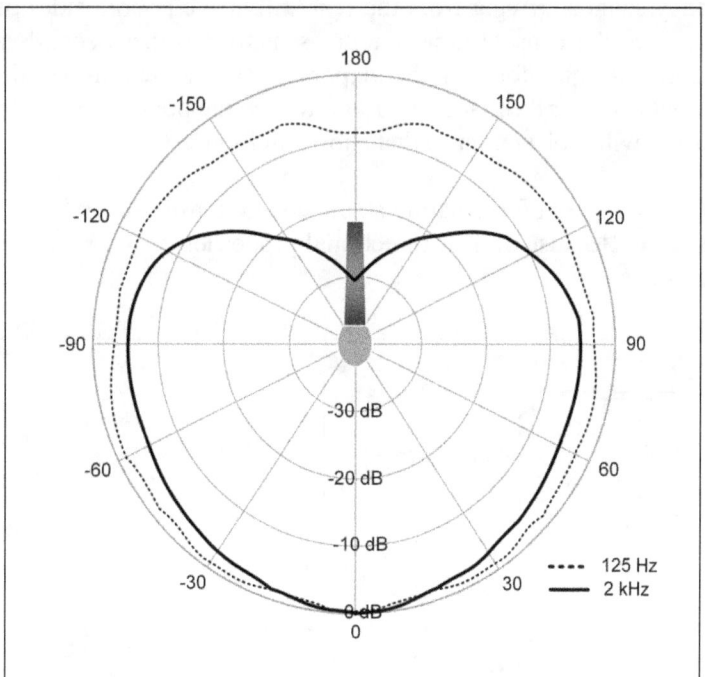

Figure 3-13   Cardioid Pickup Pattern

The cardioid microphone gets its directional properties from the strategic placement of a port on the rear or side of the microphone's capsule. The addition of this port allows sound waves to strike the diaphragm from the rear as well as the front, and so the diaphragm responds to the difference in pressure at its two surfaces. The term *pressure gradient* is therefore used to describe these capsules. The port is designed so that for sound waves arriving from the rear, the time it takes for the sound to reach the back side of the diaphragm via the port is equal to the time it takes the sound wave to travel to the front of the diaphragm via the outside of the capsule. Thus, the sound wave arrives at both sides of the diaphragm at the same time (and in phase), and so there is no pressure difference (and hence no pressure gradient) between the front and rear of the diaphragm. Since there is no pressure gradient, the diaphragm does not move and sound from the rear of the microphone is rejected. This is shown in Figure 3-14(A). For sounds arriving from the front of the capsule, the time it takes for the sound wave to arrive at the rear of the diaphragm via the port is longer than it takes to arrive at the front of the diaphragm, and so there is a strong pressure gradient established across the diaphragm. The output from the capsule is therefore high for sounds arriving from the front of the capsule. This scenario is shown in Figure 3-14(B).

The pressure gradient developed across the diaphragm is frequency dependent. Low-frequency waves develop a lower pressure gradient than high-frequency waves because high-frequency waves change the air pressure faster over a given time than low-frequency waves (the maximum gradient across the reference planes occurs at a frequency where $d = 1/2$ wavelength). This pressure response rises by 6 dB per octave.[1] If left like this, the capsule would have a low output level for low-frequency sound, and a high output level for high-frequency sound. The capsule is therefore mechanically tuned (i.e., equalized) to make it more sensitive to low-frequency sound arriving from the front of the diaphragm, thus flattening the frequency response. This tuning has a number of undesirable effects, however. First, it makes the capsule more sensitive to handling noise and low-frequency wind noise (such as plosives). Secondly, it causes the microphone to gradually lose its directional properties in the lower frequencies of its operating range, becoming more omnidirectional. Low-frequency sounds arriving from the rear still arrive at both sides of the diaphragm at the same time, but since the capsule is now more sensitive to these sounds at the front of the diaphragm, there is now unequal force on the diaphragm (i.e., a pressure gradient) and the microphone produces output even for these sounds. Thus, we see the polar response for a low-frequency sound, such as 125 Hz, will look similar to that shown in Figure 3-13.

Cardioid microphones are useful for isolating a sound source from several others on stage and for rejecting audience noise. Because of their directional properties, they are also useful in helping to control feedback on stage.

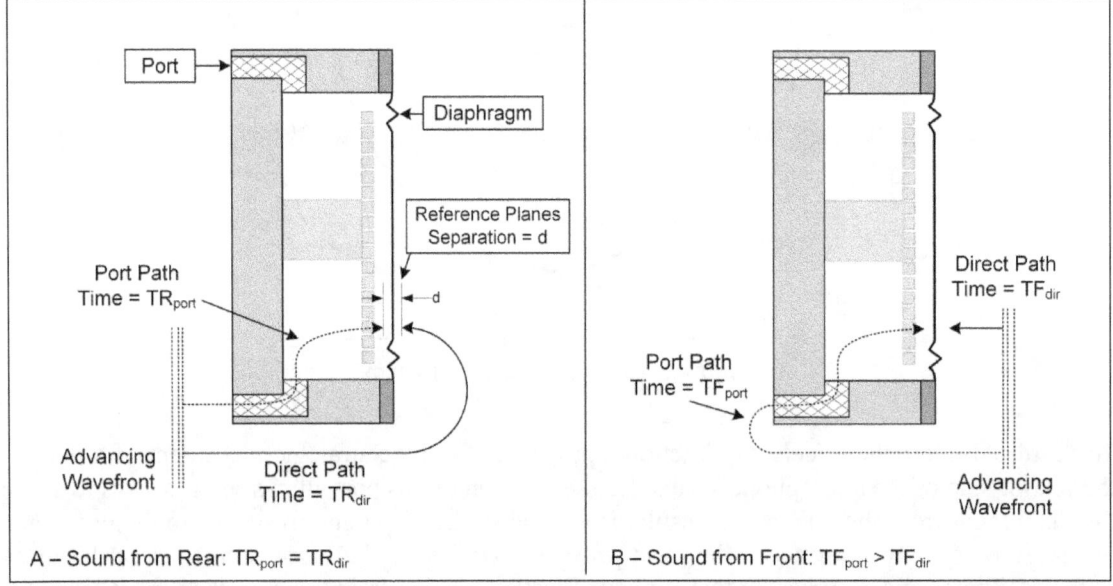

Figure 3-14 Pressure Gradient Capsule Showing Sound from the Rear vs. the Front

3.7.3   Proximity Effect

A third consequence of the tuning of the microphone's capsule to equalize the pressure gradient is the proximity effect. The proximity effect describes the increase in low-frequency response as the user moves very close to the microphone. This phenomenon is also sometimes referred to as the presence effect. Only directional microphones exhibit this effect, since this applies only to pressure gradient capsules.

At very close range, e.g., less than 3 cm, the distance from the source to the front of the diaphragm can be substantially shorter than the distance from the source to the rear of the diaphragm (via the port). Thus, we know from the inverse square law that the SPL at the front of the diaphragm will be much higher than at the rear. A large pressure gradient will therefore be established for all frequencies (including low frequencies) in this situation. Remember, though, that the manufacturer has tuned the capsule to boost the low-frequency response because low frequencies naturally produce a smaller pressure gradient. Therefore, when the low frequencies generate a large pressure gradient equal to that of the high frequencies due to the inverse square law and are also additionally boosted by the capsule's equalization, the output from the microphone takes on a bassy, boomy character.

3.7.4   Other Pickup Patterns

Omnidirectional and cardioid pickup patterns are the two most commonly encountered pickup patterns in live sound reinforcement. Microphones with other pickup patterns are used when these two common types are not adequate for the situation at hand. Some of these patterns are shown in Figure 3-15.

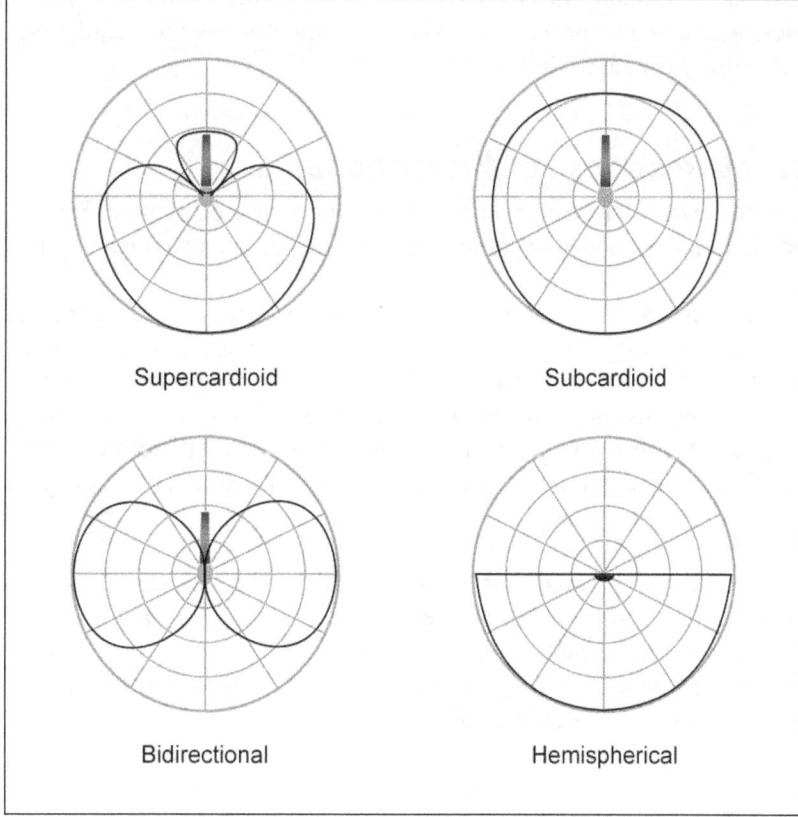

Figure 3-15   Other Pickup Patterns

Supercardioid microphones have a similar pattern to cardioid, but narrower. This pattern also exhibits a rear lobe, and so these microphones do not reject sound directly behind the microphone as well as cardioids. They are very good for sound source isolation, e.g., focusing on a soloist performing with a group. Hypercardioid and shotgun microphones offer even narrower pickup patterns, but these come with a tradeoff of even bigger rear lobes.

Another interesting pattern is the subcardioid. This is somewhat between the omnidirectional pattern and the cardioid, rather like a flattened omni, and sometimes referred to as a wide cardioid. This pattern is useful when we want to emphasize a particular area or spot while still capturing some of the ambience of the surrounding environment.

A figure-8 pattern is so called because it looks like the numeral 8 laid on its side. It is also sometimes called a bidirectional pattern. This is the natural pickup pattern of a diaphragm that is fully open at both the front and the rear, as is the case with some ribbon microphones. However, this pattern is more typically achieved by putting two diaphragms on opposite sides of a single backplate (in the case of a condenser) in the capsule and wiring them so that their signals subtract.[2] Essentially, it is like putting two cardioid capsules back to back and wiring them so that the positive output of one is connected to the negative output of the other.[3] The dual diaphragm capsule is mounted sideways in the head of the microphone, so the microphone must be addressed from the side rather than the top. With respect to this orientation (side address), the microphone therefore picks up sound equally from the front and back (bidirectional) and rejects sound from the sides. These microphones are good for users who face each other while performing. They are also used in the middle-side stereo recording technique.

A hemispherical pattern is simply like half of the omnidirectional pattern. This pattern is usually seen on what are called boundary layer microphones, used typically on large flat surfaces such as conference tables, where many speakers sit around the microphone and require equal access to it, or on podiums to help deal with phasing problems.

3.8 Frequency Response of Microphones

The frequency response of an electronic device is an important specification. Before we discuss the frequency response of microphones in particular, let us define what we mean by this term.

Unless done deliberately, a device should not change any aspect of a sound signal except its amplitude. The frequency response shows how the device changes the signal's amplitude for all frequencies of interest as the signal is processed by the device, from input to output. If changes in amplitude are equal over this range of frequencies, this is referred to as a flat frequency response and is usually desirable. The frequency response of a microphone therefore tells us how well the microphone transforms sound in relation to the frequency of the sound over its operating frequency range.

While the goal of a transducer designer is usually to have a flat frequency response, some manufacturers may deliberately emphasize or cut certain frequencies. This may even be switchable by the user (e.g., low cut to reduce noise; mid boost to emphasize voice; high boost to add sparkle). Another goal is to have a wide frequency response, i.e., it should be able to capture as much of the audio range from 20 Hz to 20 kHz as possible.

It is much easier to absorb data such as frequency response when presented in a visual format. Therefore, the frequency response is usually displayed on a graph, such as Figure 3-16. That makes it easy to see how flat the response is over the device's operating range and how wide it is with respect to the audio band of 20 Hz to 20 kHz. From this graph, we can see that the response is quite flat in the important frequency range where human speech occurs. The manufacturer has also boosted the response at the higher frequencies, which adds sizzle to higher notes and sounds.

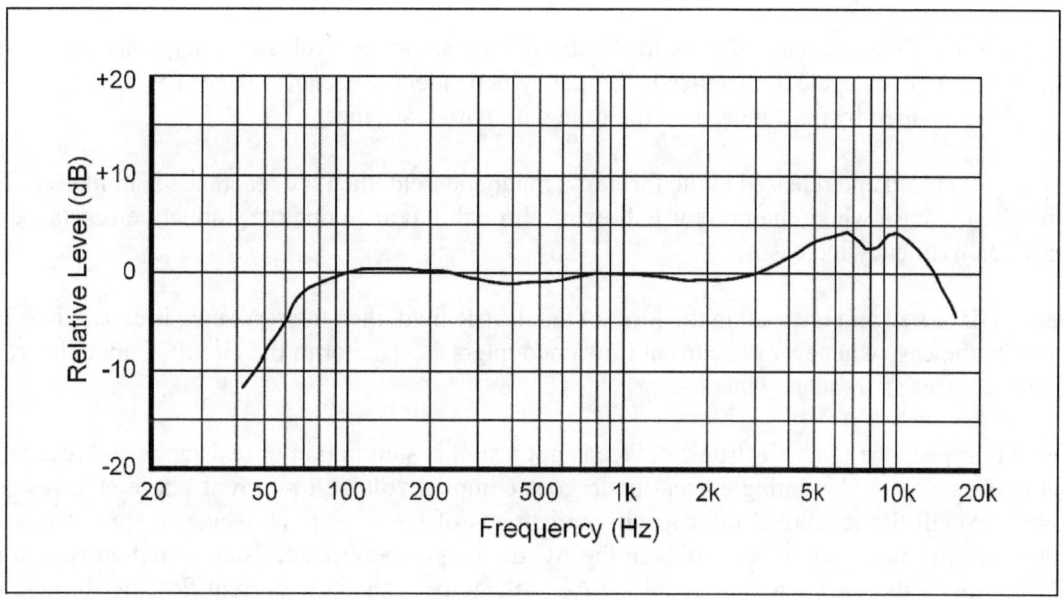

Figure 3-16   Frequency Response of Typical Live Performance Microphone

3.9 Concept of Impedance

Before we talk about microphone impedance, let us figure out what impedance really is. Suppose we have a battery connected to a device (a load) with resistance R, as shown schematically in Figure 3-17(A). Conventionally, direct current (DC) flows from the positive terminal of the battery through the load and continues on to the negative terminal of the battery. The resistance opposes the flow of direct current in the circuit. Resistance is a measure of how much the load opposes the flow of electrons, i.e., current, among its atomic structure and is measured in ohms (Ω). Resistance robs the electrons of some of their energy, as it takes a certain amount of energy to push the electrons through the atomic structure of the device, creating heat in the process. Ohm's Law tells us that if the resistance is $\boldsymbol{R}$ and the voltage of the battery is V, then the current flowing through the resistor is $\boldsymbol{I} = V/R$.

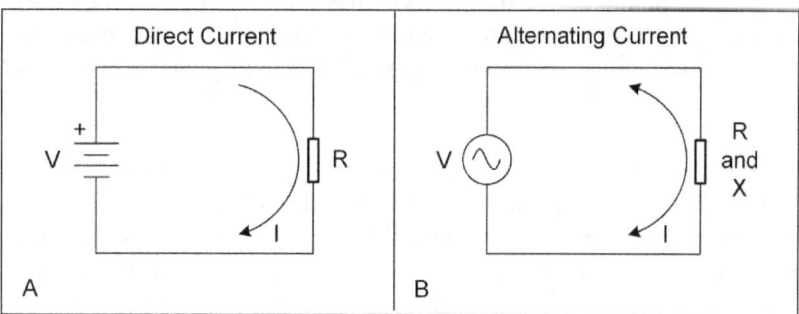

Figure 3-17   Resistance vs. Impedance

With alternating current, the generating device pushes electrons through the load in one direction, and then reverses the current flow, pushing the electrons through the load in the opposite direction. This cycle repeats for as long as there is current flowing. Impedance (Z) is the total opposition that an electronic device gives to alternating current (AC) flowing through it. Impedance is a combination of resistance and reactance. It is also expressed in ohms. This is shown in Figure 3-17(B).

Reactance (X) is a measure of how much the device stores and releases energy as the current fluctuates with each cycle. Reactance is frequency dependent. Reactance is not true resistance. No power is consumed by reactance, as is the case with pure resistance.

If energy is stored and released in the form of a magnetic field, then the reactance is inductive. This would be the case where the current is flowing through a coil (inductor). Inductive reactance increases as frequency increases.[b]

If energy is stored and released in the form of an electric field, then the reactance is capacitive. This would be the case whenever the current flow encounters a capacitor in the circuit. Capacitive reactance decreases as frequency increases.[c]

The total impedance of an electronic device is not a simple summation of resistance and reactance. This is because the alternating current undergoes complex voltage-to-current phase changes as it passes through the inductive and capacitive elements of the device. However, a similar relation between impedance and current holds in the AC domain as between resistance and current in the DC domain. If the impedance is Z and the AC voltage is V, then the current flowing through the device is $I = V/Z$.

3.10 Microphone Impedance

With the above understanding, we now know that the impedance of a microphone represents the total opposition that it presents to alternating current. We are interested in the microphone's impedance for two reasons. Firstly, the electrical current flowing between the microphone and the mixer is alternating current. Secondly, the mixer's input also has a certain impedance, and we should be aware of how that impedance interacts with the impedance of the microphone to affect our signal.

Microphone impedance is characterized as low or high. Low-impedance—or low-Z—microphones are generally less than 600 Ω, and are typically between 150 to 300 Ω at 1 kHz. High-impedance—or high-Z—microphones are generally more than 5000 Ω. There are also very low impedance microphones (~20 to 75 Ω), originally used in broadcast studios (but rarely used today), and originally designed to terminate into 150 to 600 Ω. These can be terminated by normal low-impedance mixer input stages. The low-impedance microphone input on a typical mixer has an impedance of about 2 kΩ. Generally, high-Z microphones have TS phone plugs, while low-Z microphones have XLR plugs.

The well-known maximum power transfer theorem[4] tells us that to transfer the maximum power from a source to a load, the input impedance of the load should match the output impedance of the source. However, it is neither important nor desirable that the input impedance of the mixer should match the output impedance of the microphone, since in this case, we are not trying to get maximum power transfer from the microphone, but rather efficient and clean signal transmission. The input

---

[b] For inductors, $X_l = 2\pi FL$ where π is 3.1416, F is the frequency, and L is the inductance.
[c] For capacitors, $X_c = 1/(2\pi FC)$ where F is the frequency C is the capacitance.

impedance of the mixer should be at least five times the output impedance of the microphone. Trying to match impedance exactly causes excessive signal loss and degradation. If Z_{load} is the input impedance of the mixer, and Z_{mic} is the output impedance of the microphone, then the signal loss is given by:

$$Loss\ in\ dB = 20 \log\left(\frac{Z_{load}}{Z_{load} + Z_{mic}}\right)$$

Thus, we can see that if we match impedances exactly, we would incur a loss of 6 dB, and the loss becomes greater as the mixer's input impedance decreases relative to the microphone's output impedance.

3.11 Other Microphone Specifications

Impedance and frequency response are two common elements of the microphone's specification that are well known. There are other aspects of a microphone's performance that also bear consideration.

3.11.1 Sensitivity

The sensitivity of a microphone is greatly dependent on the mass of the diaphragm. This is because the heavier the diaphragm, the greater must be the force of the air impinging on it to get it moving. Furthermore, once it is moving, a heavier diaphragm will have greater momentum. That means that it will be harder for it to change direction, and therefore it will be less responsive to high-frequency sounds which try to change the direction of travel of the diaphragm many more times per second compared to low-frequency sounds. Recall that in a dynamic microphone, the coil is directly attached to the diaphragm. This makes it relatively heavy when compared to diaphragms of condenser and ribbon microphones, which have nothing attached to the vibrating area of the diaphragm. In condenser and ribbon microphones, electrical attachments are made to the diaphragm only at its edges. Dynamic microphones are therefore typically less sensitive than condensers and ribbons.

The sensitivity of a microphone is usually given in dBV, and is a measure of the output of the microphone at a reference SPL of 94 dB and a frequency of 1 kHz. 94 dB SPL is taken to be equivalent to a pressure of 1 pascal (Pa). Also, recall from Section 1.8 that 0 dBV is referenced to 1 Vrms. Therefore, you may see the same sensitivity specification written in various forms, e.g., −31.9 dB re 1V/Pa (i.e., −31.9 dB referenced to 1V/Pa); −31.9 dBV/Pa; 25 mV/Pa; 25 mV @ 94 dB SPL. Note that some manufacturers may refer to sensitivity as output level.

3.11.2 Maximum SPL

The maximum SPL specification tells us the highest SPL the microphone is designed to operate under while still producing an acceptably clean signal. It is usually given for a 1 kHz signal with the microphone driving a 1 kΩ load, where the microphone produces no more than 1% total harmonic distortion (THD, discussed fully in Section 5.8.1).

This information is useful, as it may help us choose the right microphone for the job. For example, you would not want to take a microphone with a maximum SPL rating of 130 dB and put it right next to the beater on a kick drum. This information can also be used in conjunction with the sensitivity specification to allow us to figure out the maximum output of the microphone (useful information for predetermining preamp input voltage range compatibility).

3.12 Wireless Microphones

Wireless microphones are very popular because they allow a user the freedom to move around without the restriction of cables. The good ones are unfortunately quite expensive. Wireless microphones use the same type of capsule as found in wired microphones, but instead of connecting to the mixer via a cable, the capsule is connected to an FM transmitter. The corresponding receiver is then connected to the mixer.

The performance of wireless microphones is generally poorer than wired microphones. They are noisier due to the extra electronics involved in the signal path and face problems with signal loss and interference. The range of a typical wireless microphone is about 60 to 250 m, and they can usually operate about 8 hours on a single battery. Typically, they have transmitters of around 50 mW of radiated power.

We will look at various issues and operational characteristics of the elements of wireless microphones in the following sections.

3.12.1 Wireless Microphone Transmitters

Wireless microphone transmitters are either integrated into the same unit as the microphone capsule, as in Figure 3-18(A), in which case it is called a hand-held microphone, or may be housed in a separate unit, usually clipped on a belt (B). In the second case, the microphone capsule is housed in a small unit and may be attached to the end of a boom worn on the head, or a small cord worn around the neck (called a lavaliere microphone), or clipped to the tie or lapel (in which case it is called a lapel microphone). See Figure 14-8 for examples of these microphones. Lapel microphones can be had in either omnidirectional or directional models, but the omnidirectional capsules are generally simpler and smaller than directional ones. Similarly, they can be had in condenser or dynamic versions, and again, the condenser capsules tend to be smaller than dynamic variants. Note that because of where these microphones are normally worn (on the chest or hidden beneath clothing or hair), some equalization is usually needed to make them sound natural.

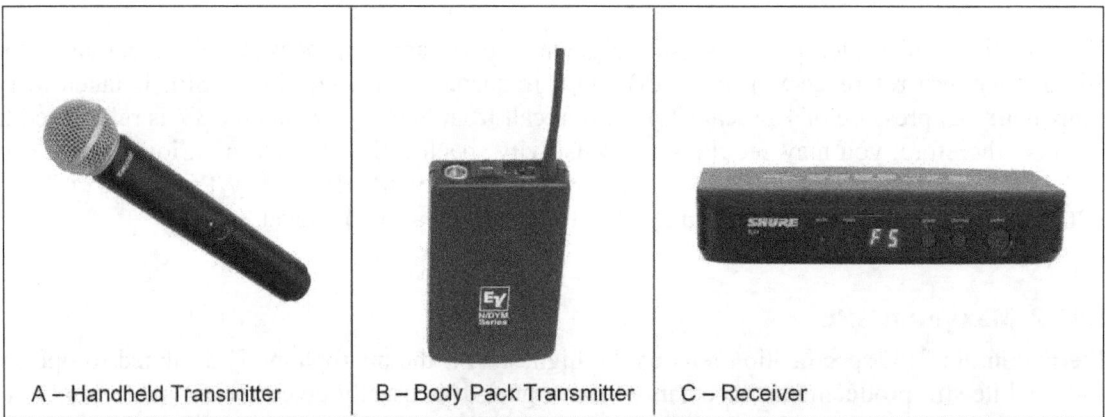

| A – Handheld Transmitter | B – Body Pack Transmitter | C – Receiver |

Figure 3-18 Wireless Microphone Transmitters and Receiver

Currently, wireless microphones operate in either the very high frequency (VHF), the ultra high frequency (UHF), or the Industrial, Scientific, and Medical (ISM) bands at 2.4 and 5.8 GHz regions of the radio spectrum.[5] In any region, the microphone will only operate in a small section of the band and will usually allow the user to select from many channels within that band. As this spectrum is shared with many other types of devices, the opportunity for interference is high. For example,

FM radio, television, land and marine communications units utilize the VHF band; television and walkie-talkies use the UHF band; and cordless phones, wireless LAN, microwave ovens, and a whole host of other devices use the ISM band. UHF systems generally have better immunity to radio frequency interference because they offer more channels from which to choose. As it seems that governments are constantly reallocating the use of radio spectrum, there is always the danger that a wireless system that is legal to operate on a given band today may well be prohibited from operating on that same band in the future.

3.12.2 Wireless Microphone Receivers

The receiver intercepts the radio signal from the transmitter and converts it into a microphone-level or line-level signal that can be processed by the mixer. Receivers come in two flavors—non-diversity and diversity. Diversity receivers, like the Shure BLX4 shown in Figure 3-18(C), were developed to deal with the problem of signal dropout.

3.12.3 Multipath Reflections and Signal Dropout

The received signal strength at the receiver is constantly changing. This variation in strength is called fading and is caused by objects in the environment attenuating, diffracting, or reflecting the signal from the transmitter (as the performer moves around) in an ever-changing way. Signal drop-out is for the most part caused by multipath reflections of radio frequency signals. This concept is shown in Figure 3-19. When the signal from the transmitter arrives at the receiver via paths of different lengths, they end up being out of phase at the receiver's antenna. In sound reinforcement work, it is reflections from walls and ceilings that usually cause these multipath signals. Signals that are out of phase will cancel each other to some extent, causing signal degradation. It is even possible that near total cancellation results, leading to dropout.

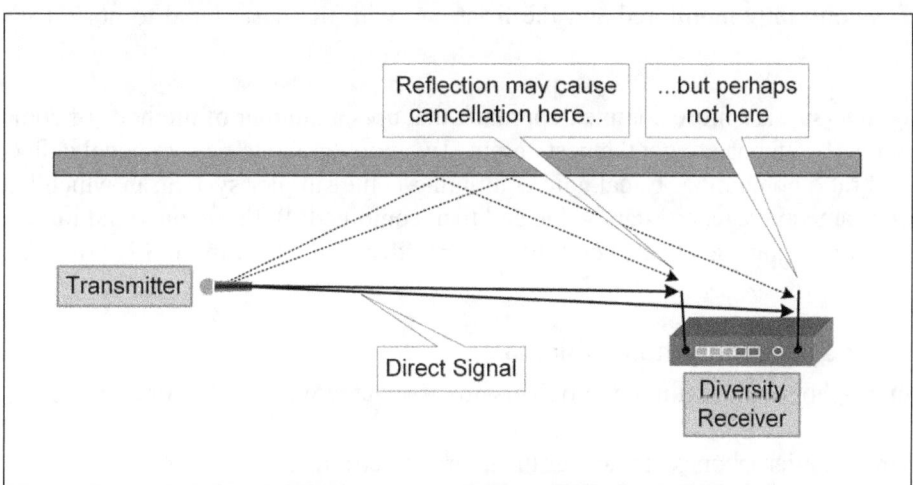

Figure 3-19   Diversity Receiver Reduces Dropouts

To try to overcome the problems with multipath reflections and dropout, receiver designers started putting multiple antennae and, in some cases, multiple receivers in a single unit. Some designs even allow the antennae to be physically detached from the receiver and placed at large distances from each other. Even if cancellation occurred at one antenna site, it is unlikely to occur simultaneously at the other antenna site. These receivers are called diversity receivers.

3.12.4 Diversity Receivers and Signal Processing

A two-step approach is used to tackle the problem of signal dropout in wireless receivers. The first, as noted above, is by using multiple antennae on diversity receivers. There are many forms of diversity, but wireless microphones used for live sound reinforcement employ some form of spatial diversity.

With simple space diversity, the antennae must be separated by at least one-half wavelength[d] of the carrier signal. The amount of antenna separation determines the degree of correlation[e] of the received signals. The antennae in a diversity receiver must be positioned so as to receive uncorrelated, statistically independent signals.

With polarization diversity, the antennae are oriented at angles to each other. This form of diversity is effective because the radio signal usually becomes polarized when it is reflected from walls. When a signal becomes polarized, it vibrates only in one plane. The angle between the antennae is set large enough so that the signals captured by each antenna are uncorrelated.

Phase diversity monitors the strength of the combined signal from the two front ends of the receiver unit. When the level of the combined signals drops below a set threshold, the phase of the signal from one antenna is shifted. Depending on the complexity of the receiver, this may simply be a 180° shift, or a smaller phase shift. This process continues until the combined signal strength is above the threshold once more.

The second step in the fight against signal dropout is in how the signals from the multiple antennae are processed in the receiver unit. Two approaches are generally used in sound reinforcement units—switching and combining.

Switching receivers are by far the simpler of the two designs. Here the output from the receiver front ends is constantly monitored and the front end with the best signal-to-noise (S/N) ratio is selected.

Combining diversity is a more complex approach that uses a number of methods of combining the signals from both antennae to get the best output. Two popular strategies are equal gain combining and maximal ratio combining. Equal gain combining is the simpler system, in which both signals are set to the same average, constant value and then combined. With the maximal ratio combining approach, the incoming signals are combined so as to derive the maximum S/N power ratio.

3.12.5 Wireless Microphone Problems

Wireless microphones present many problems not associated with wired microphones:

- Wireless microphones are noisier than wired microphones.
- Wireless microphones are prone to dropouts.
- Wireless microphones are prone to interference from other transmitters.
- The transmitter battery can go dead in middle of the performance.

---

[d] Recall that wavelength $= \lambda = c/f$ where c is the speed of light, and f is the frequency in Hz.
[e] Correlation is a measure of the similarity between two signals and may depend on their relative phase and spatial position. Here, correlation pertains to the signal fading of the channel.

Whenever possible, a wired microphone should be used. However, when wide ranging freedom of movement is required, the only real choice is a wireless microphone.

3.13 Choosing a Microphone

There are no hard rules when it comes to choosing a microphone. When choosing a microphone, the use to which it will be put must be considered first and foremost. The microphone should be matched to the application, considering such things as the SPL of the source, distance from the source, coverage area of the microphone and need for source isolation, feedback control, tone and mood, and of course, cost. Because this is such a large area, we will devote all of Chapter 14 to looking at various miking scenarios and the kinds of microphones that apply to them.

3.14 DI Boxes

DI stands for Direct Injection, or Direct Input, depending on the source of information. A DI box converts an unbalanced line or speaker-level signal to a balanced microphone-level signal. It also provides impedance matching functions.

The typical arrangement of a DI box is shown in Figure 3-20. The unbalanced, high-Z, line-level input is shown on the left of the diagram, along with a parallel output, which is a straight copy of the input. These jacks are typically connected in parallel with each other, and either one could be used as input or output (Figure 3-22 makes this clear). Because the output of a speaker-level device can be quite high, the DI usually has a switch for attenuating the input signal. This can help to prevent saturating (passive design) or overloading (active design) the DI circuitry, or overloading the preamp circuitry of the mixer. DIs also usually have a 'ground lift' switch which allows the ground plane of the input to be disconnected from the ground of the balanced output. This is provided to help alleviate problems with ground loops that cause hum in the audio output. We will discuss ground loops in detail in Chapter 13.

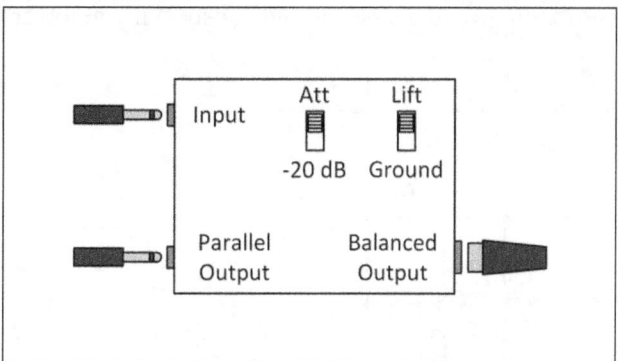

Figure 3-20   Typical DI Box

Figure 3-21 shows how DI boxes are typically connected to the output of line or speaker-level devices, such as electronic keyboards, drum machines, and guitar amps to get a signal that is then sent down the snake (multichannel shielded twisted pair cable) to the mixer. It also shows how the parallel output can be used to feed on-stage power amplifiers directly. Note that it is not really necessary to convert an unbalanced pro-audio line-level connection to a balanced connection in order to send the signal long distances, since the signal level is generally much higher that the background electromagnetic noise. Pro-audio line-level cable runs of up to 40 m can generally be used without problems. High-impedance instrument connections, e.g., guitar output, should be converted to balanced connections before sending the signal any distance over a few meters, however, as the signal level is not as high as line level, and the high-frequency capacitive losses (see Section 6.1.2) on an unbalanced instrument connection may be unacceptably large for long cable runs. Balanced lines are also needed to simplify and achieve an effective grounding scheme.

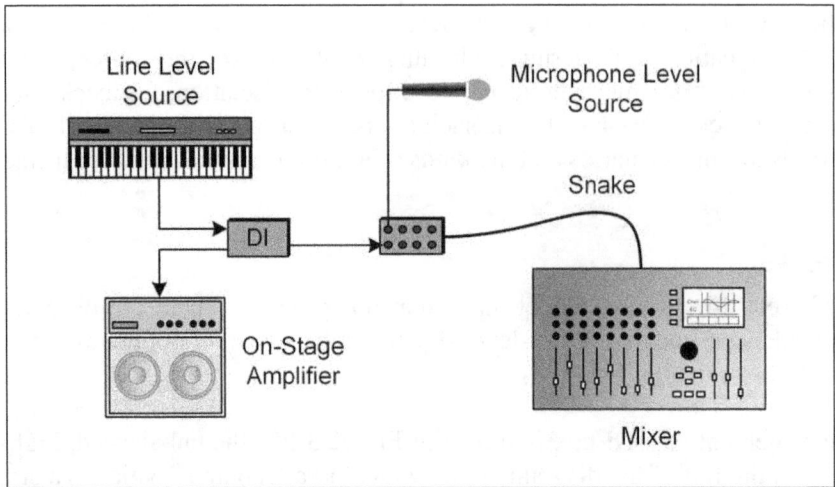

Figure 3-21 On-Stage DI Box Usage

3.15 DI Box Design

There are two general designs for DI boxes. The first type is passive. This type of DI box does not require any power to operate and consists only of a transformer and some resistors. A generic schematic of this type of DI is shown in Figure 3-22. It is very easy to make this kind of DI, but it is rather expensive to make a really good one because it is more costly to design and manufacture a transformer that does not saturate and distort the audio, especially at lower frequencies and higher input levels.

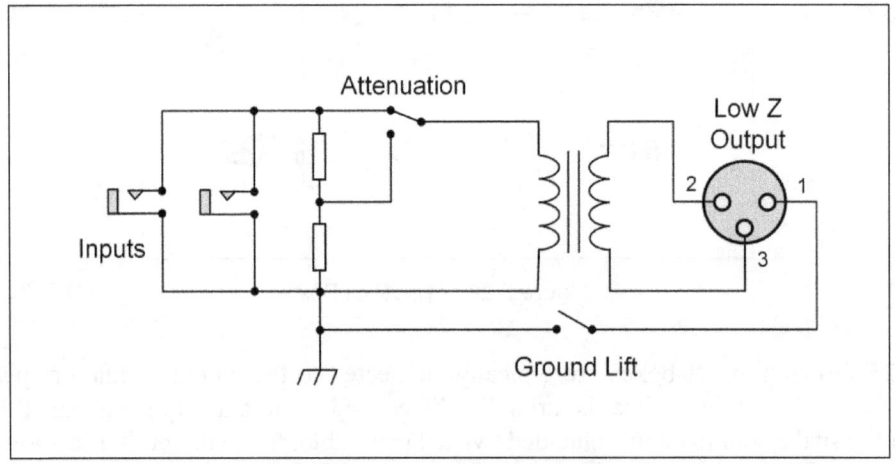

Figure 3-22 Passive DI Design

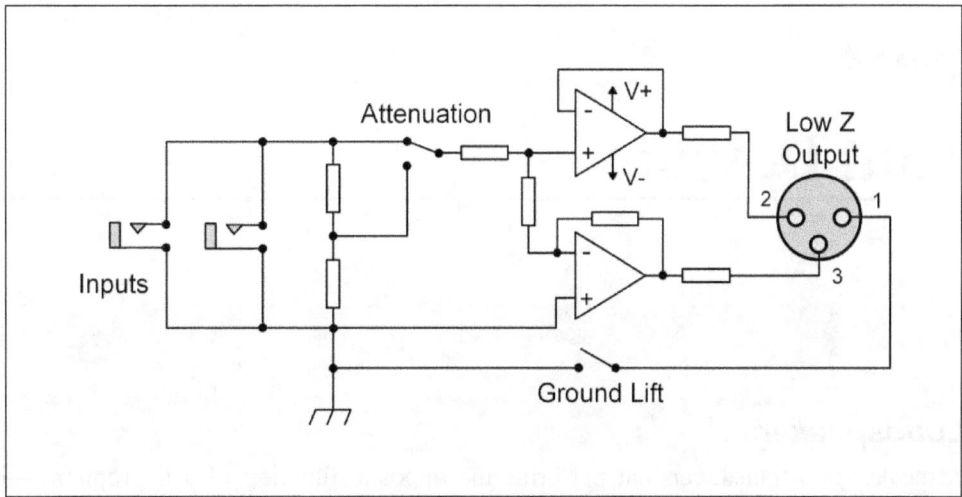

Figure 3-23   Active DI Design

The second type of DI box is active. As can be seen in Figure 3-23, this type of DI box contains active electronic components that require a power source. Power is provided either in the form of batteries, a wall power adapter, or phantom power from the mixer. It is harder to design a good active DI box, but once designed, it can be easily manufactured. A cheap active DI box may generate noise from inferior electronic components which tend to deteriorate with age.

Review Questions

1.   What are the most common types of microphones encountered in a live sound reinforcement environment?

2.   How can we increase the current generated when a wire moves in a magnetic field?

3.   Why must the diaphragm of a microphone's capsule be as light as possible?

4.   What is a capacitor (condenser)?

5.   Explain the working principle of a condenser microphone.

6.   Why are condenser microphones so much more sensitive than dynamic microphones?

7.   Explain how phantom power can be used to power a condenser microphone.

8.   What is electrostatic interference and how can it degrade an audio signal flowing in a patch cable?

9.   What kind of cable is used in a balanced connection and how does it protect against ESI and EMI?

10. Explain how a balanced connection eliminates noise picked by the connecting cable.

11. What is the difference between a cardioid and supercardioid microphone? In what situation wound you use each one?

12. Explain what is meant by the proximity effect.

13. Is it important for a transducer to have a flat frequency response? Why or why not?

14. What are some of the most common problems encountered when using wireless microphones?

15. What is a diversity receiver and how does it help solve the problem of signal dropout?

Chapter 4

Output Devices

4.1 Loudspeakers

The loudspeaker is a transducer that performs the opposite function of a microphone—it takes electrical energy and transforms it into sound energy. Its design is even similar to the microphone. In fact, for each microphone type (dynamic, condenser, etc.) there is a corresponding loudspeaker driver type.

A typical loudspeaker is really a system consisting of one or more drivers, a crossover, some protection devices, an enclosure, and other lesser components that work together to deliver the final sound to the audience. A cross-sectional view of such a system is shown in Figure 4-1. One may argue that the driver is the most important element in the system, followed by the crossover.

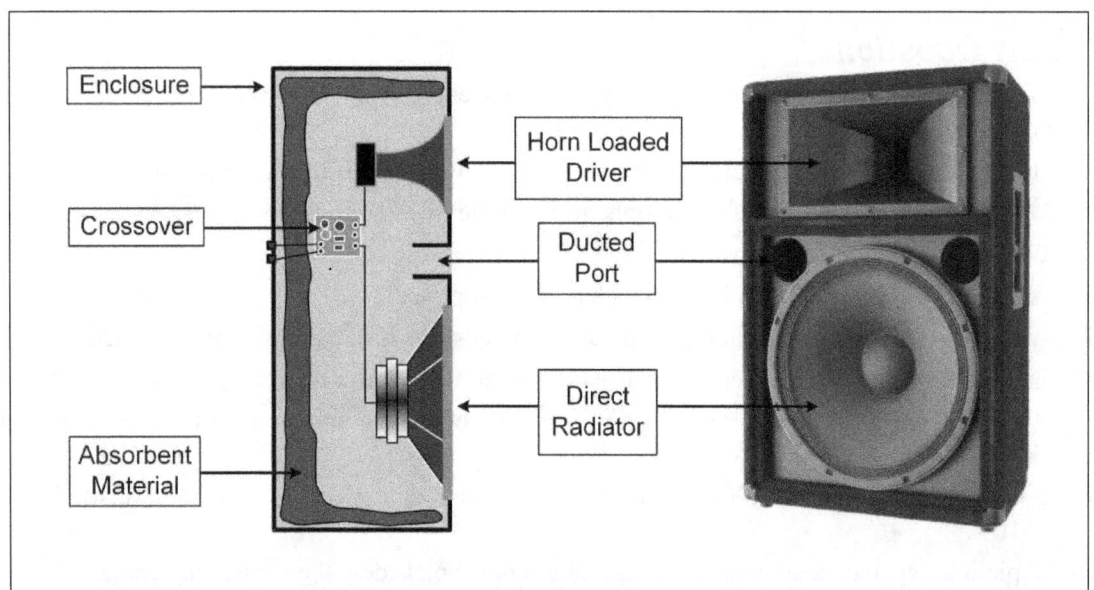

Figure 4-1   Typical Two-Way Loudspeaker System

Loudspeaker systems use different size drivers to cover separate parts of the audio spectrum (20 Hz to 20 kHz). This is because drivers with large surface areas are more efficient at reproducing low-frequency sounds, while those with small moving elements are more efficient at reproducing higher frequency sounds. The drivers used for low-frequency sound reproduction therefore tend to be large, with heavy moving structures (cone and coil sub-assemblies), while those used for reproducing high-frequency sounds need to have very light moving structures with small inertia so that they can keep up with rapidly changing waveforms. The combination of all the drivers is used to cover the entire audio range. A loudspeaker system with only two drivers that cover separate parts of the audio spectrum is called a two-way system. A loudspeaker with three such drivers is called a three-way system.

We will look more closely at the major components of the loudspeaker system in the following pages.

4.2   Dynamic Driver Principle

Most of the drivers in a loudspeaker system are dynamic drivers. Their principle of operation is shown in Figure 4-2 and can be seen to be the opposite of a dynamic microphone.

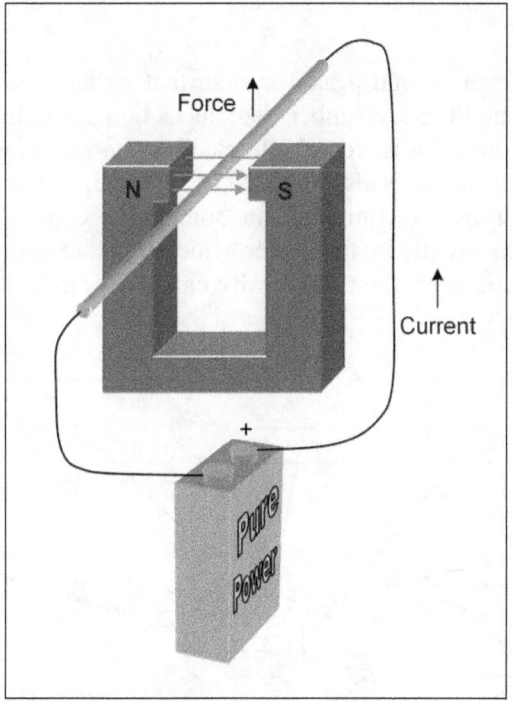

Figure 4-2   Dynamic Driver Principle

If a conductor that is carrying a current is sitting in a magnetic field, it will experience a force that is at right angles to both the direction of current flow and the lines of magnetic flux. This force pushes the conductor out of the magnetic field. Dynamic drivers work on this principle. The force on the conductor can be increased by making the magnetic flux greater (e.g., by using a more powerful magnet over a smaller area), or by increasing the length of the conductor in the magnetic field (e.g., by making it into a tightly wound coil).

4.3   Back EMF

As we mentioned above, a dynamic driver's design is very similar to that of a dynamic microphone. In Section 3.2, we showed that in a microphone, when the conductor moves through the lines of magnetic flux, a current is induced in the conductor. A similar thing happens with a loudspeaker driver (in fact, a loudspeaker could be used as a crude microphone). Normally, the driver is connected to an amplifier, which supplies current to the voice coil, causing it to move in the magnetic field. As the coil is moving through the magnetic field, however, it acts just like a microphone, and a voltage is induced across it that is opposite in polarity to the driving voltage from the amplifier. Compare Figure 3-3 and Figure 4-2. The induced voltage across the coil is called back EMF (electromotive force) because it opposes the driving current from the amplifier. The amplifier designers must give consideration to this back EMF so that the amplifier can handle it properly. We will talk more about back EMF when we look at amplifiers.

4.4 Driver Types

In the next few sections, we will take a look at the various types of drivers that are commonly used in the construction of typical sound reinforcement loudspeaker systems.

4.4.1 Direct Radiator

A direct radiator is a driver in a loudspeaker system that radiates its sound directly into the air without going through any additional chambers like ducts, horns, etc. In the dynamic direct radiator shown in Figure 4-3, the conductor is wound into a coil around a former that is then suspended in a strong magnetic field. The former is also attached to the cone, so that the cone moves as the coil moves. It is the cone that moves the air to create sound. The cone is supported at its edge by a flexible surround, and in the middle by the spider, which together act to keep the voice coil in the strongest part of the magnetic field. The chassis (also called the basket) provides a rigid support for all the elements of the driver.

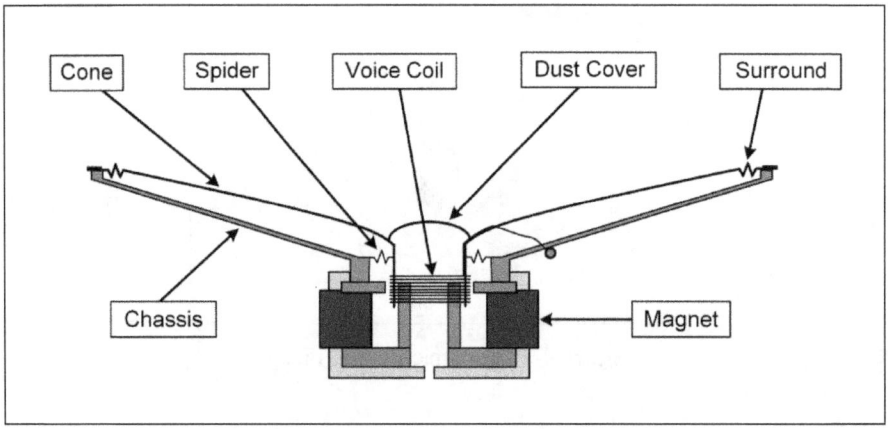

Figure 4-3 Cross-sectional View of Direct Radiator

Direct radiators are very inefficient at converting electrical energy into sound energy—only about 0.5% to 4%. This is because the moving mass of the driver (cone, coil, and former) is much larger than the mass of the air that is moved by the driver (the ratio is typically in the range of 50 to 100:1).

4.4.2 Horn Loaded Driver

When a horn is mated to a direct radiator, as in Figure 4-4, the combination is often referred to as a horn loaded driver. They are many types of acoustic horns, e.g., round and straight exponential, radial, multi-cell, and constant directivity. We will only consider general characteristics.

Horns are widely used in sound reinforcement to control the dispersion characteristics of the loudspeaker system. This is achieved by reshaping the natural dispersion of the mated driver. They are used to spread the high frequencies more, thus reducing high-frequency beaming (see Section 4.6). They are also used to reduce the mid frequency spread. The end result is the creation of a more evenly distributed and sonically balanced sound field, where the mid frequencies reach further into the audience.

Horns are used to increase the efficiency of a driver by acting as an acoustic transformer. At the driver's surface, the horn presents a high-impedance, high-pressure environment. At the horn's mouth, the impedance, or pressure, is lower. Therefore, the horn transforms little air motion at the driver's surface into greater air motion with greater speed at its mouth, and this translates into higher SPL.[1]

On the surface, horns may appear to be quite simple devices, but in actuality, they are quite complex. The difficult part of making a horn is in getting the ratio of the openings of the mouth and the throat, and the curve of the body just right so as to optimize its function.

4.4.3   High-Frequency Compression Driver

A compression driver is shown in Figure 4-4. In a compression driver, the diaphragm is coupled to a chamber that has a cross-section which, at its opening, is smaller than that of the diaphragm. Thus, the diaphragm compresses the air in the chamber when it moves. The compression chamber is in turn coupled to an expansion chamber which has a bigger cross-section at its opening. Therefore, an acoustic transformer is created, increasing the efficiency of the driver. In practically all cases, a horn is mated to the compression driver, further increasing its efficiency by increasing the transformer effect.

Compression drivers exhibit a number of problems that degrade the sound they produce. The relatively large diaphragm has many breakup modes (see Section 4.5 for further explanation) over a wide frequency range, and the compression of air and turbulence in the compression chamber causes non-linear distortion at high SPL. This causes audible sound coloration, and often leads to listening fatigue.

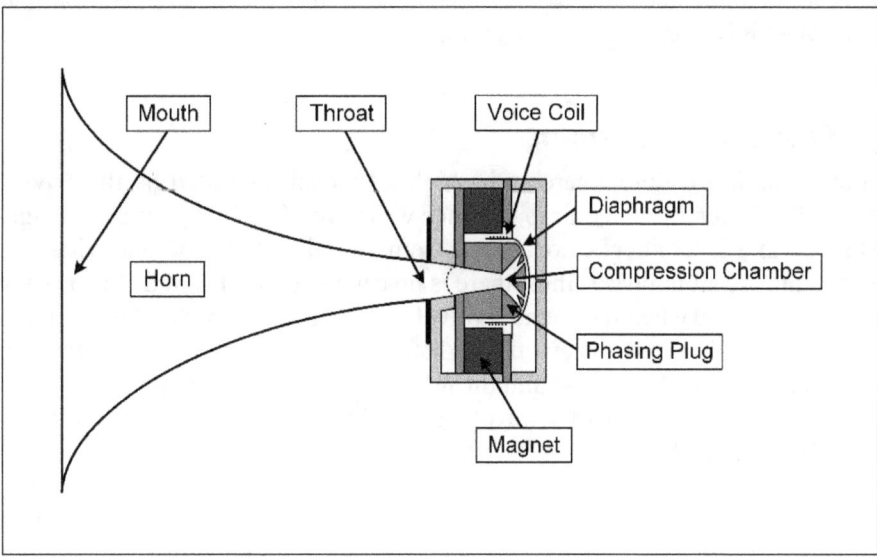

Figure 4-4   Compression Driver with Horn

4.4.4 Ribbon Driver

Ribbon drivers replace the voice coil by a ribbon conductor suspended between the poles of a powerful magnet. These drivers are used for high-frequency reproduction because the ribbon does not have enough excursion to reproduce the low frequencies. While theoretically they could be made in any size, it is difficult to build a large driver to reproduce the low bass. Because quality ribbon drivers are relatively expensive to make and are only produced in low quantities, these types of drivers are typically found in only a few higher-end loudspeaker systems.

4.4.5 Other Driver Types

There are other driver types that are not typically encountered in live sound reinforcement work. For example, there are electrostatic drivers that are simply too expensive and require additional high voltage equipment to make them practical for live sound work.

4.5 Audio Distortion from Drivers

All drivers produce some distortion of the reproduced sound. Distortion is any change in any aspect of the sound signal, other than its overall amplitude. Transducers typically produce the most distortion in a system because of the physical limitations of the device in transforming the sound signal from one form to another (in this case, electrical to acoustic).

In the case of a direct radiator, we would like the cone to act as a rigid piston attached to the voice coil. Unfortunately, this is not the case. Because the cone is not entirely rigid, different parts of it move at different rates when actuated by the voice coil. The end result is that vibrations are created within the cone itself, distorting the original sound. This is called cone breakup.

A driver will also produce a large amount of audio distortion when driven close to its operational limit. Drivers are designed to work within the linear part of both the magnetic gap and their suspension (i.e., the spider and surround). As the voice coil moves in the magnetic gap, the suspension keeps it from flying off into space. A higher-level signal will push the voice coil further out of the strongest part of the magnetic field. Thus, the same amount of current will move the voice coil less. As the driver approaches the limit of its operating range, its suspension also becomes non-linear. It therefore takes a much greater force to move the cone in this region of operation that it does to move the cone when it is closer to its rest position.

4.6 High-Frequency Beaming

Direct radiators exhibit a feature whereby the high frequencies emitted by the driver tend to be strong on axis of the driver, but fall off rapidly once we move off axis. This is called high-frequency beaming. At an on-axis listening position, sound from all points of the driver arrive at the position at the same time and are all in phase. Thus, there is no cancellation of sound. This is shown as point A in Figure 4-5. However, when the sound's wavelength is small compared to the diameter of the driver, the difference in distance from an off-axis listening position to a diametrically opposed point on the driver's cone is enough to cause significant phase cancellation at that point. The difference in distance must be between one quarter wavelength and half a wavelength for this to occur. This is shown as point B in Figure 4-5.

At long wavelengths, the difference in distance is negligible compared to the diameter of the driver, and so is the phase cancellation. Therefore, this phenomenon does not affect low-frequency sounds.

In fact, it can be seen from Figure 4-6 that low-frequency sounds from the driver will tend to be rather omnidirectional, while the dispersion narrows as the frequency increases.

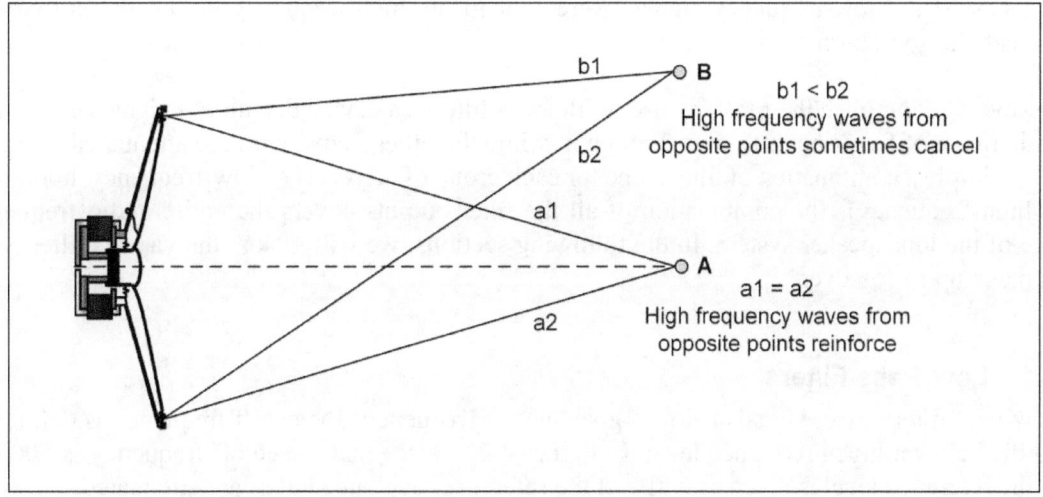

Figure 4-5   High-Frequency Beaming

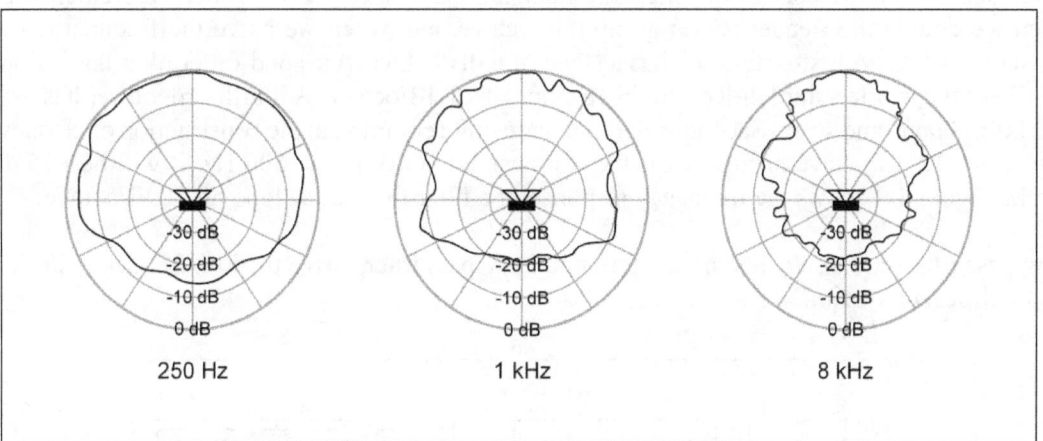

Figure 4-6   Frequency Dependent Dispersion

4.7 Crossovers

In Section 4.1, we noted that loudspeaker systems use a combination of different sized drivers to cover as much of the audio spectrum as possible. In a system such as this, we want to ensure that each driver receives only the range of frequencies that it can handle. A crossover is the device that is used to accomplish this task. We isolate the drivers in this way for two reasons.

The first is efficiency. For example, it is pointless sending very high frequencies to the bass driver. For the most part, the signal passes through its voice coil and is somewhat dissipated in the driver as heat. The real issue, though, is that the mass and inertia of the moving parts of the bass driver are so high that it simply cannot keep up with the rapidly changing high-frequency waveform. The result is a low-level, distorted reproduction of the signal.

Secondly, we have to consider driver protection. High-frequency drivers tend to have voice coils made with very thin wire. These can handle much smaller currents than the low-frequency drivers. Additionally, in typical music distribution, there is more low-frequency energy than high-frequency. If the low-frequency signals were sent to the high-frequency driver, the coil would overheat and soon burn out.

Crossovers accomplish their task by use of filters. A filter is a device that allows signals covering a certain range of frequencies to pass through it, while the other frequencies are attenuated. A cross-over is simply a combination of filters, one for each group of drivers (i.e., low-frequency, midrange, and high-frequency). The combination of all the filter outputs covers the entire audio frequency range of the loudspeaker system. In the following sections, we will look at the various filter types that make up a crossover.

4.7.1 Low-Pass Filters

A low-pass filter passes signals below a given cutoff frequency. The cutoff frequency is defined to be 3 dB below a given reference level. In Figure 4-7, we see that the cutoff frequency is 500 Hz, and the reference level is given as 0 dB. At the reference level, no signals are attenuated.

Notice that the filter does not block all signals above the cutoff frequency. The filter has a roll-off which depends on the slope of the filter. The slope is usually specified in dB/octave. Remember that when we double the frequency, we go up one octave, and when we halve the frequency, we go down one octave. A first-order filter has a slope of 6 dB/octave. A second-order filter has a slope of 12 dB/octave, while a third order filter has a slope of 18 dB/octave. A fourth-order filter has a slope of 24 dB/octave, and so on. Second-order filters are quite common in the construction of crossovers. A slope of 12 dB/octave means that if the response is −3 dB at say, 500 Hz, it will be −15 dB at 1 kHz. Figure 4-7 shows the frequency response of a filter that has a slope of 12 dB/octave.

Low-pass filters are also called high-cut filters. Low-pass filters drive the low-frequency drivers in the loudspeaker system.

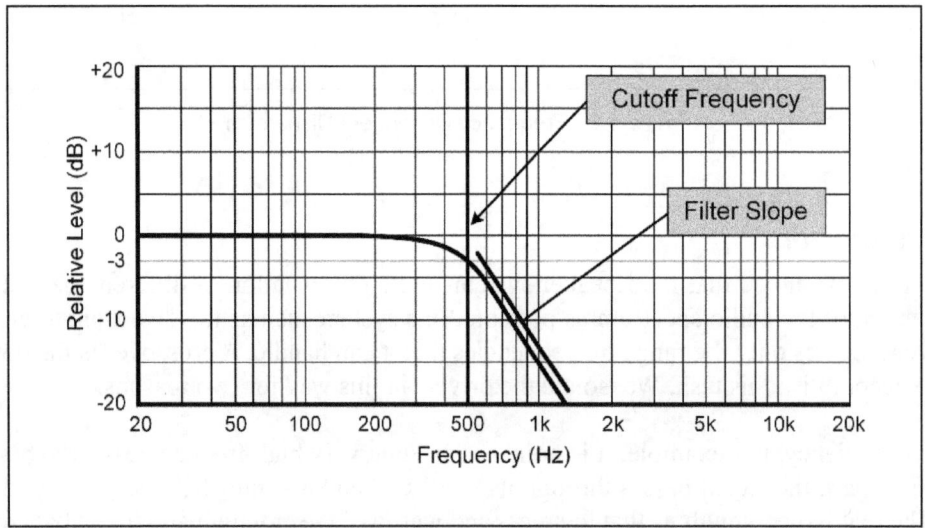

Figure 4-7 Low-Pass Filter Frequency Response

4.7.2   High-Pass Filters

A high-pass filter is the opposite of a low-pass filter, in that it allows frequencies higher than the cutoff frequency to pass. Similar terminology applies. High-pass filters are also called low-cut filters. High-pass filters drive the high-frequency drivers in the loudspeaker system.

4.7.3   Band-Pass Filters

Band-pass filters pass signals below and above a defined center frequency. The response of a band-pass filter is shown is Figure 4-8. The range of frequencies passed is determined by the bandwidth of the filter. The bandwidth of the filter is defined to be the size of the passband at the 3 dB level down from the peak. Higher bandwidth means more frequencies are passed. Band-pass filters drive the midrange drivers in the loudspeaker system.

Bandwidth is sometimes incorrectly referred to as Q. This term derives from the Quality Factor attribute of the filter. Q is actually defined as the center frequency divided by the bandwidth in hertz. The response shown in Figure 4-8 would be for a filter that has a bandwidth of about 60 Hz and a Q of about 3.3. Generally speaking, for a given center frequency, high Q implies narrow bandwidth, while low Q implies wide bandwidth.

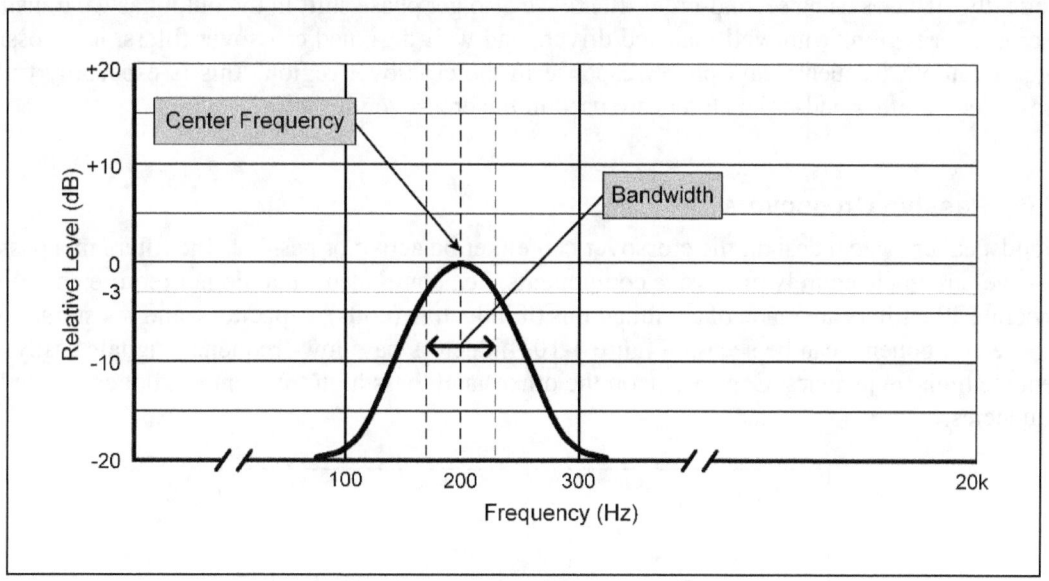

Figure 4-8   Band-Pass Filter Response

4.7.4   Combined Filter Response

In a typical three-way loudspeaker system, we would have a crossover consisting of three filters—low-pass, band-pass, and high-pass—feeding the bass, mid-range, and high-frequency drivers, respectively. In order to provide a smooth transition between the filter ranges, the output of the filters would have to overlap as shown in Figure 4-9. The point where the filters overlap is called the crossover point or crossover frequency. The drivers themselves must also have smooth response for at least an octave on each side of the crossover frequency. The combined frequency response of the three filters covers the entire audio range of the loudspeaker system.

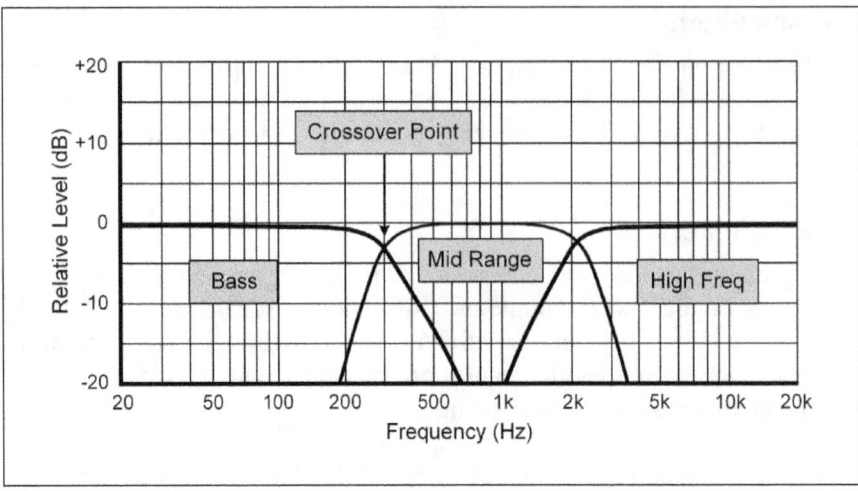

Figure 4-9 Combined Crossover Frequency Ranges

With less expensive or poorly designed systems using second-order (12 dB/octave) filters, we typically get either a deep null or slight peak in frequency response in the crossover region. In practically all cases using second-order filters, we also get phase shift in the output as we transition the crossover region. With well-matched drivers and well-designed crossover filters, it is possible to get a smooth frequency and phase response in the crossover region. This is especially true if third-order (or other odd-order) filters are used in the crossover design.

4.7.5   Passive Crossovers

In loudspeaker system design, the crossover can either be active or passive. The filters in a passive crossover are made entirely of passive components, i.e., components that do not require any power to operate. The filters are made of combinations of inductors (coils), capacitors, and resistors. Some of these components can be seen in Figure 4-10. Inductors pass low-frequency signals easily, but attenuate high frequencies. Capacitors, on the other hand, pass high frequencies, but attenuate low frequencies.

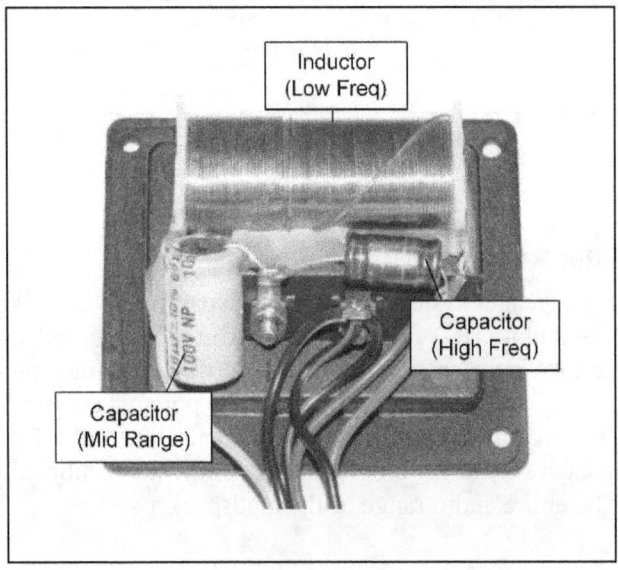

Figure 4-10 Passive Crossover

In the signal chain, passive crossovers sit between the power amplifier and the loudspeaker. Passive crossovers therefore operate at speaker-level voltages and are connected directly to the drivers. Because of the large inductors used in the better crossovers, a lot of power may be dissipated in the crossover coils themselves, thus robbing the driver and reducing the efficiency of the system.

Passive crossovers are usually buried inside the loudspeaker enclosure. There is generally no need to have access to the crossover, since its filters are usually matched to the drivers by the loudspeaker's manufacturer. The design frequencies of the filters may drift over time as the components age. Such drift would cause a slight shift in the crossover points of the system. While this may be a concern for audiophiles, it is unlikely to cause real problems for the live sound operator.

4.7.6   Active Crossovers

Unlike filters in a passive crossover, the filters in an active crossover use components that require power to operate. The filters are usually made with small amplifiers that incorporate a tuned feedback loop. The components used in an active crossover handle much less current than that of a passive crossover.

In the signal chain, active crossovers sit between the preamp and the power amplifier. Active crossovers therefore operate at line-level voltages and are never connected directly to the driver. The active crossover splits the audio into two or three bands needed for two-way and three-way systems discussed above. It then sends those band-limited frequency ranges to the power amplifier. As can be seen from Figure 4-12, a separate power amplifier is therefore needed for each band.

Figure 4-11   One Channel of Active Crossover

Active crossovers are housed in their own enclosures, with front panel controls that allow access to the filter parameters. The filters of an active crossover can thus be tuned by the user, allowing him to mix and match various drivers to achieve higher efficiencies. Figure 4-11 shows one channel of an active two-way crossover made by Behringer, with controls to set the crossover point and to set the gain of the filters for the high and low-frequency bands.

4.7.7   Multi-Amplification

The use of active crossovers in audio systems gives rise to the practice of multi-amplification, the most common manifestation of which is bi-amplification and tri-amplification.

In bi-amplification, the active crossover is used to split the music program spectrum into a low-frequency range and a mid-to-high range. These ranges are then amplified separately, as shown in Figure 4-12, with one amplifier used for the low frequencies and another for the mid-to-high frequencies. The bands are usually split like this (as opposed to a low-to-mid range and a high range) because the equal power split in the music program spectrum is around 300 Hz. In other words, if we were using a single amplifier, half of its power would be used in producing sound below 300 Hz, and the other half would be used in producing sound from 300 Hz to 20 kHz. Note that in bi-amplification, if the mid-to-high frequency unit is a two-way system, it still utilizes a passive crossover to properly route these signals. This passive crossover is much smaller than a full spectrum crossover, though, since it does not need the large inductors to handle low frequencies.

Tri-amplification takes the processes one step further and splits the music program spectrum into three ranges—low, mid, and high frequency. Each band is then separately amplified. As can be seen from Figure 4-13, we now require three power amplifiers to process the whole audio spectrum. Furthermore, the high-frequency unit still requires the use of a passive capacitor to protect it from the short surge of current as the amplifier is turned on.

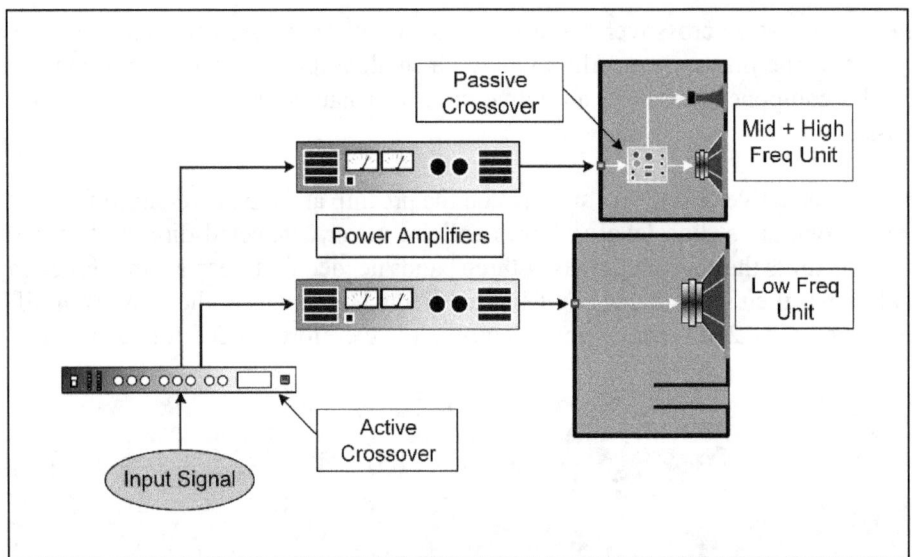

Figure 4-12 Bi-Amplification

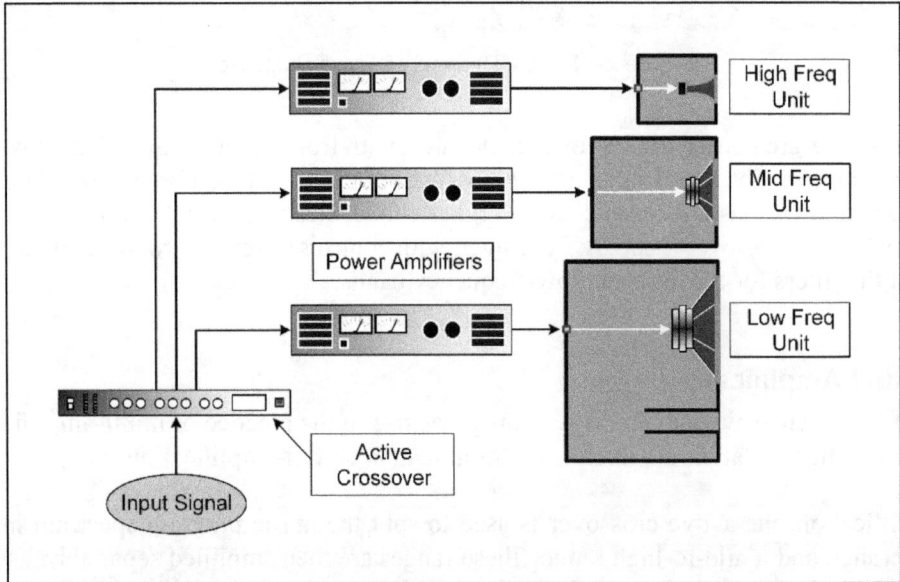

Figure 4-13 Tri-Amplification

4.7.8   Advantages and Disadvantages of Multi-Amplification

One clear disadvantage of using multi-amplification is that it requires a separate power amplifier for each band, potentially increasing the cost of the system. In some cases, multiple loudspeaker cabinets are also used, leading again to higher cost, as well as having to manipulate more pieces of equipment.

However, the disadvantages of multi-amplification are more than offset by the advantages to be gained by using this scheme. Drawing on previous work by Elliott,[2] we now detail many of these advantages here:

- Since the drivers are driven separately, and also possibly housed separately, it is easy to time-align the whole system, either by physical positioning or by electronic means. Time-alignment is discussed further in Chapter 13.

- The user has the ability to completely control the crossover points and width of passbands to exactly match the drivers in use. This leads to better loudspeaker system efficiency.

- Unlike passive crossovers where the final response of the system is determined by the interaction of the crossover and the driver's ever-changing impedance (see Section 4.11.1), with an active crossover, the frequency response can be shaped independently of the driver's impedance, since there is no interaction between the active crossover and the driver's impedance.

- The amplifier is connected directly to the driver, so none of its power is wasted in the bulky inductors of a passive crossover.

- In the audio signal chain between the amplifier and the loudspeaker, there are three impedances to consider (ignoring the connecting cables)—the amplifier's output impedance, the driver's voice coil impedance, and that of the passive crossover. When considering the shunting of back EMF generated by the driver's voice coil, the ideal would be to have zero impedance between the driver and the amplifier. In reality, the impedance of the crossover is a large factor in reducing the effective damping that can be provided by the amplifier (see Section 5.7) because it is large compared to the output impedance of the amplifier (assuming a typical solid state amplifier). With multi-amplification, the crossover is removed from the circuit, thus allowing the driver to 'see' the lowest impedance and attain maximum damping. This leads to better transient response of the system.

- Multi-amplification allows for the use of lower powered amplifiers. Consider a bi-amplified system where signals of the two bands are fed to the amplifiers. Suppose these signals are of equal strength and produce 37.5 V peak signals at each amplifier's output. If driven into an 8 Ω load, this requires 175 W of power dissipation from each amplifier (recall $W = V^2/Z$). In a system without bi-amplification, these signals are combined in the single amplifier, producing a signal that has a peak voltage of 75 V at the amplifier's output. Driven into an 8 Ω load, that requires an amplifier that is capable of delivering over 700 W. That is four times the power required for a single band in the bi-amplified configuration (or twice the combined power of the amplifiers in the bi-amplified configuration).

- Multi-amplification allows optimal amplifier usage. Recall that the equal power split of music program material is around 300 Hz. Furthermore, the energy content of typical music program decreases at a rate of about 3 dB per octave above 1 kHz. This means that less power is required to amplify the high frequencies than the low frequencies. For example, suppose 100 W is required for content below 1 kHz. At 16 kHz, the requirements would be −12 dB, or only about 7 W. If we use multi-amplification to separate and amplify the signal bands, we can use lower power (i.e., cheaper) amplifiers for the upper frequency signals. Additionally, amplifiers with poor high-frequency response can be used for the low-frequency band, while amplifiers with poor low-frequency response can be used for the upper frequency bands.

- The drivers in a multi-amplified system are isolated from each other. This allows systems that produce high SPL to be easily constructed, since it is the woofer and to a lesser extent the midrange drivers that produce the bulk of the sound energy. In a passive system, since the drives are all interconnected, the high-frequency driver would probably go into distortion when the system is driven hard, leading to quite unpleasant sound. In a multi-amplified system, the low and mid frequency drivers can be driven hard independently of the high-frequency driver, producing clean sound at high SPLs typically found in live applications.

- Intermodulation distortion (see Section 5.8.2) occurs in an amplifier when a low-frequency signal interacts with a high-frequency signal to produce new tones that are the result of the low-frequency signal modulating the high-frequency signal. Intermodulation distortion is reduced in a multi-amplified system because the high frequencies are separated from the low frequencies *before* they are sent to the amplifier.

4.8 Loudspeaker System Enclosures

The drivers in a loudspeaker system are mounted in an enclosure to facilitate handling and to otherwise change some operating characteristic, e.g., increase their efficiency or transient response. In this section, we will take a look at the most common enclosure types encountered in live sound reinforcement, and will discuss the important properties and pros and cons of each system.

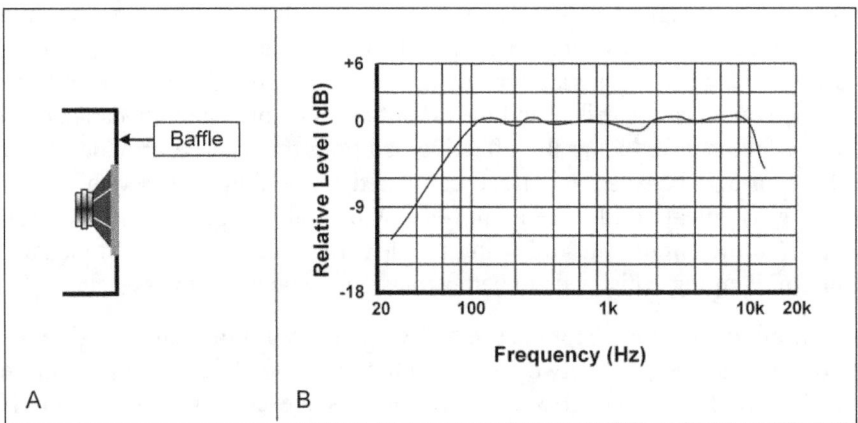

Figure 4-14 Dipole Loudspeaker System

4.8.1 Dipole

The dipole is the simplest and cheapest of all loudspeaker enclosure types. They are usually seen in guitar amplifier combos. As can be seen from Figure 4-14(A), the dipole is not really an enclosure at all, but really just a baffle on which the driver is mounted. Sometimes the baffle has sides attached to it to help control the sound. It is called dipole because the sound waves from both sides of the driver are used to produce the final sound of the system. Since the sound waves at the front of the baffle are 180° out of phase with the waves at the rear of the baffle, cancellation becomes a serious problem when the wavelength approaches the dimensions of the baffle. At these frequencies, the sound waves from the back side of the driver easily bend around the baffle and interact with the waves from the front. This creates a frequency response that resembles a high-pass filter in its lower frequency ranges. In fact, this system has a roll-off of about 6 dB/octave below its corner frequency, as shown in Figure 4-14(B).

4.8.2 Sealed Box

As the name implies, a quantity of air is trapped in a box by the seal between the drivers and enclosure structure. This is shown in Figure 4-15(A). While the system is not totally airtight, it does have very low leakage. Theoretically, the wave generated by the backside of the driver is trapped in the box and cannot interfere with the wave generated by the front of the driver, as in the case of the dipole. However, this also makes this design inefficient, as half the driver's output is lost. Additionally, in this type of system, the lower corner frequency is always higher than the driver's free air resonance. Thus, the full potential of the driver is not harnessed. It can also be seen that this system has a roll-off of about 12 dB/octave.

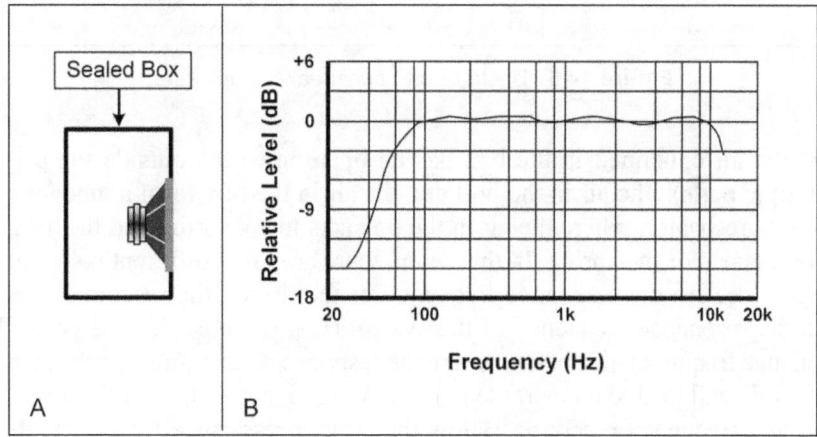

Figure 4-15 Sealed Box Loudspeaker System

There are two main variants of this design. The first is the *acoustic suspension*. In this case, the volume of the box is so small that compression and rarefaction of air trapped in the box become significant as the cone moves back and forth. This provides significant resistance to driver's cone motion as the cone moves farther away from its neutral position. The trapped air thus acts as a spring, helping to quickly restore the cone to its rest position when the signal is removed from the driver. This improves the transient response of the system. The trapped air also helps to control possible overextension of the cone when the system is driven by high-level signals. This allows it to handle more power before the driver goes into destructive overextension. The tradeoff is a steep roll-off in the bass region and reduced driver efficiency, since the spring effect is constantly working against the motion of the cone.

The second variant is the *infinite baffle*. Here the volume of air trapped in the box is very large and the driver behaves more or less as if it were operating in free air. If the sealed boxed is stuffed with absorbent material, the physical size can be reduced by up to 30% and still maintain the characteristics of the larger box. This design offers lower bass and a gentler roll-off than the acoustic suspension, but has poorer transient response.

4.8.3 Ported Box

This type of enclosure is also called a ducted box, ducted port, and a bass reflex system. It is the most common mid to large size enclosure encountered and is more efficient that the sealed box. It is shown in Figure 4-16(A).

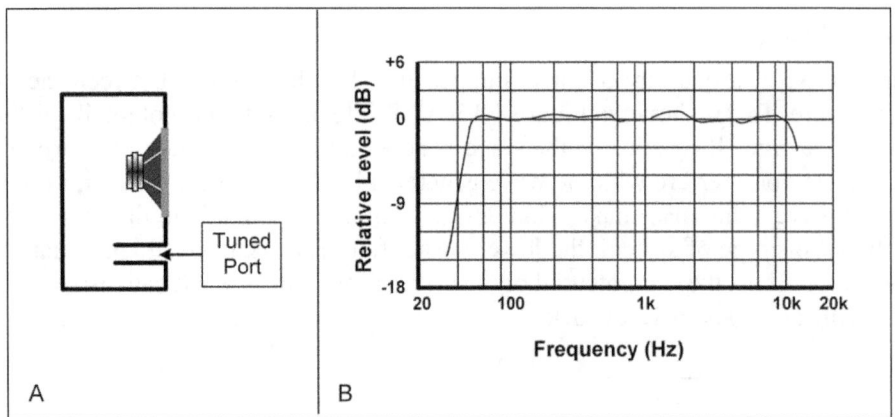

Figure 4-16   Ducted Port Loudspeaker Enclosure

In this system, the air contained in the box has an opening to the outside via a port (there can actually be multiple ports). The air in the port and the air in the box form a tuned resonant system, called a Helmholtz resonator, where the air in the box acts like a spring and the air in the port acts like the mass on the end of the spring. In this setup, there are three different resonance frequencies to keep straight: these are the resonance frequency of the driver, the resonance frequency of the port itself, and the resonance frequency of the system (i.e., the box plus the port). The system is tuned to a resonance frequency just below where the response of an equivalently sized infinite baffle starts to fall off. This will lead to an increase in the low-frequency output of the loudspeaker system and extend the low-frequency response. Below the system resonance frequency, the output from the port and the driver become progressively more out of phase, and so the response falls off rapidly, around 24 dB/octave, as shown in Figure 4-16(B).

Above the system resonance frequency, the compliance (i.e., springiness) of the air in the box keeps the cone motion of the driver in check, similar to the effect of the air in a sealed box. At the system resonance, the cone motion is lowest—most of the work is being done by the port. Below the system resonance frequency (and especially at and around the port resonance frequency) the air in the box does not offer significant resistance to the motion of the cone, and it is possible to overextend the driver with too big a low-frequency signal. This is why it not a good idea to try to boost the low bass output from a ported box with excessive amounts of EQ. In this type of system, the port's resonance frequency can easily be deduced from its impedance curve, as it coincides with the lowest frequency peak in the graph (e.g., see Figure 4-20).

4.8.4 Bass Horn

Bass horns are popular in sound reinforcement as a means of increasing the efficiency of the bass driver. Because of the wavelengths involved, bass horns tend to be rather long. In fact, as shown in Figure 4-17(A), they are usually bent or folded just to get them to fit within a reasonably sized box. Folded horns tend to have poor mid-bass response, which is not a problem for a subwoofer. However, they also typically exhibit uneven response over their working range. Bass horns have a cutoff frequency that is based on the dimensions of the opening and the flare rate, and response falls off rapidly below the cutoff frequency.

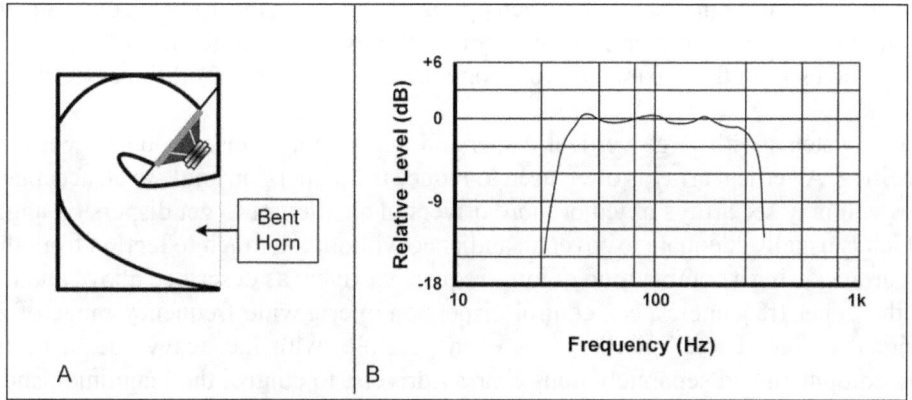

Figure 4-17   Bass Horn Loudspeaker Enclosure

4.8.5   Loudspeaker Array

A loudspeaker array is a set of drivers, in close proximity, operating together so that collectively the drivers move the same amount of air as a single larger driver whose surface area is the sum of the surface areas of all the drivers in the set. The drivers in an array may be housed in a single enclosure, or they may be housed separately to allow flexibility in stacking. Such enclosures can be sealed or ported. Arrays are thus sometimes used to produce higher SPL, especially in the low bass region, where mutual coupling occurs for frequencies where the distance between adjacent driver centers is less than half a wavelength. Two such drivers will produce 3 dB more output due purely to this coupling effect when compared to a single driver. Since the amount of low bass produced is directly proportional to the volume of air that can be moved, it may be impractical to build a single driver large enough to move the same amount of air as can be moved by even a small array.

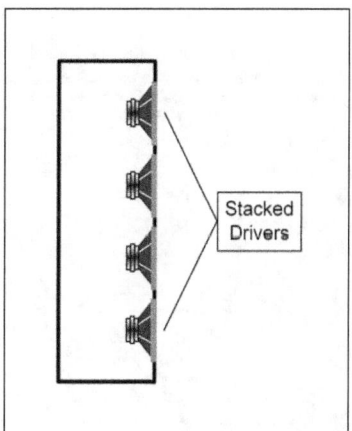

Figure 4-18   Vertical Array of Stacked Drivers

A vertical array is an arrangement where the drivers in the array are stacked vertically. This arrangement, shown in Figure 4-18, may also be called a phased array or a line array. The main use of vertical arrays is to control the vertical dispersion of the loudspeaker system in the region between 100 Hz and 500 Hz (at higher frequencies, the individual drivers already exhibit a beaming characteristic as discussed in Section 4.6). When drivers are arranged vertically, this causes the vertical dispersion angle of the system to become narrower. As we move vertically off the centerline of the array, the difference in distance between opposing drivers in the array becomes greater. Phase cancellation thus becomes more prominent over the effective frequency range of the array as out of phase sound waves from the separate drivers combine at angles further off the centerline.

Loudspeaker systems with large vertical dispersion angles may cause problems with reflections from the ceiling. A vertical array is often used to reduce the dispersion angle to an acceptable range. Sometimes you may see arrays of ten or more drivers in an attempt to get dispersion angles of 30° to 40°, which is usually adequate to cover an audience without too much reflection from the ceiling. Well-built arrays, using a combination of low-frequency drivers as described above and long throw horns for the upper frequencies, can control dispersion over a wide frequency range of 250 Hz to 8 kHz, critical for good intelligibility. It is even possible, with the heavy use of digital signal processing equipment and separately housed array drivers, to control the magnitude and phase of several key frequencies going to each driver in the array. [3] This determines, to some extent, where and how the sound from the multiple drivers cancel and reinforce each other, thus allowing the sound from the vertical array to be directed at a particular area.

4.8.6   Other Enclosure Types

There are many other types of driver and enclosure combinations that are less common, but which may occasionally be encountered in sound reinforcement use. Some of these include, Passive Radiator, Bandpass, Isobaric, and Transmission Line systems.

4.9   Passive vs. Active Loudspeakers

The passive loudspeaker system is the most common system in use. This is where the amplifier driving the speaker components is external to the loudspeaker system. This allows for a simpler and cheaper loudspeaker system, as well as greater choice in the amplifier selection. However, for the sound technician, this choice also creates the potential for amplifier to speaker mismatch.

Figure 4-19   Amplifier Module Integrated into Powered Speaker System

An active loudspeaker system is sometimes called a powered system. In this case, the amplifier driving the speaker components is built right into the loudspeaker enclosure. The active loudspeaker system may be more expensive than a passive system, but the cost may be less than a separate passive loudspeaker plus an external amplifier combination. Since the amplifier is designed into the system, it is easier for the manufacturer to match the amplifier to the driver and crossover characteristics. It is also easier to build protection into the amplifier-speaker chain. Figure 4-19 shows the amplifier module built into the Electro-Voice ZLX-15P powered speaker, which has multiple inputs on XLR/TRS combo jacks, and buffered output that is convenient for daisy-chaining additional loudspeakers fed with the same signal.

4.10 Driver Protection

Most loudspeaker systems incorporate some sort of protection that prevents the drivers from being damaged by excessively high driving currents. In some cases, the driver protection that is incorporated into the loudspeaker enclosure is only for the mid and high-frequency units. This is because the low-frequency drivers, with their larger voice coils, can handle a lot more current before failing.

The most popular protection device is the polyswitch. A polyswitch is a ceramic or polymer device with positive temperature coefficient and non-linear characteristics. A positive temperature coefficient means that the device's resistance increases with increasing temperature. The term *non-linear* means that throughout its operating range, the resistance does not increase at a constant rate with increasing temperature. At normal ambient temperatures, the device has resistance of less than 0.5 Ω. When the current flowing through the device causes its internal temperature to reach a critical level, the resistance of the device increases dramatically, causing the current flow to be reduced accordingly. The device will remain in this tripped state until the voltage present at tripping is reduced, at which point its internal resistance will return to normal levels. Polyswitches can be tripped thousands of times and can be held in the tripped state for many hours without suffering damage.

Some loudspeakers also use fast blow fuses, but these must be replaced once they are blown. Additionally, fuses tend to degrade over time with normal usage close to their operating limits. Therefore, an old fuse may blow below its peak current protection level.

Electronically controlled relays can be very effective in protecting all the drivers of a system, especially in cases where output stage failure of an amplifier may place prolonged high-level direct current onto the speaker line. Practically every modern commercially available power amplifier has some form of this protection scheme integrated with its output section, but discrete units can also be obtained and inserted between the amplifier and the loudspeaker.

Another approach to protecting loudspeaker systems is to use limiters. Limiters are inserted in the audio chain before the power amplifier. They monitor the input signal and prevent its level from going above a threshold that is set by the user, thus limiting the power amplifier's output. Under normal conditions, they work well, but cannot protect the loudspeakers from amplifier failures that cause large direct currents to be output on the speaker lines.

4.11 Loudspeaker Specifications

In this section we will take a look at the most common specification elements of loudspeakers. An understanding of these elements will help you in choosing the system that is most suitable to your particular application.

4.11.1 Impedance

The impedance of a loudspeaker is an identical concept as impedance of microphones. However, a lot more attention is usually paid to loudspeaker impedance because the speaker-amplifier interaction involves higher current flow. The lower the impedance of the loudspeaker, the greater the current the amplifier will try to deliver to the system.

Typical nominal loudspeaker impedance of 4 or 8 Ω is usually almost all determined by the bass driver. As shown in Figure 4-20, the actual impedance of the loudspeaker system varies highly with frequency. The impedance of a loudspeaker system peaks at the bass driver's in-box resonance frequency, which is typically higher than its free-air resonance (the frequency at which the driver naturally vibrates). In Figure 4-20, this is the peak around 100 Hz. At resonance, the driver generates the most back EMF, greatly opposing any driving current (hence the peak in the impedance). Bass reflex loudspeaker systems will have an additional peak in the low-frequency area, at the port's resonance frequency where the bass driver unloads (i.e., the air in the box no longer effectively provides resistance to the cone movement) and vibrates widely. Between these two peaks, the low point in the impedance occurs at the system's (i.e., box plus port) resonance frequency. As frequency rises above the driver's resonance, the impedance falls to around the driver's nominal impedance. It will start to rise again as the driver's inductance opposes the flow of current, but begin to fall off after about 2 kHz as the influence of the high-frequency components of the loudspeaker system take effect.

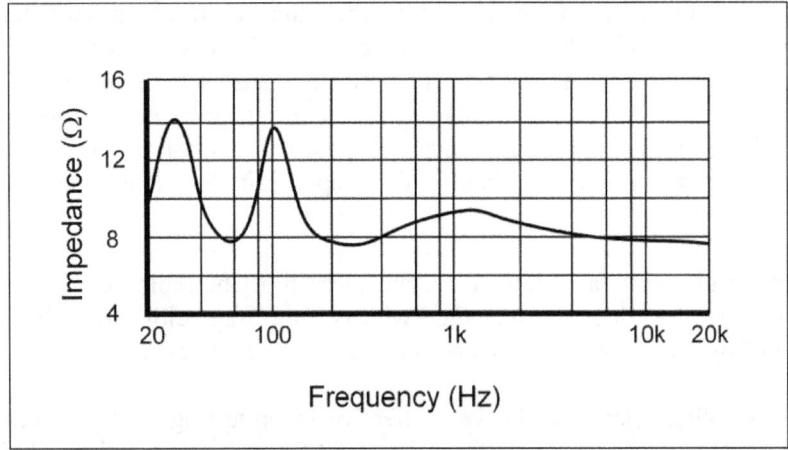

Figure 4-20   Impedance Curve of Typical Live Sound Loudspeaker System

4.11.2 Vertical and Horizontal Dispersion

Dispersion gives a measure, in degrees, of how widely the sound spreads from the loudspeaker. Figure 4-21 shows the basic concept that this specification is trying to convey. Different manufacturers may use a variety of terms to describe this specification, including *nominal dispersion*, *coverage*, and *coverage angle*. A few that are not particularly careful may also use the term *directivity*, though that term, when used in a specification, is normally reserved for more specific information described below.

We already know from our earlier discussion on high-frequency beaming that the dispersion of sound from a driver is frequency dependent. Thus, this specification is a nominal value that takes into account all frequencies emitted from the loudspeaker system. The dispersion figure tells us the included angle about the longitudinal axis beyond which the response drops 6 dB below the on-axis response. It is given for both the horizontal and vertical planes.

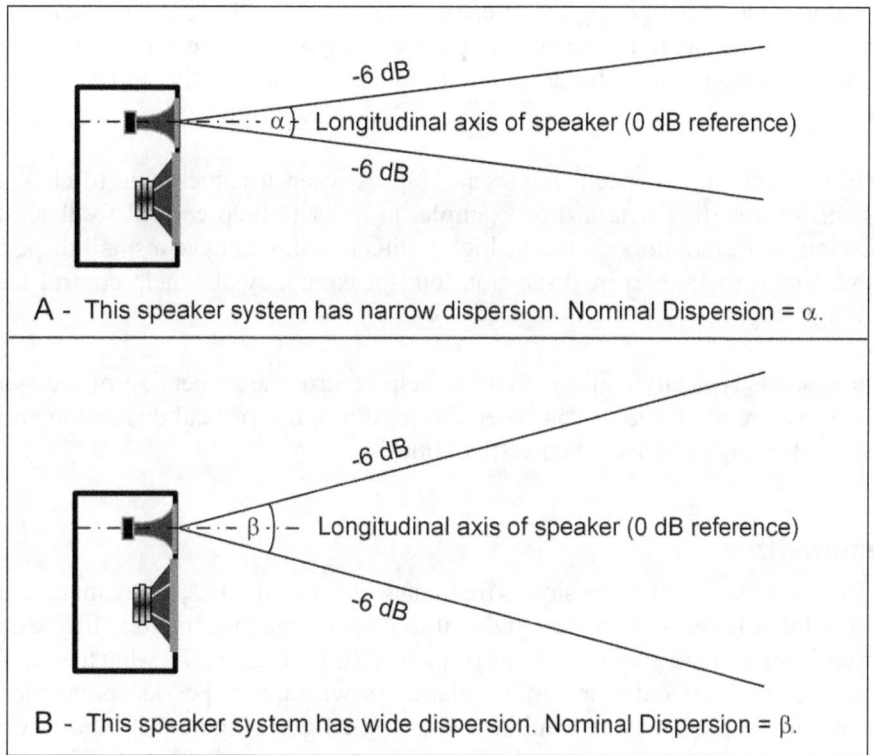

Figure 4-21   Nominal Dispersion Angles of a Loudspeaker

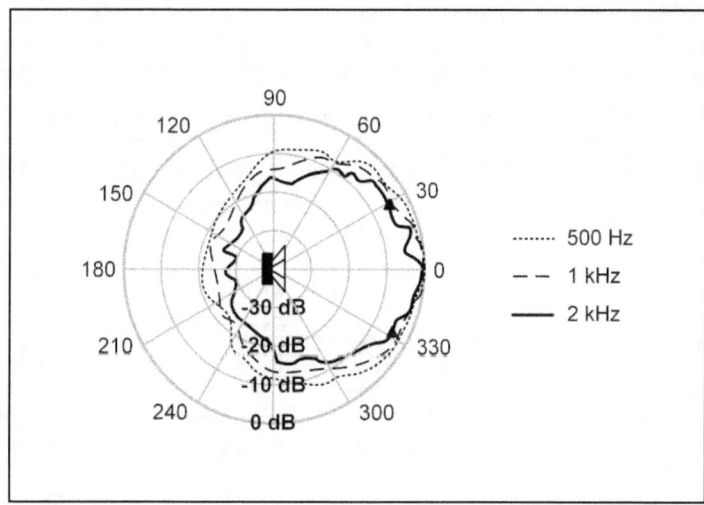

Figure 4-22   Dispersion Graph Showing Multiple Combined Plots

When a manufacturer wants to give a more complete picture of the dispersion of the loudspeaker system, they will typically publish a series of polar plots, such as shown in Figure 4-6. Each frequency of measurement may be shown as a separate plot (as in the case of Figure 4-6), or several frequencies may be combined in a single figure, as shown in Figure 4-22. Similarly, separate plots may be given for vertical and horizontal dispersion, or both planes may be combined in a single figure. Regardless of the approach taken in providing the plots, the graphs given will provide a wealth of detailed information regarding the dispersion of the loudspeaker system, far more than just the −6 dB nominal coverage angle. For example, if we take a look at the 2 kHz plot in Figure 4-22, we can deduce for ourselves what the −6 dB coverage would be for that frequency, but we can also see the reduction in level for any angle of interest, or deduce the difference between front and rear on-axis levels.

The information given in this specification can help us when it comes time to choosing a loudspeaker system for a particular task. For example, in order to help control feedback and reduce cross-monitoring, stage monitors should be highly directional, i.e., have a small dispersion figure. In a very reverberant room, narrow dispersion loudspeakers may also help control feedback and improve intelligibility.

Loudspeaker systems typically employ horns to help control the dispersion of the system. Additionally, as we saw from our discussion on enclosure types, the vertical dispersion angle can also be narrowed by stacking the drivers in a vertical line.

4.11.3 Beamwidth

As stated above, we know that dispersion is frequency dependent. Thus, many manufacturers use a different method that is particularly rich in detail to display this information. The beamwidth plot shows, for frequencies over a particular range (typically 20 Hz to 20 kHz), what the −6 dB coverage angle is for either the horizontal or vertical plane. As with the dispersion polar plots, both the vertical and horizontal plots may be combined in a single figure. Figure 4-23 shows what a beamwidth plot for a loudspeaker system looks like. We can see that below 100 Hz, this system is essentially omnidirectional. We can also see that above 500 kHz, the beamwidth remains the more or less constant for several octaves. This would likely be due to the use of a constant directivity horn in the construction of the loudspeaker system.

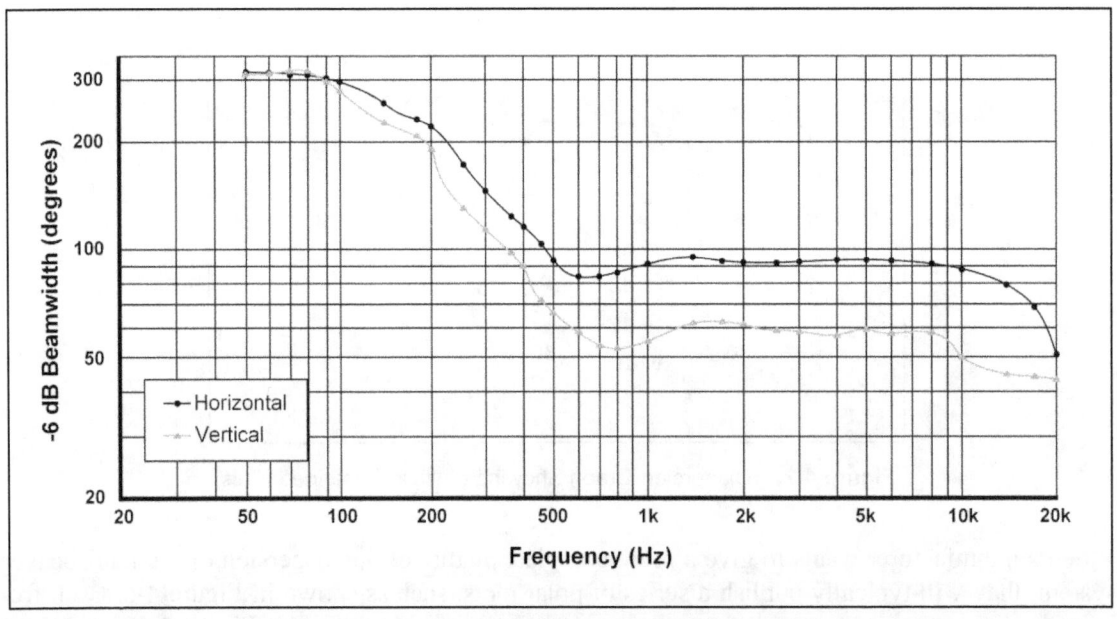

Figure 4-23 Beamwidth Plots for Horizontal and Vertical Planes

4.11.4 Directivity Index and Directivity Factor

Yet another presentation of the directional characteristics of a loudspeaker system that you will sometimes see is given in a directivity plot, shown in Figure 4-24. This one is a little hard to appreciate initially, and some explanation of the concept is in order. Basically, the directivity specification is telling us how much more effective the loudspeaker system is at concentrating its acoustic output into a beam as compared to an omnidirectional radiator, which, as its name implies, radiates its acoustic output equally well in all directions. Conceptually, imagine you have an omnidirectional source radiating at a given acoustical power and measure the SPL at some reference distance from the source. Now replace the omnidirectional source with the real loudspeaker system, point it directly at the measurement device, and measure the SPL while the system is radiating the same acoustical power as the previous omnidirectional source. The increase in SPL measured would be the directivity index (**DI**) of the loudspeaker system and is expressed in dB. The ratio of the second measurement to the first would be the Directivity Factor (**Q**) of the loudspeaker system and is dimensionless. These two measurements really tell us the same thing about the loudspeaker system and, as may be expected, there is a simple relationship between them, *viz.*:

$$DI = 10 \log(Q)$$

Of course, in reality there would be no need to actually take any measurements from an omnidirectional radiator, as the intensity at any distance from such a source can easily be calculated given the acoustical power and the use of the inverse square law. In practice, the directivity index of a loudspeaker system can actually be calculated from beamwidth measurements used to produce the dispersion plots discussed earlier, and many manufacturers define directivity factor to be the ratio of the on-axis SPL to the mean SPL of all measurements taken around the speaker system when determining the dispersion of the loudspeaker. Unlike the dispersion and beamwidth plots, manufacturers typically only publish on-axis directivity information. However, the directivity index of the speaker system may be roughly calculated for any angle in the horizontal or vertical plane by subtracting the value found on the corresponding dispersion plot for the angle in question from the on-axis directivity index.

As may be expected, the **DI** of the loudspeaker system is also frequency dependent. This can clearly be seen in Figure 4-24, where it is obvious that the concentration of energy in the beam is getting higher as the frequency increases. Again, the use of constant directivity horns in the construction of the speaker system will help to keep this attribute more or less constant over its region of control.

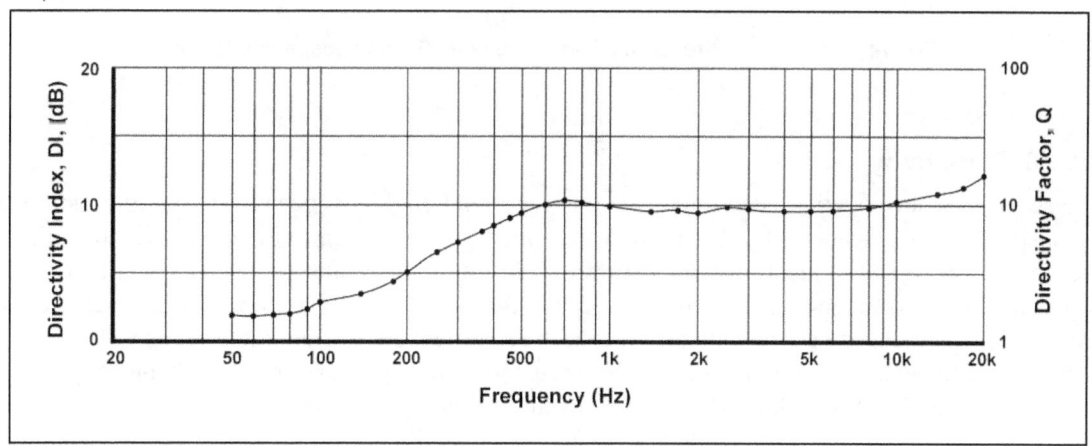

Figure 4-24   Directivity Index and Directivity Factor of Loudspeaker System

The *DI* specification is very useful in comparing different systems to determine their suitability when coverage, tight pattern control, and effective throw distance are important factors.

4.11.5 Frequency Response

The frequency response specification of a loudspeaker tells us the useful operating range of the speaker system and, more importantly, how good the loudspeaker is at reproducing sound for all frequencies over its operating range. It shows the output level of the loudspeaker when fed by a fixed level input signal over all the frequencies that the speaker system reproduces.

We generally want the frequency response of the loudspeaker to be flat over its entire operating range. That means that the loudspeaker would be able to reproduce sounds equally well for all frequencies over its operating range. It would also be desirable that the speaker system be able to reproduce sounds over the entire audio range, from 20 Hz to 20 kHz. Neither of these conditions is usually achieved, as can be seen in Figure 4-25, which shows the response of a typical 30 cm live sound speaker system. As is common for a system like this, the response falls off well before the edges of the audio range, and we find that there are significant peaks and dips in the frequency response. However, variations within ± 3 dB are often considered 'flat.' In poorly designed two-way or three-way systems, large variations in response usually occur around the crossover frequencies.

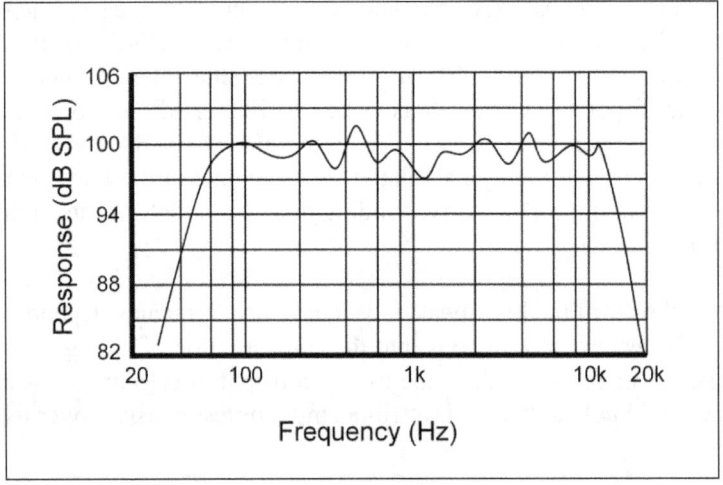

Figure 4-25 Typical Frequency Response of a 30 cm Loudspeaker System

4.11.6 Sensitivity

The sensitivity specification is an important one for sound engineers because in combination with the dispersion specification, it tells us how well a loudspeaker system will cover a given audience area. Basically, this specification tells us how good or efficient a speaker system is at converting electrical energy from an amplifier into sound. Loudspeaker systems are notoriously inefficient at converting electrical energy into sound. Most of the electrical energy input to a loudspeaker system is converted to heat. Most of what is left is converted to cone motion. Because the cone has so much more mass than the air which it moves, much of the remaining energy is given up to cone momentum. Less than 2% of the original electrical energy is actually converted to sound energy in most cases.

The sensitivity specification tells us how loud the speaker will be at a distance of 1 m from the center of the speaker when it is driven with a power of 1 W. This loudness is measured in dB SPL. For example, if such a measurement were 97 dB, then the specification would be stated as *97 dB SPL @ 1 W @ 1 m*. Sensitivity measurements are usually made with shaped (i.e., filtered) noise input that mimics the energy distribution of typical music program.

4.11.7 Power Handling Tests

As noted above, most of the energy delivered by the amplifier to the loudspeaker is converted into heat in the voice coils of the drivers in the system. Some energy is also lost in the speaker's crossover network. Fundamentally, the power handling specification of a loudspeaker should tell us how much power the system can dissipate without incurring damage.

Manufacturers use several different terms to describe the power handling capabilities of their loudspeaker systems. Terms such as 'peak,' 'continuous,' and 'RMS' need to be precisely defined by the manufacturer if we are to make sense of their power handling specification. The use of certain power handling terms can obscure the true capabilities of the speaker system and may lead a user to conclude that a speaker system can handle more power than it realistically can. Before discussing the power handling capabilities of speaker systems further, let us look at some of the terms that we will use when discussing this topic.

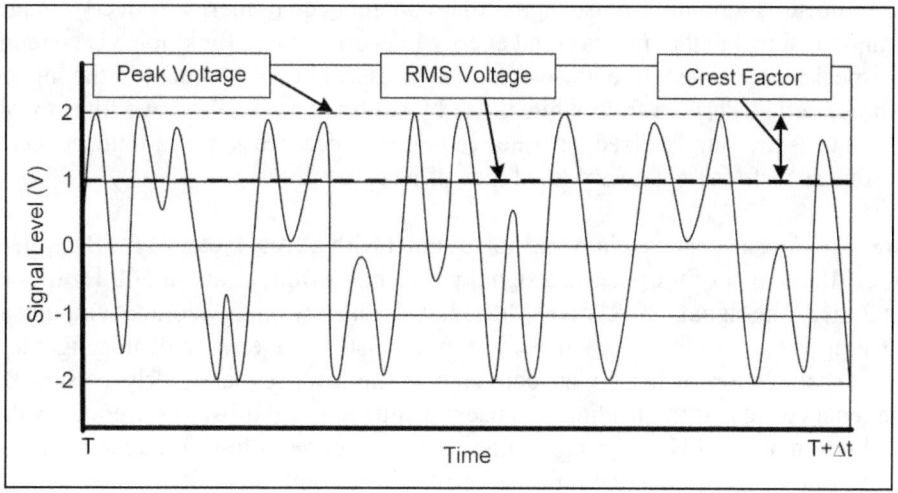

Figure 4-26 Power Handling Terms

For illustration purposes, let us suppose that Figure 4-26 shows a short burst of an alternating current (AC) sound signal. In this type of signal, the level of both the voltage and the current alternates between positive and negative values. The *peak* signal level is the maximum level reached by the signal and in this example is 2 volts. The *peak-to-peak* signal level, which is the difference between the largest negative level and the highest positive level, is 4 volts.

You will often see the term *RMS* (Root Mean Square) used by manufacturers in describing the power handing capabilities of their products. Technically, this term only applies to current or voltage, not to power. The RMS value of an AC signal is the equivalent direct current (DC) value that would deliver the same power into a given load. The RMS voltage or current of a signal is not the same as the simple average of the signal's voltage or current, respectively, which may actually be zero.

The RMS voltage of an AC signal is given by:

$$V_{rms} = \sqrt{\frac{\sum_{i=1}^{n} V_i^2}{n}}$$

That is a complicated looking formula, but what it basically tells us is this: we take the signal and sample (i.e., measure) its voltage (V) at several points ($i = 1$ to n) over a period of time. We square the value of each voltage measurement and then calculate the average (mean) of all the squared measurements. Finally, we find the square root of the mean. Hence the term *root mean square*. Clearly, since the current of the signal is also alternating, we can do a similar thing for measuring current. The power delivered to a load by that AC signal is not alternating, though, so we do not calculate an RMS value of the power.

The *crest factor* is the difference between the peak level (i.e., the crest) and the RMS level of a signal. It is usually expressed in dB. Recall that *dBV = 20 log(V_2/V_1)*. Thus, a crest factor of 6 dBV represents a 2:1 voltage ratio. Since power is proportional to the square of the voltage (remember that *P = V^2/Z*), this is equivalent to a 4:1 power ratio.

There are a number of standardized tests that reputable manufacturers can use to determine the power handling capability of their loudspeaker system. These tests generally use some form of noise signal. White noise is a random noise signal that contains equal energy at every frequency. Pink noise is a random signal that contains equal energy in every octave. Pink noise is generally used in testing loudspeakers because white noise contains excessive energy content in the higher frequencies. Pink noise can be derived from white noise by applying a −3 dB/octave filter to white noise. Because these tests are standardized, anyone reading the specification of a loudspeaker that references one of these tests can get a very good idea of its capabilities.

The **EIA RS-426-A** test consists of a white noise signal with a crest factor of 6 dB applied to a band limiting filter. The output of the filter is a signal with a bandwidth of about 500 Hz to 8 kHz with a peak about 2 kHz. The signal has relatively little low-frequency energy content and thus is not very useful in testing large low-frequency drivers. Since most of the energy of this signal is centered around 2 kHz, calculation of the power delivered to the low-frequency driver using the driver's nominal impedance value is misleading because the impedance of most low-frequency drivers rises well beyond nominal at 2 kHz. The signal does not have enough low-frequency content to cause mechanical failure of low-frequency units, so most failures with this test are thermal. The duration of the test is 8 hours.

The **EIA RS-426-B** test consists of a noise signal with a crest factor of 6 dB. The signal is flat from 40 Hz to 1 kHz. From 1 kHz to 10 kHz the signal has a 3 dB/octave roll-off. Below 40 Hz and above 10 kHz the roll-off is very steep. The test signal is a good approximation of a real music signal. The duration of the test is 8 hours. This test stresses the drivers both mechanically and thermally fairly equally.

The **AES2-1984** test is really meant for individual drivers. It consists of a pink noise signal with a crest factor of 6 dB and a bandwidth of one octave. The center frequency is varied depending on the driver being tested. The duration of the test is 2 hours.

The **IEC 60268** tests used to be called the IEC 268 tests. IEC 60268-1 is the *Sound System Equipment – General* test, but contains a section for loudspeakers. IEC 60268-5 is the loudspeaker specific test. The test consists of a pink noise signal with a crest factor of 6 dB, flat from 100 Hz to 2 kHz, with 3 dB/octave roll-off at each end. The duration of the test is 100 hours.

4.11.8 Power Handling – The Actual Numbers

The first thing an engineer should do when looking at the power handling specification is to see if the manufacturer used one of the standard tests in deriving the specification. If an in-house test was used, it is anyone's guess what the numbers mean.

The power delivered to the speaker system is calculated as $P = V^2/Z$, where V is the voltage of the test signal and Z is the speaker's impedance. When the voltage used in the calculation is RMS voltage, most manufacturers will refer to the power rating as *RMS power*. As we noted above, the term RMS power is not technically accurate. RMS is used to derive an equivalent DC value from a waveform that has both positive and negative values. Power is always positive. RMS power may also be called *average power*. Again, note that the average voltage of the test signal is generally not be the same as the RMS voltage and is not usually used in power calculations.

Program power is generally defined to be twice the RMS power, i.e., it is not a measured quantity. It is also referred to as *music power*.

If the peak voltage of the test signal is used in calculating the power, then *peak power* is the term used in stating the specification. The speaker system may only be able to withstand this level of power for a few milliseconds before something goes up in smoke. This figure is good for letting you know how hard you should not push your speakers. Peak power is also called *maximum power* or *total power* by some manufacturers.

4.12 Power Compression

Power compression is a phenomenon that is observed as more power is fed into a loudspeaker driver. Basically, the driver becomes less efficient as we push it harder, and it takes disproportionately larger amounts of power to get slightly greater volume out of the driver. There are primarily two reasons for this. The first is the rise in voice coil temperature which causes an increase in impedance. An increase in impedance will require the driving amplifier to deliver more power in order to move the voice coil. As the driver dissipates more power, its voice coil can reach temperatures of 125 °C. Practically all high-end high-frequency drivers have their voice coils suspended in ferrofluid, which helps to cool the voice coil. One manufacturer even took to blowing air over the magnet/voice coil assembly of their low-frequency driver in order to cool it, with a claimed dramatic reduction in power compression.[4]

The second cause of power compression is that as we push the speaker harder, the drivers operate nearer to their mechanical limits. In this region, the compliance becomes non-linear, i.e., it takes more force to move the cone a small distance in this region than it does to move the cone the same distance when it is operating closer to its rest position. Since the force to move the cone comes from the electrical current flowing through the voice coil, we can see that it will take disproportionately more power to get slightly more volume from the driver in this region of operation.

4.13 Choosing a Loudspeaker

Let us summarize this chapter by looking at some of the things to consider when choosing a set of loudspeakers for a sound system. As is always the case with any piece of electronic sound equipment, we should not go on specifications alone, but should also give the speaker system a good listen. Some of the key things to consider are the function of the loudspeaker (e.g., FOH, monitor, or fill); the size, layout, and reverberation characteristics of the venue; and the dominant program material to be handled.

Dispersion is one of those factors that you will want to pay attention to. Monitors require controlled, narrow dispersion, as compared to FOH loudspeakers, but in a highly reverberant space, narrow dispersion FOH loudspeakers can also be more easily aimed to help reduce problems with feedback.

Remember that sound intensity falls off rapidly with distance, especially the high frequencies, which suffer from higher absorption as well as inverse square law dissipation. You will need to weigh the options of using high-power loudspeakers with long-throw high-frequency character-istics against smaller units in a distributed system when trying to fill a large venue. The main FOH positions tend to require large powerful units, whereas fill speakers typically have more modest power requirements.

The primary function of the system, i.e., the main type of program material handled will have a key influence on selection. For example, a speech-only system does not need to reproduce sound components much below 80 Hz, nor does it need to handle frequent high-level high-frequency transients. A music system, though, is more demanding. If your program material calls for a lot of bass, then size counts.[a] Music may also contain frequent high-frequency transients, so listen to the high-frequency drivers to make sure that they are up to the task of delivering clean high-frequency sound.

Review Questions

1. Explain the principle by which a dynamic driver works.
2. What is back EMF?
3. What is the purpose of mating a horn to a driver?
4. What are some of the issues with compression drivers?
5. Explain what is meant by cone breakup in a dynamic driver.
6. What causes high-frequency beaming when sound is emitted from a direct radiator?
7. What is the role of a crossover in a loudspeaker system?
8. What is the bandwidth of a bandpass filter whose center frequency is 1 kHz and Q is 3.0?
9. How does the operation of a passive crossover compare to that of an active crossover?
10. Explain how bi-amplification works.
11. What are the advantages of using a bass reflex speaker enclosure over a sealed box design?
12. What are the advantages and disadvantages of an active vs. passive loudspeaker system?
13. What does the sensitivity specification of a loudspeaker tell us and why is it important?
14. How does a vertical line array help control dispersion in the vertical plane?
15. Discuss three approaches that are used to protect the drivers of a loudspeaker system from destruction due to excessive high current from the amplifier.

---

[a] For a driver of a given size in free air, the cone needs to move four times as much for every octave you go down in the audio spectrum if it is to produce the same SPL. Since many drivers fail to do this at low frequencies, the difference must be made up by moving more air, which can be done by using bigger drivers.

Chapter 5

Amplifiers

5.1 System Overview

If we refer back to Figure 1-1, we will see that amplifiers fall into the category of signal processors—devices that manipulate the audio signal in some way. They are essential to any live sound reinforcement system. An amplifier's job is at once conceptually simple, yet practically complex—take a low-level input signal and turn it into a high-level output signal, without changing any other aspect of the signal. No real amplifier does a perfect job of this. Fortunately, though, modern technology does an excellent job when operating in a live sound environment.

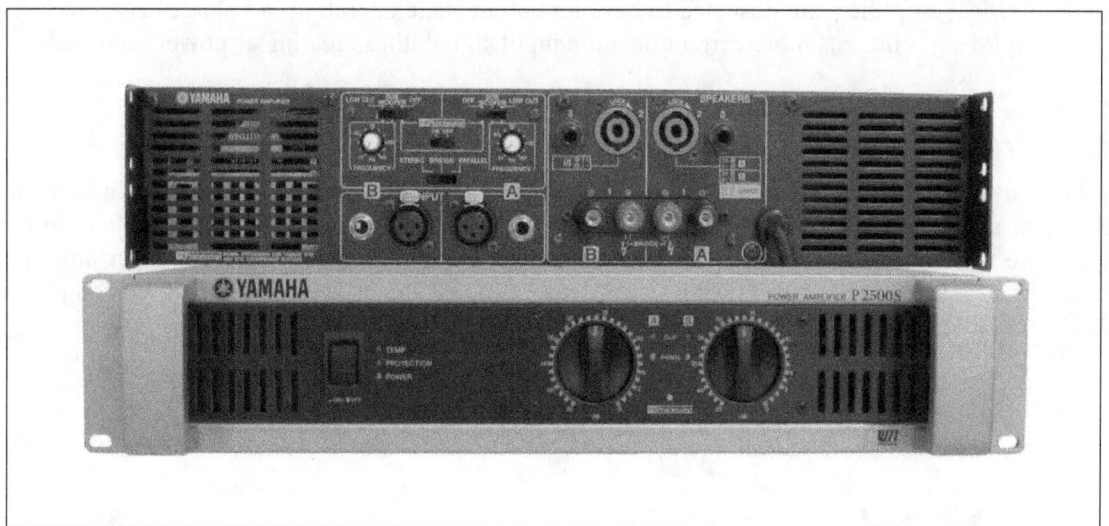

Figure 5-1   Small Live Sound Reinforcement Power Amplifier

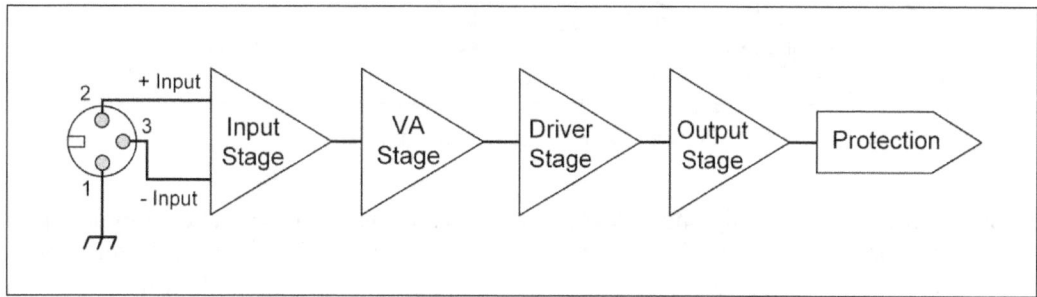

Figure 5-2   Amplifier Stages

Figure 5-1 shows a typical small live sound reinforcement amplifier. Such an amplifier usually consists of four stages, as shown in Figure 5-2. These are the input stage, which does impedance conversion and signal buffering; the voltage amplification stage, which increases the input signal to a level sufficient to drive the driver stage; the driver stage, which increases the voltage and current to a level that is acceptable to, and switches on, the output stage; and the output stage, which supplies the high voltage and current to the loudspeakers. The figure also shows that there is usually some form of protection circuitry which keeps the output devices from being destroyed in cases that would overstress these devices, e.g., output short circuit.

5.2 Current vs. Voltage Drive

An amplifier can be designed to increase the output current as the current of the input signal increases. This is called current drive. For a given fixed input signal, this kind of amplifier will try to maintain a constant current, but vary the voltage if the output load impedance changes. Current drive output stages are used for experimental setups in things like variable output impedance amplifiers and are not likely to be seen in a live sound reinforcement application.

An amplifier can alternatively be designed to increase the output voltage as the voltage of the input signal increases. This is referred to as voltage drive. In this case, for a fixed input signal, the amplifier will try to maintain a constant voltage, but vary the current as the output load impedance changes. Most amplifiers are designed to have an output stage operating in voltage drive, with the capability of sourcing lots more current than the input signal, thus creating a 'power' amplifier.

5.3 Pre and Power Amplifiers

While any amplifier can itself be thought of as consisting of several sections, including pre-amplifier and power amplifier sections, when we talk about preamps and power amplifiers in this book, we usually refer to entire self-contained units that fulfill a specific pre or power amplifier role. Figure 5-3 shows how the preamp sits between a source signal device and the power amplifier, while the power amplifier takes the signal from the preamp and feeds it to the loudspeaker.

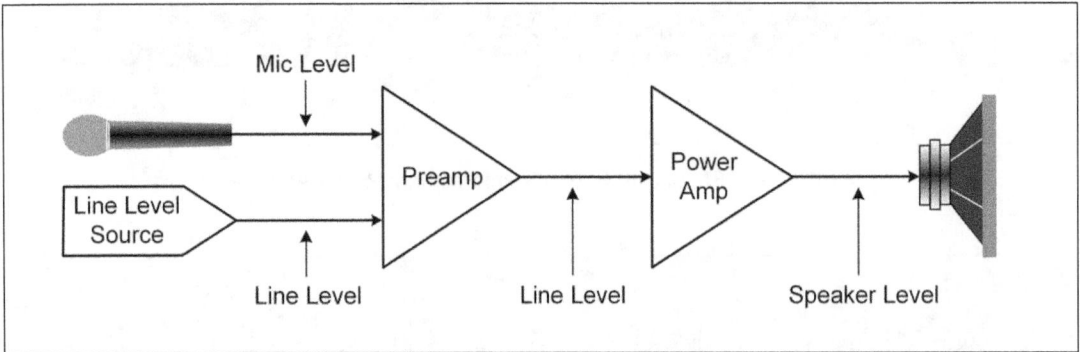

Figure 5-3 Pre and Power Amplifiers

The preamp's first role is usually one of level conversion. Typically, it will take a low-level signal, such as a microphone-level signal, instrument-level signal, or even a low line-level signal and boost it to the level required to drive the input stages of pro-audio equipment. This may be a power amplifier or some other signal processor. See Section 1.9 for a thorough discussion of these signal levels. To do a good job, preamps must therefore have high gain and produce very little noise.

Preamps may also be used as a buffer between two devices. This will be the case when the device that is being driven would draw so much current from the first device that the output voltage of the first device would fall to unacceptably low levels and the system as a whole would become too noisy. In this case, a preamp with enough output current to drive the second device would be used to interconnect the two devices.

The power amplifier's job is to take a line-level signal and boost it to a level that will drive a loudspeaker, which is generally well in excess of 20 Vrms.

Modern amplifiers use a combination of active components in their construction. In preamps, the devices that do the actual amplification are usually either a form of field-effect transistor (FET), bipolar junction transistor (BJT), integrated circuit operational amplifier (IC op-amp), or even thermionic valve (the old-style vacuum tube). In power amplifiers, FETs and BJTs may be used in the various driver stages, but the output stage is usually made up of metal-oxide semiconductor field-effect transistors (MOSFET). MOSFETs are very hardy devices that can deliver lots of current while taking the abuse a pro-audio power amplifier is likely to encounter over its lifetime.

5.4 Amplifier Classification

The output stage of almost all power amplifiers is designed using pairs of opposite polarity devices that operate in a complimentary fashion. This complimentary action allows the amplifier to produce more power more efficiently. The amplifier is classified depending on the action of these output devices with respect to one cycle of the input signal (considering positive and negative portions), i.e., how the output devices are turned on relative to each other and the input waveform. Figure 5-4 is a block diagram which shows the output devices being driven by their respective driver devices. The bias circuit determines when and for how long the drivers, and hence the output devices, are turned on. Below we will take a brief look at the most popular classes of amplifiers.

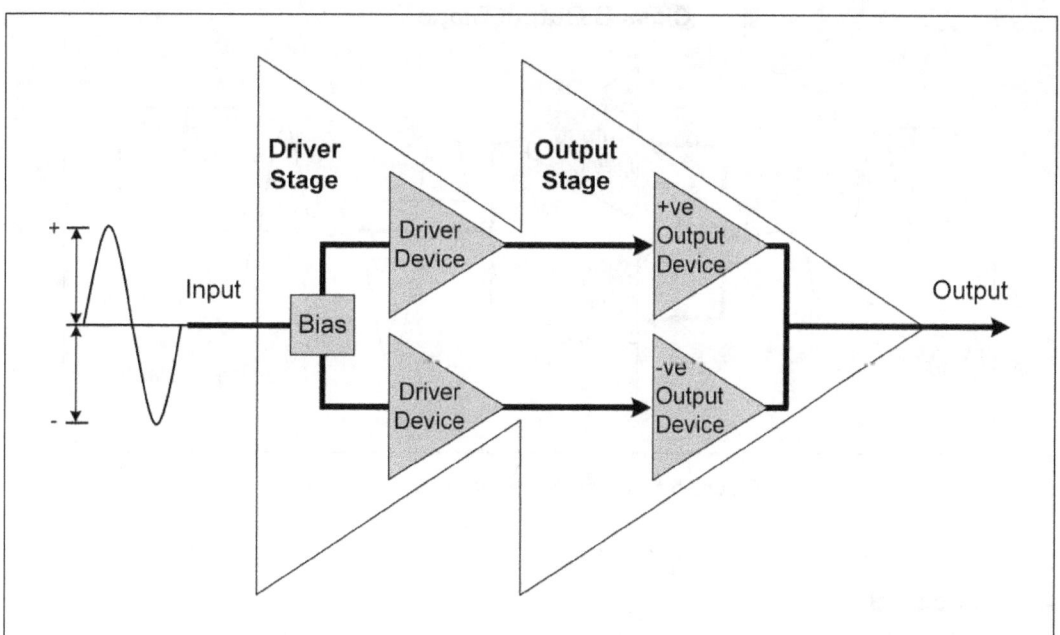

Figure 5-4 Amplifier Driver and Output Stages

5.4.1 Class A

The Class A amplifier is the exception to having complimentary pairs of output devices. In a Class A amplifier, there is only one polarity of output device. Class A amplifiers are biased so that the output devices are always conducting regardless of the input signal polarity or level, the aim being to keep the output devices in their most linear region of operation. That means that even when there is no signal present at its input, the output devices are conducting, so the amplifier wastes a lot of energy, dissipating it as heat. These amplifiers are so inefficient—achieving about 20% efficiency in reality—that you will never see a Class A power amplifier used in a live sound role. A 500 W output mono block would easily consume over 1500 W and would require a large heavy power supply and massive cooling system. A Class A amplifier may be found in a preamp, though, where signal levels are much smaller and wasted energy is much lower. The great thing about Class A amplifiers is that they produce the least amount of distortion of any amplifier type.

5.4.2 Class B

Figure 5-5 shows the operation of a Class B output stage. In a Class B amplifier, one output device conducts during the positive half of the input waveform, while the opposite polarity complementary output device conducts during the negative half of the input waveform. The two devices are never on at the same time. This is sometimes called push-pull operation.

Because one device switches off and the other switches on when the input signal crosses over from positive to negative (and vice versa), there is a glitch in the output signal, producing large amounts of crossover distortion. This distortion is so high that you will not likely see a pure Class B amplifier used in a live sound role. However, Class B amplifiers are fairly efficient, about 60% when driving real speaker loads.

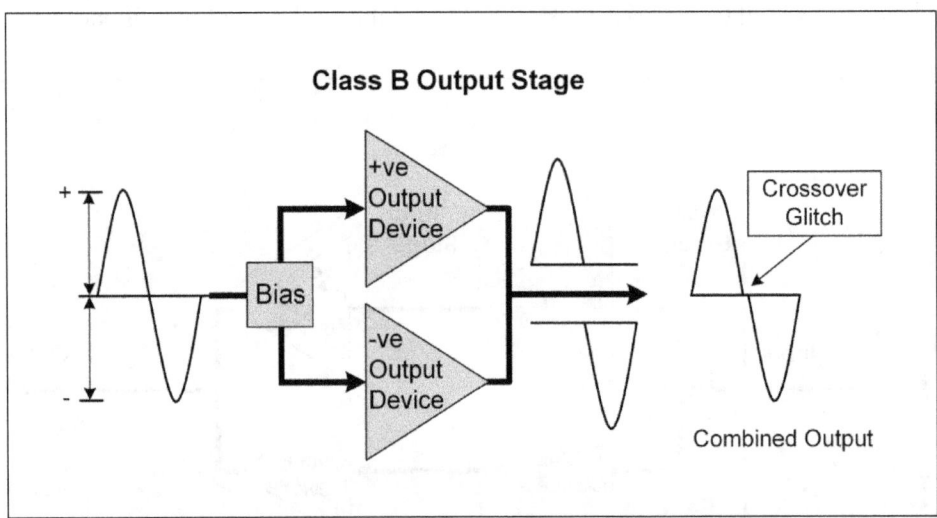

Figure 5-5 Class B Output Stage Operation

5.4.3 Class AB

In an attempt to harness the efficiency of Class B, while at the same time using the principles of Class A to eliminate crossover distortion, Class AB is used. Its operation is shown in Figure 5-6. Class AB is the most popular amplifier design for intermediate power amplifiers—up to about 500 W into 8 Ω.

Unlike Class A amplifiers which are biased so that the output device is always on and in its most linear region of operation, a Class AB device is biased so that it is barely conducting when no input signal is present. Thus, a Class AB amplifier wastes energy too, but it is very small compared to a Class A amplifier. The Class AB design is only about 50% efficient when driving real loads.

Because each complementary output device is biased to be just on, each device conducts for a little more than half the input cycle. There is thus some overlap in the crossover region, when both devices are conducting. This practically eliminates the crossover distortion of the Class B design and gives a high degree of fidelity that is suitable for live sound applications.

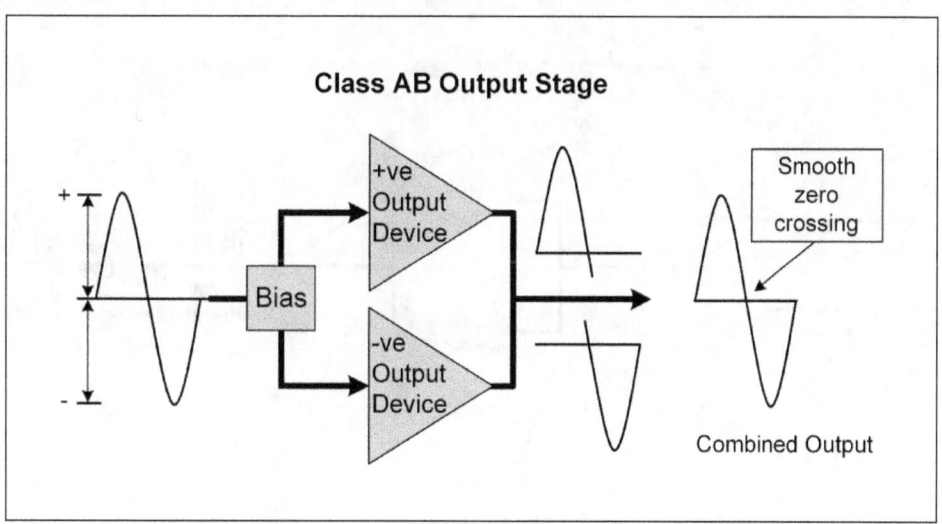

Figure 5-6   Class AB Output Stage Operation

5.4.4   Class G and Class H

Both Class G and Class H are derived from Class AB, with Class H being a refinement of Class G. These amplifier classes are popular in higher power amplifies over 500 W into 8 Ω. In Class G the output stage is connected to two different power supplies, one low voltage and one high voltage (or equivalently, one power supply with different supply voltage rails). For most of the audio program material, the low voltage power supply is used. The amplifier switches to the high voltage power supply for high-level peaks in the music program.

Class H also uses a power supply with low and high voltage rails. Like Class G, the amplifier also switches to the high voltage rail for higher-level audio input signals. In Class H, however, the output voltage of the high voltage power supply is modulated by the audio input signal. The power supply thus supplies just enough power to drive the output stage with very little wasted energy. This gives improved efficiency over the Class G design. Compared to Class AB, this configuration delivers much more power for a given weight and size.

5.4.5   Class D

Class D amplifiers work quite differently to the other amplifier classes we have been discussing up to now. The classes discussed above are all analog linear designs, where a change in voltage at the input causes a similar but larger change in voltage at the output—e.g., put a low-level sine wave at the input and you get a higher-level sine wave at the output. On the other hand, a Class D amplifier is a switching device which naturally generates a high-frequency square wave. In Figure 5-7 the comparator compares the instantaneous amplitude of the input signal with that of the generated triangle wave. If the input signal is higher, the comparator's output is positive. If the triangle wave is higher, the output is negative. The input signal is thus used to pulse-width modulate the square wave. Pulse-width modulation means that the width of the tops and bottoms of the square wave is changed based on some attribute of the input signal, in this case, its amplitude. The base frequency of the square wave is called the switching or sampling frequency, and is set by the triangle wave generator shown in Figure 5-7. The switching frequency is at least twice as high as the highest audio frequency, but usually much higher—typically hundreds of kHz.

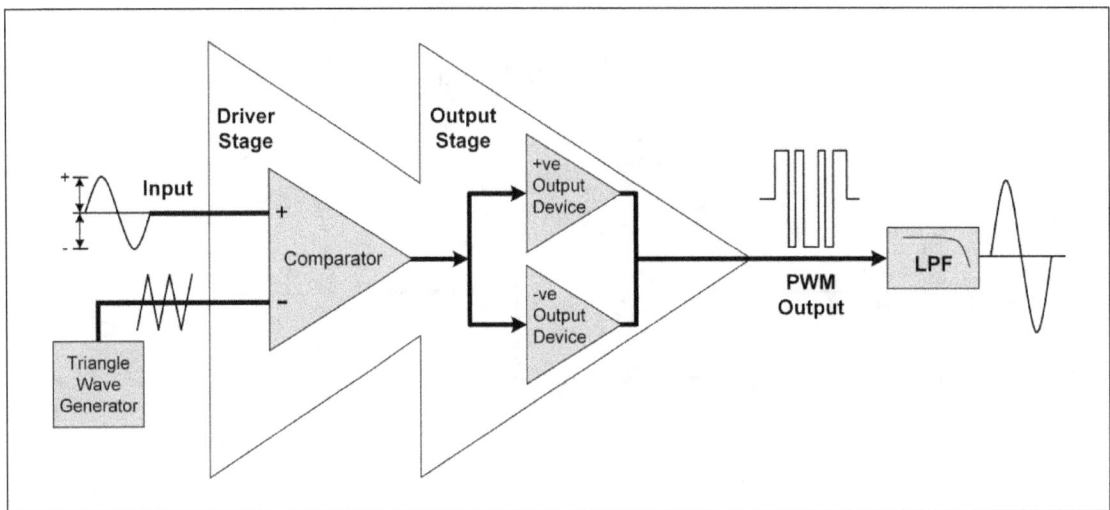

Figure 5-7 Class D Driver and Output Stage Operation

The pulse-width modulated (PWM) square wave is used to turn on and off the output devices. The output devices produce a high-level PWM square wave. As illustrated in Figure 5-7, this PWM square wave looks nothing like the original input signal. However, if the PWM signal is averaged over time, the shape of the original input signal is recovered. The way that this PWM square wave is averaged is by passing it through a low-pass filter (LPF). This filter also removes the high-frequency switching components from the signal.

In Class D operation the output devices are, theoretically, either complete off or completely on. If the output device had zero on-impedance and could be switched on infinitely fast, no power would be dissipated in the device when it is turned on. Practically, an output device still has some impedance when it is on. Furthermore, some power is lost in the finite amount of time it takes to turn on the device. These switching losses are thus greater with higher switching frequency because the devices spend more time turning on and off. Still, it is not uncommon to see Class D output stages that can achieve over 80% efficiency. Higher efficiency translates directly into physically smaller and lighter amplifiers.

With a very high switching frequency, the error in the recovered audio signal is lower and smaller output filter components can be used. Total harmonic distortion (THD) is usually higher, though, because switching transients occur more often. The distortion due to switching transients is higher than the crossover distortion in class AB amplifiers. The sampling frequency is generally chosen to be at least 10 times the highest audio frequency to be amplified. That is why early Class D amplifiers were mostly used to drive subwoofers where the maximum audio frequency is around 200 Hz (thus allowing a low switching frequency to be used, giving a good balance between high efficiency and low THD). With current designs, though, many of the issues with distortion have been addressed, and Class D amplifiers are now amongst the most commonly utilized amplifiers in full-range live sound applications (e.g., the amplifier used in the ZLX-15P referenced in Section 4.9 uses a Class D design).

5.4.6   Class D Derivatives

A number of manufacturers have created amplifier designs based substantially on the Class D principle and have marketed them as new classes, though they are not strictly new classes in the classical sense of the term. Practically all of these use proprietary techniques to get improved performance compared to the classical Class D design. Some of the more common ones that you are likely to encounter are briefly discussed here.

Class HD amplifiers are a natural advancement of Class D. Rather than using a fixed supply voltage, Class HD amplifiers utilize the Class H technique of modulating the power supply rail voltage with the input signal. This leads to even further efficiencies in power consumption.

The Class-I designation and design were developed by Crown International. It uses two sets of PWM waves in its output section, one for the positive part of the input signal and the other for the negative part. When the input signal is positive, the tops of the positive PWM wave are wider than those of the negative PWM wave, and vice versa. The final output audio wave is derived by combining the two PWM waves in an interleaved fashion, which is the source of the "I" in Class-I.[1]

Class-T was originally developed by Tripath Technology, Inc., as a proprietary extension of Class D.[2] The Class-T amplifier analyses the input signal and uses an algorithm to determine the best way to modulate the output square wave. Further, it uses feedback based on the output signal along with digital signal processing techniques to fine tune the modulation parameters (i.e., switching frequency, duty cycle) in real time.

Lab.gruppen is the source of the Class TD designation. This design uses a conventional linear output stage, such as a Class AB stage, but uses separate Class D amplifiers to modulate the supply rail voltages to track the input audio signal, similar in concept to Class H. Again, the aim is to get the benefits of higher efficiency of tracking supply voltage, with the additional low distortion of a linear output stage.[3]

5.5   Amplifier Output Power Rating

In the consumer electronic space, marketing departments have been known to push spectacular specifications that obscure the practical capabilities their products. Just as with loudspeakers, the power rating specification of the amplifier is sure to be hyped by the marketing department in trying to convince us of the stellar performance characteristics of the amplifier. Certainly, it is one which can be very confusing to the uninitiated. In the pro-audio world, things are generally not as bad as in the consumer world. Still, the power rating specification may hardly be given much real thought by the buyer as to what it really means, and this may lead to acquiring a product that may not be optimal for the application.

Before offering an amplifier for sale, the manufacturer performs certain tests to determine its power output capabilities. The marketing department takes these test results and presents them in the best light. Below we will look at some of the tests and other issues surrounding the amplifier's output power specification.

5.5.1 Power Supply Limitations – Clipping

All amplifiers have power supply limitations. The major limitation has two aspects—voltage and current. An amplifier cannot produce an output signal that is greater than what the power supply is able to deliver. That is why traditional high-power amplifiers need large power supplies.

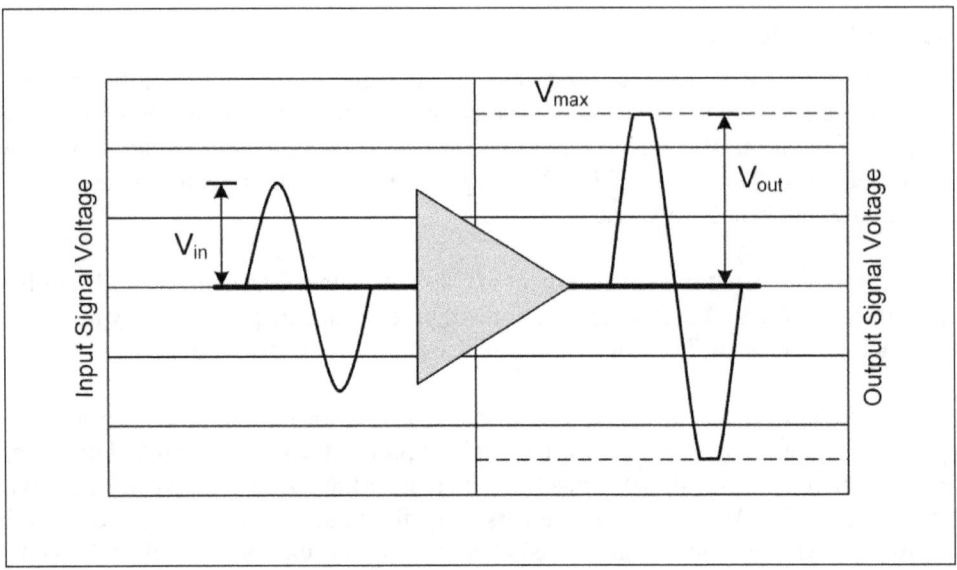

Figure 5-8 Output Signal Clipping

The output signal of an amplifier is limited to a maximum voltage that is no greater than the power supply output voltage. In Figure 5-8 this is referred to as V_{max}. Thus, if the power supply of a particular amplifier delivers a peak 50 V, the output signal of the amplifier cannot be greater than 50 V. Let's say this amplifier has a voltage gain of 40. If we drive the amplifier with an input signal of 1 V peak, the output signal would be 40 V peak. Now suppose the input signal rises to 1.5 V peak. The output signal should be 60 V peak, except that the amplifier can only deliver a maximum output of 50 V. This amplifier cannot faithfully reproduce signals where the input is above 1.25 V because the tops of the output signal are all limited to 50 V. We say that the output signal is clipped at 50 V. In Figure 5-8 we see that the top and bottom of the output signal are flat, as though they were cut off with scissors.

5.5.2 Power Supply Limitations – High Current

In the above section we looked primarily at the voltage limitation of the amplifier's power supply. However, the amplifier also has a current limitation. Even if an amplifier is operating within its designed output voltage range, if the load impedance gets too low, an attempt will be made to draw more current from the amplifier than the power supply can deliver. In this case, the power supply output voltage (and hence the output signal voltage) will drop, not necessarily uniformly, leading to distortion. An amplifier operating in this region may be perilously close to destroying its output devices, so its output protection circuitry may kick in and shut down the amplifier.

Some amplifiers, however, are built with power supplies (and output devices) that can deliver enough current to allow them to maintain the correct output signal voltage even when the load impedance drops to low levels, like 1 or 2 Ω. These amplifiers are usually referred to as high current amplifiers.

5.5.3   Power Bandwidth

When stating the power output of an amplifier, the manufacturer must also give the bandwidth over which this power output applies. For North American manufacturers, the power bandwidth is the frequency range over which the amplifier will deliver rated continuous power into a given load. European manufacturers using the DIN 45500 standard use only half the rated power for this specification, leading to a wider power bandwidth figure. Some North American and Japanese manufacturers have started using half power figures in their specifications as well. Power bandwidth may appear to be related to the frequency response, but the two are not the same thing. As we will see below, the frequency response is usually measured at a much lower power into a higher impedance, thus giving rise to results that are much better.

5.5.4   EIA RS-490/CEA-490-A Continuous Power

Pro-audio amplifiers sold in North America will usually publish power output test results derived from EIA RS-490 tests. This is a series of tests developed by the Electronic Industries Association, later known as the Electronic Industries Alliance. In 2004 this specification was revised and became known as CEA-490-A. Some manufacturers still use the earlier version, EIA RS-490, 1981. For pro-audio equipment manufacturers, these tests fulfill the spirit of the US Federal Trade Commission (FTC) rule in 16-CFR Part 432 (which actually applies to consumer audio amplifiers).

For all power output tests, the amplifier's output is terminated by resistances representing nominal values of real loudspeaker loads. Usually 8 Ω, 4 Ω, and sometimes 2 Ω are used.

For continuous average power, the test consists of driving the amplifier with a sine wave signal at various frequencies, one at a time, over the amplifier's rated bandwidth. The exact frequencies used and the number of spot checks are left to the manufacturer, who can use any frequency that best highlights the amplifier's capabilities. For each frequency, the input signal is increased until the amplifier's output signal reaches its rated total harmonic distortion (THD), usually 1%. For each frequency used, the test continues for at least five minutes, and the minimum RMS output voltage reached at 1% THD in any single test is used to determine the amplifier's continuous average power rating. See Section 4.11.7 for an explanation of RMS voltage. Note that this test allows the amplifier to be tested using only one channel at time.

5.5.5   DIN 45500

European amplifier output power tests are somewhat similar to the North American tests. Manufacturers use the appropriate subset of DIN 45500 to measure the amplifier's output power.

As with EIA tests, the amplifier's output is terminated with a load of 8 Ω, 4 Ω, or 2 Ω to represent real speaker loads. For a continuous power measurement, the input signal consists of a 1 kHz sine wave. The average voltage output by the amplifier over a 10-minute period, with no more than 1% total harmonic distortion, is used to determine its stated continuous power output.

5.5.6   EIA RS-490/CEA-490-A Dynamic Headroom

Dynamic headroom is a concept that is important in expressing how well the amplifier can handle large transients without clipping the signal. As shown in Figure 5-9, it gives the amount of power over and above the continuous power that the amplifier can cleanly produce for very short periods of time.

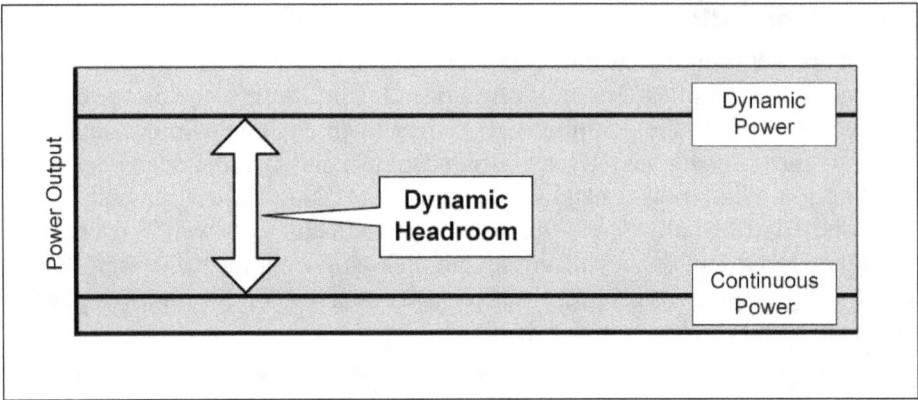

Figure 5-9 Dynamic Headroom

To determine dynamic headroom, a test signal that consists of a 1 kHz sine wave is bursted to a high level for 20 ms, then attenuated by 20 dB for 480 ms. This is done repeatedly for a period long enough to gather test results. The input signal is increased until the amplifier just starts to clip the output signal, and then it is reduced to just below clipping. The RMS voltage of just the peak portion of the output signal (corresponding to the input burst) delivered by the amplifier in this test scenario is used to determine its dynamic headroom rating.

Dynamic headroom is usually expressed in dB. Suppose the continuous power, P_{cont}, of an amplifier is determined to be 500 W into 8 Ω, while the power output from the dynamic headroom test, P_{dyn}, is 700 W into 8 Ω. Then the dynamic headroom is calculated as:

$$
\begin{aligned}
\textit{Headroom in dB} \quad &= \quad 10 \log\left(\frac{P_{dyn}}{P_{cont}}\right) \\[2mm]
&= \quad 10 \log\left(\frac{700}{500}\right) \\[2mm]
&= \quad 10 \log(1.4) \\[2mm]
&= \quad 1.5 \text{ dB}
\end{aligned}
$$

5.5.7 Amplifier Output Power Confusion

Sometimes the term *RMS power* is used to describe the output power of an amplifier. As discussed in Section 4.11.7, the term *RMS* technically applies to voltage or current. However, if RMS output voltage is used to calculate the output power, then the power rating is often referred to as RMS power.

Many terms are used to express the dynamic headroom specification of the amplifier. Dynamic power and peak power are used often for this. This is sometimes also called maximum power or total power. However, these terms are sometimes also used to express the power output calculated using the peak output voltage obtained in a dynamic power test.

Manufacturers, though, play games with all these terms. The bottom line is that if they do not provide exact definitions or state the standard test used, we may never really know just what the power rating specification means.

5.6 Amplifier Frequency Response

The amplitude response of an amplifier with respect to frequency, usually just called the frequency response in the manufacturer's publications, tells us how well behaved an amplifier is at amplifying a signal over its rated bandwidth. Over its rated bandwidth, for a given input level, the amplifier should ideally increase the signal's amplitude by the same amount regardless of the signal's frequency. This would be the ideal flat frequency response. Modern amplifiers come pretty close to this ideal over the audio range.

The best way to present the frequency response characteristic of a device is with a graph of amplitude response versus frequency, as was done for microphones in Section 3.8. This would show at a glance any deviations from a flat response anywhere over the amplifier's rated bandwidth. However, you will not likely see such a presentation of this information for an amplifier. Instead, a statement of the following form will be published:

$$\textit{Frequency Response: 20 Hz} - \textit{20 kHz, +0/−0.5 dB, 1 W/8 }\Omega$$

This statement tells us that the frequency response test was done over an audio range of (at least) 20 Hz to 20 kHz. Using a swept sine wave, the response of the amplifier is measured over this range. The output level of the amplifier at 1 kHz is taken as the 0 dB reference level. Then the maximum deviation above and below this level anywhere over this range is noted. Thus, the statement *+0/−0.5 dB* says that over the test range, there was no deviation above the 1 kHz level and the maximum deviation below the 1 kHz level was 0.5 dB. We have no way of knowing where this deviation occurred, but it is likely at one of the extremes of the stated frequency range. The specification also tells us that the test was conducted using an output power of 1 W into 8 Ω, well below the rated power of the amplifier.

5.7 Impedance and Damping Factor

Amplifiers have both input impedance and output impedance. The concept of impedance was discussed in Section 3.9.

The input impedance is a measure of the total opposition that the input stage of the amplifier gives to the flow of alternating current (AC). It gives an idea of the load that the amplifier will place on equipment that it is connected to, i.e., the equipment that is feeding the amplifier. To put as little load on this equipment as possible, the input impedance of the amplifier needs to be rather high. Typical values are in the range of 10 to 30 kΩ.

The output impedance is the impedance of the output stage of the amplifier. For a transformerless solid state output stage, typical of power amplifiers used for live sound reinforcement, this may be just a few hundredths of an ohm. This specification has no real bearing on the common loudspeaker loads that may be connected to the amplifier. However, the output impedance figures into the damping factor specification.

Damping factor only applies to power amplifiers driving loudspeakers. Damping factor is simply the ratio of the speaker load impedance to the output impedance of the amplifier:

$$Damping\ Factor = \frac{Speaker\ Load\ Impedance}{Amplifier\ Output\ Impedance}$$

Solid state amplifiers typically have a high damping factor specification (on the order of 400 to 500), while tube amplifiers have relatively low damping factors (around 10 to 20). Note that the load impedance of a speaker changes with frequency, so the damping factor is usually stated at the nominal impedance of the speaker (typically 4 or 8 Ω).

Damping factor is only an issue for the lower frequencies, i.e., an issue for the woofers in the loudspeaker system. The problem is that the cone and voice coil of woofers have a lot of mass, and hence inertia, and tend to continue moving even after the driving signal from the amplifier is removed. Thus, the driver continues to produce sound when none is intended. As was discussed in Section 4.3, when the voice coil of the speaker moves, it also generates back EMF, a voltage that is of opposite polarity to that of the driving voltage from the amplifier. By shunting this back EMF, the unwanted motion of the speaker cone is damped, and the lower the impedance into which this back EMF is shunted, the more the motion is damped. This implies that for an amplifier, a bigger damping factor is better, as it may be able to tame unwanted woofer oscillations and improve the low-frequency sound of a poorly damped loudspeaker.

However, the situation is not so straightforward. There are three (ignoring the loudspeaker cables for now) impedances to consider in the signal chain between the amplifier and the loudspeaker, viz. the amplifier's output impedance, the impedance of the crossover, and the driver's voice coil impedance. Since the crossover's impedance is much larger than the output impedance of the amplifier (assuming a typical solid state output stage), the real effective damping of the amplifier is greatly reduced from what the typically high damping factor specification might have you believe. Even if we had an amplifier with zero output impedance (suggesting an infinite damping factor), given a typical 8 Ω nominal loudspeaker with a voice coil resistance of 6.5 Ω, and a crossover resistance of 1.5 Ω, we would have a more realistic damping factor of about 4. Of course, since the impedance of the crossover will vary with frequency, there is plenty of opportunity to achieve even lower damping.

5.8 Distortion Characteristics of Amplifiers

All amplifiers introduce some distortion into the signal they are processing. Distortion occurs anytime the output signal is different in any way, other than amplitude, from the input signal. Distortion generally falls into two categories—non-linear and linear.

Non-linear distortion occurs when new audio waves, at frequencies not present in the input signal, are added to the amplifier's output signal. This is caused by non-linearities in the transfer function of the amplifier. The transfer function of the amplifier is simply a mathematical way of describing how the amplifier derives its output signal from the input signal. It is partially visualized by a frequency response curve, as was described in Section 5.6.

Linear distortion results in shifting the relationship of certain attributes of the signal, without adding new frequencies to the signal. For example, the amplifier may shift the phase of some frequencies of the signal while not shifting others, resulting in distortion.

Figure 5-10 shows how a waveform can be distorted by an amplifier. There is overshoot and undershoot visible at the top and bottom of the waveform. There is settling, causing a rippling of the waveform top and bottom, and there is a transient distortion because the amplifier is too slow to keep up with the input signal. In the following sections, we will take a look at some of the things that cause signal distortion in amplifiers and their presentation in the amplifier's specification.

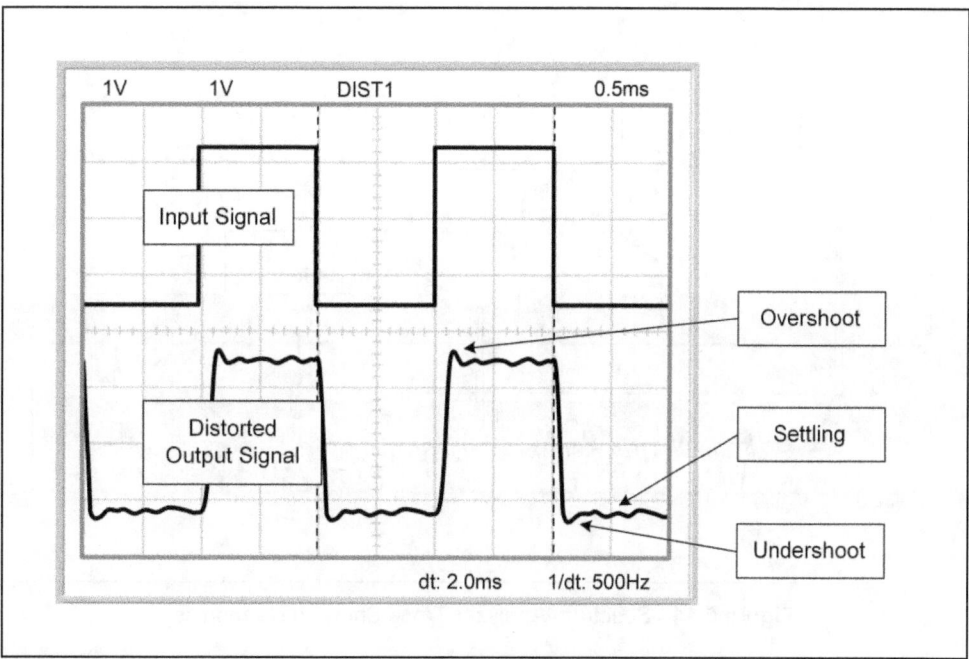

Figure 5-10   Distortion of Waveform by Amplifier

5.8.1   Harmonic Distortion

Harmonics are replicates of an original waveform at whole number multiples of its original frequency. The frequency of the original waveform is called the fundamental. The second harmonic would be twice the fundamental frequency, i.e., an octave higher than the fundamental. Similarly, the third harmonic is three times the frequency of the fundamental, and so on for higher-order harmonics. Figure 5-11 shows a spectrum analyzer trace where a 2 kHz fundamental, the third harmonic at 6 kHz, and the fifth harmonic at 10 kHz are highlighted. Other harmonics are also visible in the trace. Since harmonics are whole number multiples of the fundamental, they are musically related to the original waveform, i.e., they combine with the original wave to form a chord that is not unpleasant to the ear. This is in contrast to enharmonic frequencies, which add to form very dissonant, harsh, and unpleasant sounds.

With harmonic distortion, we have new frequencies added to the output that were not present in the input signal. These new frequencies are harmonics of the original frequencies. Total harmonic distortion (THD) can be measured by applying a pure sine wave to the amplifier input and then examining the amplifier's output for harmonics. A spectrum analyzer or FFT analyzer is used to show the harmonics and to measure the level of each harmonic. THD is then calculated by obtaining the ratio of the square root of the sum of squares of the RMS voltage of each harmonic to the RMS voltage of the fundamental. So, if V_n is the RMS voltage of the nth harmonic and V_1 is the RMS voltage of the fundamental, then:

$$THD = \frac{\sqrt{V_2^2 + V_3^2 + \cdots + V_n^2}}{V_1}$$

As per CEA-490-A, only harmonics that are bigger than 10% of the strongest harmonic would be included in this calculation.

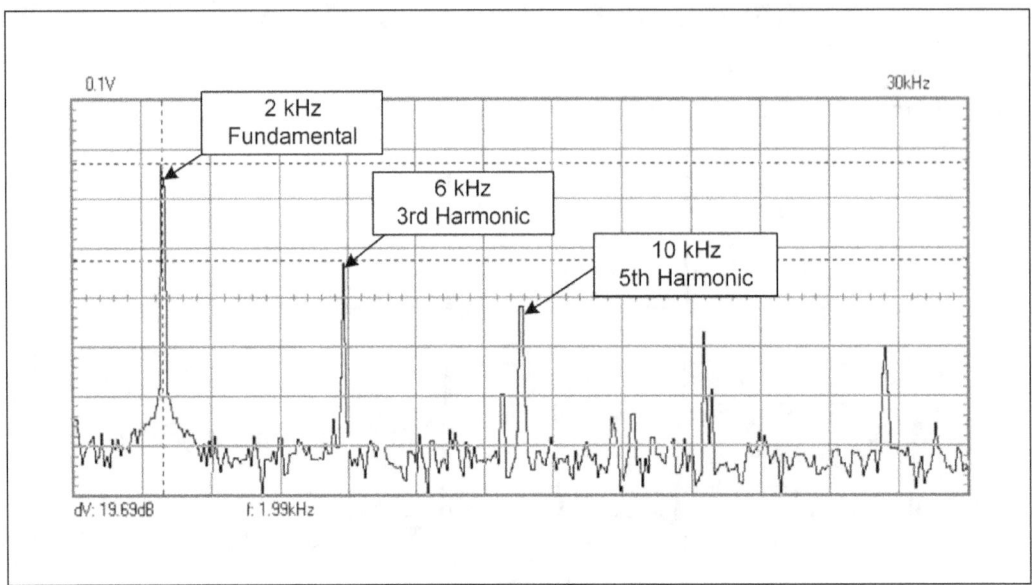

Figure 5-11 Spectrum Analyzer Trace Showing Harmonics

There is, however, a better test than simple THD. It is a measurement of total harmonic distortion plus noise, or THD+N. This is a better test because it measures everything that is added to the input signal—harmonics, hum, active component noise, etc. THD+N values are bigger than straight THD values for the same amplifier, and so some manufacturers who want to make their equipment look better may only give THD values. THD+N is measured by applying a pure sine wave to the amplifier. At the output, a very narrow notch filter removes the fundamental. What is left—harmonics plus anything else added by the amplifier—is measured by an accurate RMS volt meter.

It is generally agreed that most humans cannot hear THD below 0.5%. Since practically all modern audio amplifiers spec well below this, THD can most likely be disregarded when considering this aspect of amplifier performance.

5.8.2   Intermodulation Distortion

When two signals are applied to an amplifier, the output should consist of only these two signals, only bigger in amplitude. If these two signals interact with each other inside the amplifier in any way to cause the output signal to deviate from the above ideal, then we have intermodulation distortion, or IMD. There are two aspects to IMD. In the first case, a strong signal modulates a weaker signal, causing its amplitude to vary in accordance with the stronger signal's frequency. A spectral analysis would show the modulation components as sideband tones equally spaced about the modulated tone, as shown in Figure 5-12. In the second case, input signals whose frequencies are close to each other interact to produce new beat tones. In Figure 5-12, the input tones are shown as f_L and f_H. As can be seen, the new tones are not harmonically related to either input signal, but rather to the sum and difference of both signals, with frequencies $f_H + f_L$, $f_H - f_L$, etc. This type of distortion therefore produces harsh sounding output.

Different standards bodies have slightly different procedures for measuring IMD. The amplifier's manufacturer should therefore state which approach was used in their documentation.

In the SMPTE and DIN tests, a low-frequency sinusoidal signal and non-harmonically related high-frequency sinusoidal signal are summed together in a 4:1 amplitude ratio and applied to the amplifier. Typical frequencies used for the SMPTE test are 60 Hz and 7 kHz. Nonlinearities in the amplifier will cause the higher frequency signal to be modulated by the lower frequency signal, and distortion products will be seen spaced at multiples of 60 Hz on each side of the 7 kHz tone. The IMD test equipment measures the average variation in output amplitude of the high-frequency signal. The ratio of this variation to the amplitude of the high-frequency signal is given as the IMD specification.

In the ITU-R test, two non-harmonically related high-frequency sinusoidal signals, f_L and f_H, are summed together in equal amplitude and applied to the amplifier. In this scenario, the frequencies of the test signals are close to each other, typically spaced only 1 kHz apart. Nonlinearities in the amplifier will cause the tones to modulate each other, again producing distortion products at $f_H - f_L$, $2f_H - f_L$, and $2f_L - f_H$. Only these first three distortion components below 20 kHz are taken into account in this measurement of IMD. While the ITU does not specify the actual frequencies to be used in the test, many manufacturers use 19 kHz and 20 kHz for the two input signals. For these two particular frequencies, only the distortion product at 1 kHz is typically measured (i.e., only the $f_H - f_L$ product is used). In this case, the IMD is calculated as the ratio of the voltage of this 1 kHz product ($f_H - f_L$) to twice the voltage of f_H at the amplifier's output.

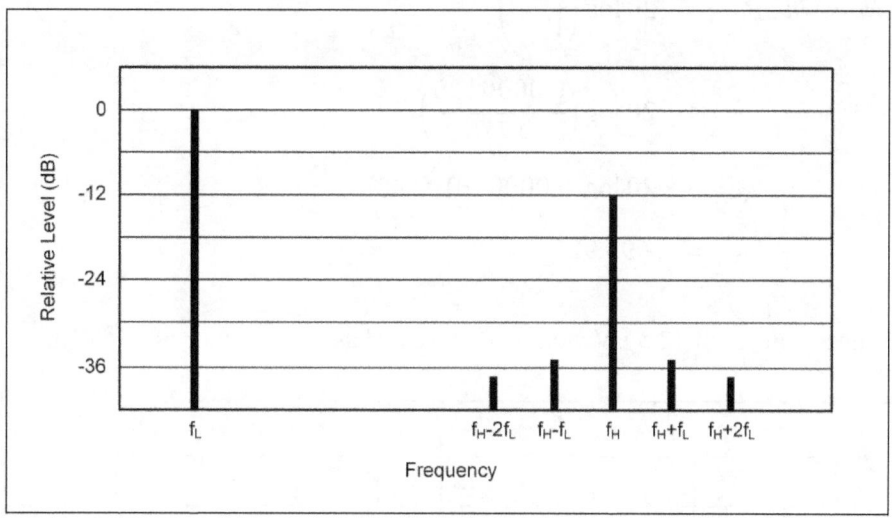

Figure 5-12   Sum and Difference Tones in SMPTE IMD

5.9 *Other Amplifier Specifications*

There are other specifications which you will see published for amplifiers. Many of these are so loosely stated that they are unclear in their implications. Left to the interpretation of the user, they do not provide much useful information about the amplifier. Below we will try to demystify some of the more common ones.

5.9.1   Amplifier Noise and Signal to Noise Ratio

Manufacturers usually publish some sort of noise specification for their amplifiers. Sometimes they list the noise floor, or residual noise, of the device. Oftentimes, an even more confusing signal to noise ratio (S/N or SNR) specification is listed.

All electronic devices produce noise due to the random motion of electrons within the device. Even some passive components are known to be noisier than other types (e.g., carbon resistors are noisier than metal film resistors). As components age, they may become noisier. This inherent noise produced by a device when no signal is connected to its input is called the noise floor of the device. It is always present and cannot be removed.

Measuring the noise floor of an amplifier is simple enough. It is in its presentation that things get tricky.

To measure the noise floor of an amplifier, some standards (e.g., CEA-490-A[4]) specify that a 1 kΩ resistor be connected across the input of the channel being measured. However, sometimes a signal generator set to zero output is connected to the input instead (e.g., 60268-3[5]). The output of the amplifier in the chosen test configuration is then measured by an RMS voltmeter. The voltage measured is typically converted to a dB referenced value and presented as the noise floor. For example, suppose a test of an amplifier gave a noise reading of 12.5 µV. This would be converted to a dBu value, which has a reference voltage of 0.775 V, as follows:

$$Noise\ in\ dBu = 20\log\left(\frac{V_{out}}{V_{ref}}\right)$$

$$= 20\log\left(\frac{0.0000125}{0.775}\right)$$

$$= 20\log(0.000016)$$

$$= -95.92$$

So, a noise measurement of 12.5 µV would be presented as a noise floor of −96 dBu.

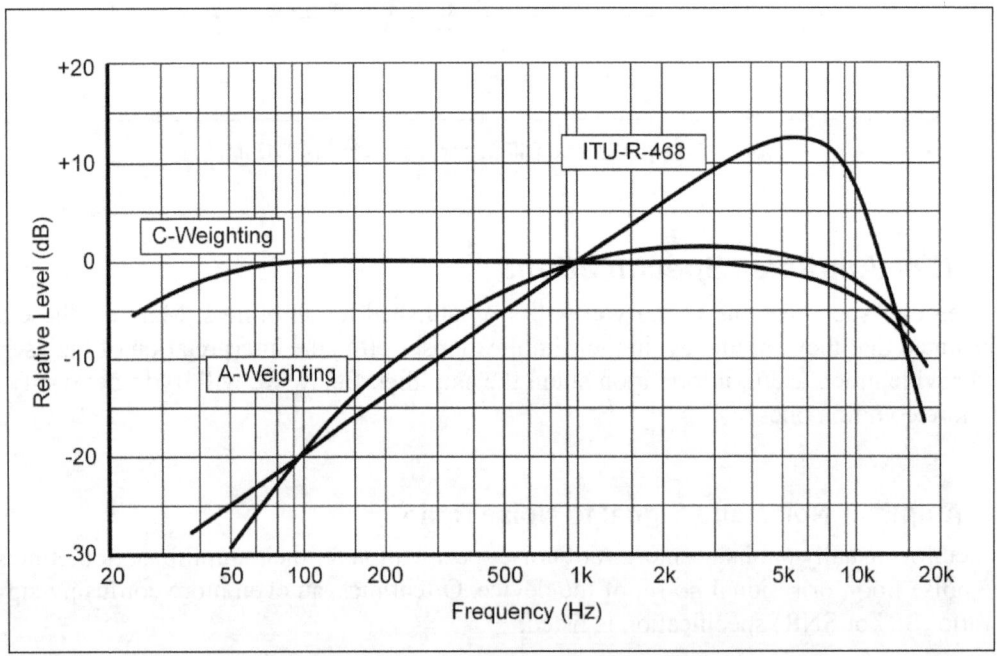

Figure 5-13 Weighting Filters Used in Noise Measurements

Noise has a very wide bandwidth, with significant energy content outside of the normal audio frequency range of 20 Hz to 20 kHz. In order to more closely correlate the noise measurement to what humans can hear, the bandwidth of the noise measurement is usually constrained to below 22 kHz. Standard weighting filters with known envelope characteristics are typically used to shape the noise, usually to the manufacturer's advantage, before the output noise reading is taken. Few manufacturers will present unweighted noise specifications because a weighting filter can improve the specification quite a bit. Figure 5-13 shows the shapes of the common weighting filters that may be used to shape the noise signal.

The A-weighting filter rolls off the low end quite a bit (down 20 dB at 100 Hz) and somewhat the very high end as well (10 dB at 20 kHz). The use of this filter drastically reduces low-frequency noise and can hide things such as 50/60 Hz hum and its harmonics from the measurement. This leads to much improved noise specifications and probably accounts for the popular use of this weighting filter by many manufacturers.

The C-weighting filter is generally flat, being down 3 dB at about 30 Hz and 8 kHz. It is seldom used by manufacturers.

The ITU-R 468 filter rolls off the low end just as drastically as the A-weighting filter, but also emphasizes the midrange noise (up by 12 dB at 6.3 kHz). This correlates better with how humans actually hear. However, because of this midrange emphasis, it leads to worse noise specifications than the A-weighting filter, so is also seldom used by manufacturers.

The SNR specification of an amplifier is the ratio of a reference signal to the weighted output noise level described above. Therefore, the other crucial piece of information that must be given in order to make sense of the SNR specification is the level of the reference signal. Clearly, the higher the level of the reference signal, the better the specification will be. For example, the CEA-490 specifies that a reference signal level of 0.5 Vrms be used. However, many manufacturers of pro-audio equipment use the pro-audio line level of +4 dBu, i.e., 1.23 V instead. To see what a large difference in the SNR specification this makes, let us look at an example.

Suppose two manufacturers each has an amplifier with a noise floor of 12.5 μV. If manufacturer A uses a line reference of 0.5 V, his SNR would be calculated as follows:

$$SNR = 20 \log\left(\frac{0.5}{0.0000125}\right)$$
$$= 20 \log(40000)$$
$$= 92.04 \text{ dB}$$

That gives us an SNR of 92 dB. Manufacturer B, using a line reference of +4 dBu, would publish an SNR of 100 dB for the same amplifier. This is a much better looking specification for sure, but the amplifiers are of equal quality. However, we have no way of knowing this unless they state the reference level and weighting used in the test.

5.9.2   Channel Separation

Channel separation is sometimes called crosstalk. Normally, one would not expect the signal in one channel of an amplifier to be heard at the output of another channel. However, because of inductive coupling of the circuit board traces, stray capacitance coupling of the outputs into the inputs, and channel power supply interaction, it happens in real devices. Crosstalk measures how much one channel leaks into another channel. It is the ratio of the output of a channel that is not driven to the output of a driven channel. Poor channel separation may not be a big problem for a power amplifier as a whole, but it would be very bad for a preamplifier. For example, it is especially annoying on some low-end mixers.

5.9.3   Common-Mode Rejection Ratio

For a balanced input stage where one input is non-inverting and the other is inverting, any signal that is applied simultaneously to both inputs is called a common-mode signal. In this case, the output should be zero, i.e., the common-mode signal should be rejected, as shown in Figure 5-14. However, in reality the output is not usually zero for the stage. The common-mode rejection ratio, or CMRR, is the ratio of the output voltage to the input voltage for a common-mode signal.

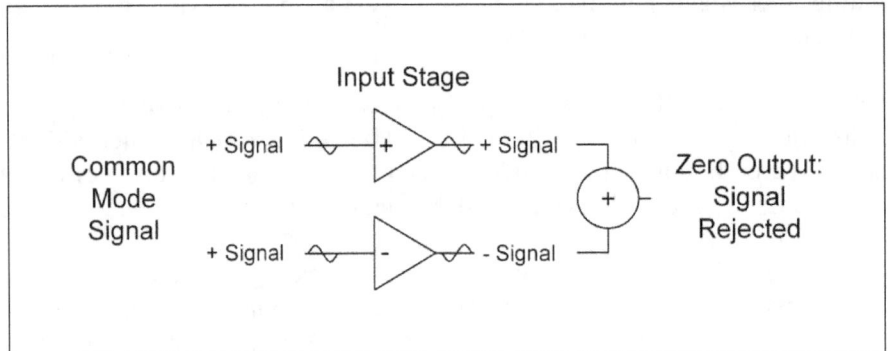

Figure 5-14   Common-mode Signal Rejection

As will be discussed in Chapter 6, balanced lines and input stages are used in an effort to eliminate noise in the interconnection of various stages of the sound system, leading us to conclude that the lowest possible CMRR is desirable in the amplifier. A non-zero CMMR can arise if the impedance of each input is not identical to the other (i.e., the input stage is not actually balanced), if one input is amplified more than the other in the first part of the input stage, or if the phase of one input is slightly shifted by the input device with respect to the other input. If it is the case that the phase of one input is shifted, then even a signal on a perfectly balanced connection will have some distortion introduced when the inputs are summed in the amplifier.

5.10 Bridge Mode Operation

In stereo mode, the two channels of the amplifier operate independently of each other (except for unintended and undesired crosstalk). Figure 5-15 illustrates that when properly connected, the positive output terminal of the amplifier is connected to the positive terminal of the loudspeaker, while the negative terminal of the loudspeaker is connected to the audio ground plane of the amplifier. This causes a positive output voltage from the amplifier to move the speaker cone forward, while a negative voltage from the amplifier will move the speaker cone backwards.

Figure 5-16 shows the amplifier in bridge mode operation. Here the input from a single channel (usually channel 1 or A) is used to feed both channels. It is sent as usual to channel 1, but it is also inverted and sent to channel 2. If the signal output from channel 2 is not exactly opposite in polarity to the original signal on channel 1, then a measure of distortion will be introduced to the audio output. The positive output terminal of channel 1 is connected to one terminal of the loudspeaker, while the positive output terminal of channel 2 is connected to the other terminal of the loudspeaker. The loudspeaker is not connected to the amplifier ground. The two channels thus operate together in a push-pull fashion, effectively doubling the voltage across the loudspeaker as compared to stereo mode operation.

Now recall that if V is the voltage across the loudspeaker and Z is the speaker impedance, then the current, I, flowing through the loudspeaker is given by:

$$I = \frac{V}{Z}$$

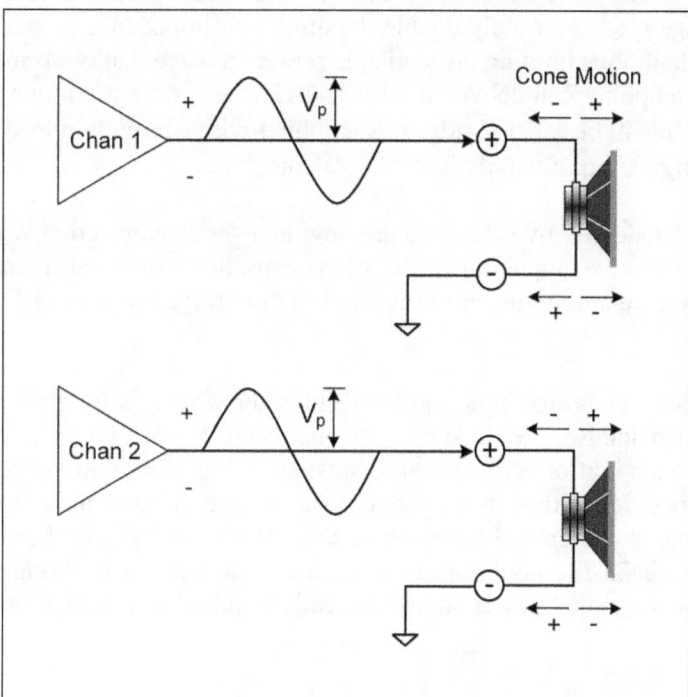

Figure 5-15   Stereo Mode Operation

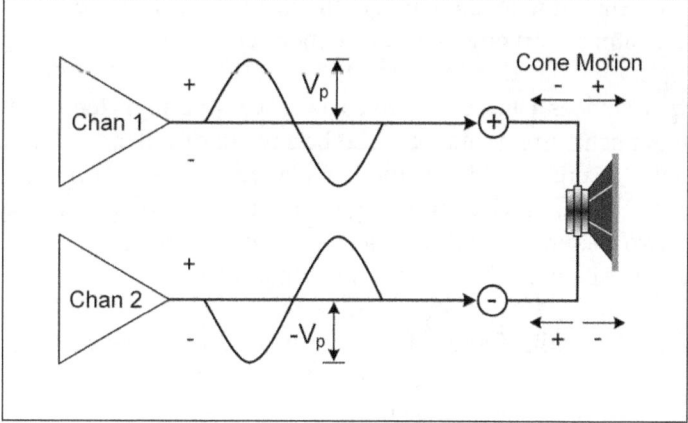

Figure 5-16   Bridge Mode Operation

Therefore, if the load impedance remains the same, the amplifier will try to deliver twice the amount of current in bridge mode as in stereo mode. Compared to stereo mode then, the power, W, potentially delivered by the amplifier is:

$$W = 2I \times 2V$$

That is four times the available power, or a 6 dB increase, as compared to a single channel in stereo mode. However, the output devices of the amplifier cannot likely sustain that much current flow, so the power actually available is greatly reduced.

Because the output devices cannot handle the large amounts of current, an attempt to bridge the amplifier driving the same low impedance as in stereo mode will likely trip its protection circuitry. Therefore, in order to bring the current flow back to allowable levels, the lowest acceptable load impedance for bridge mode is usually double the minimum impedance of stereo mode. This cuts the current flow in half, thus limiting the available power to twice that of normal stereo operation. For example, if an amplifier can deliver 1000 W per channel into a minimum load impedance of 4 Ω in stereo mode, it will be safe to bridge this amplifier with a minimum load impedance of 8 Ω. In this bridged configuration, it will deliver 2000 W into 8 Ω.

Note that in bridge mode, the two channels are now in a series connection with the loudspeaker. This means that the effective output impedance of the amplifier is doubled. Therefore, from Section 5.7, we can see that the damping factor of the amplifier in bridge mode is half of the normal value in stereo mode.

So, when would you ever bridge your amplifier? In other words, is it better to have two 500 W channels driving two loudspeaker systems, or one 1000 W channel driving one loudspeaker system? There is no straightforward answer to this, as it depends on many factors. Is the single speaker system more efficient than two separate systems? Is it more compact and easier to handle? Perhaps we need lots of power to drive a sub and no other suitable amplifier is available. It may also be cheaper to obtain the double peak-to-peak output voltage by bridging two low-power channels than by the upgraded power supply typically required in the single-channel high-power amplifier.

5.11 Output Device Protection

As we mentioned above, bridging an amplifier into too low a load impedance will likely trip its protection circuitry. We should note, though, that there are other instances where this circuitry will also be engaged, e.g., output short circuit, or even poor ventilation.

The amplifier's output devices can be destroyed by excessive high temperature. This is usually caused by passing too much current, but may also be a result of failed cooling fans or restricted air flow over the heat sinks. To prevent this destruction, manufacturers incorporate circuitry into power amplifiers to protect the output devices from high temperature and current over extended periods. There are generally two methods used to protect the output devices, *viz.* reduce the gain of the channel, or stop the flow of current by opening an output relay.

Gain reduction is used in mild forms of output stress, such as excessive clipping where the amplifier is operating at full power for long periods. Gain reduction may also be activated when the amplifier is operating into very low-impedance loads, such as 2 Ω. It would be activated when an output device is initially short circuited. The aim is to reduce the amount of current that the output devices are passing.

Device disconnection is the more drastic form of protection. The bottom line here is to stop the current flowing through the output device. This is usually done by opening a relay in series with the output device, disconnecting the speaker load. This circuitry may be triggered when excessive current flow is sensed, such as with a short-circuited output. The trigger may also be based on temperature, for example, if an output device gets too hot. This itself may be due to excessive current flow, but may also be due to poor ventilation.

In many amplifiers, device disconnection is also triggered when DC voltage is detected at the output. Under normal circumstances, DC is not seen at the amplifier's output, and its presence is usually a sign that the output stage has failed in some way. In this case, tripping the amplifier's protection circuitry protects the loudspeaker's low-frequency drivers from burning up, and it may even protect the output stage from further damage.

5.12 Choosing an Amplifier

There are a number of things to consider when choosing an amplifier. While there are no hard and fast rules, a number of guidelines can certainly be laid out.

First, we should note that for live sound work, some things can be ignored in a modern amplifier. For example, distortion and noise specifications are usually so good in a modern amplifier that they tend to be of no consequence in the noisy environment of live sound reinforcement.

An amplifier should not be chosen in isolation from the loudspeaker system, since the efficiency of the speakers will play a big role in determining the amount of power required from the amplifier. Remember that a speaker system that is only 3 dB less efficient than another speaker system will require twice as much power to produce the same SPL. Bear in mind too, that if the loudspeakers have very low impedance, i.e., less than 4 Ω, a high current amplifier will be required to drive them properly, otherwise the amplifier's protection circuitry may be constantly kicking in.

When using an amplifier in a bi/tri-amplification application, the amplifiers used for the high-frequency drivers do not need to be as powerful as the one used for the low-frequency drivers and tend to have opposite requirements in terms of frequency response extension. In a similar vein, in a speech-only system, the low end is usually cut off below about 80 Hz, so an amplifier with a poor low end could be used. However, for a system used for music, the amplifier should cover the entire spectrum.

Amplifiers used for live sound reinforcement should always have output protection built in. Accidents will happen and, for practically any power amplifier, an output short circuit would destroy the output stage without it. However, one must also be careful not to get an amplifier that is too restrictive for the application. For example, some amplifiers include non-adjustable infrasonic filters as part of their protection scheme. If a subharmonic synthesizer is used to generate very low frequencies and such an amplifier is part of the signal chain, the final sound may be disappointing as the amplifier will filter out most of the generated material.

Something that is not often considered is the efficiency of the amplifier, i.e., how much power the amplifier consumes while it is working. We really want to choose the most efficient amplifier that fits with the rest of our requirements. Efficient amplifiers tend to run cooler, last longer, and may be easier to haul around. In smaller venues, more efficient amplifiers put less strain on the electrical system and could mean the difference between successfully performing a show and a tripped electrical breaker.

A live sound reinforcement power amplifier is all about power and should be sized properly for the venue. One rule of thumb that has been used in the industry for a long time is that the amplifier's output wattage should be twice the power handling capability of the loudspeaker being driven. We will look at detailed methods of calculating the power requirements of a system in Chapter 13.

The ability to monitor and control the amplifier remotely via a software application running on a computer should be considered if the amplifier is installed in a location that is far removed from the engineer's mixing location or if there are several amplifiers employed in the implementation of the sound system (both of which are typically the case, especially if multi-amplification is used). As the adoption of Ethernet digital networks for audio distribution continues at a rapid pace (we discuss this in Chapter 12), this has become a viable option even for smaller systems, since the monitoring and control application can run over the same network infrastructure. When there are multiple amplifiers used in the system, having the ability to remotely observe and control the operating parameters of the amplifiers aids significantly in troubleshooting problems that may arise during a performance, and eases the burden of managing the system as a whole.

In the end, how the amplifier sounds with your chosen loudspeakers should factor heavily into the decision. After all, it's all about the sound, and all the great specs won't mean much if the system does not satisfy the requirement for great sound.

Review Questions

1. Discuss the similarities and differences between pre and power amplifiers.
2. Why are pure Class A power amplifiers not used in live sound deployments?
3. Compare and contrast the operation of a Class A output stage with that of a Class AB output stage.
4. Explain the operation of a Class D amplifier.
5. What happens to the output signal of an amplifier when the input signal is so high that the output signal would exceed the maximum voltage that the power supply is able to deliver?
6. What is meant by a high current amplifier?
7. What is dynamic headroom of an amplifier and why is it important?
8. What is the difference between the power bandwidth and the frequency response of an amplifier?
9. What is the damping factor of an amplifier and how can it improve the sound reproduced via connected loudspeakers?
10. What is harmonic distortion of an amplifier and how important is it in modern pro-audio equipment?
11. Explain the key differences between harmonic distortion and intermodulation distortion.
12. What voltage does a level of +4 dBu correspond to?
13. What is the SNR of an amplifier that has a noise floor of −95.5 dBu, and the test input signal level is +4 dBu?
14. What is common-mode rejection in a differential amplifier and how does it work?
15. Explain how a two-channel amplifier can be operated in bridge mode.

Chapter 6

Interconnecting Devices

6.1 Balanced vs. Unbalanced Interconnections

The focus of this chapter is primarily on the cables we use to interconnect the audio devices we typically find in our racks with each other, our instruments, and loudspeakers. However, before we get to the specific cables themselves, it is useful to discuss the two broad input and output connection schemes found on many of the devices, *viz.* balanced and unbalanced. In many cases, there is some uncertainty as to just what constitutes a balanced connection. We will therefore now discuss in detail what makes a balanced connection, and how we can make interconnections between the various types of input and output configurations.

6.1.1 Input and Output Stages

Before talking about balanced and unbalanced connections, let us first take a brief look at the input and output stages that allow us to get a signal into or out of the devices themselves. No doubt you have seen the terms *balanced* and *unbalanced* applied in the marketing literature to inputs and outputs on your favorite processor or mixer. While this is a very common use of these terms, you will see shortly that this is not strictly the most accurate use of these terms.

If we take a high-level look at the output stage of a preamplifier, such as may be found in an effects processor or mixer, it may be represented as shown in Figure 6-1. Part (A) on the left side of the figure shows what marketing generally calls an unbalanced output. This is more properly called a single-ended output configuration. Here the signal from the earlier stage is amplified and provided on a single output line. The amplifier is depicted as having an output impedance with respect to ground of Zo.

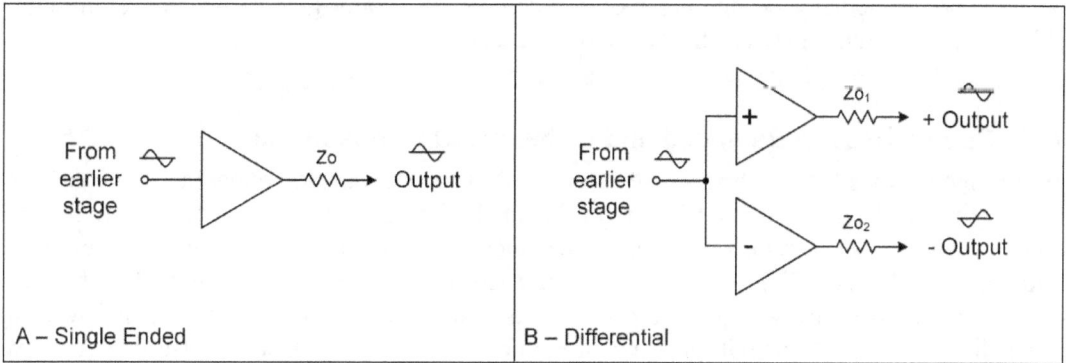

Figure 6-1 Amplifier Output Stage Configurations

In Figure 6-1(B) we again show the same signal from the earlier stage being amplified. This time, though, it is amplified by two stages in parallel. One side keeps the polarity of the signal as it was when it entered the stage (shown by the + sign at the amplifier and the +*Output*), while the other inverts the signal (shown by the − sign at the amplifier and the −*Output*) before output. Here the output impedance of the non-inverting side is depicted as Zo_1, while that of the inverting side is shown as Zo_2. This is conceptually the arrangement that is used to create signals for balanced connections, and indeed this arrangement may be referred to as a balanced output in the literature. This output arrangement is also called double-ended and sometimes also called differential.

On the input side of the coin, there are also similar concepts in design. Figure 6-2(A) shows a high-level design of the input stage of an amplifier where the signal is sensed on a single wire. The signal level on the input is referenced to ground, and the input stage amplifies the voltage found on this single input. In this case, the input impedance of the amplifier with respect to ground is represented as Zi. In the marketing literature, this type of input configuration is often called unbalanced.

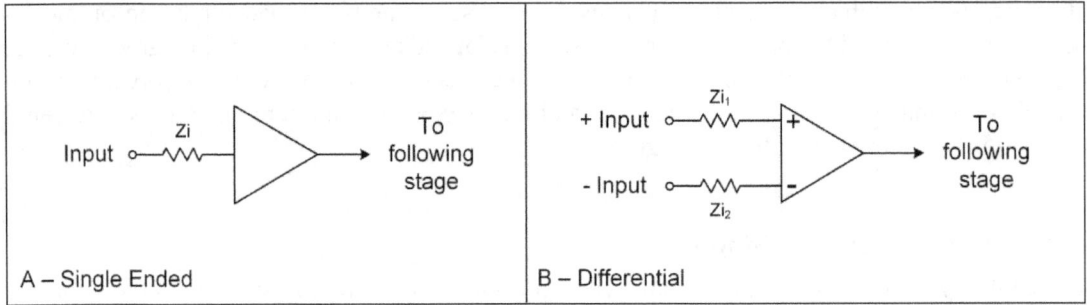

Figure 6-2 Amplifier Input Stage Configurations

In Figure 6-2(B) we show a common arrangement where the input signal to the amplifier is accepted on two ports, shown as +*Input* and −*Input* in the diagram; the inputs are also called non-inverting (shown by the + sign at the amplifier) and inverting (shown by the − sign). The input impedance of the non-inverting input is depicted as Zi_1, while that of the inverting input is shown as Zi_2. In this arrangement, the input stage amplifies the *difference* between the two inputs. Thus, this type of input configuration is called differential. Another way to think of this configuration is that the signal on the −*Input* port is inverted before being added to the signal on the +*Input* port. Under normal operating conditions, the expectation is that the signal present at the −*Input* is equal, but opposite in polarity, to that present at the +*Input* port. While these inputs are also commonly referenced to ground, they can actually be floating, i.e., not referenced to any ground, since the amplifier is looking at the difference between the two input signals.

6.1.2 Cables Used in Balanced and Unbalanced Connections

Figure 6-3(A) shows the construction of cable used in an unbalanced connection. One conductor forms a cylindrical shield around the core conductor. The shield is typically connected to the signal ground and also forms the return signal path. The other conductor sits in the center of this cylindrical shield. This is called a coaxial cable, or simply coax. Coaxial cable provides acceptable performance over reasonable distances (less than 40 m) when used with line-level connections, and over much smaller distances (e.g., 5 m) with instrument connections. In the past, it has also been used with high-impedance microphones connected to preamplifiers with very high input impedance, though this is no longer seen in contemporary professional applications.

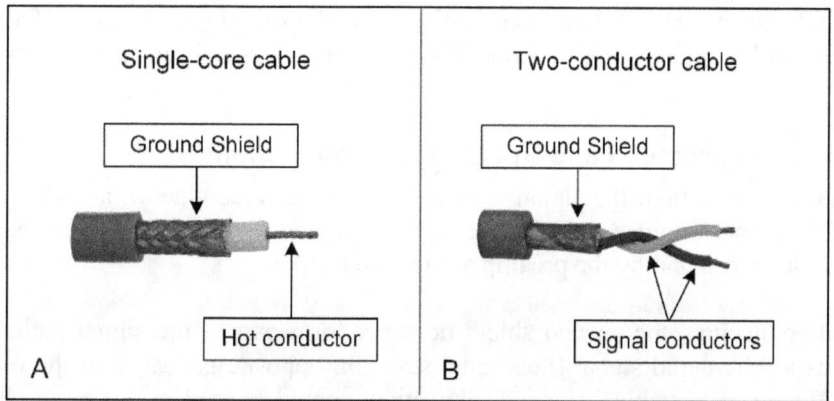

Figure 6-3   Cables Used for Audio Connections

The grounded shield of the coax cable is an effective defense against electrostatic interference (ESI). ESI occurs when unwanted current is induced in a conductor via capacitive coupling with another nearby conductor. This can happen because charged particles are attracted to, or repelled from, the conductor based on the voltage difference between the conductor and its environment. When an insulator prevents the charged particles from freely flowing between the two conductors, a charge is built up in their vicinity. As the voltage on the conductor changes, the charge also changes, and this sets up a changing electric field. The shield, connected to ground, works to block that changing electric field from inducing a current in the core conductor.

The use of coax cable causes loss in high-frequency response over long distances because the cable itself acts like a capacitor, with the shield acting as one plate, the core conductor acting as the other plate, and the insulator between them acting as the dielectric. Since high-frequency signals easily pass through a capacitor, this forms a simple low-pass filter, allowing high frequencies to be conducted to ground.

Balanced connections use a 2-conductor cable consisting of a twisted pair of wires surrounded by a shield. This is shown in Figure 6-3(B). Only the twisted pair forms the signal path. The shield does not form part of the signal path and is usually grounded so as to protect the signal carrying lines from ESI.

The twists in the conductors help to reduce the effects of electromagnetic interference (EMI), both as a defensive strategy and as a preventative measure. EMI occurs when a varying magnetic field induces an unwanted current in the conductor. Twisting the conductors reduces the distance between them and distributes them evenly around a central axis. Defensively, when the conductors are tightly twisted, this makes them behave as one wire in an interfering magnetic field. Thus, any varying magnetic field induces the same current in each conductor, and so the induced interference can be easily canceled by the circuitry of the balanced connection as described in the next section. If the conductors were not twisted, the induced current would very likely be different in each conductor, making it impossible to eliminate the interference even with the balanced circuitry. Preventatively, when differential signaling is used to send the audio signal along the twisted pair, the magnetic fields created by the (equal but opposite) current flowing in each conductor will be equal but opposite, and thus should cancel each other out. Thus, potential interference created by sending a signal along the cable should also be practically eliminated.

As you can imagine, not all cables are created equally. In low quality cables, the shield may consist of very loose stranded braid, and the signal conductors in 2-conductor cables may only be loosely twisted (or not be twisted at all). In higher quality cable, the shield will be made of a tight braid or aluminum foil that has no gaps and the conductors will be tightly twisted (e.g., 1 turn every 2 cm).

6.1.3 Signals on Balanced and Unbalanced Connections

In an unbalanced connection, the signal is sent down the core conductor, and the ground shield forms part of the signal circuit. Any noise picked up along the way by the cable is simply amplified along with the desired signal by the preamp at the mixer input.

In a balanced connection, the ground shield does not form part of the signal path; the signal is usually sent as a differential signal (i.e., equal strength, opposite polarity) on the twisted pair of signal wires. Whenever possible, balanced connections should be used, as this allows higher signal-to-noise ratios to be achieved throughout the entire system. To understand why this is so, look at Figure 6-4. In a differential circuit, the signal is sent on one wire (positive or hot) of the twisted pair and an inverted copy of the same signal is sent on the other wire (negative or cold) of the pair. Any noise that is picked up by the cable affects both signals equally. At the processor input, the positive signal is processed normally; however, the negative signal is inverted during processing. This means that the two source (original) signals are now of the same polarity, while any noise picked up in the cable will now be of opposite polarity. The two signals are then added together, causing the noise to cancel and giving a signal that is twice the strength, or 6 dB higher, as in either conductor alone.

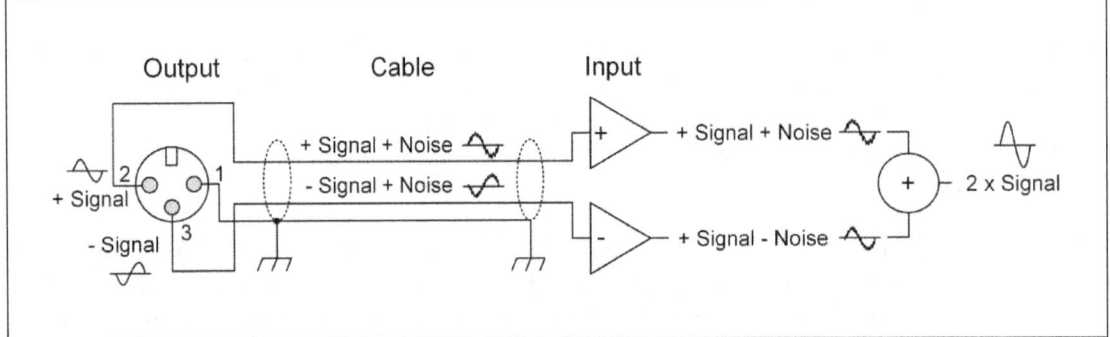

Figure 6-4 Noise Cancellation on Balanced Connection

6.1.4 What Makes a Connection Balanced?

Figure 6-5 shows the output of one device (Device A) connected to the input of a second device (Device B). The devices are depicted using the differential output and input stages discussed in Section 6.1.1. The ground connection between the two devices that is typically present has been left out for clarity. The figure shows the connection that is typically called balanced. Why is it referred to as *balanced*? To answer that question, we will take a look at what happens when noise current is induced in the connectors (Leg 1 and Leg 2) between the devices. Of course, these connectors are typically twisted, as discussed earlier, so as to ensure that any externally varying magnetic field induces a current in each leg equally.

First, let us look at one leg of the connection, say Leg 1. The total impedance, Z, of that leg with respect to ground is comprised of the output impedance of Device A, the input impedance of Device B, and of the wire connecting the two devices. A disturbing magnetic field will induce a current in the leg, and that current will flow to ground. The voltage appearing at the +*Input* port of input amplifier of Device B due to this current flowing in Leg 1 is strictly dependent on the total impedance, Z, of the leg. If the current flowing in the leg is I, then the voltage across the $+Input$ port is given by $V_1 = IZ$, and this is the voltage that will be amplified due to noise induction. Now, assuming that the same current is induced in Leg 2 by the same magnetic disturbance, we can see that a voltage V_2 will also be impressed across the $-Input$ of the amplifier. It should also be clear that the only way V_1 will be equal to V_2 is if the impedances on the two legs are the same. When the impedances of these two legs are the same with respect to ground, the connection is said to be balanced, and the same voltage (due to external interference) will be presented at the inputs of the amplifier of Device B. This is the common-mode voltage that is discussed in Section 5.9.3, and this condition is necessary for the input amplifier to eliminate noise due to interference.

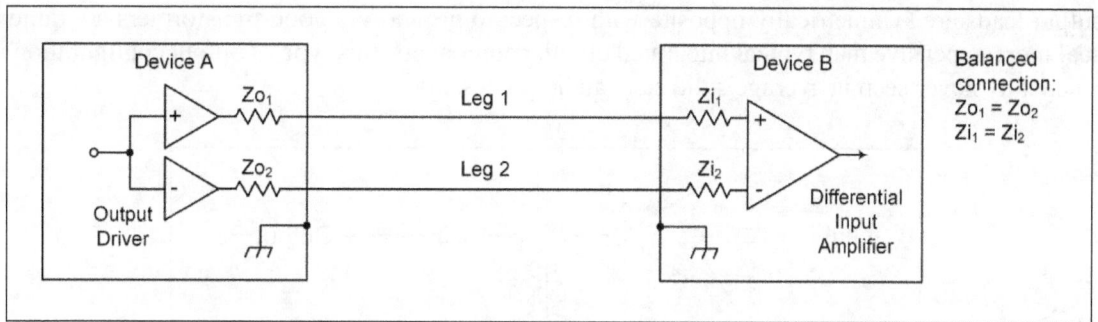

Figure 6-5   Differential Connection can be Balanced or Unbalanced

When the impedances of Leg 1 and Leg 2 are different, an induced current will cause different voltages to appear across the inputs of the differential amplifier. Because the voltages are different, the amplifier will not be able to eliminate the noise caused by this voltage—remember, the amplifier amplifies the *difference* in voltage at the two inputs, and that difference is only zero when the voltages are the same. When the impedances of Leg 1 and Leg 2 are different, the connection becomes unbalanced. If the connection of one leg is broken, this becomes an extreme case of an unbalanced connection, i.e., we can consider that the impedance of the leg is infinite, no current flows, and no voltage is impressed across the input of the amplifier. This, in a sense, is what we have with single ended connections, and so we can say that single ended connections are unbalanced.

Manufacturers generally take great care to match the input impedances of differential amplifiers. This matching is important to maintaining a balanced connection, and the common-mode rejection ratio (CMRR) specification discussed in Section 5.9.3 can tell us how well they have achieved this match. However, as we have seen, the line output impedances of the source device also must be matched in order to maintain a balanced connection. Unfortunately, though, output impedances are much less tightly controlled.[1] This means that in reality, it would be almost impossible to create a truly balanced connection between devices. However, the effects of mismatched impedances can be minimized by making output impedances very small while making input impedances very large. That way, variations in output impedance have only a very small impact on the overall impedance of the line, while the input impedance, which is tightly controlled, has the greatest impact.

6.2 Balanced Output Drivers

In Section 6.6, we will be talking about the many ways to interconnect balanced and unbalanced input and output stages. As you will see, how we do this depends greatly on the design of the balanced output stage. While the input stage of the preamplifier is either transformer-balanced or based on a variation of the differential amplifier discussed above, there are several forms of the balanced output stage that can be used, and in many cases the terms used by the manufacturer in their marketing literature or specifications do not clearly describe the output circuitry. In the following paragraphs, we will take a closer look at some of the output stages that you may find in many devices. Understanding the differences will help later in choosing the correct type of cable when it comes time to interconnect devices.

6.2.1 Transformer-Balanced Output

Figure 6-6 shows the circuit for a typical transformer-balanced output. This is one of the simplest but most effective means of providing a balanced output. The signals appearing at the transformer output leads are symmetrically opposite with respect to ground. As good transformers are quite a deal more expensive than typical integrated circuit components, this type of output configuration is practically never seen in average solid state audio processors.

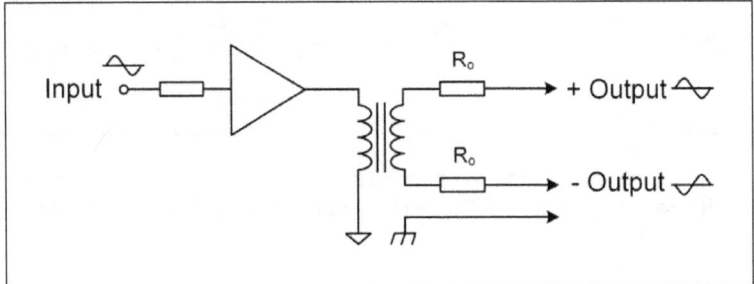

Figure 6-6   Transformer-Balanced Output

6.2.2 Active Balance Output

Active balanced outputs are popular because the circuit components are relatively inexpensive and can be provided in a single integrated circuit. Figure 6-7 shows one way such a circuit can be realized. Here the input to the stage is simply fed to two op-amps. One is configured as a non-inverting amplifier, while the other is configured as an inverting amplifier, and the gains of the two amplifier modules are set to be identical. This circuit thus delivers identical but opposite polarity signals on its two output ports. Resistors *Ro* are used to adjust the output impedance of the op-amp modules, but these are typically not critically matched.

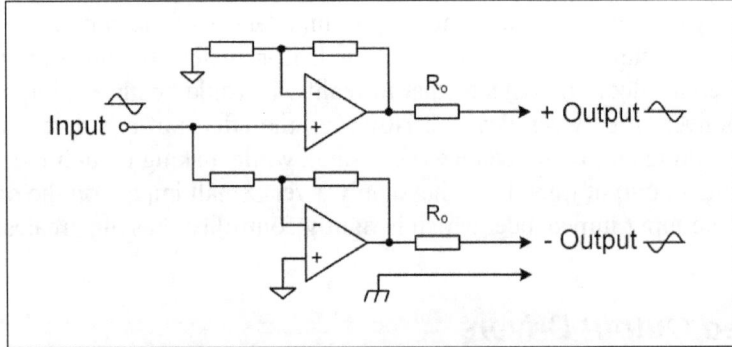

Figure 6-7   Active Balanced Output Stage

6.2.3   Impedance-Balanced Output

A simpler way of providing a balanced output is to use what is termed an impedance-balanced output configuration. This is regularly seen on line-level outputs of many processors. Some manufacturers may call this a *balanced/unbalanced* output in their literature. Figure 6-8 shows what the circuit may look like. As you can see, only one half of the output (i.e., only one leg of the connection) is actually driven! The other half is simply connected to signal ground. Of course, this is not an issue for either the balanced condition of the connection or the differential amplifier at the input of the other device. As we saw earlier, the balanced state of the connection depends only on the impedances of the two legs being matched. Thus, if Ro_2 is adjusted so that it matches the output impedance of the driven line, the connection will still be balanced. Since at the input of the other connected device the differential amplifier amplifies the difference in the signal on the two lines, it will be perfectly happy with no source signal on one line. The only negative impact of this is that the output level from the differential amplifier stage will be half what it would be if both lines were driven with equal but opposite polarity signals. From this discussion we also see that signal symmetry is not a prerequisite for a balanced connection.

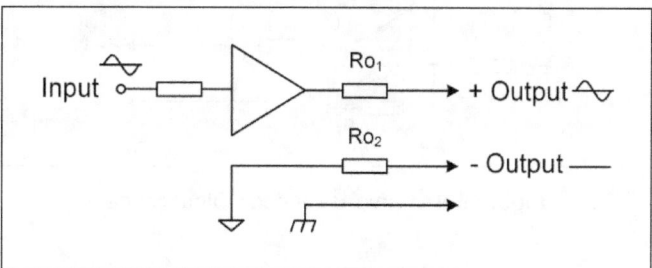

Figure 6-8   Impedance Balanced Output Stage

6.2.4   Servo-Balanced Output

A servo-balanced output is one of the harder configurations to understand, as many manufacturers state that they use this type of output on their devices, but they do not explain what it is or why they use it. This configuration is sometimes called a cross-coupled output, and in Figure 6-9 we can see why—the output from each leg is fed back to the input of the other leg. This feedback is negative for the circuit, i.e., it works in completely the opposite way from the familiar acoustic feedback that we are all familiar with—the greater the feedback signal, the lower the output from the circuit. Take a closer look at the upper half of Figure 6-9 which drives the positive (non-inverted) output. We see that the input signal to the op-amp consists of the original input to the whole circuit as well as a signal tapped from the negative (inverted) output of the lower side. Note that the feedback signal is opposite polarity to the original input signal, and so when the feedback signal is higher, the overall output of the op-amp for the positive side is lower. The circuit is symmetrical, so it works the same way for the lower half, which drives the negative output leg. When properly trimmed, the outputs of this circuit will be equal, but opposite in polarity.

So why go to the trouble of all this complexity? Well, when connecting a balanced output to an unbalanced input, as we will see below, we usually need to consider the construction of the cable used for the interconnection. The best approach would be to use the simplest cable without worrying about the circuit, and this is what a servo-balanced output allows us to do. If we use a TS plug on the output of a servo-balanced output stage, it will short to ground the negative output of the circuit. This removes the negative feedback to positive side, and so the output signal level of the positive legs rises to double its typical level. At the same time, the feedback from the positive leg into the negative output side drives the output of the negative side to practically zero. Thus, there is no harmful result from shorting the negative output line to ground. This behavior mimics that of a transformer-balanced output, but at a smaller cost, size, and weight. At the same time, since the output level of the positive side is practically doubled, there is no loss of signal level at the output of the input differential amplifier on the other connected device (as stated in the previous section when there is a signal on only one leg of the balanced input). This form of operation may cause overloading of a single ended input, so we must be mindful of this when interconnecting a servo-balanced output to a single ended unbalanced input.

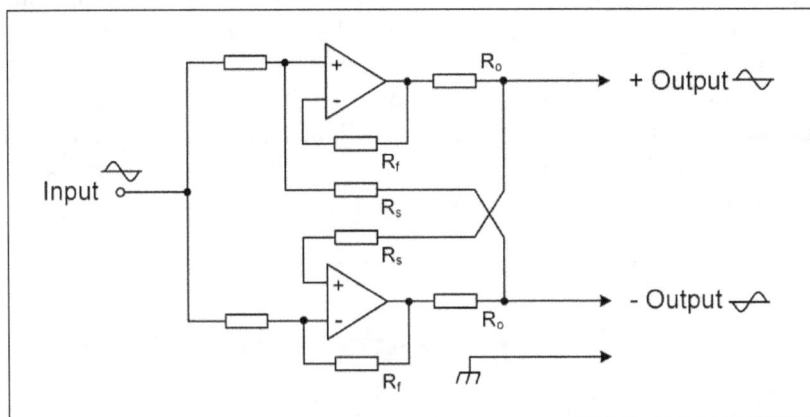

Figure 6-9 Servo-Balanced Output Stage

6.3 Simple Patch Cables

No doubt about it—sound reinforcement involves handling a lot of cables of many different types and functions. The more we handle them, the more likely that sooner or later we will cross paths with one that does not behave as we expect. Having a sound knowledge of what is under the cover of the most common cables along with the signal flow paths and entry-exit points of the connectors not only increases confidence in dealing with such issues, but will prove invaluable if we ever need to do an emergency repair in the field. Patch cables are the straightforward interconnects used to get a signal from one device to another. Below we will review the most common ones that you will encounter on practically every gig.

6.3.1 TS to TS Cable

This is the simplest cable in the box, with 1/4-inch TS (Tip-Sleeve) plugs at both ends, connected by a single-core shielded cable. The signal runs along the core, from tip to tip. The ground connection is on the sleeve. This cable is used for unbalanced connections, as in line-level interconnects, guitar cables, keyboard cables, and other instrument cables.

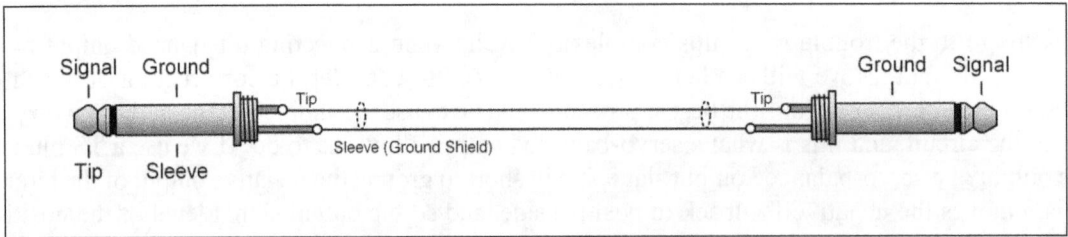

Figure 6-10 TS-TS Patch Cable

6.3.2   TRS to TRS Cable

This cable is used for line-level balanced connections, e.g., between effects processors and the mixer. This cable has 1/4-inch TRS (Tip-Ring-Sleeve) plugs at both ends, with the positive signal on the tip, the negative (inverted) signal on the ring, and the ground connection on the sleeve. The cable wiring itself consists of a twisted pair surrounded by a ground shield, like the one discussed in Section 6.1.2.

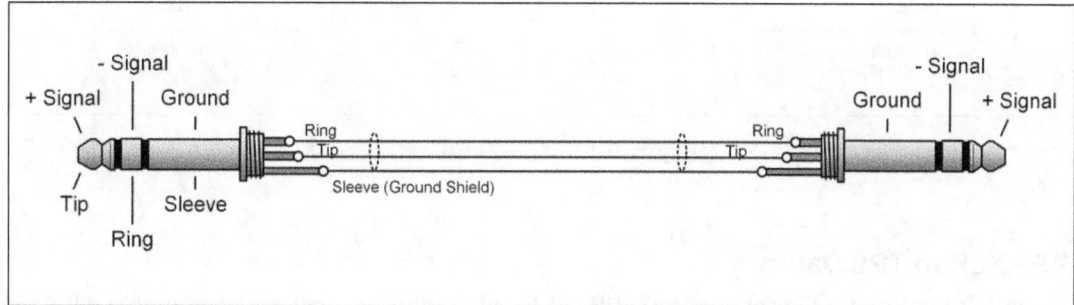

Figure 6-11   TRS-TRS Patch Cable

6.3.3   TRS to RCA Cable

With the popularity of mini MP3 players and cell phones with player software, one of these cables should probably be in your collection. This cable takes the stereo signal from the player and breaks it out on to two separate RCA plugs. The left channel is sourced from the tip, while the right channel is sourced from the ring of the TRS plug. Many mixers have RCA jacks that allow direct connection using this kind of cable, but if not, an RCA to 1/4-inch adapter may be used. Of course, the TRS plug on this cable is the miniature 3.5 mm (1/8-inch) type.

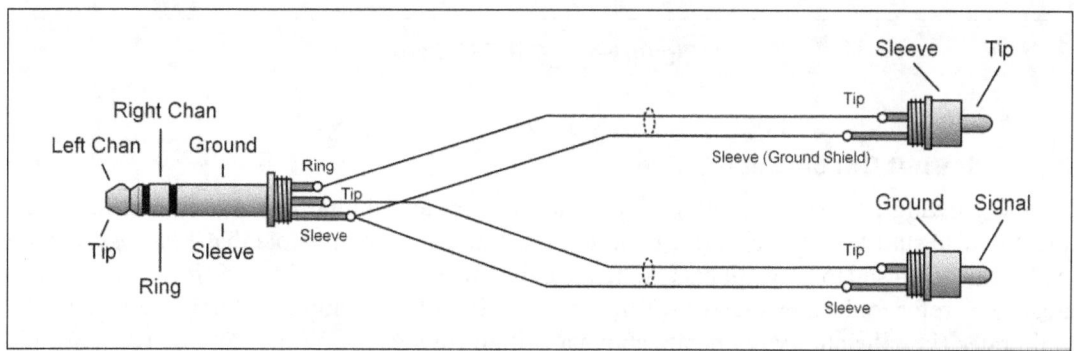

Figure 6-12   TRS-RCA Patch Cable

6.3.4   XLR to XLR Cable

This is the common microphone cable used to connect low-impedance microphones to the mixer. This cable can also be used as a line-level patch cable where the equipment provides a line-level signal on XLR connectors (e.g., most mixers provide main left/right outputs via XLR). This cable has a female connector at one end (which connects to the output of one device—the signal source) and a male connector at the other (which connects to the input of the second device—the signal sink). The cable wiring itself consists of a twisted pair surrounded by a ground shield.

Conventionally pin 1 is connected to the ground shield, while pin 2 carries the positive signal and pin 3 carries the inverted signal.

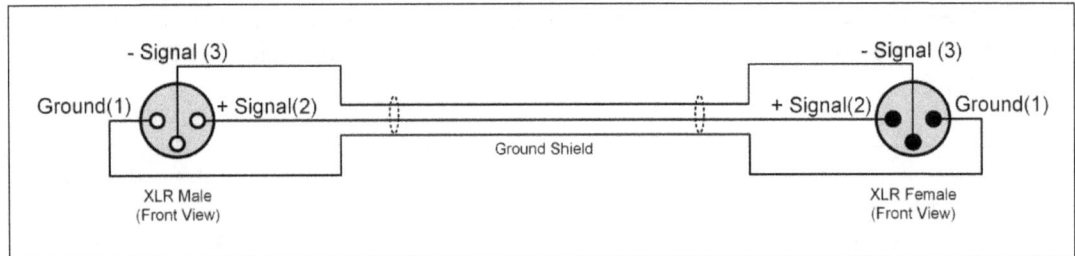

Figure 6-13 XLR-XLR Cable

6.3.5 XLR to TRS Cable

This cable is essentially the same as an XLR-XLR cable with one end replaced with a TRS plug. The XLR end of the cable can be either male or female depending on whether it is intended to mate to the input or output, respectively, of the processor. This cable is typically used to carry line-level signals over a balanced connection between processors.

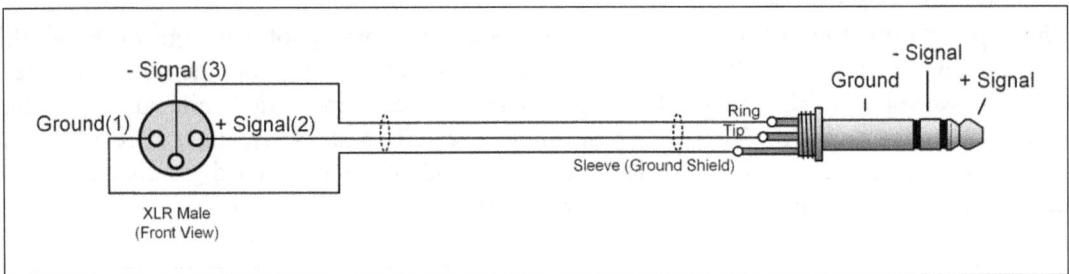

Figure 6-14 XLR-TRS Cable

6.3.6 Ethernet Cat 5e/Cat 6 Cable

With the growing popularity of affordable digital mixers and digital remote control, this kind of cable will also start to become a more common sight in the gig bag. Note that *Ethernet* refers to a specific protocol used to transmit data along the cable, while *Cat 5e* and *Cat 6* (for Category 5e and Category 6) refer to the specification of the cable itself and its ability to reliably carry data above certain rates (i.e., the ability to operate above a certain frequency). While Cat 5e or Cat 6 cable is used to carry digital audio data, e.g., between the onstage digital snake head and the FOH mixer, the actual protocol used to carry such data may or may not be based on standard Ethernet. However, in general, all such cable is commonly called Ethernet cable, regardless of the actual protocol used to send data over the cable.

Ethernet cable consists of eight conductors terminated at each end by an 8-position 8-contact (8P8C) modular plug, commonly referred to as an RJ45 plug. The eight conductors are arranged as four twisted pairs, where the twist pitch varies slightly between pairs to help reduce crosstalk. To increase interference immunity, Cat 6 cable used in audio applications may also be shielded, and is sometimes referred to as shielded twisted pair, or STP. The unshielded variety (called unshielded twisted pair, or UTP), though, is more common and works perfectly well in most environments.

The twisted pairs are numbered 1 to 4, but depending on which standard is referenced, the numbering scheme varies. This makes no difference to its application in transporting digital audio, since the important aspect of this interconnect is that the pins on the RJ45 plug at one end of the cable are connected to the correspondingly numbered pins on the other end of the cable. Figure 6-15 shows an Ethernet cable that follows the T568A pair-numbering scheme.

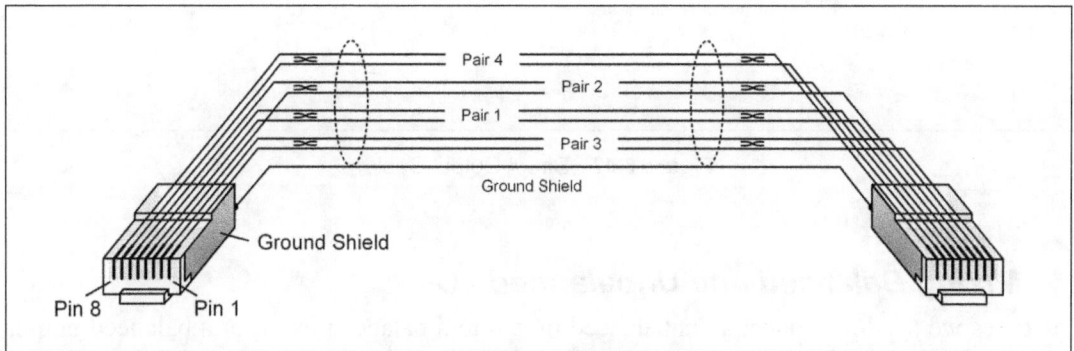

Figure 6-15   Ethernet Cable

6.4 Insert Cable

Insert cables are generally used when we want to break the signal path and insert a signal processor in the signal chain. The process is discussed extensively in Section 9.12, so we will not repeat the details here. Figure 6-16 shows the cable that is used to accomplish the task.

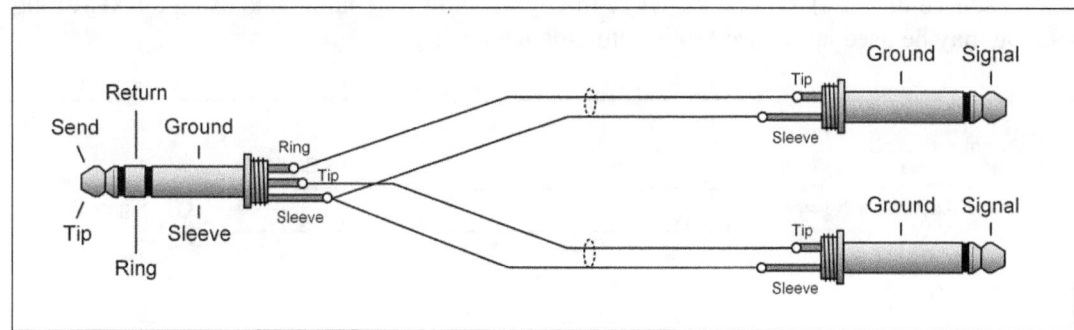

Figure 6-16   Insert Cable

6.5 Direct Output Cable

Some mixers feature onboard microphone splitters that allow the channel's input to be sent right back out the mixer to be used for other purposes, e.g., to feed a multitrack recorder. Most mixers, however, do not have this capability. On the other hand, practically every mixer has an insert point on every microphone channel. One old trick that can be used to provide a copy of the channel's input signal is to use a cable like the one shown in Figure 6-17. This is commonly referred to as a direct out cable, and it uses the insert point of the channel to tap the signal. As can be seen, the left TRS plug is the same as in any insert cable. However, we see that the tip is immediately connected to the ring, thus essentially restoring the link and sending the signal right back into the mixer unchanged. At the same time, the tapped signal is also sent to the tip of the TS plug and thus made available outside the mixer.

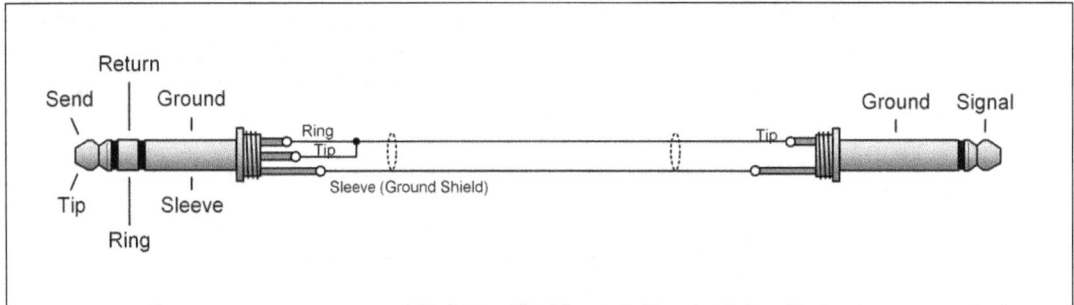

Figure 6-17 Direct Output Cable

6.6 Mixing Balanced and Unbalanced I/O

Sometimes we need to connect an unbalanced output to a balanced input, or a balanced output to an unbalanced input. Some I/O ports take care of this condition by themselves, but it is really useful to know how to make a cable to do this if you ever need to.

6.6.1 Unbalanced Output to Balanced Input

Most effects processors have unbalanced outputs, while many mixers, amplifiers, and other processors have balanced inputs. Figure 6-18 shows the cable that can be used to connect devices in this case. The key things to note here are that the positive input is fed the signal from the device while the negative input is connected to ground. This avoids leaving the negative input floating (a potential source of noise). Of course, while the figure shows the input side using an XLR plug, a TRS plug may be used just as well if the situation demands it.

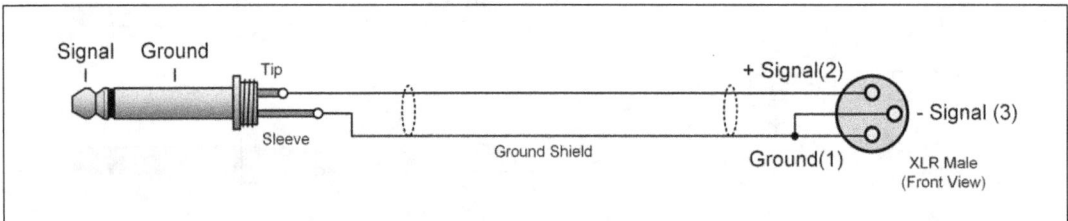

Figure 6-18 Unbalanced to Balanced Connection

6.6.2 Balanced Output to Unbalanced Input

If we are connecting from say, the balanced output of a mixer to the unbalanced input of an effects processor, we would need a cable like the ones described here. In this case, we have a number of options and must also choose the proper approach depending on whether the output device is transformer-balanced or not.

As discussed earlier, most equipment encountered will likely use electronically buffered output. Even here we have a choice of the actual cable construction. Figure 6-19 compares two approaches. In each case, the positive signal from the output is connected to the tip connector of the TS plug going to the unbalanced input. In (A), we see that the negative signal from the output is connected to the ground (sleeve) of the TS plug, while the ground shield from the output is left disconnected at the input at the TS end of the cable. In Figure 6-19(B), we use a single-core cable and the negative signal from the balanced output is not used. The ground at the output is connected to the ground of the input at the TS plug via the ground shield. Both of these configurations will work well, but the approach in (A) has some advantages. First, because it is using both the positive and negative output signals, the signal level at the input will be twice as high as in (B) (see Section 5.10 for an explanation). Additionally, notice that in (A) the ground connection between the two connected devices is broken, and that may help in eliminating hum caused by a ground loop.

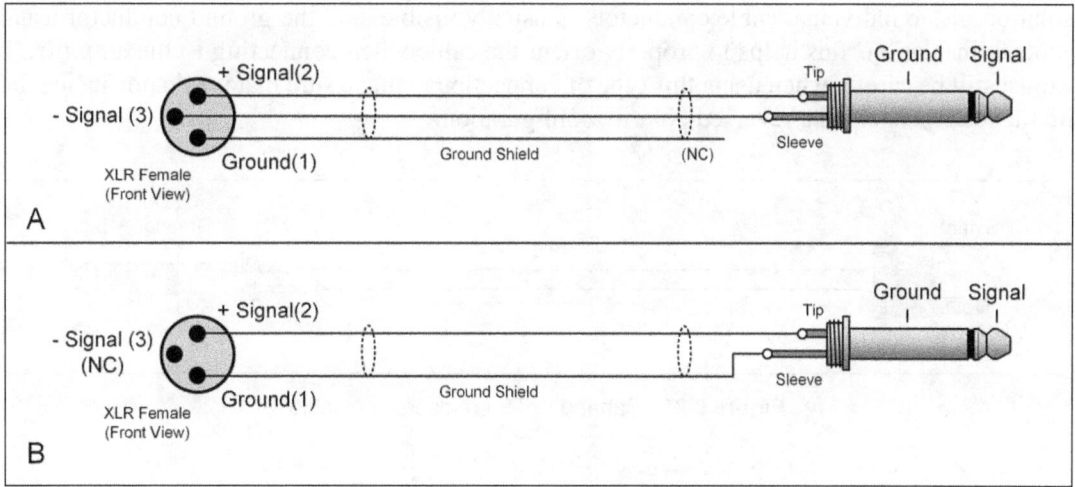

Figure 6-19   Balanced to Unbalanced Connection

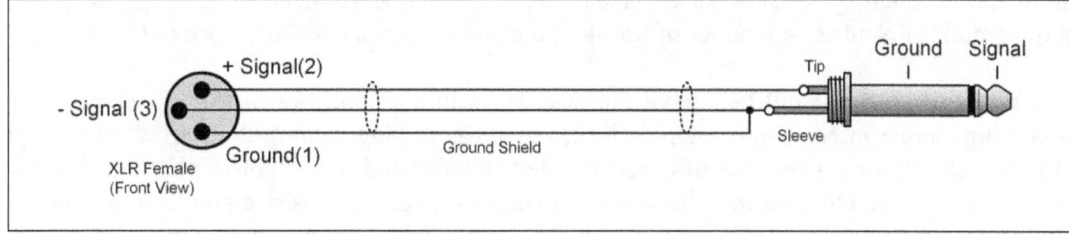

Figure 6-20   Balanced to Unbalanced Connection - Transformer Output

When the output is transformer-balanced, we can also use the cable shown in Figure 6-19(A), which will again allow the full output voltage to be available at the input on the unbalanced side of the connection while also taking advantage of the broken ground. The cable shown in Figure 6-19(B) cannot be used for a transformer-balanced output stage, since in this case, one end of the transformer secondary is left floating, and the signal at the input side will be very low (essentially leakage). It is also possible to use a cable like the one in Figure 6-20, which essentially shorts the negative output to ground while maintaining the ground connection between the two sides of the connection. Both of these approaches may also be effectively used with servo-balanced outputs.

6.7 *Loudspeaker Cables*

Loudspeaker cables come in many shapes and sizes, depending on the power requirements and on the connector type available at the amplifier and speaker. The one common thing that you will see with loudspeaker cables is that there is no ground shield. Of course, they also tend to be quite hefty to enable handling of the high current that typically flows through them.

6.7.1 Banana to TS cable

Many power amplifiers have banana jacks for the loudspeaker output connection, and many live sound reinforcement loudspeakers have 1/4-inch jacks for audio inputs. Thus, the banana-TS cable is a very common loudspeaker cable arrangement. Banana plugs are very easy to dislodge from their jacks, so it is imperative that the amplifiers are secure when using this type of connection. Similarly, unless the loudspeaker uses a locking 1/4-inch jack, the TS plug is also fairly easy to remove from the speaker, which is something that you do not want to happen while driving the loudspeakers during a performance. There is no keying on the banana plug, but the color of the insulation of the individual cable conductors is usually visible, and the ground conductor usually has black insulation. This helps to properly orient the cable when connecting to the amplifier, but we must still be careful when using this type of connection to make sure that we do not accidentally wire the loudspeakers in a reversed polarity configuration.

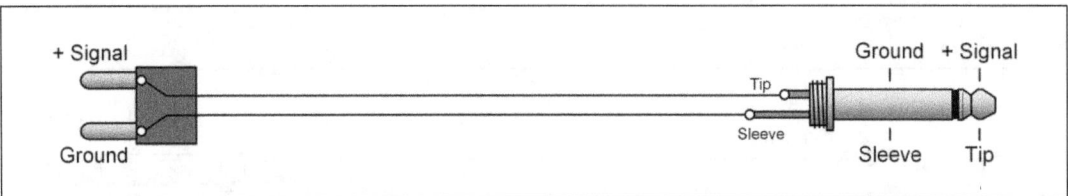

Figure 6-21 Banana to TS Loudspeaker Cable

6.7.2 2-Pole Speakon to TS cable

A step up from the banana plug is the Speakon connector. The Speakon plug is a locking connector that fits into a matching Speakon jack. Once mated, it cannot be accidentally disengaged. Speakon is a trademark of Neutrik. A number of Speakon connectors are shown in Figure 6-22.

Many pro-audio power amplifiers have Speakon jacks that accept 2 or 4-pole Speakon plugs (a 2-pole plug can be inserted into a 4-pole jack, but a 4-pole plug cannot be inserted into a 2-pole jack). The Speakon plug consists of a keyed outer cylinder and a concentric plug. The electrical speaker contacts are recessed into the space between the two. The positive signal is delivered via contacts (referred to as 1+ and 2+) on the central plug, while the return ground connectors (referred to as 1− and 2−) are on the inside of the outer cylinder. In the 2-pole Speakon to TS cable shown in Figure 6-23, the 1+ pin is connected to the tip of the TS plug.

Figure 6-22 Speakon Connectors – 4-Pole Plug, 2-Pole Plug, and 4-Pole Jack

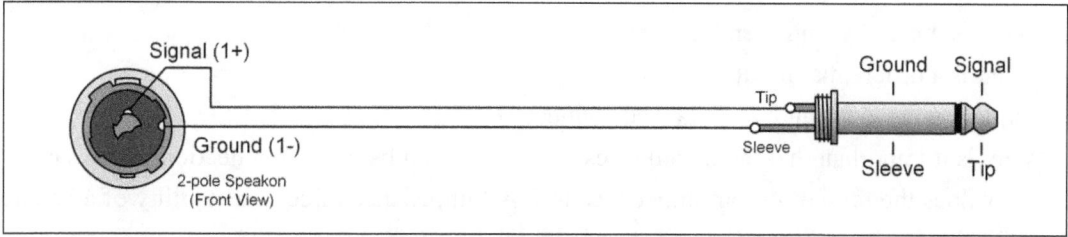

Figure 6-23 2-Pole Speakon to TS Cable

6.7.3   2-Pole Speakon to 2-Pole Speakon cable

When both the amplifier and the loudspeaker have Speakon jacks, this is the preferred cable to use, as the connectors are secured on both ends. The cable, shown in Figure 6-24, is connected symmetrically at both ends, i.e., pin 1+ is connected to pin 1+, and 1− is connected to 1−.

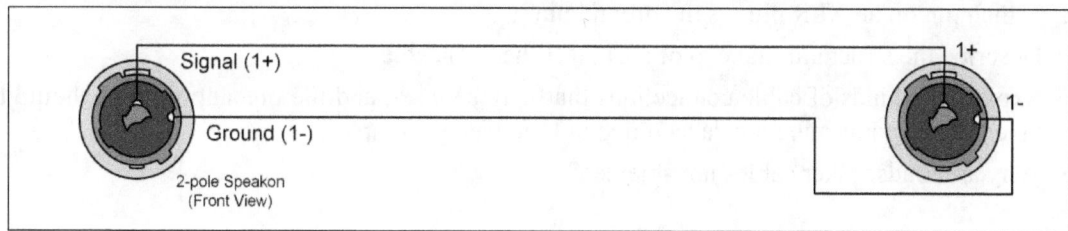

Figure 6-24 2-Pole Speakon to 2-Pole Speakon Cable

6.7.4   4-Pole Speakon to 4-Pole Speakon cable

If the loudspeaker is wired for bi-amplification (see Section 4.7.7) and also has a properly wired Speakon jack, then the 4-pole Speakon cable connector is very convenient to use. The wiring arrangement is shown in Figure 6-25. In this case, the high-frequency signal could be sent along pins 1+ and 1−, while the low-frequency signal utilizes pins 2+ and 2−.

Figure 6-25 4-Pole Speakon to 4-Pole Speakon Cable

Review Questions

1. What is meant by single-ended output?
2. What is a differential input amplifier?
3. Explain what is meant by a balanced connection.
4. Why is it important that the impedances on the legs of a balanced connection be the same?
5. How does the ratio of output impedance to input impedance affect the stability of a balanced connection?
6. Compare and contrast a transformer balanced output stage with an active balanced stage.
7. What is a balanced/unbalanced output stage?
8. What is a servo-balanced output stage?
9. Explain why signal symmetry on the legs of a connection is not required for a balanced connection?
10. Why are the wires in a two-conductor patch cable normally twisted?
11. On a stereo to TRS patch cable, which contact of the TRS plug carries the left channel signal?
12. Which pin on an XLR plug is the ground pin?
13. Describe the structural makeup of a Cat 6 (Ethernet) cable.
14. Explain the kinds of cable connections that may be used, and the precautions that should be taken, when connecting a balanced output to unbalanced input.
15. Why are loudspeaker cables not shielded?

Chapter 7

Digital Audio

7.1 Digital Audio

These days, every effects rack is likely to contain at least one piece of digital audio processing equipment. As the price of digital signal processing technology has fallen, the popularity of digital audio equipment has increased to the point where they are used for everything from reverbs to synthesizers to equalizers. How are digital processors different from their analog counterparts and what actually goes on inside these little boxes? In this chapter, we will shed some light on the general concepts of digital audio processing.

7.2 Numbers and Numerals

Digital audio processing is all about numbers—numbers that are manipulated inside of the digital signal processing device. Before we go any further, though, we need to clarify the difference between a number and its representation in a form that is usable by humans.

A number is the actual physical quantity or measure of something. If I wanted to show a quantity of three, I may display three beads, for example. If I then wanted to show a quantity of five, I may add two more. What if I wanted to show a quantity of 4564138? Using my system of beads, this becomes an impractical scheme. A more practical scheme is to use numerals. A numeral is a symbol that is used to represent a number. A numeral is not (necessarily) equal to a number because as we will see shortly, depending on the context, the actual number it represents can change.

Different numerals can be used to represent the same number. For example:

- **1, 2, 3, 4** are Western Arabic numerals
- **I, II, III, IV** are Roman numerals
- ***one, two, three, four*** are word numerals

In the decimal number system these numerals all represent the same set of numbers.

7.3 The Decimal Number System

Before we delve into digital audio processing, it will be useful to review some concepts about number systems. We will use the familiar decimal number system as a bridge to understanding the system used by digital audio processors. The decimal number system is the one we use in everyday life. It is a positional system that uses the set of ten numerals $\{0, 1, 2, 3, 4, 5, 6, 7, 8, 9\}$ to represent all the numbers we need to describe. We say it is a positional system because the value associated with a digit in a number representation depends on its position within that representation.

Let us recall some things that we learnt in elementary school. Each decimal digit is associated with a multiplier that is tied to its position. The least significant (i.e., lowest value) position is the position at the extreme right of the number. The multiplier associated with this position is 1. The multiplier of the next position to the left is 10. The next three positions, going right to left, have multipliers of 100, 1000, and 10000, respectively. Each new value is 10 times more than the last. Recall from Section 1.8.2 that these are all powers of 10. The sequence 1, 10, 100, 1000, 10000 is the same as $10^0, 10^1, 10^2, 10^3, 10^4$.

To clarify this further, look at Figure 7-1, where in the decimal number system we represent the number *one thousand one hundred and one* as **1101**. We can see that when we break the representation down into its constituent digits that each digit has an individual value that is a multiple of a power of 10, starting with the least significant digit to the right with a value of 1, and finishing with the most significant digit to the left with a value of 1000.

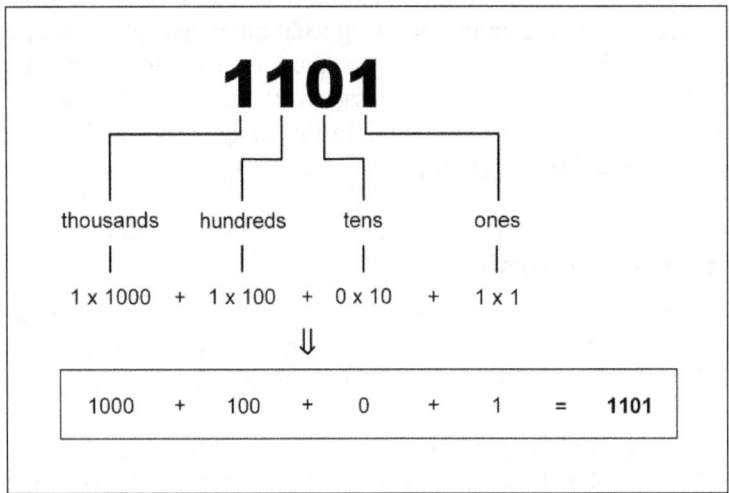

Figure 7-1 Numeric Example in Decimal Number System

In summary, there are two important things to note about this system:

1. Each decimal digit is taken from a set of ten fixed numerals.
2. In multi-digit numbers, each decimal digit has a value that is ten times more than an identical digit one place to its immediate right.

Because of this, the decimal system is called a **base-10** system.

7.4 Number Space in the Decimal Number System

Two interesting questions now arise in our look at the decimal number system. The first question is this: How many different numbers can be represented with a given number of digits? If we have one digit, we can represent 10 different numbers, *viz.* 0 to 9. If we have two digits, we can represent 100 different numbers, *viz.* 0 to 99. For every digit that we add, we can represent 10 times as many numbers as before. We can thus represent 10^n different numbers with *n* digits.

The second question is this: What is the biggest number that can be represented with a given number of digits? If we have one digit, the biggest number is 9. If we have two digits, the biggest number is 99, and so on. The biggest number is one less than the maximum amount to numbers that can be represented because zero is the first number in the list. The biggest number that we can represent is therefore $10^n - 1$.

So, for the decimal, or base-10, number system we can state the following:

- With n digits, we can represent 10^n different numbers.
- With n digits, the biggest number we can represent is $10^n - 1$.

7.5 Digital Audio Processors

Practically everyone is familiar with the modern personal computer—the PC. The PC is a generalized digital computer that has certain elements that allow us to interact with it and perform useful tasks. For example, it has a user interface consisting of a keyboard, mouse, video display, etc. It has data storage devices like solid state drives and digital memory. It also has some sort of central processing unit (CPU) that allows it to perform complex manipulation of data.

Well, a digital audio processor is really a specialized digital computer, just like your PC, but optimized to perform audio related tasks. Just like your PC, it has a user interface that consists of knobs, buttons, and perhaps a small display that allows the user to enter data into the unit. Figure 7-2 shows the user interface of the MX400 digital processor made by Lexicon. Just like your PC, the digital audio processor also has some form of digital memory that stores its programs and data, and it has a specialized microprocessor that executes those programs to manipulate that data. As we will discuss later in more detail, the digital audio processor also has a means of getting an audio signal into and out of the unit. The data used by the digital audio processor consists of this audio signal in a numeric form, and a set of numbers, perhaps entered by the user, telling the digital audio processor how to change that audio.

Figure 7-2   User Interface of a Digital Audio Processor

The digital audio processor stores its data in memory which consists of special electronic switches. These switches can have only two states, *viz. off* and *on*. It is common practice to represent these states by using *0* and *1*, respectively. Remember, all our data entered into the processor is stored as numbers in this memory. Therefore, it is natural to use a number system based on just two numerals—*0* and *1*—to represent these data. This number system is called the *binary*, or *base-2*, number system.

7.6 The Binary Number System

When compared to the decimal number system, the binary system is very similar. The concepts are the same. A digit in the binary number system is called a *bit*—a contraction of **b**inary dig**it**.

Each bit is associated with a multiplier that is tied to its position. The least significant bit (LSB) is the position at the extreme right of the number. The multiplier associated with this bit is 1. The multiplier of the next bit to the left is 2. The next three positions to the left have multipliers of 4, 8, and 16, respectively. Each new value is 2 times more than the last. These are all powers of 2. The sequence 1, 2, 4, 8, 16 is the same as $2^0, 2^1, 2^2, 2^3, 2^4$.

Let us look at an example to help clarify this. Figure 7-3 shows the binary number 1101. This number consists of the same numerals as used in the decimal example of Figure 7-1. However, because the number system is different, the value of the number is quite different. As we did with the example in the decimal number system, we break the number down into its constituent bits. We show each bit's equivalent decimal value so that we can easily compare the value of the two numbers. In this example, the least significant bit has a value of 1. The next higher bit has a value of zero, while the most significant bit has a value of eight. The equivalent decimal value of this number is 13.[a]

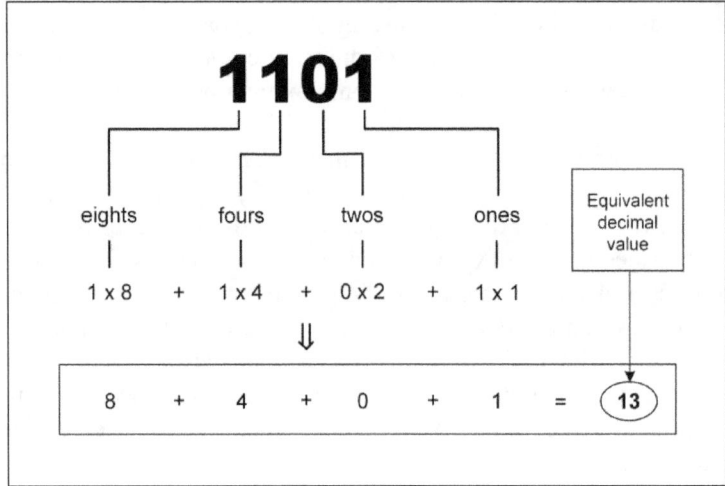

Figure 7-3   Numeric Example in Binary Number System

For reference, Table 7-1 summarizes the differences between the familiar decimal number system and the binary system.

Base 10 (Decimal)	Base 2 (Binary)
Uses {0, 1, 2, 3, 4, 5, 6, 7, 8, 9}	Uses {0, 1}
Each digit changes by a multiple of 10	Each bit changes by a multiple of 2

Table 7-1   Comparison of Decimal and Binary Number Systems

7.7   *Number Space in the Binary Number System*

Let us now answer for the binary number system the same two questions that we answered for the decimal number system concerning the number space. First, how many different numbers can be represented with a given number of bits? If we have one bit, we can represent 2 different numbers, *viz.* 0 and 1. If we have two bits, we can represent 4 different numbers, *viz.* (in binary notation) 00, 01, 10, and 11 (equivalent to decimal 0, 1, 2, and 3). For every bit that we add, we can represent twice as many numbers as before. We can thus represent 2^n different numbers with *n* bits.

---

[a] The representation of numbers in binary format can get overly complicated when we consider negative or fractional numbers. In order to keep our discussion simple, we will consider only positive whole binary numbers, without any loss of insight into the concepts presented in this section.

The second question is: What is the biggest number that can be represented with a given number of bits? If we have one bit, the biggest number is 1. If we have two bits, the biggest number is 3, and so on. Just as we saw with the decimal system, the biggest number is one less than the maximum amount to numbers that can be represented. The biggest number that we can represent is therefore $2^n - 1$.

We can thus summarize the number space of the binary system with following statements:

- With n bits, we can represent 2^n different numbers.
- With n bits, the biggest number we can represent is $2^n - 1$.

7.8   Word Size in Digital Audio Processors

Why do we even care about the number space in the binary number system? Well, here's why. We know that digital audio processors use numbers internally to represent and manipulate audio. In digital audio, each such number is called a word. The number of bits making up the word is called the *word size* or *word length*. Sometimes the term *bit depth* is also used for this. The word size is fixed for any given digital processor. That means that a given digital processor is limited in the number of different numbers it can represent internally. If its word size is n bits long, we know that it can only represent 2^n different numbers. As we will discuss in detail later, in order to accurately represent audio in digital format, we need to be able to represent many different numbers. This will allow us to represent very small changes in the audio signal. The term *resolution* is used to describe this. High resolution means that very small changes can be discerned. In order to have high resolution, we need many bits in each word. More bits in the word give us more resolution, i.e., more accuracy, higher fidelity.

Manufacturers often use the word size as a selling point of their devices. Figure 7-4 shows the labeling on a digital audio processor. We can see that the manufacturer proudly displays that this processor is a 24-bit device. With 24-bit words, this processor can represent 2^{24} different numbers. That is over 16.7 million numbers (16777216 numbers to be exact)—good for high resolution audio. As a comparison, the common audio CD uses 16-bit words to store its audio. With 16 bits, we can only represent 65536 different numbers. Many pro-audio processors use 20 or 24-bit words internally.

Figure 7-4   Label on Digital Processor Showing Word Size and Sample Rate

7.9   Analog vs. Digital Signals

Analog signals are what occur naturally. Even when we take sound energy and convert it to an electrical signal, such as with a microphone, that electrical signal is still analog. The analog signal is continuous. That is, it has a value at every possible instant in time. That also means that it is infinitely precise in its representation of the amplitude of the signal. No matter how small a period in time we look at, if our instruments were accurate enough, we could measure the change in amplitude from the previous instant. When we draw the representation of an analog signal on a graph with *Time* on the horizontal axis and *Amplitude* on the vertical axis, we draw it as a continuous line, as in Figure 7-5, which shows a classic representation of one cycle of a sine wave.

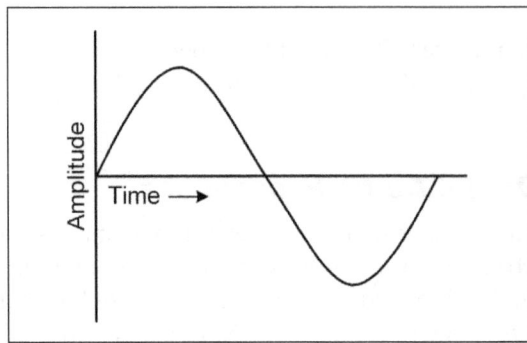

Figure 7-5   Graphical Representation of Analog Sine Wave Signal

A digital signal is a representation of a natural (i.e., analog) signal using a sequence of numbers that are chosen from a *finite* and *fixed* set of numbers. The set is finite because the quantity of numbers in the set is limited to a certain amount. The set is also fixed because the actual numbers used to represent the signal are also fixed. If the set were {0, 1, 2, 3} it would always be {0, 1, 2, 3}. It could never be changed to {1, 1.5, 2.5, 3}. In a 16-bit digital processor, you could think of this set as containing 65536 numbers.

Digital signals are derived from analog signals by measuring—or sampling—the analog signal at specific, fixed instants in time, e.g., once every second. This sequence of measurements is simply a set of numbers. Each measurement is called a sample, and you should realize that this set of samples *is* the digital signal. We can see that a digital signal is discontinuous in time, i.e., it has no value in the interval between samples. In sampling the signal, the value of the signal's amplitude, i.e., its voltage, at the instant that it is sampled is rounded up or down to the nearest value in a given set of numbers. This is the set described in the previous paragraph. Thus, a digital signal is also discontinuous in amplitude. Furthermore, because we have to round the measurement to fit the values in the set, the digital signal also has finite precision. If we were to represent the digital signal on a graph similar to Figure 7-5, it would look like just a series of dots, as shown in Figure 7-6. Here we see the same signal represented by 25 discrete samples.

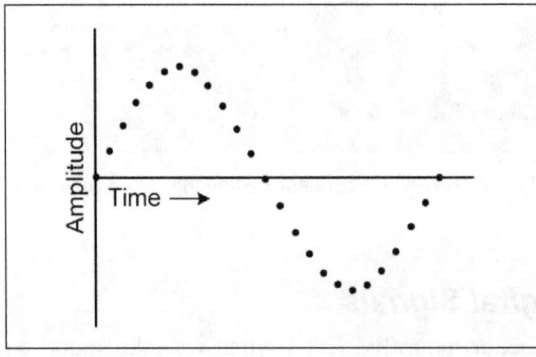

Figure 7-6   Graphical Representation of Digital Sine Wave Signal

Since digital signals exist within the digital audio processor, and analog signals exist outside of the processor, the signals must be converted from one type to the other when digital signal processors (DSP) are used. Analog to digital converters (ADCs) are used to get the signal into the processor and digital to analog converters (DACs) are used to transform the digital signal back to analog when the processing is done. Figure 7-7 shows this relationship of the signal converters and DSP to the overall layout of the digital audio processor. Let us now look at the details of how these conversions are done.

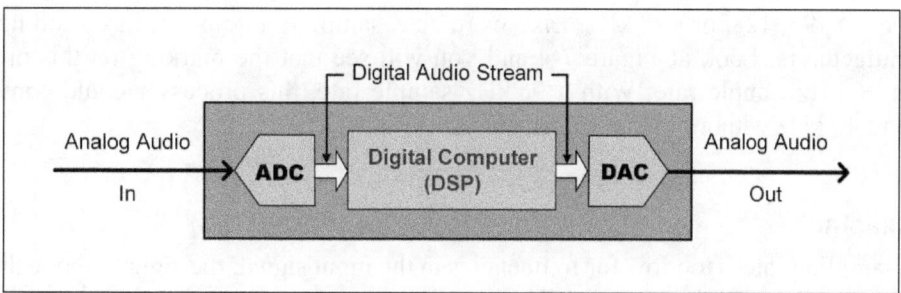

Figure 7-7   ADC and DAC in Digital Audio Processor

7.10 Analog to Digital Conversion

Analog to digital conversion is at least a three-step process. The first two steps—sampling and quantization—were briefly mentioned above, but we will now discuss them in more detail. The third step is encoding, where the data from the second step is represented in a format that is used by the digital processor. Encoding takes data in one format and represents it in another format without changing its meaning or value.

7.10.1 Sampling

Sampling is the act of measuring the level of the audio signal at regular intervals. It converts the analog signal into a series of voltage measurements, i.e., from a continuous signal into a discrete-time, continuous-amplitude signal. Continuous amplitude means that the measured amplitude of the signal is as accurate as our measuring device will allow, and it can have any value within the range of the instrument. Figure 7-8 shows this graphically, where each sample is represented by a dot. Notice how each dot occurs at a discrete instant in time. If we sample faster, these dots will be closer together. If we sample at a lower rate, they will be further apart.

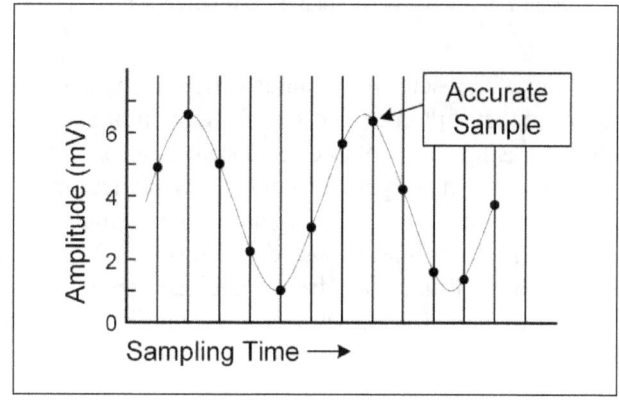

Figure 7-8   Sampling a Signal

The rate at which the samples are taken is called the sampling rate, measured in hertz (Hz), and is a key consideration in the sampling process. For periodic signals such as audio signals, if the sampling rate is bigger than twice the frequency of the signal, no information will be lost when the signal is recreated in analog form by the DAC. Half of the sampling frequency is known as the Nyquist frequency (after Harry Nyquist, who worked on sampling issues in the 1920's). For audio, since the highest frequency that humans can hear is taken to be 20 kHz, the sampling rate must be bigger than 40 kHz in order to reproduce all the frequencies in the audio spectrum. The common audio CD system uses a sampling rate of 44.1 kHz, and many digital audio processors use a sampling rate of 48, 96, 128, or 192 kHz. Like word size, sampling rate is used as a selling point by many manufacturers. Look at Figure 7-4 and you will see that the marking on this processor is boasting a 96 kHz sample rate. With a 96 kHz sample rate, this processor could convert audio signals up to 48 kHz without loss of information.

7.10.2 Aliasing

When the sampling rate is too low for frequencies in the input signal, the signals above the Nyquist frequency cannot be reproduced and will be lost from the reconstructed output signal. Moreover, another signal, lower in frequency, will be created in its place! This is because there are not enough samples to follow all the changes in the signal's shape, and when the DAC 'connects the dots' a lower frequency signal is the result. This new signal is called an alias of the original signal and introduces distortion into the audio signal. Figure 7-9 shows a signal that is sampled too slowly and how the alias will appear in the output instead of the original signal.

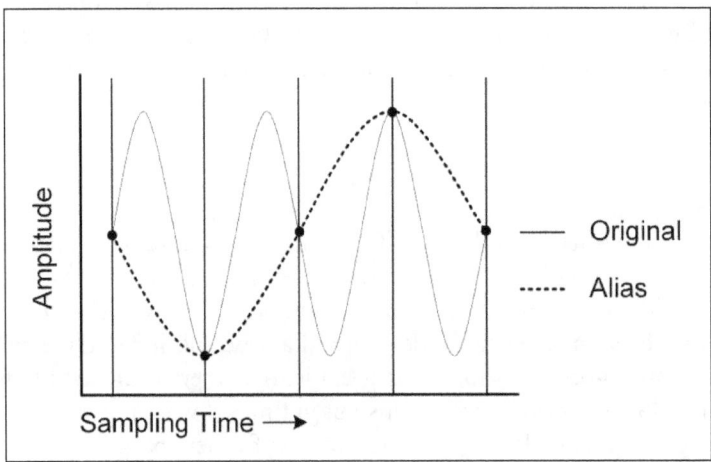

Figure 7-9 Aliasing Resulting from Low Sample Rate

Aliasing will occur for all signals present in the input which are above the Nyquist frequency. In 1949, Claude Shannon showed that if there are no signals in the input with a frequency greater than the Nyquist frequency, then aliasing does not occur. Therefore, to prevent aliasing, we could filter out from the input all signals above the Nyquist frequency while leaving lower frequency signals untouched. Unfortunately, such a filter would have to have an extremely steep roll-off. Such analog filters are costly and typically have nonlinear phase characteristics. Nonlinear phase means that as the audio signal passes through the filter, some frequencies are delayed more than others. No one has yet found a way to make the perfect analog brick wall filter that would be required in this case. To get around the problem, we use a two-step approach.

First, we make the sampling rate really high. Sampling at a rate higher than the minimum required by the Nyquist theory is called oversampling. For example, if we take the base sampling rate to be 48 kHz, this gives us a Nyquist frequency of 24 kHz, just above the upper limit of the audio band. If our sample rate is raised to 192 kHz, this is referred to as 4x oversampling. With oversampling, the Nyquist frequency also becomes quite high, and aliasing only occurs for frequencies that are far outside the audio band. With a processor that is sampling at 192 kHz, aliasing will only occur for frequencies above 96 kHz.

The second step is to low-pass filter the audio input with a cutoff frequency just above 20 kHz. This removes most of the signal between 20 kHz and the Nyquist frequency and effectively solves the aliasing problem in the audio band. This filter is commonly called an anti-aliasing filter, and it is placed before the ADC in the processor's signal chain. It is shown as AAF in Figure 7-10.

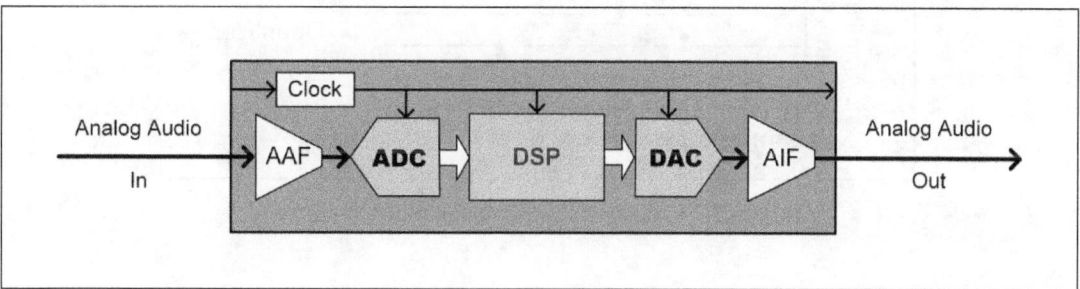

Figure 7-10   Digital Audio Processor Components

7.10.3 Quantization

The measured voltage of a sample can be any value within the input voltage range and has the maximum precision afforded by the measuring device. The ADC, however, has two limitations. It has limited precision, and it has a limited set of numbers that it can represent, dependent on its internal word size. Remember that if the word size of the ADC is n bits, it can only represent 2^n different numbers. Therefore, every measured sample must be converted, i.e., approximated, to the closest number that matches what the ADC actually uses. This process is called quantization and generally adds error to the sampled signal. This error is called quantization error. With a constantly varying analog input signal, the quantization error also varies, up to plus or minus half the value that the least significant bit (LSB) represents, and is called quantization noise. This noise adds distortion to the original audio signal.

As a simple example, suppose we quantize the signal shown in Figure 7-8. This signal has an input voltage range of 0 to 7 mV. If we use an ADC that has an internal word size of 3 bits, we are only able to represent eight different numbers. With the ability to represent eight numbers, we could divide up the input voltage range equally and represent voltage points of 0, 1, 2, 3, 4, 5, 6, 7 mV. Thus, in this case the LSB represents a value of 1.0 mV and our maximum quantization error would be ±0.5 mV. Once we have decided to divide the input voltage range like this, intermediate voltages, e.g., 2.29 mV, could not be represented. Therefore, a measurement of 2.29 mV in the sample would be quantized to 2 mV by the ADC. Table 7-2 shows how the first 10 samples are quantized for this example, and the result is shown graphically in Figure 7-11.

Our original sample set, when quantized, now gives us a discrete-time, discrete-amplitude signal.

Sample	1	2	3	4	5	6	7	8	9	10
Measured	4.91	6.56	5.01	2.29	1.08	3.00	5.69	6.40	4.22	1.62
Quantized	5	7	5	2	1	3	6	6	4	2
Error	−0.09	−0.44	0.01	0.29	0.08	0.00	−0.31	0.40	0.22	−0.38

Table 7-2   Quantized Samples of Input Signal

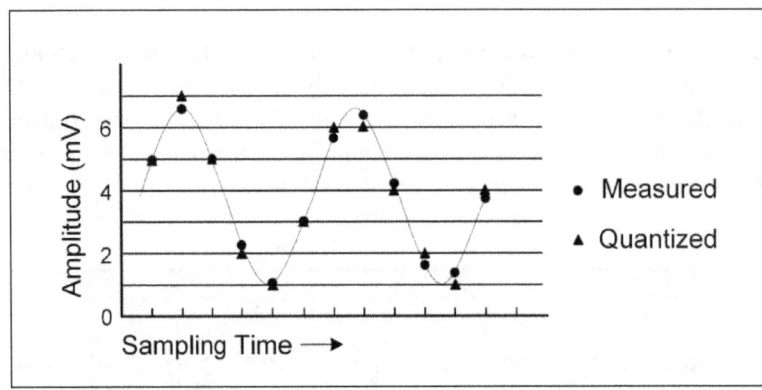

Figure 7-11   Samples Showing Quantization Error

7.10.4 Encoding

At this stage in the analog to digital conversion process, we have a set of numbers that represents the original audio signal. How we choose to represent these numbers in binary form to be used internally by the processor is called encoding. There are many ways in which the numbers could be encoded. For example, in the binary number representation, do we store the least significant bit first or the most significant bit first, or is the order of the bits changed in some other way? If we have multiple channels of audio, do we store samples taken at the same time in the same block of binary data or do we store them separately? We could use some standard encoding scheme or we could make up a completely unique one. From the perspective of the processor, it really does not matter, so long as it understands how to correctly interpret the digital data.

The process we just described in the foregoing sections, i.e., sampling, quantization, and encoding, is very common in converting an analog signal to digital, and is called pulse-code modulation (PCM), because we measure the value of a sample (pulse) and convert it to a binary number (code). PCM is the scheme used in practically all forms of digital audio processing.

7.11 Digital Signal Processing

Once the ADC has done its job, the digitized samples are delivered to the DSP core. The DSP implements algorithms that manipulate the digital signal in a variety of ways by using a combination of a specialized digital processor and software. An algorithm is simply a set of steps that describe what actions are performed to achieve a particular result. These algorithms create delays, filters, feedback loops, phase shifting, etc., to achieve many different audio effects. In the next chapter, we will look at how some of these effects are achieved.

7.12 Digital to Analog Conversion

Once the digital signal has been manipulated inside of the DSP, it must be converted back into analog form to be heard by humans. This is the job of the digital to analog converter (DAC). The DAC takes each sample and, at precisely the same sampling rate as the ADC, generates and holds a voltage corresponding to the encoded sample value. Conceptually, this is like connecting the dots of the samples with lines at right angles to each other. The generated signal looks like the one shown in Figure 7-12, here shown with a low sample rate for clarity.

7.12.1 DAC Sampling Rate and Clocking

The DAC must use the same sampling rate as the ADC to generate the analog signal. If they are different, the output signal will have a different frequency than the original signal. A faster DAC sampling rate will give a higher output signal frequency. Sometimes the ADC and the DAC use the same clock within the digital audio processor to generate the sampling period. In this case, it is easy to assure that the ADC and the DAC use the same sampling frequency. Many times, though, they use separate clocks. For example, a digital signal may be sent from one processor to another. In this case, the ADC is in one processor and the DAC is in another. In order to maintain the strict timing requirements between the ADC and the DAC, processors have the ability to derive a clock signal from an incoming digital bit stream. The clock of the receiving audio processor must be accurately synchronized to the incoming signal, otherwise the output signal from the DAC will be mostly garbage and unintelligible audio.

The clock frequency must also not jump around from one tick to the next, i.e., the duration of each tick should be the same. This variation in tick length is called jitter and is usually very small in modern equipment. Jitter causes distortion of the output signal, since the generation of the signal occurs at each clock tick.

On some audio processors, provision is made to use an external clock source for all internal timing requirements. This allows an entire rack of equipment to be synchronized to one master clock. These processors also usually provide their internal clock signal to external equipment. Figure 7-10 shows how the clock relates to the other components of the digital audio processor.

7.12.2 Sample Rate, Word Length, and Fidelity

Since the DAC only generates a voltage at every sample interval (i.e., clock tick), the output signal is very jagged—like a set of stairs. This is a form of distortion. If the sample rate is too low, this distortion is very obvious. At high sampling rates, the distortion is still there but is harder to detect.

In Figure 7-12, we consider the case where we have a large word length, but a low sample rate. As we know, if we have a large word length, we will be able to generate very accurate samples in the ADC. However, with a low sampling rate, we will only generate relatively few of them. The DAC will only have a small number of samples to work with, and the output signal will be very jagged—highly distorted. The distortion is approximately the percentage of the average signal value that the least significant bit represents. That is why in digital audio processors, distortion increases as signal levels decrease.

The dynamic range of the DSP is also tied to the word length. The theoretical maximum dynamic range of the DSP is equal to $6.02n + 1.76$ dB where n is the word length. For large word lengths, this is approximately equal to the number of bits times 6 dB, and is a good rule of thumb.

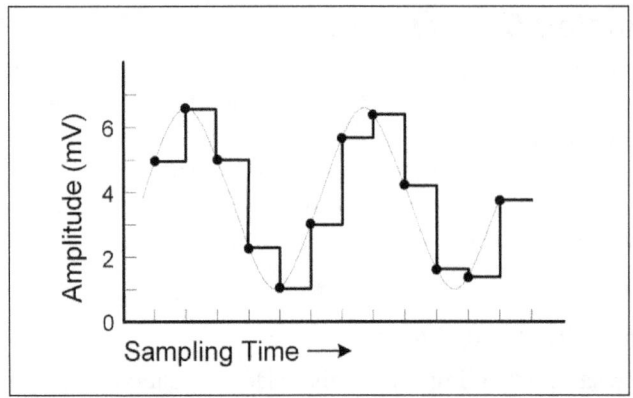

Figure 7-12 DAC Output with Large Word Length and Low Sample Rate

In Figure 7-13, we have the same input signal as previous. This time, it is digitized with a high sampling rate, but the word length is only 3 bits. Many samples will be generated, but because the number of choices that each value can have is low, the signal is still very jagged. Notice, though, that it is still much better than the case in Figure 7-12. All things being equal, this tends to be generally true, i.e., an increase in sampling rate gives better fidelity than an increase in word size.

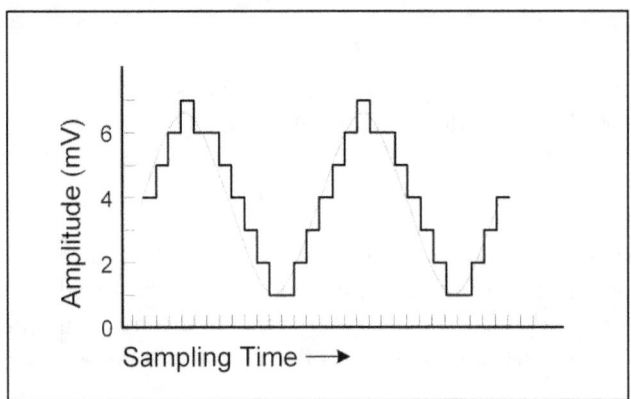

Figure 7-13 DAC Output with Small Word Length and High Sample Rate

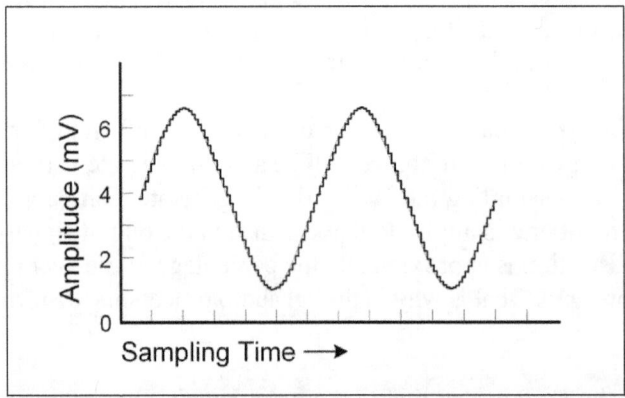

Figure 7-14 DAC Output with Large Word Length and High Sample Rate

To really have an accurate representation of the original signal, though, we must have both a large word size and a high sample rate. Though the output signal is still jagged, it is not as obvious as before. The signal now follows the original signal very closely. We can see this in Figure 7-14.

7.12.3 DAC Output Filtering

No matter what the sample rate and word size, the output signal from the DAC is still jagged. This signal looks like a low-level, high-frequency signal superimposed on a high-level, low-frequency signal. As a comparison, Figure 7-15(A) shows a 30 kHz sine wave being added to a 2 kHz sine wave. The resulting superimposed wave is shown in Figure 7-15(B). It looks very much like the wave in Figure 7-13.

The output signal from the DAC does indeed consist of unwanted high-frequency energy superimposed on the desired audio signal. Part of this energy is high-frequency noise from the digital to analog conversion process, which causes the output to look like Figure 7-15(B). The output signal also contains images of the original audio signal clustered around multiples of the sampling frequency. To get rid of this high-frequency content, the output from the DAC is passed through a low-pass filter, called an anti-imaging, or reconstruction, filter. This filter is shown as AIF in Figure 7-10. The filter removes information in the output signal above 20 kHz.

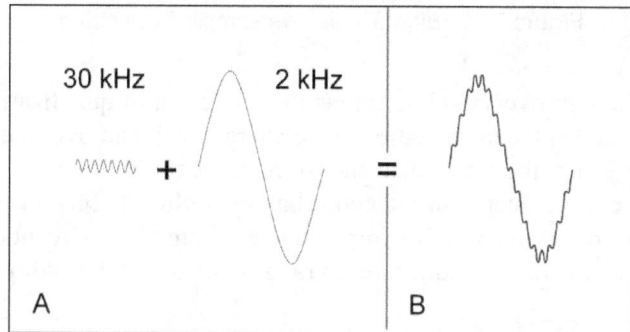

Figure 7-15   30 kHz Sine Wave Superimposed on a 2 kHz Sine Wave

7.13 1-Bit Audio Processors with Oversampling

The above discussion of digital audio processors implied a parallel n-bit system from start to finish. However, this is not always the implementation used in modern digital audio processors. Parallel 16-bit or 24-bit converters that operate at high sampling rates are expensive to manufacture. Compared to parallel converters, 1-bit delta-sigma converts are fairly cheap, and their use has become popular in digital audio processors. As their name implies, 1-bit delta-sigma ADCs do not generate an n-bit word for each sample, but rather generate a 1-bit data stream where the density of ones in the stream is equal to the average level of the input signal. Similarly, a 1-bit DAC will take a 1-bit data stream and—in combination with the output filter—generate an analog signal whose level is the average of the density of ones in the 1-bit stream.

Delta-sigma converters operate at very high oversampling rates, typically 64x, 128x, and even higher. As we saw above, increasing the sampling rate also pushes the Nyquist frequency up, and this has a few advantages. The first one is that the analog AAF can now be made with a gentle roll-off, with linear phase characteristics. Building analog filters with steep roll-offs is expensive, and they typically have nonlinear phase characteristics (see Section 8.2.1). Using a delta-sigma converter with high oversampling rate, a digital filter with steep roll-off can be implemented after the ADC to attenuate frequencies above the audio band and well below the Nyquist frequency. Compared to analog filters, well behaved digital filters are easy to implement in the software of the DSP. Similarly, a digital filter can be implemented before the DAC to eliminate the high-frequency images, and a cheap analog AIF with gentle roll-off can be used after the DAC to eliminate high-frequency noise. For illustration purposes, Figure 7-16 shows an example of 8x oversampling. This figure makes it clear how the audio band now takes up only a small part of the overall spectrum of operation, and the Nyquist frequency, at $1/2$ f_s, is well above the high end of the audio band. In this case, the analog filter slope is gentle compared to the digital filter.

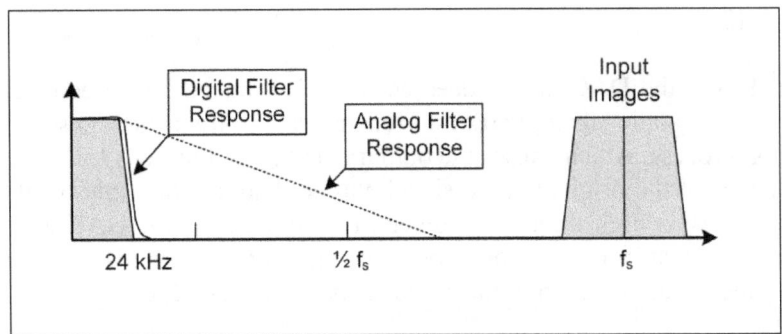

Figure 7-16   Eight Times Oversampling Spectrum

Another advantage of a high oversampling rate is the reduction of quantization noise in the audio band. With the ADC process, the quantization noise energy is spread over the entire Nyquist band. With a higher sampling rate, the total noise energy remains the same, but is spread over a wider frequency range, so the noise energy in the audio band is reduced. This improves the SNR of the ADC. Not only that, but delta-sigma ADCs implement a scheme whereby much of the quantization noise is also shifted from the lower frequencies to ranges outside of the audio band, further increasing the SNR.

One disadvantage of high oversampling rates of the delta-sigma converter is the large amount of data that it generates. Therefore, before sending data to the DSP core, the digital low-pass filter of the ADC converts the 1-bit data stream to an n-bit stream. Another digital filter stage, called a decimation filter, then simply throws away samples that it gets from the low-pass filter until an equivalent sample rate no lower than twice the Nyquist frequency is obtained. This prevents the DSP from being choked with too much data to process and maintains enough data so that the analog signal can be accurately reconstructed later.

7.14 Advantages of Digital Audio Processing

Digital audio processors seem like a lot of trouble—there is conversion and quantization noise produced in the conversion process, and aliasing and jitter problems to overcome, to name a few. These things work to distort the original audio signal. So why use digital audio processors? Well, the answer is simple. There are cheap solutions to these problems available in modern technology that allow us to get very good performance from our digital audio equipment. Below, we summarize the benefits of using digital technology to manipulate the audio signal.

- The core processors are cheap, readily available computer technology.
- The converters and filters are also cheap and readily available COTS technology.

- The processors can easily be programmed to manipulate the digital signal.
- There is no degradation of the signal when moved from one step to the next.
- High dynamic ranges are achievable with large word lengths.
- The DSP performance is stable with changes in temperature.
- There is virtually no change in performance of the DSP due to aging components.

So, we see that there are several advantages to using digital technology to process audio signals, and judging by the trends in the new audio processors on the market, manufacturers fully understand and are capitalizing on these advantages.

Review Questions

1. What is the difference between a number and a numeral?
2. What are the essential differences between the decimal number system and the binary number system?
3. What is meant by word size of a digital audio processor and why is it an important specification?
4. What does it mean to sample an analog signal?
5. Compare and contrast an analog signal with its digital counterpart.
6. What is the purpose of the analog to digital converter in a digital processor?
7. Explain the phenomenon of aliasing and how it can be avoided when performing analog to digital conversion.
8. Explain the role of quantization in the process of analog to digital conversion and its effect on audio quality.
9. What is jitter and how does it affect the audio quality of a digital processor?
10. What is the approximate maximum theoretical dynamic range of a 24-bit digital processor?
11. Generally speaking, which gives the greater increase in fidelity: an increase in sample rate, or an increase in word size? Why?
12. What is the purpose of an anti-imaging filter?
13. If the LSB of an ADC represents 0.5 mV, what is the biggest quantization error that the converter will produce?
14. What is oversampling and what are the advantages and disadvantages of implementing it?
15. Discuss the pros and cons of using digital audio processors as compared to analog processors.

Chapter 8

Audio Effects Processors

8.1 Role of Effects Processors in the Signal Chain

Take a look at Figure 1-2 and you will see a block of devices labeled *Effects*. Effects processors sit in the middle of the signal chain and are a subset of signal processors that manipulate the signal after it is delivered from the input devices. The job of the effects processor is to change the audio signal in very specific ways to meet the requirements of the user. In one sense, they allow the audio engineer to sculpt the audio signal as an artist sculpts a block of marble. In this case, the engineer enhances the audio signal to suit a particular taste. From another point of view, effects processors allow the sound engineer to change the audio signal to compensate for deficiencies in the environment. In this case, the engineer enhances the audio signal to mitigate problems caused by phenomena like absorption or reflection. In the following sections, we will take a closer look at several of the effects processors that you are likely to find in a typical effects rack of a live sound reinforcement system.

8.2 Equalization Filters

The equalizer is the most common effects signal processor encountered, from basic tone control on small mixers to 1/6 octave per band standalone systems. The equalizer is used to adjust the frequency response of the system to suit the taste of the sound engineer and to compensate for system or environmental problems that cause feedback and adversely alter the sonic characteristics of the loudspeaker system.

The equalizers available today may use a variety of filter types. We first discussed filters in Section 4.7. There we talked about low-pass, high-pass, and band-pass filters. These filter types were discussed in the context of allowing only a certain portion of the audio spectrum to pass while attenuating the rest. However, equalizers not only attenuate signals, but may also boost a particular region of the spectrum without attenuating anything. Other filter types may be employed for this, as discussed below.

At the low end of the range of the equalizer's control, it may employ a filter that boosts or cuts all frequencies below a certain frequency. Similarly, at the high end, it may use a filter that boosts or cuts all frequencies above a certain frequency. Figure 8-1 shows the response of the filter used to control the low end of the spectrum, and while it looks like a low-pass filter, it is actually a bit different. This filter has a turnover frequency at the usual 3 dB point, and a stop frequency around which the response levels off to a flat shelf. Thus, this is called a low shelving filter. On the high end, a similar filter, call a high shelving filter can be used to boost or cut all frequencies above the turnover frequency. Because these filters boost everything beyond their turnover frequencies, energy content outside the range of human hearing will also be boosted. Therefore, the sound system must take care to filter out infrasonic and ultrasonic frequencies that can waste our amplifier's power and possibly damage the drivers in the system. Knowing this limitation, many manufacturers either bound their equalizers with high-pass and low-pass filters, or simply do not use these types of filters in their equalizers. These filters, though, are still quite popular in the EQ sections of many mixers.

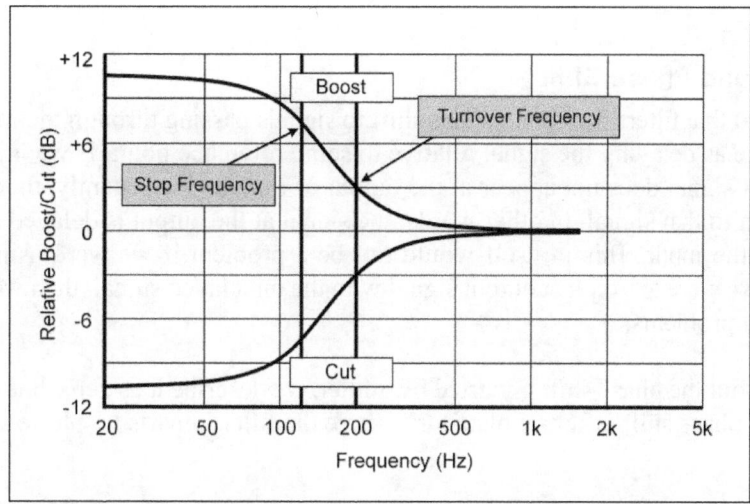

Figure 8-1   Frequency Response of Low Shelving EQ Filter

To gain better control over the range of frequencies that we boost or cut, we can use a class of filter whose response looks like Figure 8-2. If the shape of this response looks familiar, that is because we have seen a similar shape when we talked about band-pass filters in Section 4.7.3. In fact, we describe this filter using the same parameters of center frequency, bandwidth, and gain, as we use for band-pass filters. The difference here is that this class of filter can boost a particular range of frequencies without attenuating any other frequencies. We call these filters peaking filters.

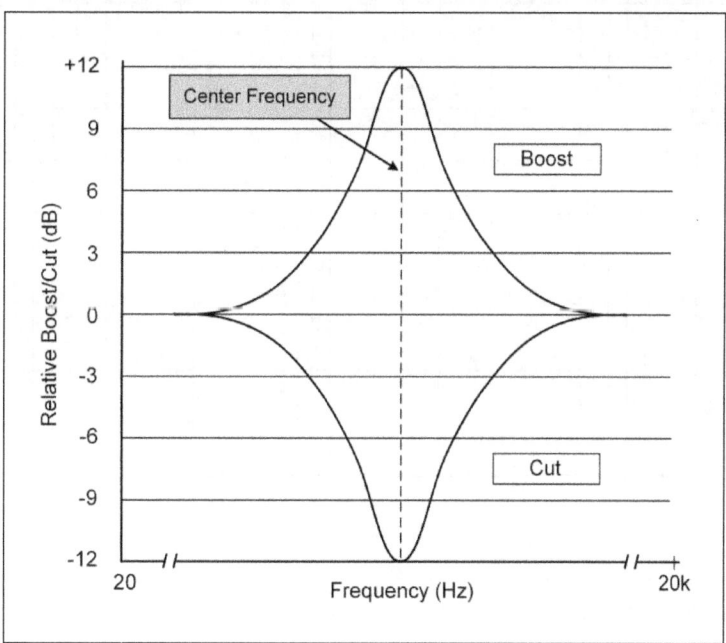

Figure 8-2   Frequency Response of EQ Peaking Filter at Maximum Boost/Cut

Since peaking filters boost frequencies within a limited bandwidth, they can be used at the low and high ends of the equalizer's range of control without the risk of overly boosting energy content well outside the range of human hearing. Thus, if an equalizer uses this type of filter, we do not have to worry about using infrasonic and ultrasonic filters to limit the energy content of the signal just because we added such an equalizer to the signal path.

8.2.1 Filters and Phase Shift

We have all heard that filters introduce phase shift to signals passing through them. Phase shift in a signal is the same as delaying the signal relative to some reference point. If we inject a signal into a device and that signal does not appear at the output of that device instantly, then we have introduced phase shift to that signal. In other words, the signal at the output is delayed some amount of time relative to the input. This in itself would not be a problem if we were using that signal in isolation, but if we were to mix that output signal with the undelayed signal, then we may encounter some unintended problems.

When we talk about the phase shift imparted by a filter, we describe it as zero, linear, or non-linear. The case of zero phase shift is self-explanatory—here the filter imparts no phase shift to the audio signal.

If the filter has a linear phase response, then it delays all frequencies by the same amount of time. In this case, the absolute phase shift angle increases linearly with frequency. For example, a 1 kHz sine wave has a period of 1 ms. A filter that imparts a 0.25 ms delay to all frequencies would cause a 90° phase shift to such a 1 kHz signal, since 0.25 ms is a quarter of the period and a whole cycle

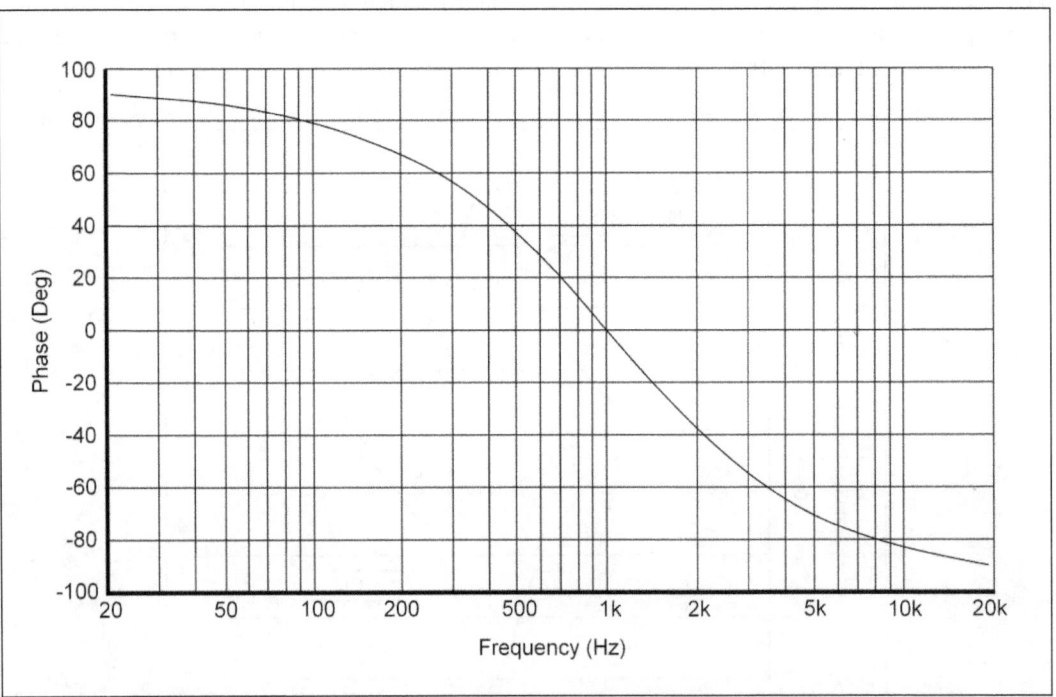

Figure 8-3 Phase Response of a Typical Band-Pass Filter with 1 kHz Center Frequency

takes 360°. On the other hand, the same filter would impart a 180° phase shift to a 2 kHz wave because such a wave has a period of only 0.5 ms, and 0.25 ms represents half of its 360° cycle. While it is beyond the scope of this book, it can be shown that any complex waveform is actually a summation of several sine wave components of various frequencies (e.g., see Appendix 1 in Howard[1]). This means that for a complex waveform passing through a linear phase filter, the relative time relationship of these individual components, and hence the shape of the waveform, are preserved. This is important in eliminating distortion of transients in dynamic music signals.

A filter that has a non-linear phase response delays different frequencies by different amounts of time. Many analog filters fall into this category. Figure 8-3 shows the phase response of one common design of an analog band-pass filter with a center frequency of 1 kHz. Notice that relative to the center frequency, low frequencies are sped up by increasing amounts of time, while high frequencies are delayed by increasing amounts, up to a maximum of 90°, or a quarter of a wavelength.

Analog filters are made using some combination of reactive components (we first looked at these in Chapter 3, when we talked about capacitors and inductors) and typically exhibit a non-linear phase response. On the other hand, digital filters can be made to have any phase response desired, including (if we ignore the issue of processor latency for the moment) zero phase shift.[2] The digital filter is really a small computer program that is executed on the hardware of the digital audio processor. The program simply manipulates the bits that represent the audio signal to give the desired result. We discussed digital audio signals and processors in Chapter 7.

Some processors allow us to connect filters together, and so an idea of the composite phase shift of multiple connected filters would be useful. Suppose we connect three filters with phase responses similar to the one shown in Figure 8-3, and center frequencies of 1 kHz, 1.25 kHz, and 1.6 kHz in series. The resulting summed phase response is as shown in Figure 8-4. At first glance, this graph looks discontinuous at about the 300 Hz mark, but that is only because we are constraining the display of the phase angle to between −180° and +180°. Since −180° and +180° are identical points on the phase circle, the phase seems to abruptly jump from −180° to +180°. We can get a clearer picture if we unwrap the phase trace and display it as in Figure 8-5, which gives us the same information, but with continuously increasing phase angle. Now we can easily see that the high frequencies are shifted by roughly 360° relative to the low frequencies. This picture gets worse as we add more filters, i.e., the delay of the high frequencies keeps getting bigger compared to the low frequencies, and this could have an audible effect on fast transients.

Now suppose we connect these same three filters in parallel and look at the combined phase response, presented in Figure 8-6. We can see that it looks almost the same as the case for the single filter because the final result is not the cumulative sum of all three stages, but rather it is the incremental difference in shifts among the three filters. Even when we add more filters, the extreme phase shifts do not grow much beyond that shown. Thus, a parallel connection will give better performance when processing transients.

Many effects processors used in the live sound arena employ phase shifting in their implementation. As practitioners, we don't usually give much thought to how filters and phase shifting are used in the effects, but rather that we get the final sound that we want from utilizing the effects. On the other hand, we care about phase and phase shift very much when it comes to locating microphones, loudspeakers, and listening positions. This is a completely different matter, discussed in Chapter 13 and Chapter 14.

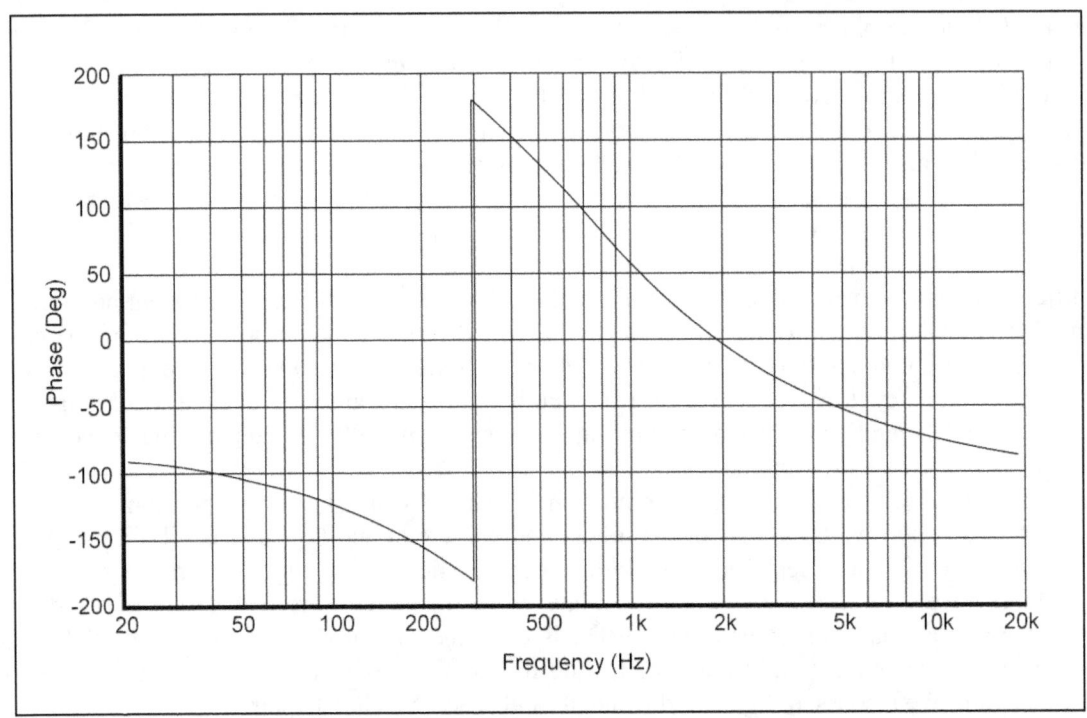

Figure 8-4 Phase Response of Three Filters Connected in Series

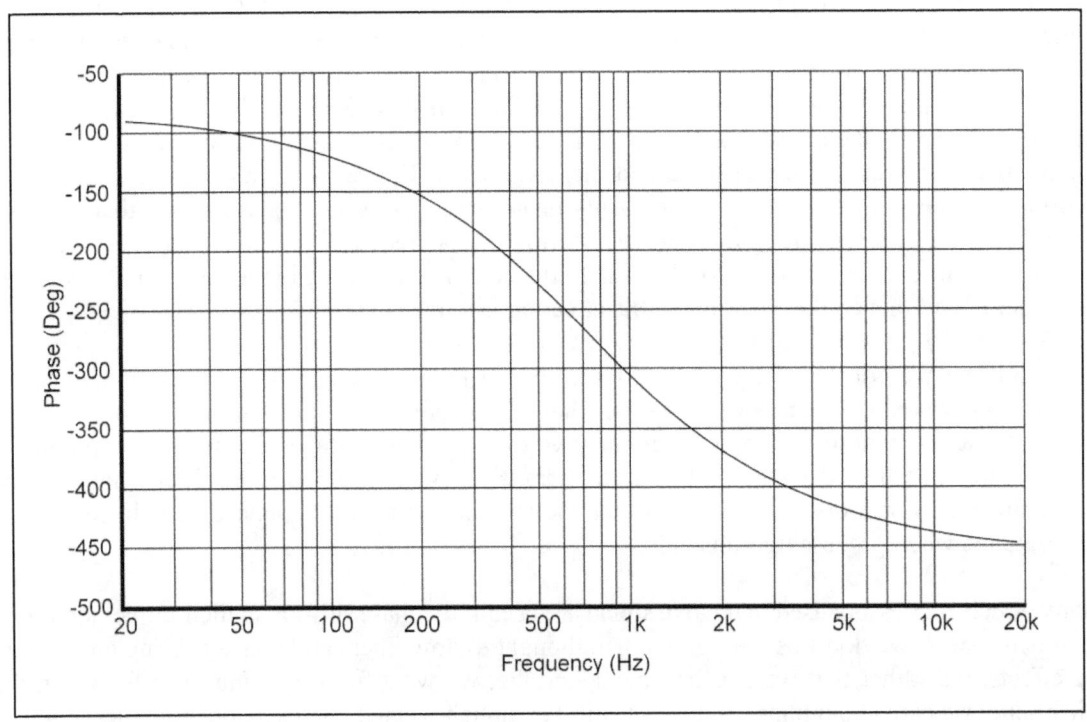

Figure 8-5 Unwrapped Phase Response of Three Filters Connected in Series

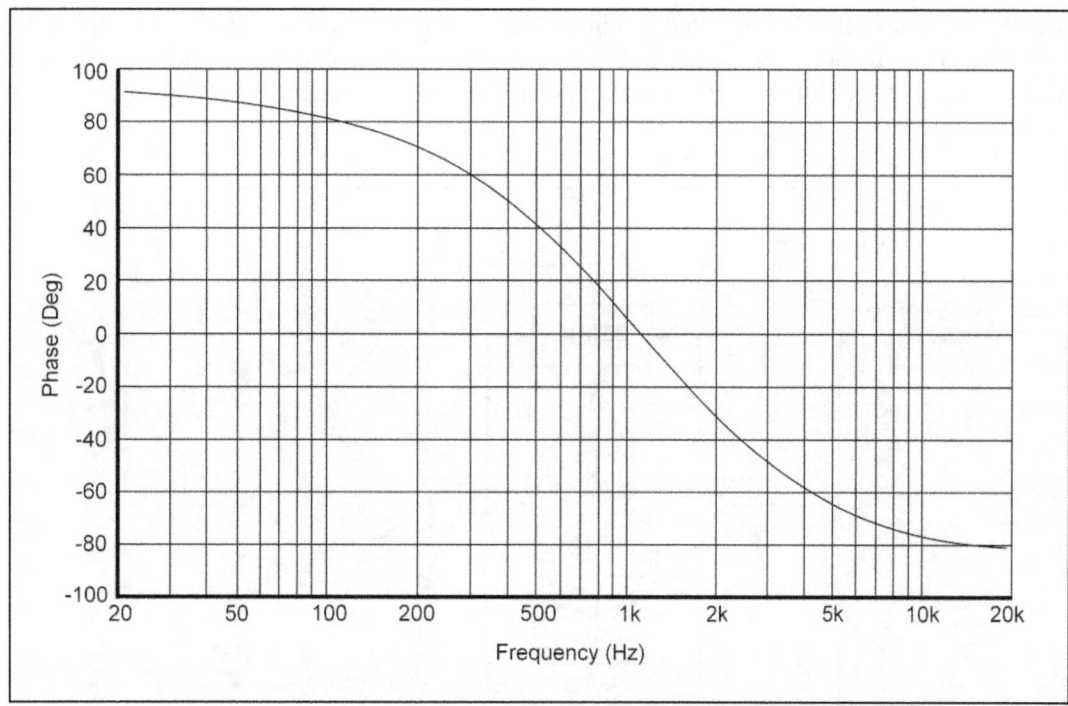

Figure 8-6   Phase Response of Three Filters Connected in Parallel

8.3 Graphic Equalizers

Graphic equalizers are made by connecting a set of peaking filters or band-pass filters in parallel, as shown in the overview schematic of Figure 8-7. Recall from Section 4.7.3 that two important attributes of a band-pass filter are its center frequency and bandwidth. In a graphic equalizer, the center frequencies of the filters are set by the International Organization for Standardization (ISO) and cannot be adjusted by the user. Additionally, the bandwidth of the filters cannot be adjusted by the user. The functionality built around each filter allows it to be used to boost or cut the audio signal level. The typical response of a single filter is shown in Figure 8-2. The audio band of 20 Hz to 20 kHz forms a range of nearly ten octaves. In the most common graphic equalizers, this range will be divided into 31 bands of 1/3 octave each.

When the equalizer uses peaking filters, the filters could be connected in series or parallel. Generally, the filters of the equalizer are connected in parallel. As we saw in the previous section, connecting the filters in parallel gives better overall phase response as compared to connecting them in series. If the equalizer uses band-pass filters, though, the filters must be connected in parallel in order to be able to pass the entire audio spectrum. If they were connected in series, the first filter would pass only its passband, which would then be blocked by subsequent filters in the chain. Furthermore, filters down the chain would have no signal to process, since their passband will have been blocked by earlier filters.

By adjusting the gain of each filter, the center frequency (and others close about the center frequency) can be boosted or cut. The gain is adjusted by a slider control on the front of the equipment, usually over a range of 6 or 12 dB. Note that the actual control of frequency response, even with a 31-band equalizer is quite imprecise. The filters actually have a fairly wide bandwidth at moderate boost/cut levels—more like an octave rather than 1/3 octave. The bandwidth is only 1/3 octave at maximum boost or cut. This causes a particular control to affect many frequencies on each side of its center frequency. This is illustrated in Figure 8-8, where it can be seen that at a boost of 4 dB, the filter has a bandwidth of two octaves. Figure 8-9 shows the graphic nature of the graphic equalizer. By looking at the sliders, a very rough view of the overall frequency response of the unit can be seen.

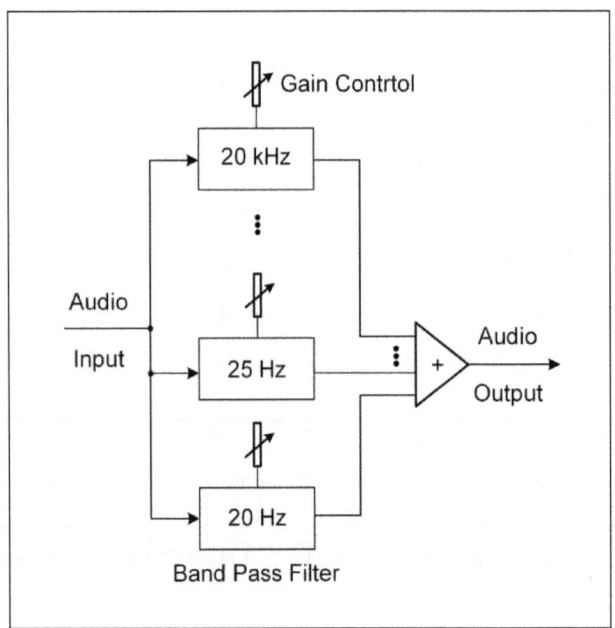

Figure 8-7 Graphic Equalizer Schematic Overview

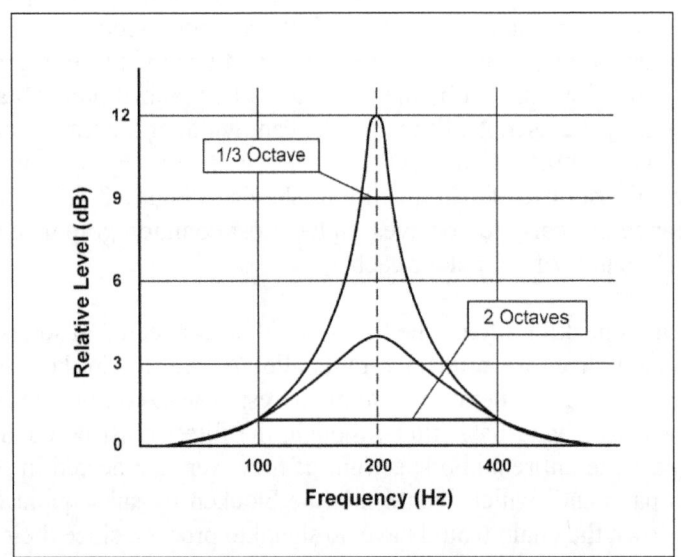

Figure 8-8 Bandwidth of Filter Varies with Boost

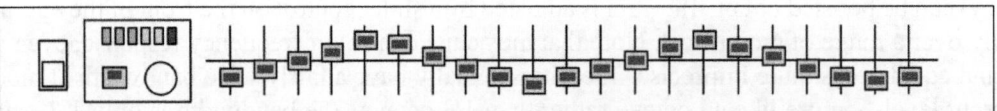

Figure 8-9 Graphic Equalizer Sliders Show Frequency Response

8.4   Constant-Q Equalizers

The Quality Factor, or Q, of a band-pass filter is closely related to its bandwidth. Recall from Section 4.7.3 that Q is defined as the center frequency divided by the bandwidth of the filter. Therefore, for a given frequency, the narrower the bandwidth of the filter, the bigger the Q. High Q filters affect a smaller range of frequencies.

As we saw above with a conventional graphic equalizer, the bandwidth of any particular filter associated with a given frequency band changes as the slider is moved. At less than full boost or cut, the bandwidth can be over an octave wide. It is therefore not possible to have very accurate and subtle adjustment of the frequency response of the sound system using a conventional graphic equalizer. For adjusting the overall tone of the sound system an equalizer with wide bandwidth filters is fine. However, for countering problems with room resonance and other system anomalies, narrow bandwidth filters are really required.

In a Constant-Q graphic equalizer, the bandwidth of the filter associated with a frequency band does not change when the slider is moved. For a 31-band equalizer, the bandwidth remains 1/3 octave throughout the entire boot/cut range. This means that the Q of the filter also remains unchanged—thus the term, *constant-Q*. Compare Figure 8-8 with Figure 8-10. In each case, the filters are shown with gains of 12 dB and 4 dB. The effect of the constant-Q filter is clearly seen. An equalizer that utilizes constant-Q filters can achieve more accurate and subtle adjustments of the frequency response of the system. This type of equalizer was made popular by Rane Corporation in the mid 1980's.

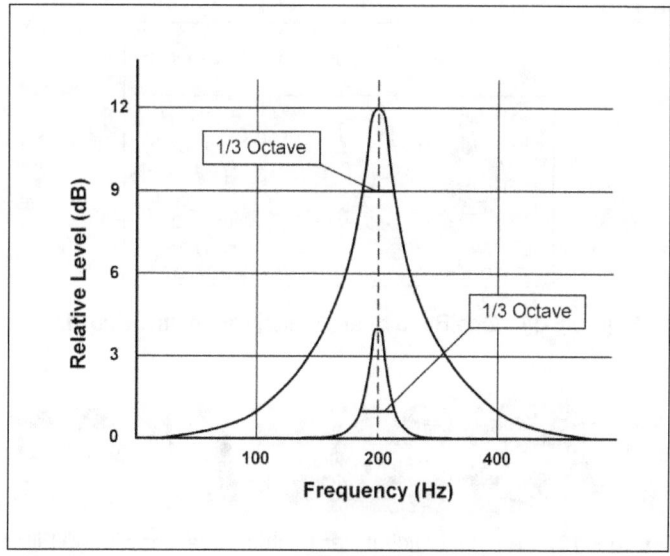

Figure 8-10   Bandwidth is Unchanged with Constant-Q Filters

8.5   Parametric Equalizers

In a graphic equalizer, only the gain of the filter is adjustable via the boost/cut slider control. The center frequency and bandwidth are not adjustable by the user. With a parametric equalizer, all three parameters—center frequency, gain, and bandwidth—of the filter are adjustable by the user. This gives the users the greatest flexibility in precisely controlling the frequency response of the system. However, this control comes at higher complexity and cost.

The better parametric equalizers use band-pass filters based on a state-variable design. In a state-variable filter, all three parameters mentioned above are separately and independently adjustable. The drawback of this design is that it uses more components and is thus more expensive to realize than other designs. In other filter designs, the center frequency is usually independently adjustable. However, the gain and bandwidth may not be independently adjustable of each other. The work-around for this is usually to have the gain function implemented separately from the filter section (i.e., the actual gain of the filter is fixed) and then use feedback to control overall gain of the filter block.

Parametric equalizers are very useful in countering feedback. In this application, a very narrow bandwidth notch filter is set on the specific problematic frequency without affecting surrounding frequencies. This type of precise control is not possible with graphic equalizes which, because of the wide bandwidths of their filters, also cut out some of the sound that the user wants to let through. Other very useful applications are removing 60 Hz hum, or helping a specific instrument cut through a mix.

Analog parametric equalizers are limited in the number of bands that they can support in a single unit because each band must have the full complement of controls, and each control takes up a fair amount of space. A typical band configuration is shown in Figure 8-11, and a typical unit may have five bands. On the other hand, digital parametric equalizers can have many more bands because all the filters share the same set of front-panel controls. The number of filters is limited only by the circuit board space and the computing power of the unit. For example, Figure 8-12 shows a Behringer FBQ2496 which has 40 independent filters (20 per channel).

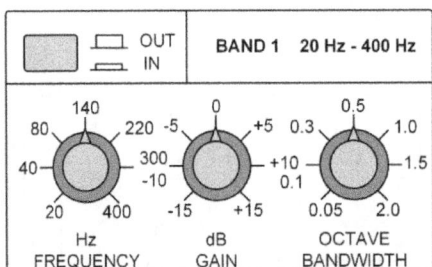

Figure 8-11 One Band of an Analog Parametric Equalizer

Figure 8-12 Behringer Digital Parametric Equalizer with 40 Filters

8.6 Compressors and Limiters

The dynamic range of a music program signal is the difference in level between the loudest and softest sounds in that program. Compression is the reduction in dynamic range of a signal. A compressor can be thought of as an automatic volume control. It reduces the level of a signal only when that level crosses a threshold set by the user. When the signal level is below the threshold, the gain of the compressor is unity, and the signal level is unchanged.

Limiting is the same as compression, but with a very high compression ratio, usually over 10:1. The time it takes the compressor to respond to the input signal—referred to as attack time—is also very short in a limiter application. This is used when we do not want the signal to go above a certain level. Some compressors have a separate limiter section built in to the unit, where the only parameter that the user has to (or is able to) set is the threshold.

8.6.1   System Overview

Figure 8-13 shows a block diagram of a compressor based on a feedforward design. There are three key building blocks in the heart of the compressor. The first is the voltage controlled amplifier, or VCA. The VCA is special because its gain can be set and adjusted by an external control voltage. This control voltage is provided initially by the second key block of the compressor—the above-threshold level detector. The above-threshold level detector's output is derived from the level of the input signal and a threshold setting determined by the user. This voltage is passed to the third block—the compression ratio control—which scales the voltage and uses it to set the gain of the VCA. These blocks form the actual compression stage of the compressor. This structure is called feedforward because we are measuring the input signal and deriving a control signal that is applied in the same direction as the input signal flow through the compressor to control the gain of the VCA. Other designs, based on a feedback model, may measure the output signal of the VCA and derive a control signal that is sent to the above-threshold level detector and ultimately back to the VCA to control its gain.

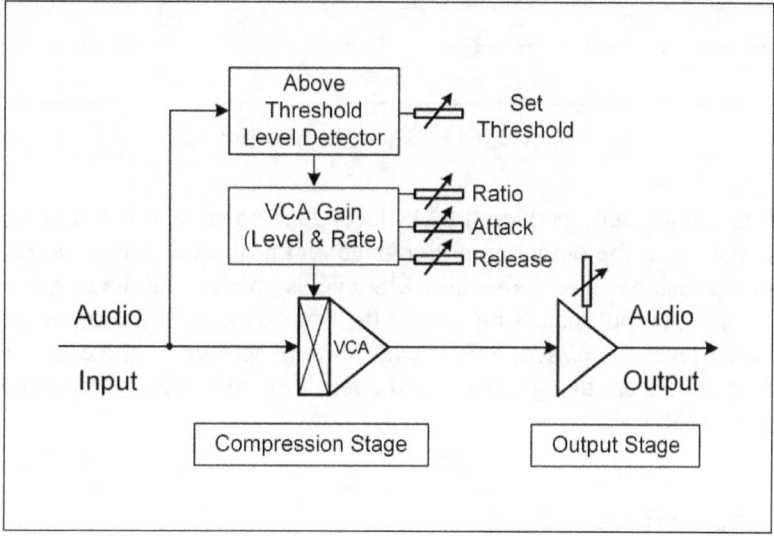

Figure 8-13   Overview Schematic of Compressor

After a signal is compressed, its overall loudness is less than before compression. Manufacturers therefore incorporate some means of increasing the output level of the signal to where it was before compression, e.g., they may add a small amplifier to the back end of the compression stage. The result is a signal that sounds as loud as, or louder than, it did before compression, but with reduced dynamic range.

8.6.2   Signal Processing

In Figure 8-14, we see a few milliseconds of a signal being compressed. This diagram is divided into three parts. On the left is the input signal before it is compressed. In the center is the signal just after it is compressed, while on the right is the compressed signal after its level is boosted for output from the compressor. The compression for this signal is quite high. In the center diagram, notice that the signal level is reduced only when it crosses the threshold. Notice that even though the level is reduced, the peaks are still somewhat above the threshold. As we will see below, this is because of the compression ratio and attack time settings of the compressor. Also, do not confuse the threshold level with the peak level of the output signal shown at the right of the figure. The level of the output signal does not depend on the threshold, but rather on how much it is boosted after compression.

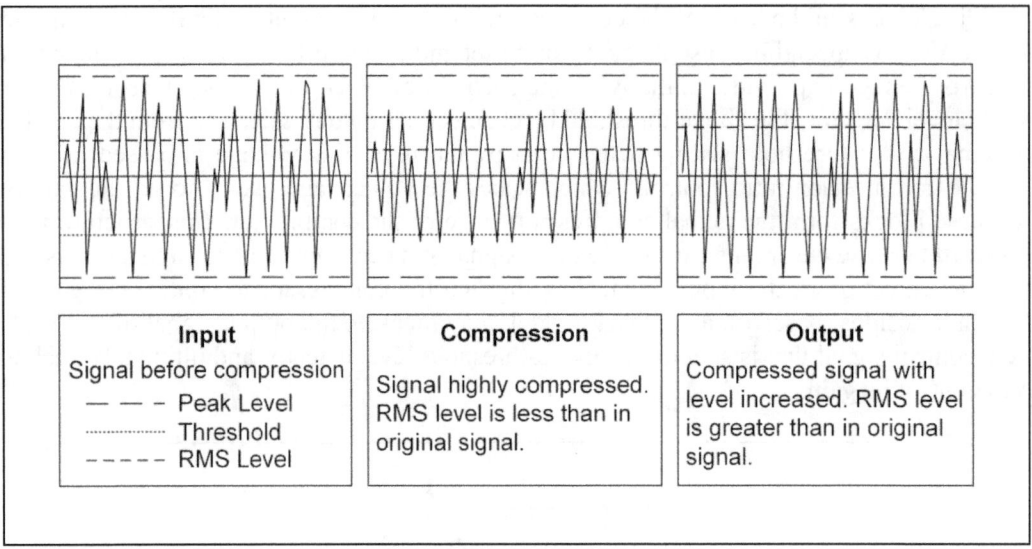

Figure 8-14 Signal Compression

When the signal is compressed, its overall RMS level goes down, and it sounds softer. The signal level is brought back up in the output amplifier stage, but now since the dynamic range has been reduced by the compression stage, its overall RMS level is greater, and the signal sounds generally louder. Note also in the output that, compared to the original signal, the low-level sections of the signal have had their levels increased. This is why some people say that a compressor makes soft signals louder. As can be seen, though, this is just a result of amplifying the compressed signal and not the compression itself.

8.6.3 Compression Ratio

The compression ratio is one of the key parameters that can be set for the compression process. You are telling the compressor by how much the input signal has to increase (in dB) in order to cause the output signal to increase by 1 dB. This only applies after the input signal crosses the threshold, of course. There is no compression below the threshold. For example, if the compression ratio is set to 2:1, this means that the input signal would have to increase by 2 dB beyond the threshold in order for the output signal to increase by 1 dB. This is shown graphically in Figure 8-15. In this example, the threshold is set at 3 dB. We can see that for the case of 2:1 compression, when the input signal level increases from 3 to 5 dB, the output level only increases from 3 to 4 dB. A ratio of 1:1 means that no compression will take place, no matter what the threshold is set to. The maximum setting of ∞:1 (infinity to 1) means that the output signal will not go beyond the threshold, no matter what the input signal level. This is sometimes referred to as brick wall limiting.

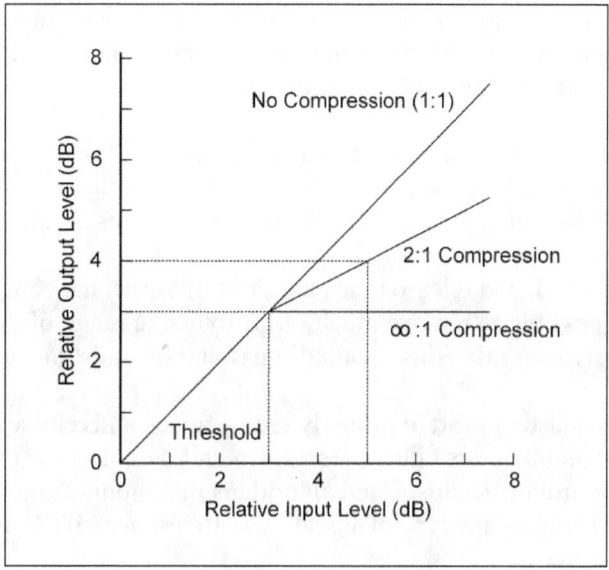

Figure 8-15 Compression Ratio

8.6.4 Attack and Release Times

Depending on its mode of operation, the compressor will take more or less time to reduce or restore gain in response to the changes in level of the input signal. The attack time is the amount of time it takes after the input signal crosses the threshold for the compressor to reduce the gain to the amount dictated by the compression ratio. When the input signal is no longer above the threshold, the compressor will restore the gain to unity (i.e., no compression). The time it takes to do this is called the release time. In Figure 8-16 we can see that this process can take a few cycles of a waveform to complete. Because the response is not instantaneous, the output signal overshoots the gain reduction target on the start of compression, and even after the signal falls below the threshold, the output signal level is still reduced for a number of cycles. The attack time is usually much shorter than the release time. Attack time is on the order of 0.1 to 200 ms, while release time is on the order of 50 ms to 3 seconds.

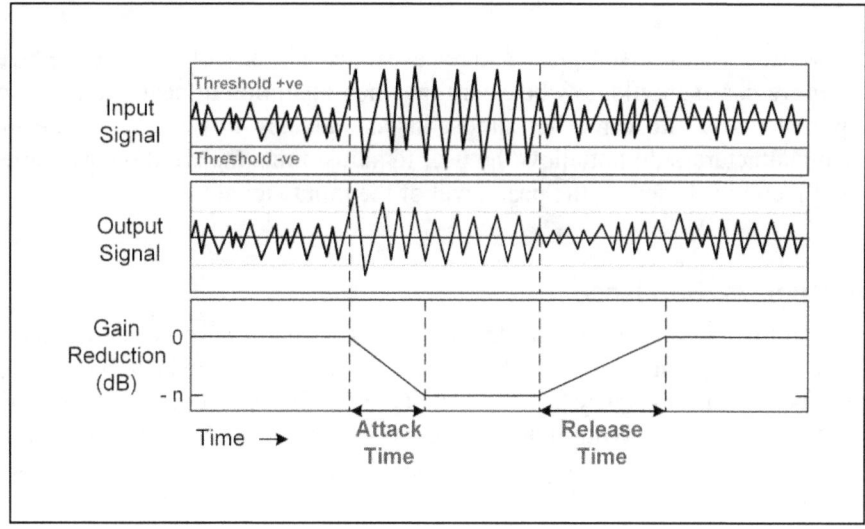

Figure 8-16 Attack and Release Times

When the release time is very short and the compression ratio is high, the background noise of the system may be heard quickly rising as the compressor restores gain on soft passages and then falls as it compresses loud passages. This is called breathing.

When the release time is long and the compression ratio is high, it takes a long time to restore unity gain after the input signal drops below the threshold. This means that soft signals may be lost and you may actually hear the level of the signal rising and falling. This is called pumping.

Now suppose that both attack and release times are short. If we set the compression ratio high and the threshold low, it is possible to remove much of the dynamic range of the music program. This results in a monotonously loud mix. This is called squashing or squishing.

It takes much practice to master the art of properly setting attack and release times on a compressor. For this reason, some manufacturers either preset them and do not allow the user to change them (e.g., Rane DC 22S), or provide sophisticated algorithms that monitor the input signal and set the attack and release time based on the type of signal (e.g., Behringer MDX1600).

8.6.5 Level Detection

The compressor needs to follow the changing level of the input signal to determine if it has crossed the threshold. This is the function of the block labeled *Above Threshold Level Detector* in Figure 8-13. There are actually two separate functions in this block. The first is the actual signal level detector and the second is the threshold crossing detector. The signal level detector sends its output to the threshold crossing detector. The level detector can typically be set to function in one of two modes—*viz.* peak or RMS.

In peak mode the output of the level detector follows the instantaneous changes of the input signal—sort of like riding the waveform of the signal. In this case, the threshold detector knows instantly when the input signal crosses the threshold. In RMS mode, the output of the level detector follows the RMS level of the input signal averaged over a short period of time (see Section 4.11.7 for a definition of RMS). This means that the threshold detector does not follow the instantaneous changes in the input signal, and high-level transients and short burst in the input signal will get through the compressor without being compressed. This can help preserve some of the dynamic range of the program material.

In the previous section we discussed attack and release times without reference to the level detection mode of the compressor. It should now be clear that when operating in peak mode, the compressor tracks the input precisely and can react almost instantaneously to changes in the input signal. Indeed, many manufacturers do not allow the user to adjust the attack and release times when the level detection function is based on the peak level of the input signal.

8.6.6 Hard Knee vs. Soft Knee

If we look again at Figure 8-15, we can see that compression starts as soon as the signal crosses the threshold. This mode of operation is called hard knee compression—the knee being the point in the graph where it bends sharply away from the unity gain line and compression begins. This is highlighted in Figure 8-17(A). With high compression ratios, this can cause an abrupt change in level of the processed signal. This change may be annoyingly audible in some program material. It is possible to achieve more subtle compression results, even with high compression ratios, by employing a more gradual onset of compression. In this case, the area around the knee in the compression graph is no longer a sharp point, but a smooth curve, as shown in Figure 8-17(B). This is commonly called soft knee compression and is useful for vocal compression, especially with high ratios.

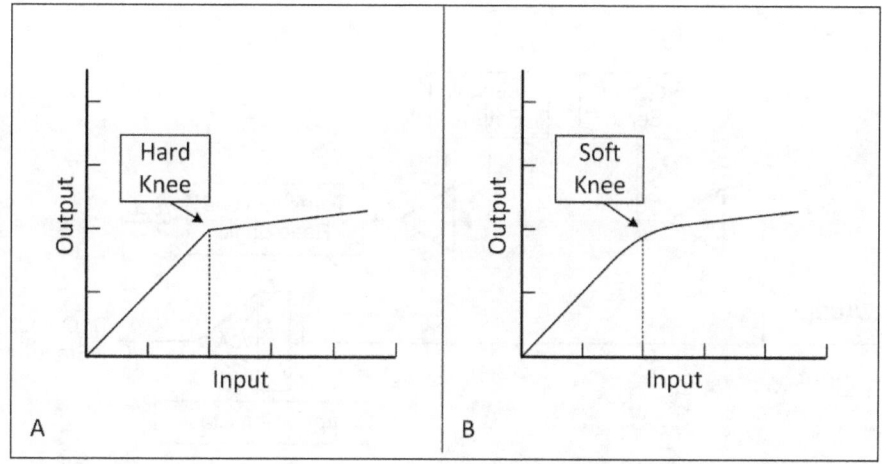

Figure 8-17   Hard Knee vs. Soft Knee Compression Characteristic

8.6.7   Sidechain

In our discussion of compressors until now, the gain of the VCA has been derived from the input signal via the signal level detector and the above threshold detector. However, compressors generally provide the user another way to vary the gain of the VCA. This is via the sidechain return of the unit and is shown in Figure 8-18. The sidechain loop is formed by tapping a signal off the audio input and providing it outside of the unit via the sidechain send or sidechain output. The sidechain return or sidechain input allows an external signal to be fed to the gain reduction circuitry of the compressor. Thus, sidechaining allows us to use a signal that is totally unrelated to the main audio input to control the compression of the input signal. This is a very useful feature, and examples of its use are easy to spot. For example, DJs use it to perform ducking. In this case, they patch a copy of their microphone signal into the sidechain return (since the sidechain input is line level, the microphone signal must be amplified before being connected to the sidechain), while the FOH signal goes through the compressor input. Now when the DJ speaks, the FOH signal will be compressed, and the louder he speaks, the more the FOH signal is compressed.

Another more interesting use of sidechaining is to allow an instrument to cut through a mix. For example, in a mix, the bass guitar and the bass drum often compete for the same space. If the engineer wants the thwack of the bass drum to punch through the mix, he may put the bass guitar through the main input of the compressor and feed the bass drum signal into the sidechain return. Now when the bass drum is kicked, the bass guitar is compressed, and the bass drum gets to the front.

The sidechain send signal can be used with the sidechain return to alter the compression characteristics when we want to use the input signal to drive the compression. One such application is de-essing. When a performer speaks into the microphone and articulates words starting with **SH** or **TH**, or having an **S** sound somewhere in the word, there is sometimes excessive hissing sound produced. This hissing sound is called sibilance and can be fairly annoying. A technique for removing this sibilance is to compress the audio signal when excessive high-frequency energy is detected in the signal. This can be achieved by inserting an equalizer in the sidechain path of the compressor—the sidechain send is connected to the equalizer input, while the output from the equalizer is connected to the sidechain return. The equalizer controls are then set to boost the high frequencies while cutting the low frequencies.

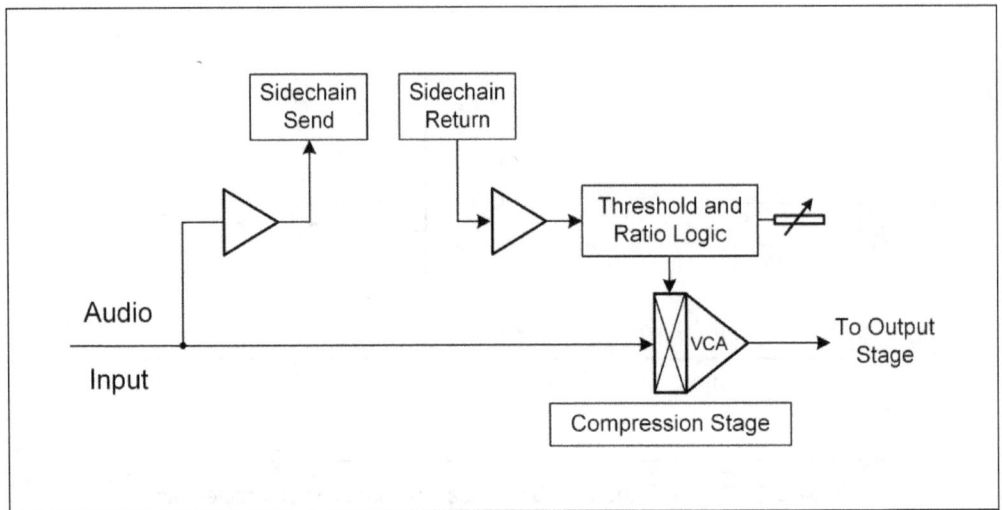

Figure 8-18 Compressor Sidechain Send and Return

8.6.8 Multiband Compression

Multiband compressors are not usually seen in a live sound reinforcement environment, probably because of their cost and because they require a fair bit of tweaking to set up properly. Their home is typically the recording and mastering studios. However, if you are compressing the final FOH mix, one of these compressors can be very useful. Recall from our discussion on active crossovers in Section 4.7.7 that the bass frequencies carry nearly half of the total energy of typical pop music program. That means that the bass end dominates the compression results when using a full range compressor—the type we have been discussing up to now. When a bass note comes through, the entire mix is compressed, and some mid and high-frequency content may be lost. Multiband compressors allow you to get around this problem by dividing the input signal into separate frequency bands—much like how we use an active crossover. Three bands are common in typical units. Each band is then compressed separately and finally summed before being output from the device. Now when a high-level bass note comes through, only the low end is substantially compressed, and the mid to upper regions of the mix are barely touched. On better units, the crossover points are fully adjustable by the user, and all the usual compression parameters—threshold, ratio, attack, release—are separately applicable to each band. Lower end units, though, force the user to pick from a set of presets that are mapped to a particular musical style or instrument. In this case, the crossover points are fixed and the compression parameters, though adjustable, cannot be applied to each band separately. Here again, preset algorithms determine the distribution and nature of compression among the bands when the user makes an adjustment to any compression parameter.

8.7 Expanders and Noise Gates

Expanders function in an opposite fashion to compressors. They are used to increase the dynamic range of a signal. Expanders work by reducing the level of the input signal when that signal falls below a threshold that is set by the user. When the signal level is above the threshold, the gain of the unit is unity, and the signal level is unchanged. Thus, whereas a compressor makes loud signals softer, an expander makes soft signals even softer.

Noise gating is the same as expansion, but with a very high expansion ratio, usually over 10:1 (and usually used in conjunction with a low threshold). When used as a noise gate, the unit will completely attenuate the input signal, producing silence (or rather, residual gate noise) as the output when the input signal level falls below the threshold. The action of a noise gate is very abrupt and the effect on the output signal is very obvious, so care must be taken when using noise gates.

Because the circuitry and function of an expander is so close to that of a compressor (both are VCAs, both reduce the level of the input signal, only the gain reduction logic is different), they are usually packaged in the same unit.

The photo in Figure 8-19 shows a one channel of a Behringer MDX2200 compressor, which also has an expander/gate section to the extreme left of the unit.

Figure 8-19   One Channel of the MDX2200 Compressor/Limiter/Expander/Gate

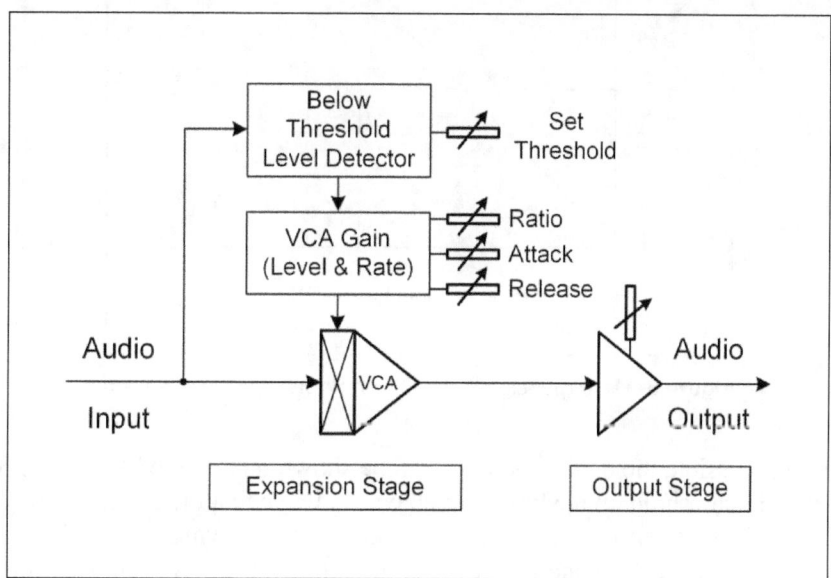

Figure 8-20   Overview Schematic of Expander

8.7.1   System Overview

Figure 8-20 shows the block schematic of an expander. If you compare this diagram with Figure 8-13 you will notice that they are almost identical. In fact, the only real difference is that the *Above*

Threshold Level Detector of the compressor has been replaced by a *Below Threshold Level Detector* in the expander. Similar to compressors, expanders use a voltage controlled amplifier, the gain of which is derived from the input signal via the level detector. The user sets the level below which expansion will begin via the threshold control on the unit.

As with compressors, manufacturers may add a small amplifier to the back end of the expansion stage so that the user can adjust the output level of the signal. In fact, in many combo units, the expansion stage and the compression stage share the same back-end amplifier.

8.7.2   Expansion Ratio

The expansion ratio is the reduction in output signal relative to the reduction in input signal. For example, an expansion ration of 2:1 means that the output will be reduced by 2 dB for every 1 dB decrease in level of the input signal. This reduction only happens when the input signal level is below the threshold. When the input signal level is above the threshold, the gain of the VCA is unity, so its output level matches the input level. (Note that some manufacturers express the expansion ratio as 1:X, where *X* is some number. This means that for every 1 dB reduction in input signal level, the output is reduced *X* dB).

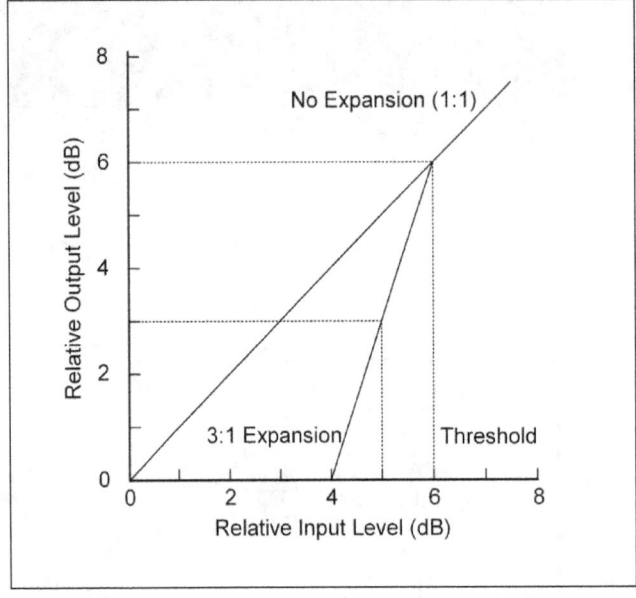

Figure 8-21   Expansion Ratio Graph Showing 3:1 Expansion

As we did for compression, the expansion ratio can be shown on a graph like Figure 8-21. To read the graph, start at the threshold on the Input Level axis and note that the corresponding output level is the same. Now if we go left on the Input Level axis—thus showing a reduction in input signal level—we can see that the corresponding reduction in output signal level on the Output Level axis is greater. With the 3:1 expansion line on the graph, we can clearly see that the reduction in output level is three times the reduction in input level.

8.7.3   Other Expander Parameters

As with compressors, expanders have user adjustable parameters for attack and release times. Additionally, hard knee and soft knee threshold crossing may be selectable. For both compressors and expanders, there is a trend among many manufacturers to automate much of the parameter setting process by implementing sophisticated algorithms that analyze the program material and adjust critical parameter values on the fly. This mode is usually user selectable by the press of a button on the unit.

Expanders, along with compressors, are used in fixed mode and built in to some systems without the user even being aware of their presence. An example is an FM wireless microphone transmitter/receiver package. The signal is compressed and amplified in the transmitter to raise the low-level signals high above any noise that is picked up during transmission. In the receiver, the signal is expanded back to the original dynamic range, lowering the noise level. This process is called companding.

8.8 Digital Delays

A delay processor simply makes a copy of the input signal and sends it to the output after a user adjustable delay interval. Practically every delay available today uses low cost digital technology. While there are some standalone units available, most delay processors come packaged in a multi-effects unit, such as the Behringer DSP2024 Virtualizer Pro shown in Figure 8-22. The delay time is usually adjustable from one millisecond up to several seconds.

Figure 8-22   Digital Multi-Effects Processor with Delay

Besides the delay time, the delay unit may also have other controls and features, such as feedback, multiple delay engines and taps, and stereo mixing and routing. Delay engines are also at the heart of many other effects processors, such as chorus, reverb and flanging.

Note that all digital (as opposed to analog) processors of any kind have some inherent delay, referred to as latency, because digital processing is a serial process of several discrete steps where each step takes a small amount of time. This latency is a separate issue and should not be confused with the deliberate use of delay processors. Latency in a digital processor is an undesirable attribute, and manufacturers work to reduce it to as small a value as possible. In your typical digital effects processor, a latency value of around 2 to 3 ms is normal.

8.8.1   System Overview – Delay Engine

In a simple digital delay line, the delay engine makes a copy of the input signal and stores it for a short time (the delay time) before sending it to the output. The core of a digital delay is a very basic store and forward circuit shown in Figure 8-23. The major components are the ADC which takes the analog audio and produces digital samples; the buffer, which is simply a block of digital memory where the actual delay happens; and the DAC, which converts the digital samples back into an analog audio signal. For clarity, the buffer in Figure 8-23 is shown to be only 9 samples long and is shown containing six samples in positions 2 to 7. In reality, this buffer may be over 3 million samples long. The concepts surrounding the operation of the ADC, DAC, and sampling were discussed in Chapter 7.

Let us look a bit more closely at how the delay processor works. Every sampling interval the ADC produces a digitized sample which is stored in the buffer. Exactly where it is stored is determined by a pointer to the next free memory location (NFL). After storing the sample, the NFL pointer is advanced to the next memory location. Additionally, every sampling interval, the DAC takes a sample from the buffer and converts it to analog to be output from the processor. Exactly where it is taken from is determined by the next output sample (NOS) pointer. After reading the sample, the NOS pointer is advanced to the next memory location. It is the 'distance'—called the offset—between these two pointers that determines the delay. This offset is calculated when the user adjusts the delay time. Figure 8-24 shows how the pointers advance from one sampling interval to the next.

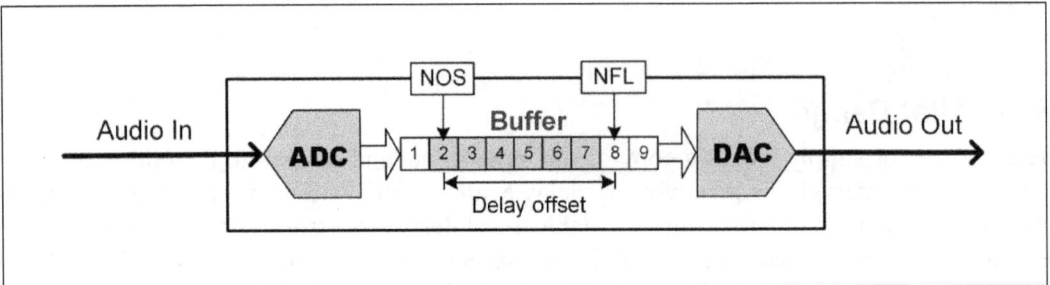

Figure 8-23   Digital Delay Line with 9-Sample Buffer

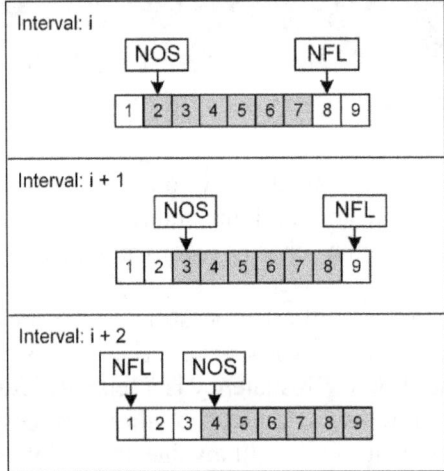

Figure 8-24   Delay Pointers Move Each Sampling Interval

While the user sets the delay time in milliseconds or seconds, all the action at the core of the processor happens in terms of the sampling interval. If the delay unit uses a 96 kHz processor, each sampling interval will be approximately 0.0104 ms long. That means that a 100 ms delay, for example, is equivalent to 9600 sampling intervals. Remember, though, that the NOS and NFL pointers are advanced once per sampling interval. If we wanted to output a sample after a 100 ms delay, all we would need to do is set the NOS pointer 9600 samples after the NFL pointer. That way, any sample that is written at the NFL pointer location will be output by the DAC 9600 sampling intervals, or 100 ms, later. That is how the pointer offset is calculated. While Figure 8-23 shows the buffer as a linear block, it is logically implemented in a circular fashion. When a pointer gets to the last location of the buffer, it simply loops around to point to the first location, as shown in Figure 8-24, for sampling interval *i+2*.

It can be seen that using this approach, the maximum delay time available depends strictly on buffer size. For a 96 kHz processor, a 30 second delay would require a buffer size of 2.88 million samples. With a word size of 24 bits, that would require just over 8.6 MB of buffer memory—not a lot by today's standards.

8.8.2   System Overview – Delay Processor

If we add a few simple components around the delay engine described above, we arrive at the basic delay processor shown in Figure 8-25. Here we have simply mixed a variable amount of the delayed signal with the original undelayed signal at the output of the processor. For delay times of less than 60 ms or so, the effect is a widening, or fattening, or creation of lushness to the sound. Above 100 ms, a distinct echo can be heard.

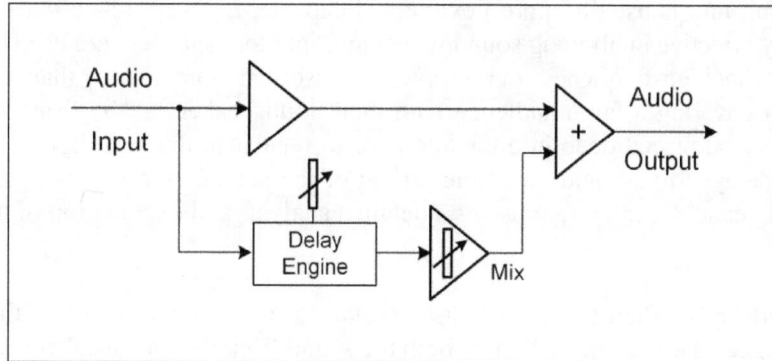

Figure 8-25   Basic Delay Processor

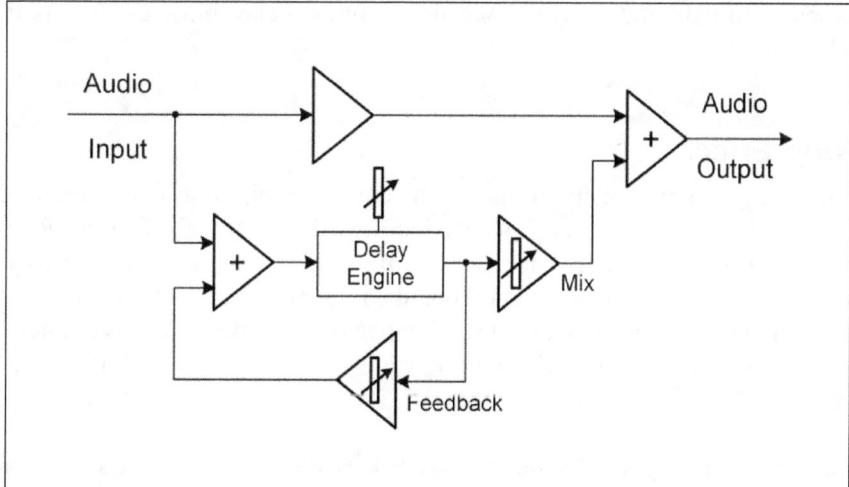

Figure 8-26   Delay Processor with Feedback

To get more utility from the delay unit, a variable feedback loop is added to the delay line, whereby some of the delayed output is mixed with the input signal and fed back through the delay engine, as shown in Figure 8-26. This can create greater ambience and a reverb like effect.

You will also find other delay processor configurations, some of which are more suited to the studio rather than live applications. For example, there is the multi-tap delay, or tapped delay line, which consists of a series of delay engines, where the output of one delay engine is connected to the input of the next. The output signal is tapped from the output of each delay and summed to give the final output. This type of delay can create many interesting rhythmic effects and has many applications in other processors. Then there is the cross-channel delay configuration, where the output of the delay engine on each channel is fed back to the input of the other channel. This can give interesting ping-pong effects. To get just the effect that is desired, these processors require a fair bit of tweaking to setup. They therefore have limited use in a live environment, where a lot of things happen on the fly, and there is maybe just one chance to get it right.

8.8.3 Delay Usage

In a live sound reinforcement environment, delay units are best used in their simple forms, i.e., in a real delay function, and not to produce complex effects, where a dedicated processor such as a chorus or flanging unit is usually more flexible and capable. In its role as a pure delay unit, the processor is very effective in aligning sound where multiple loudspeakers are used to cover a large audience. Recall that high-frequency sound waves are absorbed more quickly than low frequencies. Therefore, when covering a large audience with main loudspeakers at the front, high-frequency speaker units may be placed deep into the audience to rebalance the sound. This high-frequency signal must be delayed to coincide with the arrival of the sound from the main loudspeakers to prevent annoying echo. Chapter 13 presents a detailed analysis and explanation of how this can be implemented.

Another use of delay is when the main house speakers are well forward (more than 5 m) of the performers on stage. The audience will hear both the sound from the loudspeakers, and then a short time later, the direct (i.e., non-reinforced) sound from the performers on stage. In this case, the entire house mix can be delayed to sync with direct stage sound.

Of course, when used deliberately with long delay times, delay units can be used to produce interesting echo effects.

8.9 Phasing Effect

Many effects use the interaction of audio signals that have been phase shifted to produce interesting sounds. Phase shifting, or phasing, is probably the most fundamental of these. Phasing may be applied to individual instruments or voices in the mix to give them a kind of spacey, swooshing effect. It is a popular effect for guitars and keyboards. In Section 1.13 we introduced the notion of relative phase between waves and discussed what happens when identical waves interact with each other. Before we go further, we will quickly review this notion of relative phase and interaction of phase shifted audio.

Figure 8-27 shows a single cycle of a sine wave used for illustration. It starts out at an amplitude of zero, rises to a positive maximum level, falls back to zero, continues falling to a negative minimum, and then rises back to zero amplitude. The sequence we just described constitutes a single *cycle* of this sine wave. The sine wave is periodic, i.e., the cycle noted above repeats itself over and over for as long as there is energy to sustain the wave. The time it takes for the wave to complete one cycle is referred to as the *period* of the wave, while the number of cycles of this wave that occurs in one second is called the *frequency* of the wave. The period and the frequency of a periodic waveform are related by the simple equation:

$$Period \ = \ \frac{1}{Frequency}$$

Notice also, that in Figure 8-27 we have taken the single cycle of this wave and fit it to 360°. This will make it more convenient for us when we talk about relative phase. It does not matter what the frequency of the wave, one cycle is always made to fit into 360°. Take a look at the figure and note where the 90°, 180°, and 360° positions fall. Notice too, that 360° and 0° actually represent the same position along the waveform.

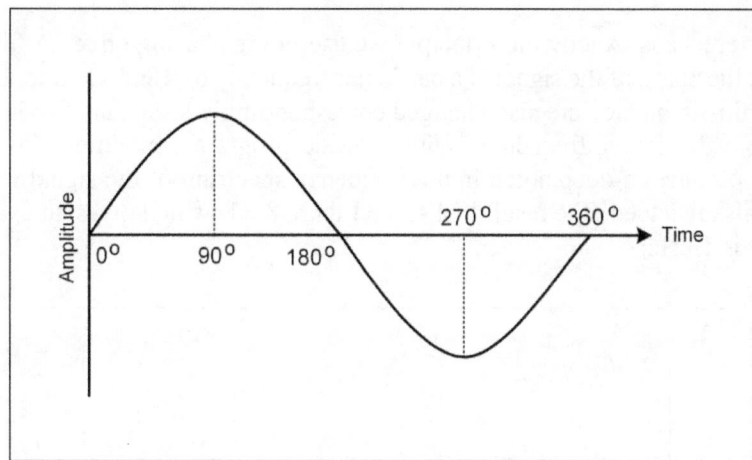

Figure 8-27   One Cycle of a Sine Wave fitted to 360°

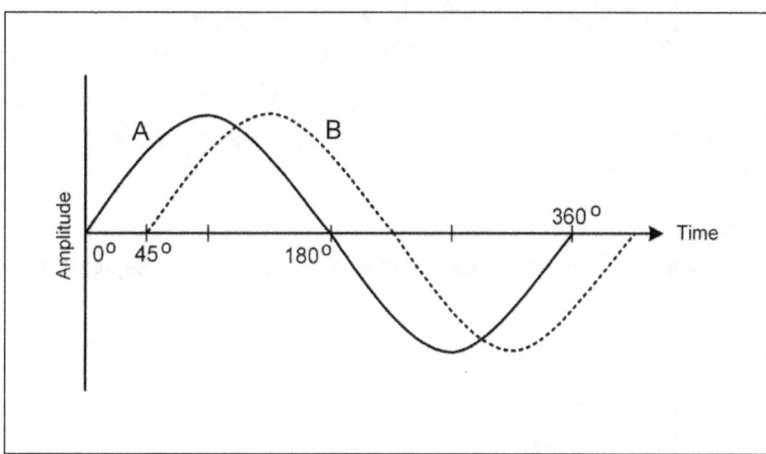

Figure 8-28   Sine Waves 45° Out of Phase

Now if we want to talk about relative phase, we need to compare two waves. In Figure 8-28 we see two identical sine waves, *A* and *B*, and we see that wave *B* starts a little bit later in time than *A*. In fact, using the notion of fitting a cycle into 360°, we can say that wave *B* lags *A* by 45° (which is the same as saying that *A* leads *B* by 45°). We can also say that relative to wave *A*, wave *B* is 45° out of phase, or simply that these two waves are 45° out of phase. The interesting thing for us is what happens when these two waves combine or interact. For the case where the waves are 45° out of phase, the end result is shown in Figure 8-29. Here we see that adding waves *A* and *B* gives us wave *C*, which is bigger than either wave *A* or *B* alone. This is sometimes called constructive interference. In fact, when adding two equal amplitude sine waves, if the relative phase is between

0° and 119°, or 241° and 360°, the resultant wave will be bigger than either one. At 120° or 240°, the resultant wave will be the same amplitude.

Let us suppose now that waves *A* and *B* were 180° out of phase. The result of adding these waves would be total cancellation as shown in Figure 8-30, because for every point on wave *A*, the corresponding point on wave *B* occurring at that exact time would be its exact opposite value. Thus, if a point on wave *A* were +0.5 V, the corresponding point on wave *B* would be −0.5 V. When adding equal amplitude sine waves, phase differences between 121° and 239° result in a smaller wave than either one. This called destructive interference.

Destructive interference is exactly the principle we use in the phasing effect. A filter combination is used to change the phase of the signal at a particular frequency by 180°. Components of the signal on either side of this frequency are also changed correspondingly less than 180° in phase as we get further away from the center frequency. With a music program containing a wide spectrum of frequencies, this produces a deep notch in the frequency spectrum of the signal when the original and filtered signals are added. The result looks like Figure 8-31, which gives an example where the center frequency is 1 kHz.

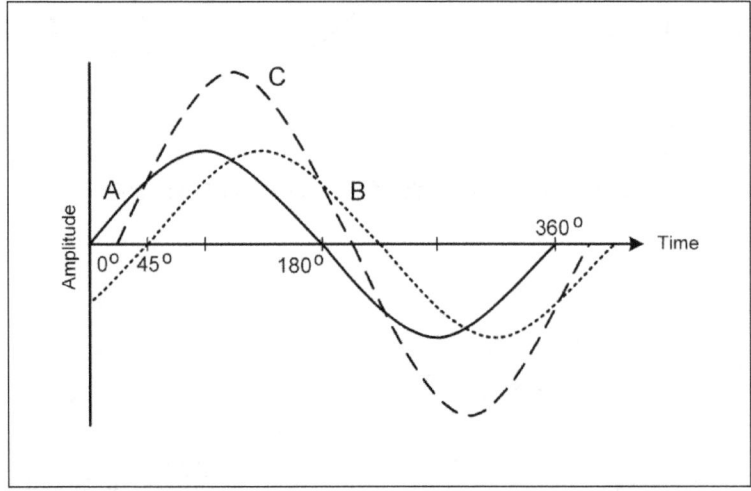

Figure 8-29   Sine Waves 45° Out of Phase Add to Give a Bigger Wave

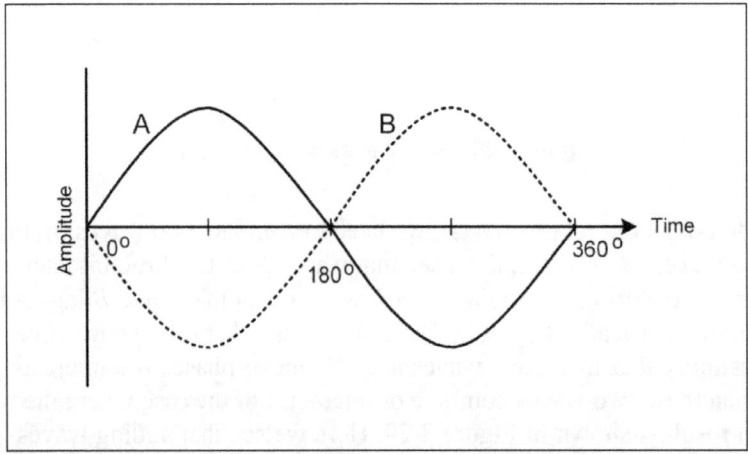

Figure 8-30   Sine Waves 180° Out of Phase Cancel Each Other When Added

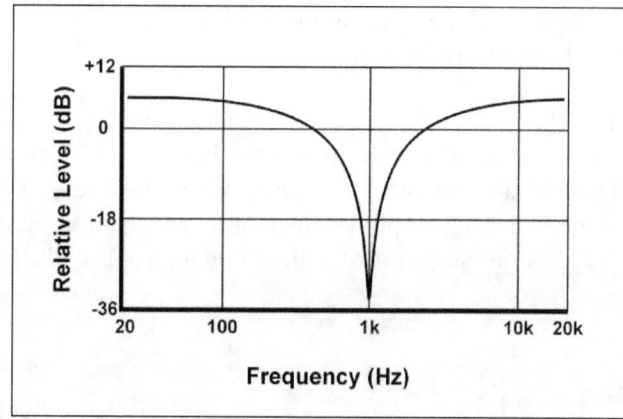

Figure 8-31   Notch in Frequency Spectrum Resulting from Adding Signals

8.9.1   System Overview – Phasing

Until now, when we discussed filters, we have talked about them mostly in the context of attenuating or boosting certain parts of the audio spectrum. That is, the frequency response of the filter was our primary concern, while the phase response of the filter was only secondary. In Section 8.2.1, we discussed the fact that a filter has a phase response in addition to a frequency response. Since we now know that all filters have a phase response as well, we can use that knowledge in the design of a phase shifter effects processor. What we need is a filter that allows us to manipulate the phase of the signal while leaving the amplitude at all frequencies intact. Such a filter does indeed exist and is called an allpass filter. An allpass filter has a flat frequency response—it passes all frequencies equally without attenuating any of them. This filter would not be much use as the primary filter in a loudspeaker crossover, but is perfect for making a phaser.

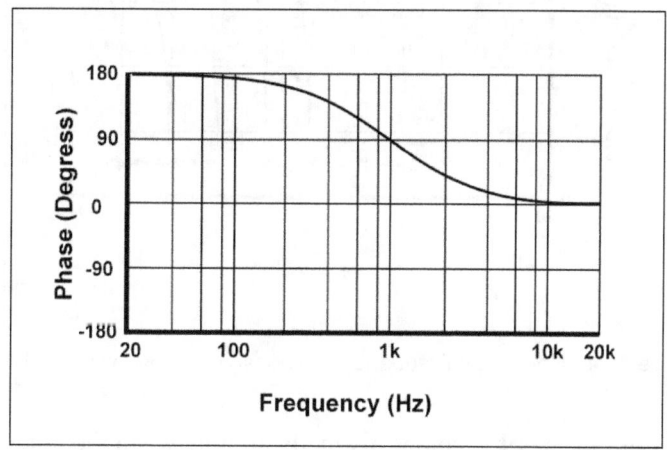

Figure 8-32   Phase Response of 1 kHz First-Order Allpass Filter

Traditional analog first-order allpass filters have a phase response such that the output is shifted 90° at the corner frequency of the filter, such as shown in Figure 8-32 for a filter with a corner frequency of 1 kHz. To shift a signal by 180° at a particular frequency would thus require two allpass filter sections—or stages—in series, with each stage having a corner frequency equal to the frequency of interest. A digital allpass filter, however, can be designed to have any phase response.[3] For the purposes of our discussion, we will refer to the allpass filter section as a block, where a block produces a 180° shift in the phase of the signal at the frequency of interest. In the first-order analog case, a block would be made up of two filter stages.

Figure 8-33 shows how in the most basic design of a phaser, the input signal is tapped and sent simultaneously to the output stage and to the allpass filter bank. Each block in the filter bank shifts the phase of the signal by 180° at a particular frequency. When the filtered signal is mixed with the original signal in the output summing amplifier, deep notches, equal in number to the number of allpass filter blocks, will appear in the output signal, giving a response like Figure 8-34.

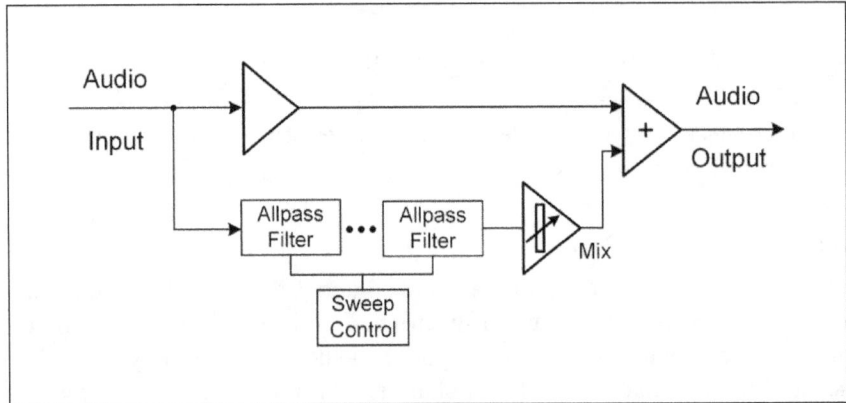

Figure 8-33 Basic Phase Shifter Block Diagram

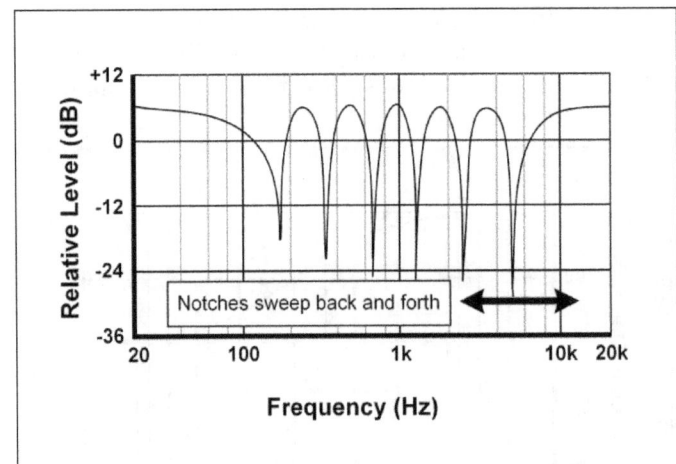

Figure 8-34 Notches in Frequency Response of Phaser with Six Blocks

The swooshing effect of the phaser becomes distinctly audible when the notches are swept up and down the in frequency. This is achieved by modulating the allpass filters together, assuring that their corner frequencies—and thus the notches—move in a predetermined, lockstep fashion. The corner frequency of each allpass filter is swept back and forth by a sweep control function, which can be viewed as an advanced form of low-frequency oscillator with an exponential ramp output. In reality, this function is performed by a microprocessor in modern digital effects processors.

Phasers usually come packaged as an effect in a multi-effects unit. There is not much to control in a phaser. The manufacturer internally controls the spacing between the notches and allows the user to control the modulation depth and speed of the sweep function. The modulation depth determines the width of the frequency range across which the notches sweep, and the speed determines how fast the notches sweep up and down this frequency range. The mix control determines the ratio of the filtered signal to the original in the output and controls the depth of the notches. When this ratio is 1 (i.e., equal amounts of filtered and original signal), the notches will be deepest. To make things more interesting, the phaser effects processor typically includes an adjustable feedback loop that allows the user to send some of the processed output back to the input, as shown in Figure 8-35.

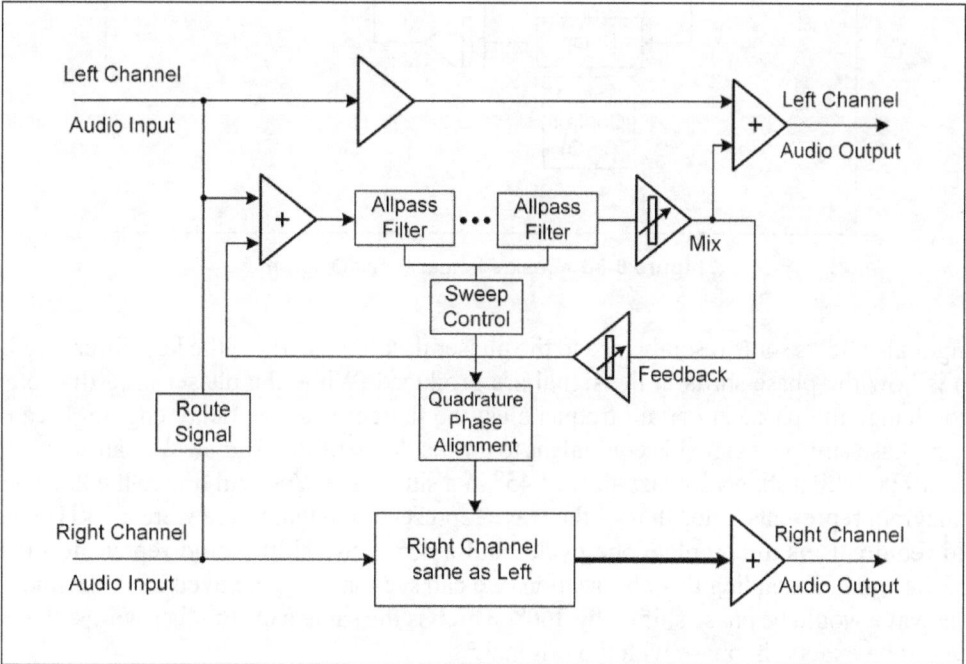

Figure 8-35   Stereo Phase Shifter with Feedback

Most phaser implementations also provide stereo output. Figure 8-35 shows that even with a single mono input signal routed to both channels, a stereo output can be generated by running the sweep control of the two channels 90° out of phase (called quadrature phase). This modulates the allpass filters in a way that ensures that the corresponding notches in the output signals of each channel do not continuously occur in the same place, and adds to the sense of spatial separation of the two channels.

8.10 Flanging Effect

The flanging effect sounds very similar to the phase shift effect described above. Indeed, the spacey, swooshing effect of the flanger is also produced by the sweeping action of deep notches in the spectrum of the audio signal. The main difference between the phaser and the flanger is the number of notches and the way these notches are produced. Like phasing, this effect is popularly applied to guitars, keyboards, and drums.

8.10.1 System Overview – Flanger

At the core of the flanger is a delay engine like the one described in Section 8.8. In fact, when we look at the block diagram of the flanger in Figure 8-36, we see it looks very much like the delay effect shown in Figure 8-25. The difference here is that the flanger includes a means of modulating the delay time.

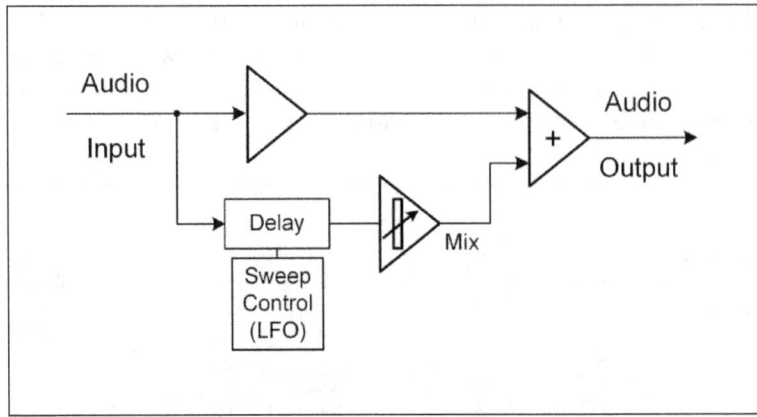

Figure 8-36   Basic Flanger Block Diagram

The flanger also bears some resemblance to the phaser described earlier. The key difference between the two is how the phase shifts in the signal are produced. While the phaser used discrete allpass filters to change the phase at certain frequencies, the flanger uses the delay engine. It can do this because a phase shift in a signal is equivalent to a time delay of that signal. We can see this clearly in Figure 8-28, which shows a phase shift of 45° in a sine wave. You will notice the horizontal axis of that diagram represents time. Thus, if the wave represented in that figure were a 1 kHz sine wave, it would require 1 ms to complete one cycle, and a 45° phase shift would represent a 0.125 ms[a] delay of the wave. Extending this observation, we can see that if we delayed a 1 kHz sine wave by 1 ms, the wave would be phase shifted by 360°, which is the same a 0°. In other words, the resulting wave would be exactly in phase with the original.[b]

But what happens to waves of other frequencies when we delay a signal by 1 ms? A 2 kHz wave, for example, has a period of 0.5 ms. In this case, a 1 ms delay will cause this wave to be phase shifted by 720°, or two whole cycles. Again, the end result is a wave perfectly in phase with the original. In fact, any simple periodic wave that has a frequency such that the delay is a whole number multiple of its period will come out perfectly in phase. This is the same as saying that a whole number of cycles of the wave will fit precisely in the delay time. Similarly, any simple periodic wave where the number of cycles that fit the delay is a whole number plus a half (i.e., 0.5, 1.5, 2.5, etc.) will be 180° out of phase with the original. Waves of other frequencies will have various phase shifts between these two extremes. Thus, we can see that if we delay a complex music program signal containing a wide spectrum of sound and add the delayed signal to the original signal, as is done in our basic flanger, there will be a series of deep notches in the output signal resulting from the cancellation of the signal at the frequencies where the waves are 180° out of phase. Figure 8-37 shows the frequency response of the flanger when the actual delay is about 11 ms. Compared to the phaser, we can see that there are hundreds more notches produced by the flanger. The response shown in Figure 8-37 is commonly called a comb filter response because the

---

[a] Since a single cycle of a wave occupies 360°, and 360° of a 1 kHz sine wave takes 1 ms to complete, the 45° position along a sine wave occurs at a point that is (45/360*1) ms from the start. This is equal to 0.125 ms.
[b] To figure out how much phase shift a particular delay causes, just find what fraction of the period it is and multiply by 360°. For example, a 1 kHz wave has a period of 1 ms, and a delay of 0.5 ms causes a phase shift of 0.5/1*360°, or 180°.

series of deep notches resembles the teeth of a comb. Correspondingly, the basic design of a flanger is sometimes called a feedforward comb filter.

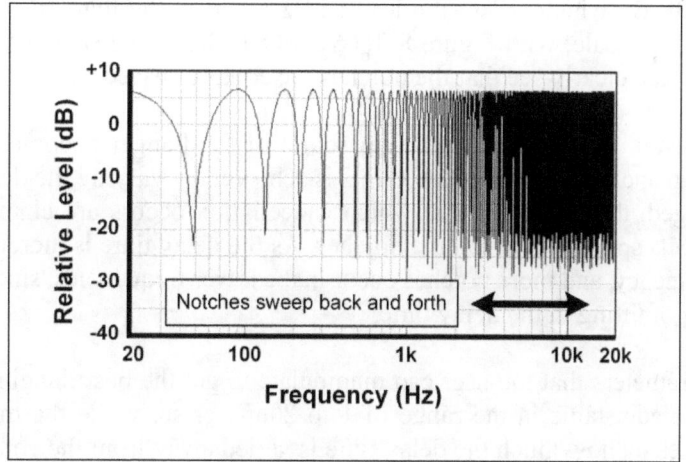

Figure 8-37   Frequency Response of Flanger

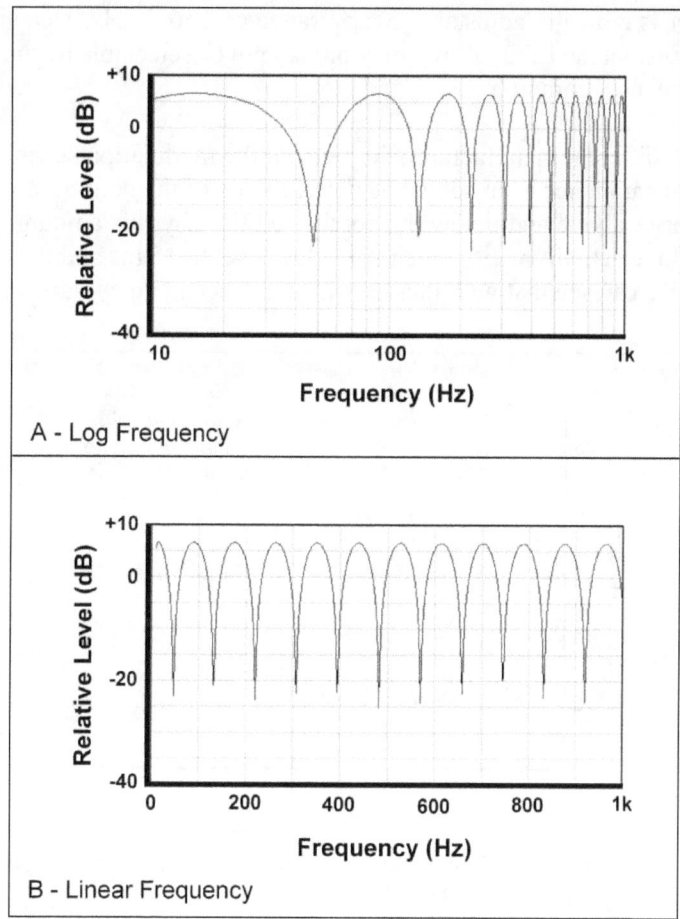

Figure 8-38   Comparison of Same Trace Using Log and Linear Frequency Scales

Since the frequencies at which the nulls occur are determined by a simple linear function of the delay time, the notches in the flanger's output signal are equally spaced. The notches do not appear to be equally spaced here because the frequency is shown on a logarithmic scale. However, if we compare Figure 8-38(B), which shows the low 1 kHz range of the flanger's frequency response with a linear frequency scale, with Figure 8-38(A), which shows the same data on a logarithmic frequency scale, we can clearly see that the notches are equally spaced.

As is the case with the phaser, the swooshing sound of the flanger becomes audible when the notches are swept up and down in frequency. This is achieved by varying the delay time. When the delay time is changed, the frequencies at which cancellation occurs are changed. Changing the delay time also affects spacing between the notches. As the delay time is increased, the first notch moves down in frequency, and more notches occur in the lower frequencies, since more cycles now have the possibility of fitting in the delay time.

There are a few parameters that the user can manipulate to get the best flanging effect. The delay time is usually user adjustable in the range of 1 to 20 ms or so, while the modulation or depth parameter determines by how much the delay time is varied away from the configured delay time. A third parameter, typically called speed or rate, determines how fast the delay time varies between the extremes determined by the depth parameter. The speed parameter actually controls the frequency of a sweep control function, which can be thought of as a low-frequency oscillator (LFO). The speed parameter is typically adjustable over a range of 1 to 50 Hz. Depending on the manufacturer, the waveform that the LFO follows may or may not be selectable by the user, but triangular and sinusoidal are the most common.

As may be expected, different manufacturers implement the modulation scheme in different ways. Sometimes the depth parameter is presented as a percentage of the configured delay time, and the actual delay time varies above and below the configured time by this amount. Figure 8-39 shows how the total delay time actually varies when the delay is set to 15 ms and the depth is 33% with a triangle waveform. We can see that with these settings, the actual delay varies from 10 to 20 ms.

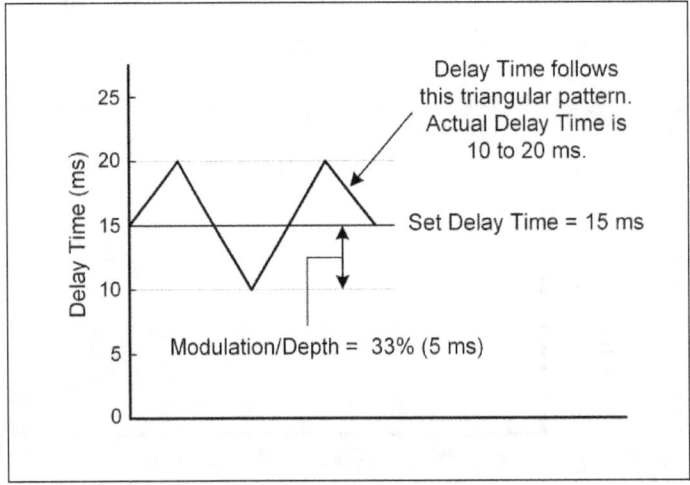

Figure 8-39   Delay Time Modulation Based on Percentage of Delay

Alternatively, the depth parameter may be presented as an actual time shift. Here the configured delay time is taken as the minimum delay that will be applied to the signal. The actual delay time will vary between this minimum and the maximum determined by the sum of the configured delay time and the depth parameter time. Figure 8-40 shows how the total delay time actually varies when the delay is set to 5 ms and the depth is set to 10 ms with a triangle waveform. With these settings, the actual delay varies from 5 to 15 ms.

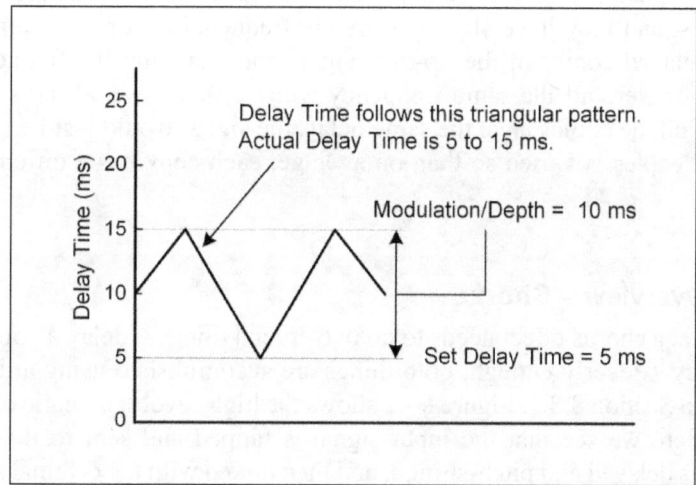

Figure 8-40   Delay Time Modulation Based on Fixed Time Shift

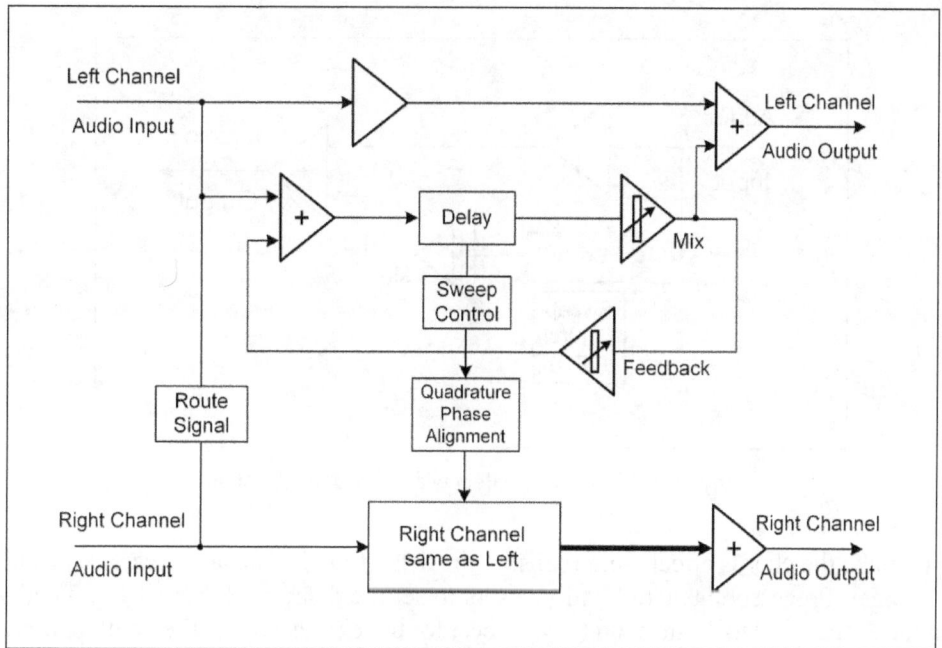

Figure 8-41   Stereo Flanger with Feedback

As with the phaser, flangers may include a user adjustable a feedback loop. This gives the user more flexibility in sculpting the sonic properties of the applied effect by playing with certain resonances created when some of the delayed output is mixed back into the input signal. Additionally, as is the case with most modern effects processors, the flanger implementation will likely be packaged as a stereo unit. In this case, the delay modulation of the two channels is linked, but operated 90° out of phase with each other (quadrature phase). Figure 8-41 shows a typical arrangement of a stereo flanger with feedback.

8.11 Chorus Effect

A chorus effect tries to mimic a real chorus by taking the audio signal from a single instrument and turning it into a signal from several instruments. The thing that gives a real chorus its body, richness, or lushness is the fact that the voices are not in perfect unison. The voices are somewhat delayed amongst themselves, and they have slightly different frequencies. A chorus effect does the same thing by making delayed copies of the original signal and changing the frequency of the copies slightly. If the copies were all the same frequency with no delay, we would simply get a louder signal. Similarly, if all the copies have the same delay, the signal would just be louder, so the delay between individual copies is varied so that, on average, each copy has a different delay from the rest.

8.11.1 System Overview – Chorus

As mentioned above, a chorus effect needs to do two crucial things—delay a copy of the signal and change its frequency. Cleverly enough, both things are accomplished using a delay engine of the type we discussed in Section 8.8.1. Figure 8-42 shows the high-level schematic of the simplest form of chorus effect. Here we see that the input signal is tapped and sent to the delay core of the processor where it is delayed and pitch-shifted, and then mixed with the original signal in the output section of the processor. You will notice that this arrangement is identical to that of the basic flanger, and that is because the fundamental processes of delaying and mixing the signals in both processors are the same.

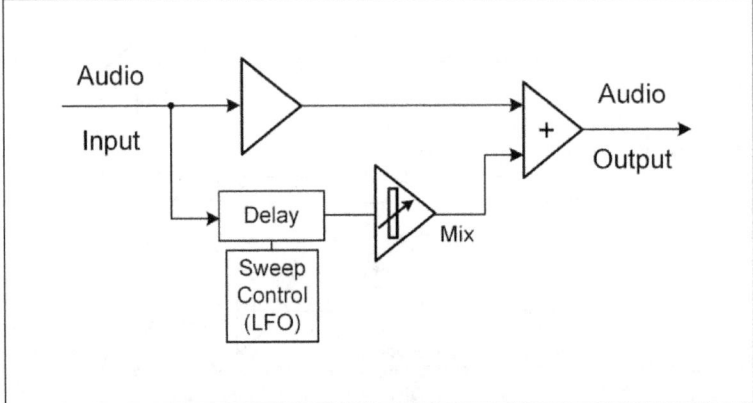

Figure 8-42   Chorus Effect with One Delay Section

In order to make the chorus effect more realistic, the delay time is constantly varied, just as we saw with the flanger. Processors generally allow users to set the delay time from 1 to 50 ms or so, but values in the range of 20 to 30 ms tend to work best for the chorus effect. The depth parameter (also called modulation or sweep) determines by how much the effects processor varies the set delay time. The rate at which the delay time is varied is controlled by a sweep control function in a similar fashion to the way it is done in the flanger discussed in Section 8.10.1. Again, this parameter is typically user adjustable in the range of 1 to 50 Hz, with values in the range of 1 to 5 Hz giving the best effect. Most processors also allow the waveform of the LFO to be switched among a number of choices, with sinusoidal and triangular being most common.

The same delay core and LFO are used to vary the pitch of the delayed copy of the signal. There are many schemes for accomplishing this, but the most common use either the slope or the amplitude of the LFO signal to vary the rate at which the delay buffer is read. If the delay buffer is read more slowly than it is written, then the pitch of the signal is lowered. If the buffer is read more quickly than it is written, then the pitch of the signal is raised. When we change the modulation depth or frequency of the LFO, the variation in pitch is correspondingly affected. If the pitch change scheme is based on the slope of the LFO waveform and we use the triangle wave, then we can see that we will have only two pitch shifts—one where the wave is ramping up, and another when it is ramping down. If the scheme is based on the amplitude of the wave, then we can see that the pitch shift will be constantly changing at a fixed rate. When we use the sine wave for the LFO, the amount of pitch shift will be constantly changing and so will be the rate of change of pitch, regardless of whether the slope or amplitude of the wave is used for control.

Another trick that is used to make the chorus effect more realistic and enhance its lushness is to add more delay sections in parallel. Figure 8-43 shows a chorus effect with two delay sections. Each section may have its own LFO modulator, or may share a common modulator. If they share a common LFO, then the delay of each section will have to be shifted out of phase with other sections so as not to be in unison (and simply create a louder signal). Even greater realism may be achieved by operating the effect in a stereo configuration, also shown in Figure 8-43. Each channel's LFO is operated out of phase with the other. Additionally, because the reproduced sounds emanate from different physical locations in the stereo field, the effect is greatly enhanced. To make the chorus even more realistic, randomly changing delay times may be used, as well as changing the loudness of the delayed signal slightly over time by using an LFO.

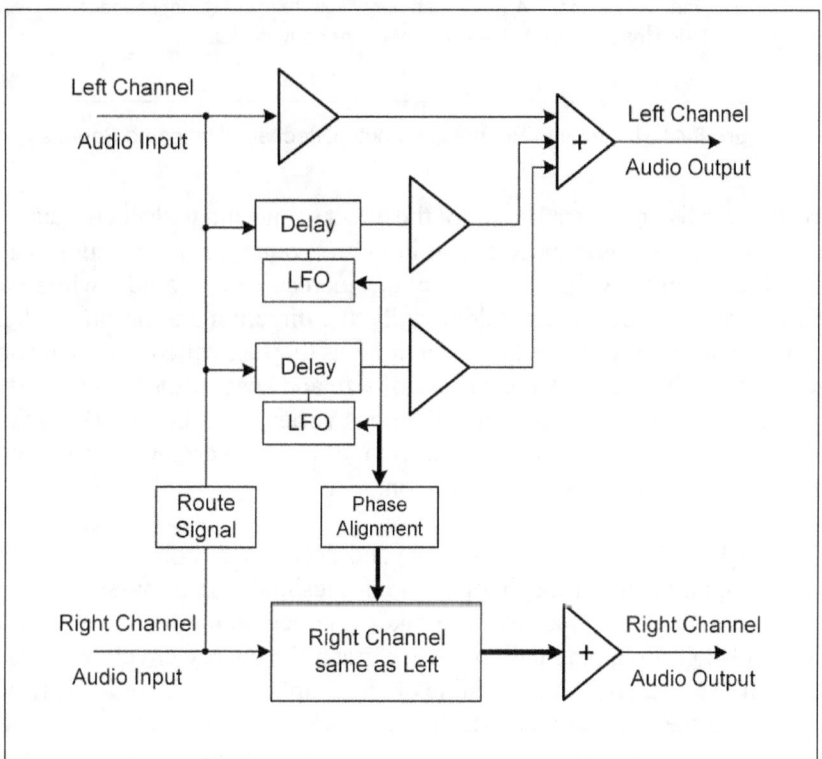

Figure 8-43   Stereo Chorus Effect with Two Delay Sections

8.12 Reverberation Effect

When we first discussed reverberation in Section 2.5, we learned that if we have a sound source in an enclosed space, a listener will hear the direct sound from the source and, a short time later, reflected sound from the enclosure's structure. Figure 8-44 illustrates this. Sounds that arrive at the listener within the first 80 ms of the direct sound add to the perceived loudness of the sound and are called early reflections. These sounds are directionally well defined and correlated to the room's geometry. They may have bounced off the enclosure's structure only once or twice and are shown as the dotted lines in Figure 8-44.

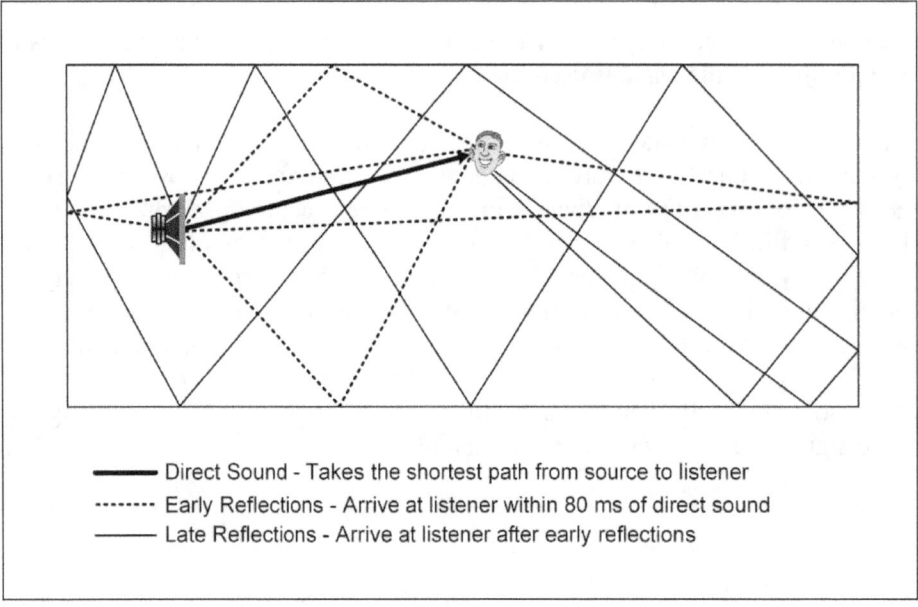

Figure 8-44   Listener in Room Hears both Direct and Reflected Sounds

Sounds that arrive at the listener after 100 ms of the direct sound add to the persistence of the sound. That is, even after the source has stopped transmitting the sound, these reflections cause the sound to 'hang around.' The rate of arrival of these reflections increases very rapidly with time and is much higher than that of the early reflections. Additionally, the direction and timing of these reflections appear more chaotic and random. These late reflections, as they are called, are what constitute reverberation. Reverberation gives the listener a sense of size and spaciousness of the room. The actual transition between early reflections and late reflections varies with the room's size and geometry, and research puts this region over a wide range from 70 ms to over 250 ms.[4] We will refer to the duration of early reflections as *ER* in our discussion.

One way to find out how these reflections behave and decay in a room is to measure its impulse response. An ideal impulse can be thought of as an infinitesimally short burst of sound, existing for a mere instant in time. Practically, an impulse takes a larger finite amount of time and could be realized by a hand clap or a popping balloon. If we graph the energy envelope of an impulse over time, it would be shown as a single vertical line on the graph, as in Figure 8-45. If we generate an impulse in a room and then measure and display the intensity of the reflections over a period of time, we get the impulse response of the room. Devices that measure the impulse response may actually generate a swept frequency tone or a burst of specially formulated noise (e.g., maximum length sequence) and then mathematically process the reflected sound to derive the impulse response. The measured impulse response generally looks like Figure 8-46. For the sake of discussion, we can simplify the impulse response as shown in Figure 8-47, and match the regions in this graph to the types of reflections depicted in Figure 8-44. We will use Figure 8-47 as our simplified model of reverberation in an enclosed space.

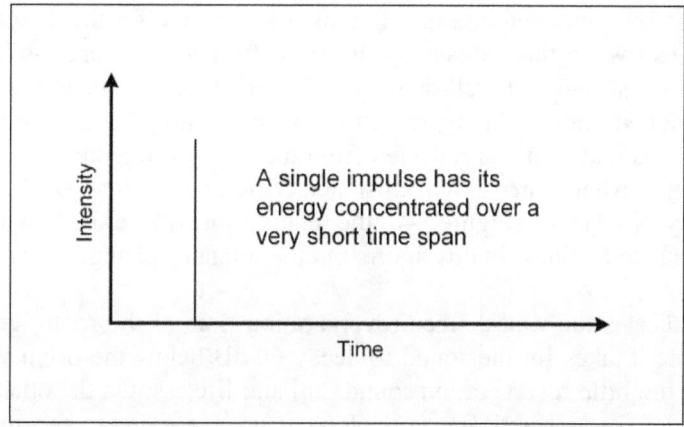

Figure 8-45   An Ideal Impulse Lasts Only an Instant

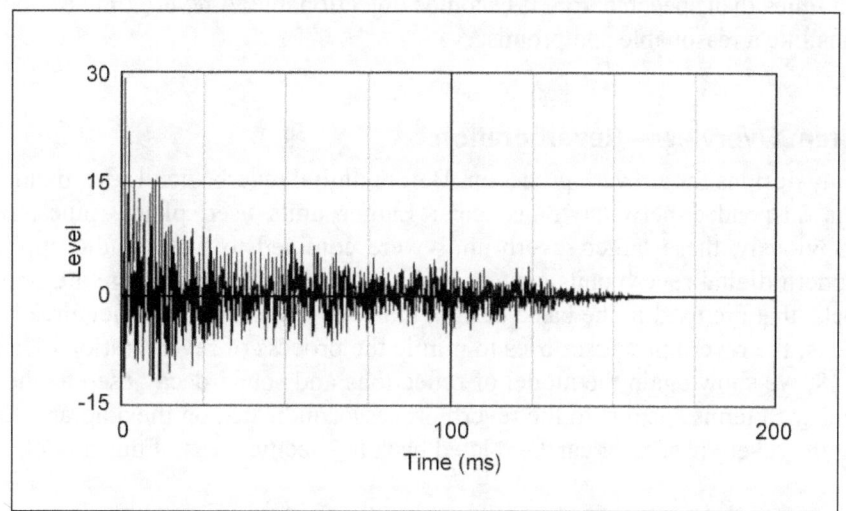

Figure 8-46   Measured Impulse Response Shows Chaotic Dense Reflections

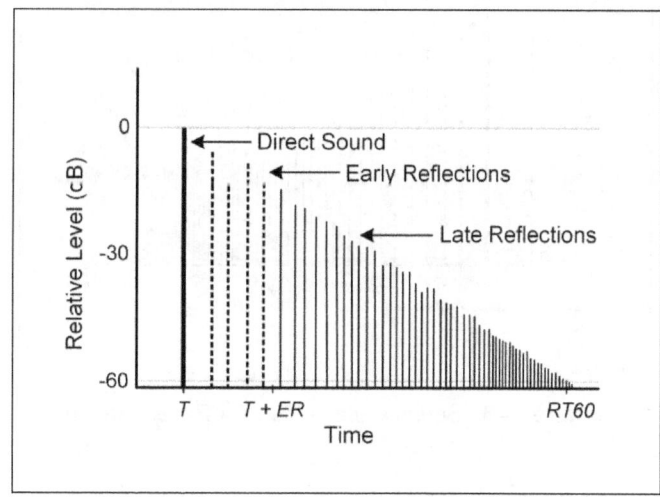

Figure 8-47   Simplified Sound Decay in Enclosed Space

Once an impulse has been initiated in a room, it takes some time for the direct sound to arrive at the listener. The instant when this happens is shown as time T in Figure 8-47. Over the next ER milliseconds, relatively strong and well-defined early reflections arrive at the listener's position, represented by the dotted lines in the figure. After this, the dense chaotic late reflections build up rapidly. The more times that sound is reflected from the walls, floor, and ceiling of the room, the less the intensity will be when it arrives at the listener. Generally, the intensity of the reflected sound decays exponentially. Note that in Figure 8-47 the relative sound level is shown on a dB scale, and thus the decay appears to be linear in this figure, but the intensity change is actually exponential.

Figure 8-47 also indicates the well-defined reverberation time of the room, generally designated RT_{60}. This is the time it takes for the sound to decay 60 dB below the original SPL of the direct sound. Spaces with too little reverberation sound dull and lifeless. On the other hand, rooms with too much reverberation, i.e., where RT_{60} is very long, cause the sound to become muddy and make speech difficult to understand. This is because newly created sounds blend into the reflections of the old sounds which are still bouncing around the room. Generally, music can tolerate longer reverberation times than speech before it becomes objectionable. A reverb time of about 0.5 to 1.5 seconds seems like a reasonable compromise.

8.12.1 System Overview – Reverberation

There are many designs for a reverb processor. Before digital reverb, small reverb units were made using springs suspended between transducers. Larger units used plates, pipes, or even real chambers. Obviously, these larger reverb units were confined to fixed studio applications. The design of modern digital reverb units can get highly complex. Even so, there are only a few core building blocks that are used in the basic design, many of which we have met already. With these building blocks, the reverb processor tries to mimic the process of reverberation discussed above. In Figure 8-48, we show again the model of reflections and sound decay used as the basis of the reverb design, with terms specific to the reverb processor indicated on the diagram. Each block in the design of the reverb processor can be related back to specific areas of this model.

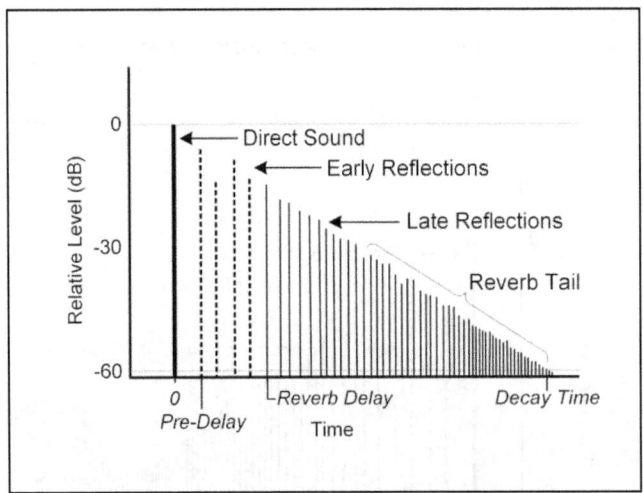

Figure 8-48 Sound Decay Model for Reverberation

The time between the direct sound and the arrival of the first early reflection is generally called Pre-Delay, and we will use it as such in our discussion. Some manufacturers refer to this as Reflections Delay. Pre-Delay is usually user adjustable. The time of the onset of late reflections may be referred to as Reverb Delay. However, some manufacturers also use the term Pre-Delay to refer to the time between the direct sound and the start of the late reflections. Some reverb processors allow the user to directly adjust this time as well, but usually this parameter is influenced by other parameters such as enclosure size and type. The Decay Time parameter generally maps directly to the RT_{60} time. Manufacturers do not appear to follow any standard naming convention for reverb parameters, so the only way to know for sure what the parameters of any reverb processor refer to is to read the manufacturer's manual.

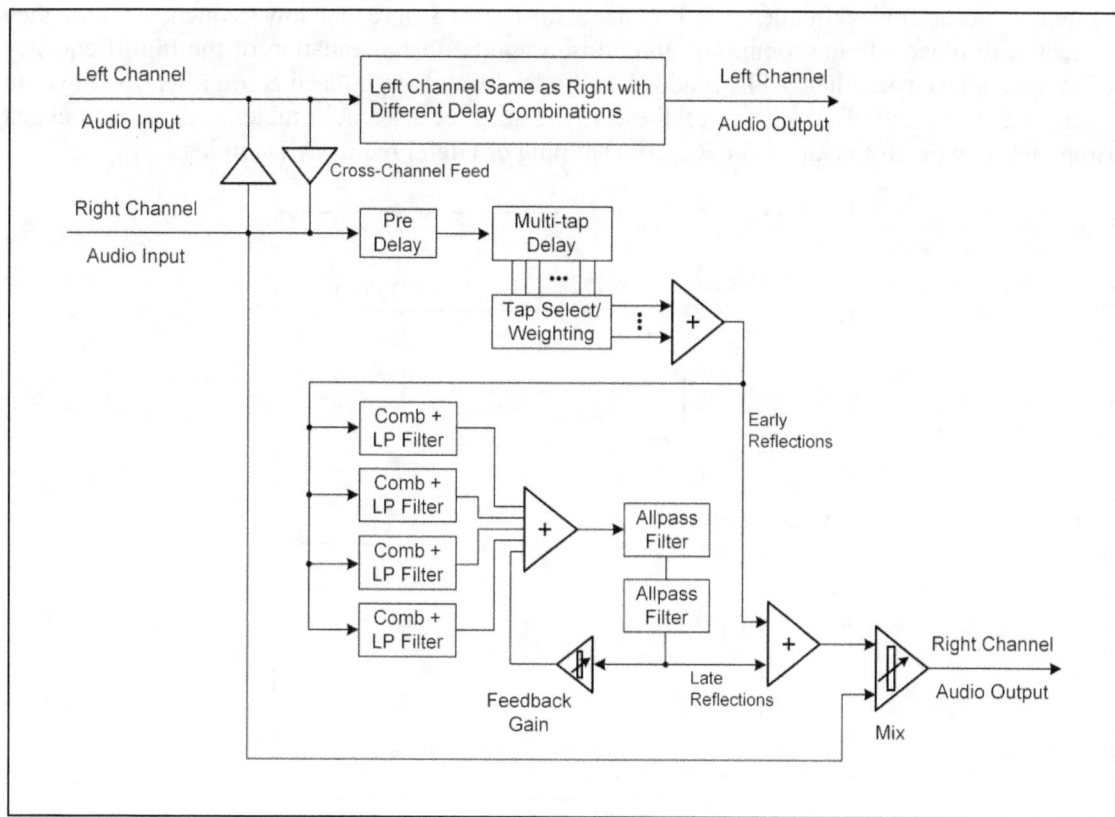

Figure 8-49   Block Diagram of Reverb Processor

Figure 8-49 shows a block diagram of a reverb processor. Pre-Delay is realized by passing the input signal through a user-adjustable delay line. The audio then goes into a multi-tap delay block, which has outputs at several delay points. There may be between three thousand and four thousand taps in this block. By selecting and weighting (i.e., applying a specific gain to the signal) varying numbers of these taps, the character and density of the reverberation can be changed. Processors may allow this initial selection to be made via a parameter like Reverb Name, which may include such values like 'Hall,' 'Cathedral,' etc. The weighted outputs from the multi-tap delay are summed and used to model the early reflections.

The summed output from the multi-tap delay is also fed to a bank of feedback comb filters to create the dense late reflections. The design of the feedback comb filter is shown in Figure 8-50(A). We can see that this device merely sends an attenuated, delayed copy of the output back to the input. The gain in the feedback loop needs to be less than 1, otherwise we will get the same effect as runaway feedback in our microphone-amplifier-speaker chain. This device is called a comb filter because its frequency response looks similar to Figure 8-38, only inverted. However, the real interest for us with this device is its impulse response. Suppose d is the delay time of the delay core, then if we input a single impulse to the feedback comb filter, its output will consist of a series of exponentially decaying impulses, spaced *d* ms apart, as shown in Figure 8-50(B). Thus, we can use a bank of these comb filters to rapidly increase the density of the reflections. In order to avoid overlapping impulses in the summed outputs of the comb filters, their delays must not be multiples of any single number and are usually chosen to be relatively prime. Overlapping impulses would emphasize a particular frequency and add unwanted coloration to the audio.

As you may recall from Section 2.7, as sound waves travel through the air, high frequencies are attenuated more than low frequencies. They are also absorbed more than low frequencies when they interact with objects in the room. To more closely model this attenuation of the high-frequency reflections, a low-pass filter (LPF) is added to the feedback loop of each comb filter, as shown in Figure 8-51. The cutoff frequency of these filters may be user adjustable, and the associated parameter may be given names like Reverb Damping or High Frequency Damping.

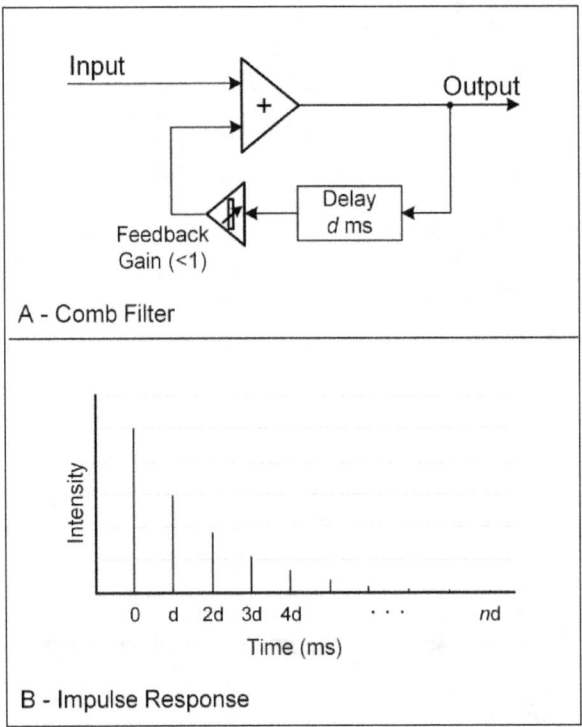

Figure 8-50 Feedback Comb Filter and its Impulse Response

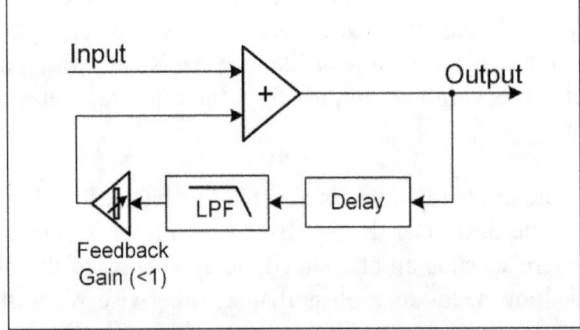

Figure 8-51 LPF Inserted in Feedback Path Models High-Frequency Absorption

The impulse density is further increased, and the phase alignment of the impulses is shifted, by feeding the summed output of the comb filters through a series of allpass filters. We have seen allpass filters used in the phaser effect to change the phase of the input signal without changing the amplitude at any frequency in the audio spectrum. However, as with the comb filter, we are also interested in its impulse response here. The design of an allpass filter is shown in Figure 8-52(A). As we can see, it is a combination of a feed forward comb filter and a feedback comb filter in a single unit. The impulse response of the allpass filter is also a series of decaying impulses spaced d ms apart. However, the initial impulse is negative, i.e., its polarity is inverted compared to the other impulses. Thus, the allpass filters add to the realism of the reverberation and allow us to more closely model the reverb tail in a real enclosure where the impulses have every conceivable phase value. By controlling the gains in the feedback and feedforward loops and the delay time of the comb and allpass filters, the decay time of the reverb tail can be manipulated. Manufacturers typically provide a Decay Time parameter to allow the user to adjust these elements of the reverb processor.

The output from the comb filter-allpass block models the late reflections depicted in Figure 8-48. This output is summed with the early reflections, and a variable amount of this combined signal is mixed with the original input signal to produce the final output of the reverb processor. The user can usually control this balance via the Mix parameter of the processor, with values ranging from 0% (input signal only—dry) to 100% (processed reverb only—wet).

Figure 8-49 also shows that stereo reverbs are made by cross tapping small amounts of the left channel input signal into the right channel, and vice versa. Additionally, the channels are further decorrelated by ensuring that the delays in opposite channels have slightly different delay times.

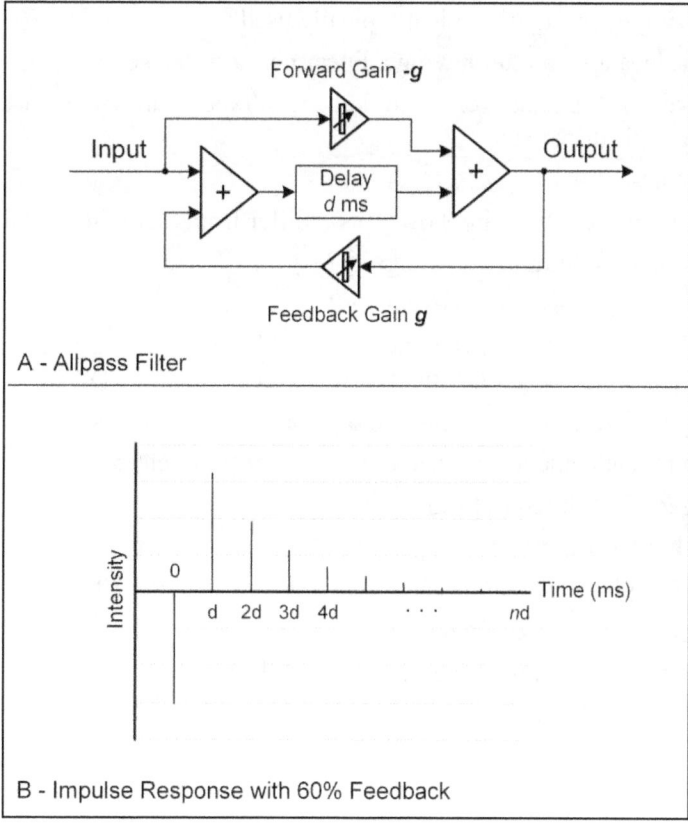

Figure 8-52 Allpass Filter and its Impulse Response

There are a number of tricks that manufacturers incorporate into their basic processors to extend their functionality. One such trick is the gated reverb, popular on percussion instruments. In this case, the reverb tail is abruptly chopped off at a user adjustable time. The actual duration of the reverb tail is determined by a combination of a gate threshold and the gate hold time. Once the reverb level falls below the gate threshold, the processor will continue putting out the signal for the duration of the hold time and then abruptly cut it off.

Another trick is to reverse the amplitude envelope of the reverberation, so that the impulses start out at a low level and rise to a high level before being abruptly cut off. It is basically the amplitude envelope of Figure 8-48 in reverse, so this feature is usually called reverse reverb. This feature should not be confused with studio applications where the entire process of reverberation is reversed and recorded. In the reverb processor used in live applications, it is only the level envelope that is reversed—the early reflections still come first, but at a low level, followed by the late reflections, increasing in level until they are cut off.

A reverb processor is used to add reverberation to the sound field when there is too little. Note that with reverb processors, we can only add reverberation to our sound field. To reduce reverberation in an enclosed space requires the use of physical acoustical treatments. Judicious use of reverb can add a sense of spaciousness and depth or distance to the sound. Adding too much reverb will cause the sound to lose clarity and become muddy. Very little, if any, reverb should therefore be added to plain speech, whereas singing, especially in a mix, can tolerate quite a bit. We will discuss this issue further when we look at mixing in Chapter 16.

Review Questions

1. What is a shelving filter and where is it typically used?
2. Explain what is meant by a filter having a linear phase response.
3. On a 1/3 octave graphic equalizer, a single control (slider) may have a bandwidth of a few octaves. Explain this discrepancy.
4. What is a constant-Q filter?
5. What is a parametric equalizer and how does it differ from a graphic equalizer?
6. What does a compression ratio of 5:1 signify?
7. What are some practical uses of an expander?
8. What is latency in a digital processor and how could it affect interaction with other processors in an analog signal chain (e.g., the effects loop of an analog mixer)?
9. What is an allpass filter? Give an example of where one may be used.
10. Explain how a phaser achieves its characteristic swooshing effect.
11. How is a flanger effect different from a phaser?
12. What is a comb filter response?
13. Explain how multiple delay engines are used to create a chorus effect.
14. What is the impulse response of a room?
15. What is meant by reverberation time of an enclosed space?

Chapter 9

Analog Mixers

9.1 Central Role of Mixers in the Sound System

For the audio engineer, the mixer is where it all comes together; the glue of the sound system; the center of the live sound universe. Indeed, in a very real sense, the front of house mixer is the center of the live sound reinforcement system. The term *front of house* (FOH) refers to equipment or resources whose primary function is to provide audio coverage to the main audience area. Figure 1-2 shows very well how central the mixer is to the whole sound system. In addition to its FOH role, the mixer may also provide sound sources to other locations, such as stage monitor systems, show monitor stations, and feeds to recording and live broadcast systems.

The functions of the mixer—in order of importance—are to mix, route, and process the input signals. Mixing can be defined as the act of combining multiple input signals into one or more output signals with control over the level of each input signal in the output signal. Routing is the ability to send various combinations of input signals to different locations, both within the mixer itself, and outside of the mixer (e.g., to external effects devices). Practically all mixers offer some basic effects processing in the form of EQ. Some mixers, however, have extensive effects processing on board, eliminating the need for external processors. In the following sections, we will be looking in depth at all these functions of a typical analog live sound reinforcement mixer.

Depending on its size and locale, a mixer may also be called a mixing board (smaller mixer), console (larger mixer), or desk.

9.2 Powered Mixers

Some mixers have a power amplifier built right into the mixer cabinet, eliminating the need for external power amplifiers. These are generally referred to as powered mixers. These mixers tend to be on the smaller side—typically 24 channels or less—as large amplifiers can get quite bulky and heavy. However, for small shows, these powered mixers can be quite convenient, as it means that there is less equipment to haul around, and the mixer to amplifier cabling is already done in the box. The majority of mixers, though, have no built-in power amplifier. Since the function of the amplifier is quite separate from than of the mixer and we have already dealt with amplifiers in Chapter 5, we will not consider powered mixers any further.

9.3 Functional Layout

Figure 9-1 shows the traditional layout of a small to midsized FOH mixer. The traditional layout of the mixer places the channel input controls on the left of the board, and the master output and monitor section on the right. The channel input section is divided into channel strips for mono and stereo channels. A stereo channel is like two mono channels managed by one set of controls. While a mono channel generally has an XLR jack for microphone input and a 1/4-inch TRS (Tip-Ring-Sleeve) jack for line-level input, as well as an insert jack that allows placing an external processor in-line with the input signal, a stereo channel usually only has two TRS jacks for line-level signal input. Additionally, stereo channels may have fewer filter and EQ controls than mono channels, and they have no phantom power feed. Other input sources on the mixer are the auxiliary returns and tape inputs. These channels usually have input controls located to the right of the input channels section.

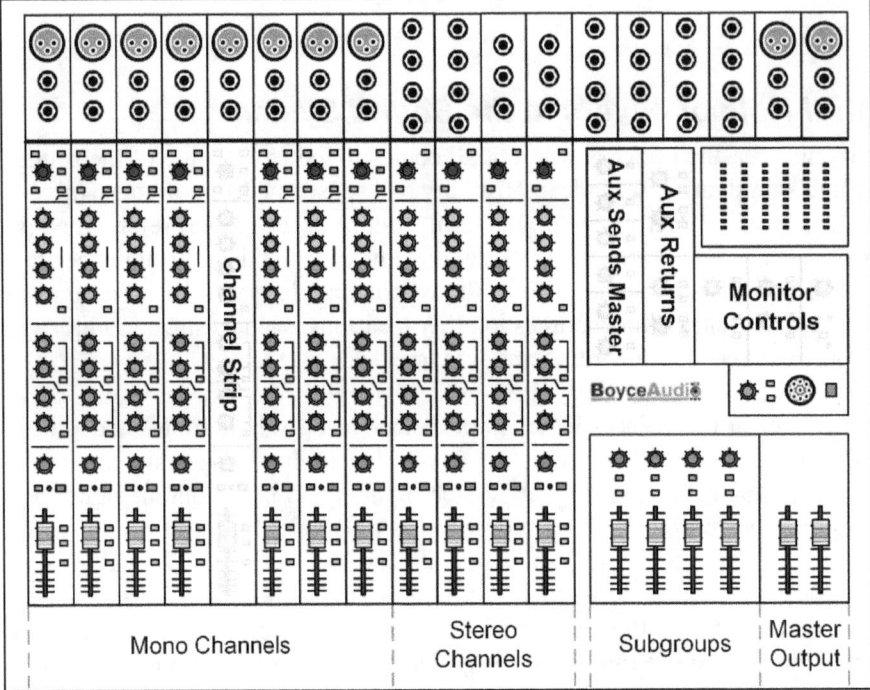

Figure 9-1   16-Channel Mixer with 4 Subgroups

The output controls are divided into an auxiliary send master section, a monitoring section, a subgroup section, and a master output section. On midsize to larger mixers, controls for a talkback microphone are also located in this section of the mixer. This usually has flexible routing options and allows the engineer to talk to the performers on stage via the monitors.

Each auxiliary send (referred to as *aux send*) channel has an output from the mixer. These are used to send various combinations of the input signals to off-board (thus the term *auxiliary*) processing equipment, such as effects processors, or to provide a stage monitor mix. The subgroups are used primarily to help simplify the control and routing of similar or related channels. However, each subgroup also has its own output from the mixer, and these can be used to feed a multitrack recorder, special feeds, or in a pinch to provide additional monitor mixes. The monitor output is used to provide a feed to the audio engineer, either via headphones, or an amplifier and speaker fed from the control room output on the mixer. The term *control room* is really used in a studio setting, but this term is found on many live sound mixers, as manufacturers try to target their mixers to both the live and studio markets. The master output is the last stop on the mixer before the signal exits the board. These outputs feed the FOH amplifiers. We will discuss all of these functional groups of the mixer in detail in later sections of this chapter.

The layout described above is generally used for mixers with up to 32 channels. For bigger mixers, manufacturers start with the above layout and add additional input channel strips to the right of the output section. This keeps the output section conveniently located in the middle of the console and distributes the input channel controls in a way that makes it easier for the engineer to access them without too much lateral movement.

9.4 System Overview

Let us now take a high-level look at the signal flow through the mixer and examine the role of various processing and routing blocks in manipulating and managing the audio signals.

9.4.1 Input

To start, we will consider the input side of the mixer. Figure 9-2 shows the block diagram of the most common subsystems found in a typical mono input channel.

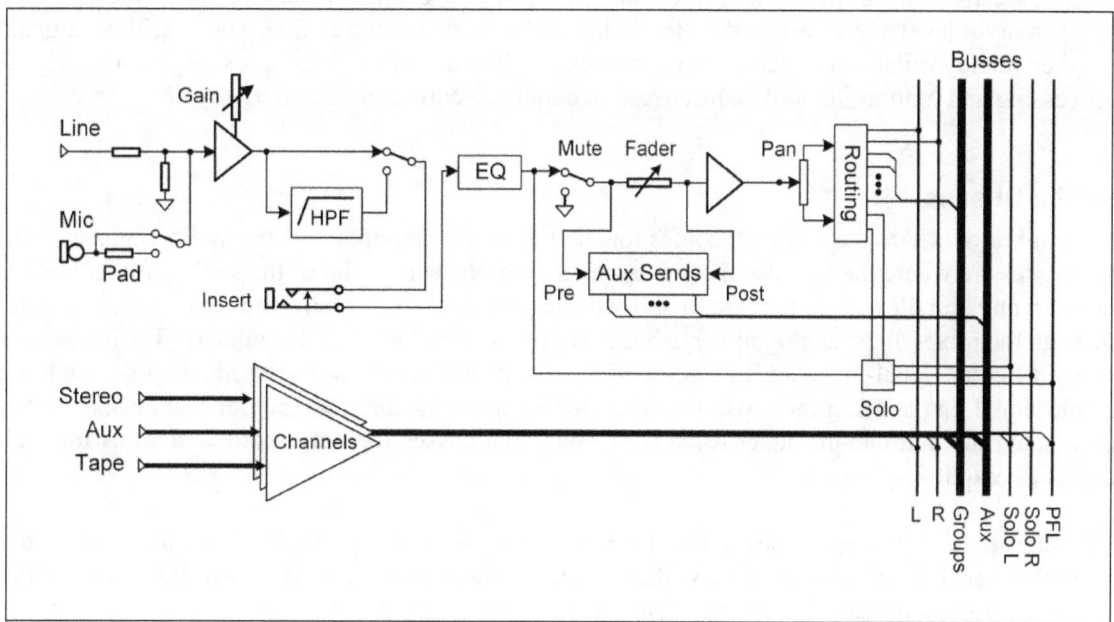

Figure 9-2 Input Side of Mixer

Starting at the left, we have both the line input and the microphone input feeding into the channel preamp. To maximize noise rejection, both of these inputs are usually balanced. We discussed balanced vs. unbalanced connections in Chapter 6. A signal is not normally provided to both the line and microphone inputs simultaneously, and so on some mixers, a switch selects one or the other of these inputs to be the active source. On some mixers, a switchable resistive network, called a pad, is inserted before the preamp to reduce the level of the input signal. The pad reduces the input level typically by 20 to 30 dB, and thus prevents overloading the channel preamp and subsequent clipping when the input signal is very high. The channel preamp has a gain control to set the gain for the amplifier and the optimum signal level for downstream subsystems. A switchable high-pass filter follows the preamp and is used to remove unwanted low-frequency content from the input signal, e.g., rumble picked up by stage microphones or low-frequency self-masking sound from vocals.

A normalized insert point is typically placed in the signal path of mono channels and is used to allow the insertion of external processing devices into the signal path. We will take a closer look at how insert points are used in Section 9.12. Note that on some mixers, the insertion point follows the equalizer section.

On smaller mixers, the equalizer section may be as simple as two bands consisting of high-frequency and low-frequency shelving controls. Larger mixers have more complex arrangements, with at least two bands of fully or semi-parametric mid frequency and perhaps semi-parametric high and low-frequency shelving controls.

After the equalizer subsystem, we have the channel mute switch followed by the channel fader and the channel buffer amplifier. The buffer amplifier feeds its output signal into the panoramic potentiometer (usually called the *pan pot*, or simply *pan*), which sends the signal to the left and right, even and odd routing network. This network consists of a collection of switches which simply direct the signal onto the main left and right, even and odd group, and solo busses.

Signals are also tapped from before (pre) and after (post) the channel fader and sent to the auxiliary subsystem for level adjustment and buffering before being sent to the auxiliary busses. These signals are then made available to external signal processors. The use of the auxiliary subsystem for signal processing and monitoring will be discussed in depth in Sections 9.13 and 9.14.

9.4.2 Busses

If the mixer is where the audio all comes together from the standpoint of the audio engineer, then the busses are where the signals all come together from the standpoint of the analog mixer. Sooner or later, any and all signals that are mixed with other signals eventually end up on a bus running through the mixer. In an analog mixer, a bus is simply a shared electrical conductor. The bus allows many different signals to be added together, i.e., mixed, and then to be accessed and processed as a single signal. In the mixer there will be busses for the main left and right output channels, a bus for each subgroup, a mono pre-fader listen (PFL) bus, two busses for stereo solo, and a bus for each aux send output.

The number of subgroup busses in the mixer is usually a selling point that is highlighted by the manufacturer. This is because, as we will see later, the subgroups are a very powerful feature of the mixer and form a mixer-within-mixer concept. Thus, for example, a mixer with four sub-groups is referred to as a four-bus mixer, and one with eight subgroups will be referred to as an eight-bus mixer. The number of subgroup busses even sometimes makes it into the mixer's designation, with a name like BX2442 suggesting that the mixer has 24 input channels, 4 subgroups, and 2 main output channels.

9.4.3 Output

Let us now take a look at the output side of the mixer. We will highlight only the most common output channels. Figure 9-3 shows us the overview of the arrangement of these channels.

First, we have a representative aux send output channel which gets its signal from its corresponding aux send bus. While Figure 9-3 shows the aux busses collapsed, remember that each aux send channel has its own dedicated bus. From the bus, the signal first goes into a summing amplifier. This is the same for all output channels; whenever a combined signal is taken off a bus, it always goes into a summing amplifier. Care must be taken that this amplifier is not overloaded by the sum of all the signals on the bus. The signal then passes through the aux send master attenuator that controls the level of the signal going into a buffer amplifier before sending it to the outside world. Buffer amplifiers are used to isolate one part of a system from another, especially when they are changes in certain electrical characteristics, e.g., impedance, voltage levels, or drive current requirements.

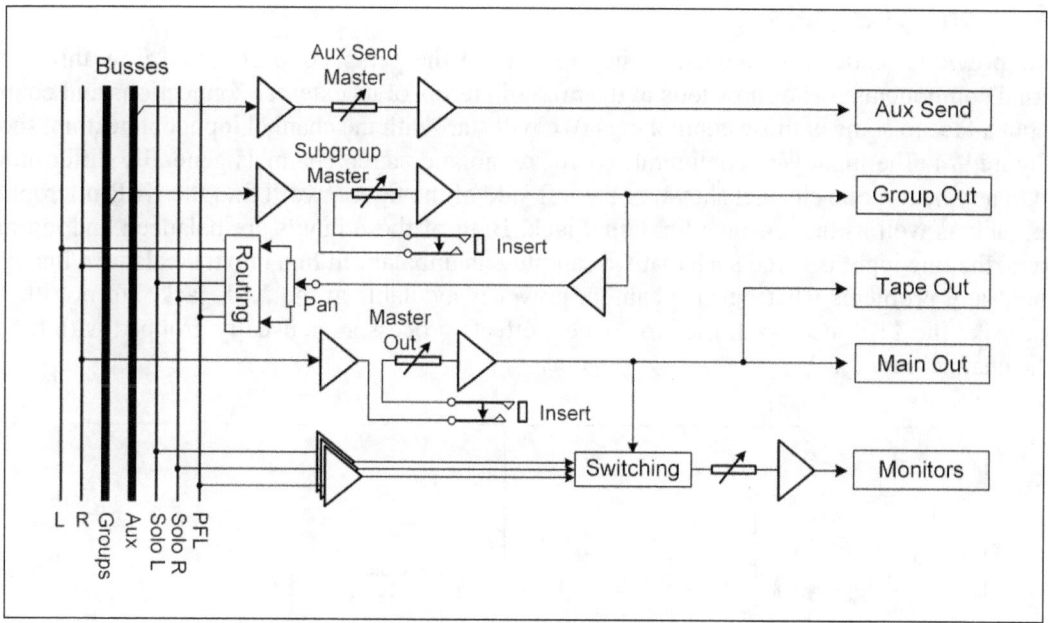

Figure 9-3   Output Side of Mixer

The next channel type shown in the figure is the subgroup output. The subgroups form a sub-mixer within the mixer and so tend to have a lot of associated controls. The subgroup output is the most complex output channel and actually has many characteristics of a simplified input channel. There is a dedicated subgroup bus for every subgroup channel in the mixer. Like the aux send channel, the subgroup output gets its signal from its dedicated subgroup bus, which feeds first into the summing amplifier, then through the subgroup master fader and finally the buffer amplifier. The subgroup channel, though, has an insert point that can be used in the same way as the one on an input channel. The subgroup channel also has routing options, usually only to main left and right, solo left and right, and PFL busses. On more complex mixers, an equalizer section will sometimes be found on each subgroup channel as well.

On FOH mixers, there are two main output channels, *viz.* Left and Right. Each output channel gets its signal from its corresponding bus. The main output channel also has an insert point before the master output fader, between the summing amplifier and the buffer amplifier. Tape outputs are tapped off of the main outputs.

The Monitor outputs are used to feed headphones and also provide a line-level signal for amplified monitor loudspeakers. Since the monitor channel is meant to allow the engineer to sample signals flowing through the mixer, it typically has a wide choice of inputs from which to take its signal. For example, it may take its signal from the PFL bus, left and right solo busses, aux send busses, tape return inputs, as well as the main output lines. A complex switching network determines what is heard on the monitors.

Whatever mixer you may have, you can usually find the manufacturer's detailed block diagram for it. This will be similar to the input and output diagrams we have looked at above, but with much more detail. Studying this diagram will go a long way towards helping you understand the signal flow through your mixer.

9.5 Input Channels

In the previous sections we looked at the mixer from the perspective of signal flow through its internal components. Let us now look at the mixer in terms of its external connections and controls and put a face to some of those components. We will start with the channel input connectors, shown in Figure 9-4. The input jack configurations for mono and stereo channels generally differ quite a bit. On a typical mono channel shown in the left side of the figure, we'll find the XLR microphone input jack as well as the 1/4-inch line input jack. Both of these inputs are balanced and, on most mixers, the line input is wired such that we can plug an unbalanced line into the balanced line input without any problems whatsoever. Phantom power is available at the XLR jack, but not the line input jack. The TRS insert jack used to connect effects processors is usually grouped with the rest of the channel input jacks.

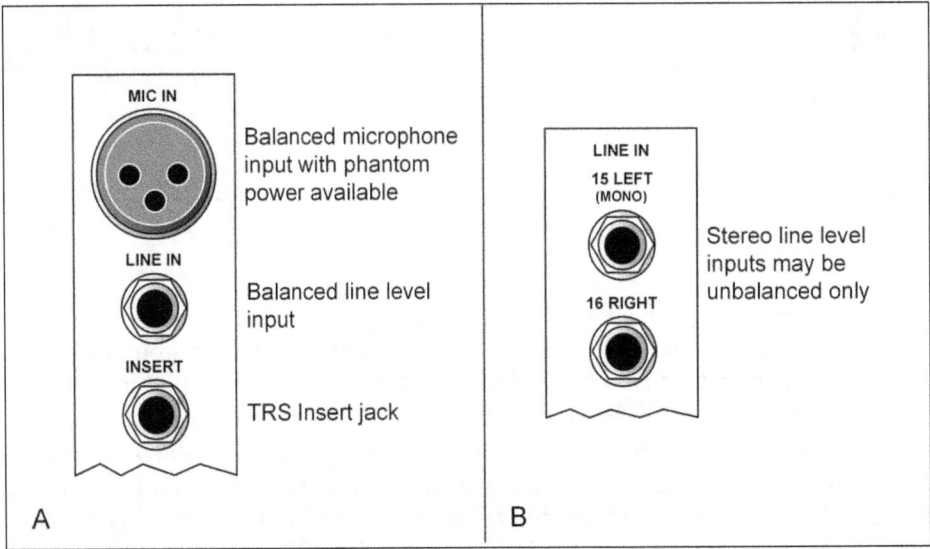

Figure 9-4 Mono and Stereo Channel Input Jacks

The stereo input jacks, shown in Figure 9-4(B), feed two adjacent channels. Depending on the quality of the mixer, these inputs may or may not be balanced. These inputs are usually wired so that if a line is plugged into the left input only, the signal is fed to both the left and right channels. As these are line-level inputs, there is no phantom power available at these jacks.

The input channel strip is shown in Figure 9-5. For any simple mixer, this can typically be divided into four or five sections. At the top of the strip is the preamp section. The main control here is the gain potentiometer that sets the gain for the channel preamp. Depending on the complexity of the mixer, an array of different switches may also be found here. The microphone/line input selector, phantom power, and pad switches as discussed earlier usually live here. The high-pass filter switch may also be found here, but on some mixers, it may be located in the EQ section. A switch for reversing the polarity of the input signal is sometimes seen as well and is useful in certain miking situations, e.g. opposed miking of drums or piano, where one microphone is placed above the drum or strings, and the other is placed below, with the microphones pointed towards each other. In such a case, the signals from the two microphones may be out of phase and significant cancellation may occur when the signals are added in the mixer unless the polarity of one of the signals is reversed.

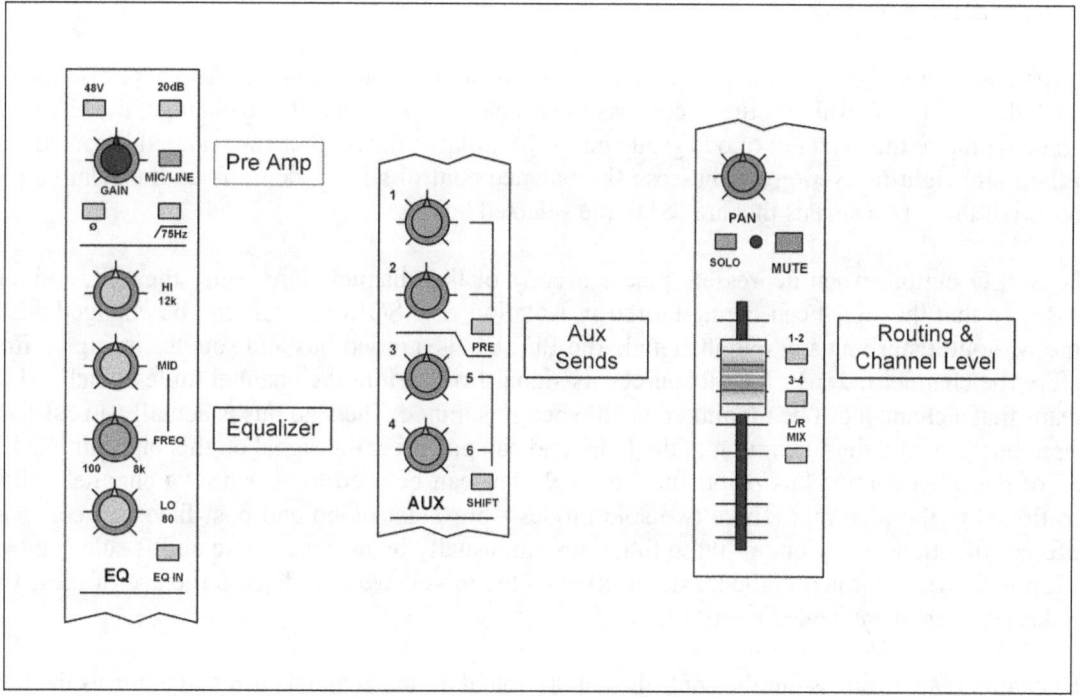

Figure 9-5   Input Channel Strip (In Three Sections)

An equalizer section usually follows the preamp section. This one shows a simple three-band section with a sweepable mid band. The high and low bands are usually shelving controls, while the mid band is peaking, as discussed in Section 8.2. Bigger mixers will have multiple, fully parametric midranges, and may have semi-parametric high and low bands. This EQ is the one used by the engineer to set the tone and mood of the channel's sound.

Below the EQ section are the auxiliary sends controls. The first two in the example shown are switchable between the pre-fader and post-fader signals. As the name implies, pre-fader signals are taken from a point in the signal chain that is before the channel fader, while post-fader signals are tapped from a point after the fader. Figure 9-2 makes this more obvious, where signal lines feeding the Aux Sends block are labeled *Pre* and *Post*. Pre-fader signals are useful for feeding monitor mixes for the stage performers, since the signal is unaffected by what the engineer does with the house mix via the channel fader. Figure 9-5 also shows a situation that is common in midsize mixers. This mixer has six aux send channels in all, but to save cost it has only four aux send controls per input channel strip. The last four aux sends are selectable two at a time, via the Shift button. For any given channel, we can either send a signal to the aux send 3 and aux send 4 busses, or the aux send 5 and aux send 6 busses, but not all four at the same time. These signals are post fader and are usually sent to effects processors.

Following the auxiliary sends section we have the pan control that is used in conjunction with the routing switches to route the signal onto the bus pairs. For example, depressing routing switch *1-2* and turning the pan control all the way counterclockwise will route the signal to subgroup 1 bus only, while turning it fully clockwise will route the signal to subgroup 2 only. Placing the pan pot at any position between these two extremes will send the signal to both busses. The level of the signal sent to each bus will then be proportional to the position of the pan pot, with that level being equal when the pan pot is centered. This works the same way for all the bus pairs that are selectable via the routing switches. By depressing more than one bus switch, the signal can be sent to multiple bus pairs at the same time.

The pan pot is found only on mono input channels. For stereo input channels, the pan pot is replaced by a balance control, which actually consists of two ganged pots. In a stereo channel, the left signal is routed only to the main left or odd group busses. Similarly, the right channel signal is routed only to the main right or even group busses. The balance control simply determines the relative proportion of these two signals that are fed to the selected busses.

The SOLO button, when depressed, places a copy of the channel signal onto the PFL and solo busses so that the signal can be monitored in isolation. The SOLO switch can be engaged at any time without disturbing any output signal. The PFL bus is a mono bus and sources its signal from before the channel fader. In fact, it sources its signal from before the channel mute switch, which means that a channel can be monitored even when it is muted. The solo bus is actually a post-fader stereo bus pair, and thus operation of the fader and pan pot affect the signal on this bus pair just like any of the other routing bus pairs. Thus, the solo bus can be used to hear how a channel will be positioned in the final mix. These two solo modes—pre-fader mono and post-fader stereo—have different functions. Only one of these functions can usually be used at a time and is selected by a PFL mode switch located in the master section of the mixer. We will discuss the use of these two modes later when we look at monitoring.

The channel fader rounds out the controls that are found on the channel strip and controls the level of the signal going onto the routed busses. It is located at the bottom of the strip, close to the engineer so that it is easily accessible. On the smallest compact mixers, the fader may be a rotary potentiometer, but it is usually implemented as a slider potentiometer. The slider type of control gives the engineer more room to work with, is easier to control, and gives instant visual feedback on where the channel level is set.

9.6 Subgroups

The subgroup controls are found in the master section of the mixer. This section can be thought of as a mixer-within-mixer. The subgroups are sometimes referred to as submasters because in the progression of the mixdown process, they are just one level above the master outputs. The subgroup faders are also sometimes referred to as the bus faders because there will be one subgroup strip for each routing bus in the mixer. Thus, in an eight-bus mixer, there will be eight subgroup faders. Figure 9-6 shows the controls for a four-bus mixer.

Subgroups are used primarily for organizing and simplifying the mixdown process. For example, we could route all of the individual pieces of a drum set to a single subgroup. We would use the individual channel faders to fine tune the mix of the individual drums, but if we needed to change the overall level of the drums in the FOH mix, only the single subgroup fader would need to be manipulated.

Each subgroup has a dedicated insert point. This makes them very useful in sharing scarce resources like effects processors across a group of channels. For example, suppose we had four backing vocal channels and we wanted to pass each vocal through a de-esser. If we inserted a de-esser on each backing vocal channel, we would need four of them. This would be an expensive approach. Alternatively, we could route all the backing vocals to a subgroup and insert a single de-esser in the subgroup channel—a much more cost-effective approach to using a scarce resource. This approach also saves time, since only one set of connections and one set of adjustments have to be made on the inserted processor.

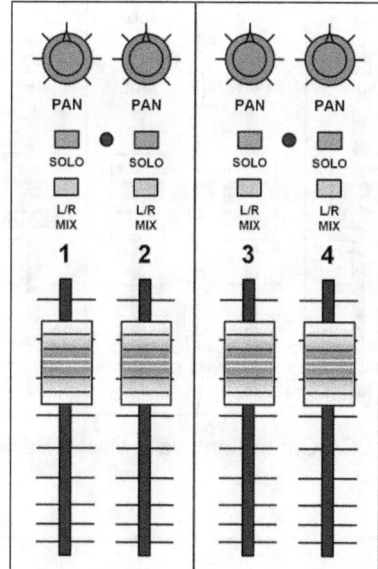

Figure 9-6   Subgroup Controls for a 4-Bus Mixer

Notice that just like the input channel, the subgroup channel also has a routing assign switch and a pan pot, and these controls work the same way as on any input channel. Given their hierarchy in the mixdown process, though, subgroups can be routed only to the main L and R mix busses. In larger mixers, each subgroup may have its own EQ controls, just like an input channel strip. Remember, too, that each subgroup has its own output jack. This means that the signal on each subgroup is available to feed a multitrack recorder, to provide additional stage monitoring, or to feed another location.

9.7   VCA Groups and Mute Groups

As we move up the mixer scale and the mixers become more complex and expensive, we may find a grouping mechanism, similar to subgroups, called VCA groups. A representative VCA master is shown in Figure 9-7.

VCA stands for voltage controlled amplifier. We first discussed this type of amplifier in Section 8.6 when we talked about compressors. The key thing to remember about a VCA is that its gain is controlled by a voltage external to the amplifier. In a mixer with VCA groups, the control of the signal levels of the channels is quite different than with a traditional non-VCA mixer. Figure 9-8 shows the difference. In Figure 9-8(A) we see the fader control of a channel on a regular (non-VCA) mixer. Note that the audio signal actually flows through the fader on its way to the subsequent circuitry. In Figure 9-8(B) we have the fader control of a channel on a mixer that has VCA groups. Notice now that the simple fader is replaced by a VCA, the gain of which is controlled by the fader which now adjusts the VCA's control voltage (V_c). No audio actually flows through the channel fader in this case. This design adds to the complexity and cost of the mixer because every channel must have this circuitry.

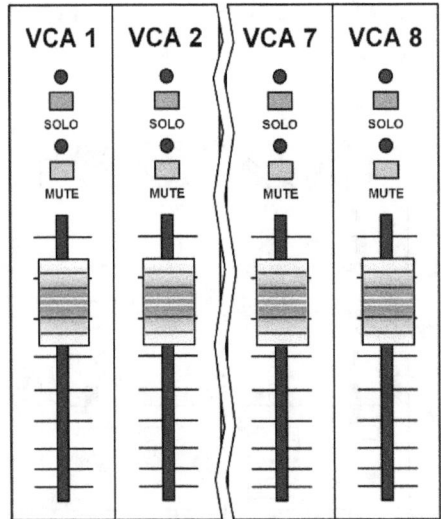

Figure 9-7 VCA Group Controls in Master Section of Mixer

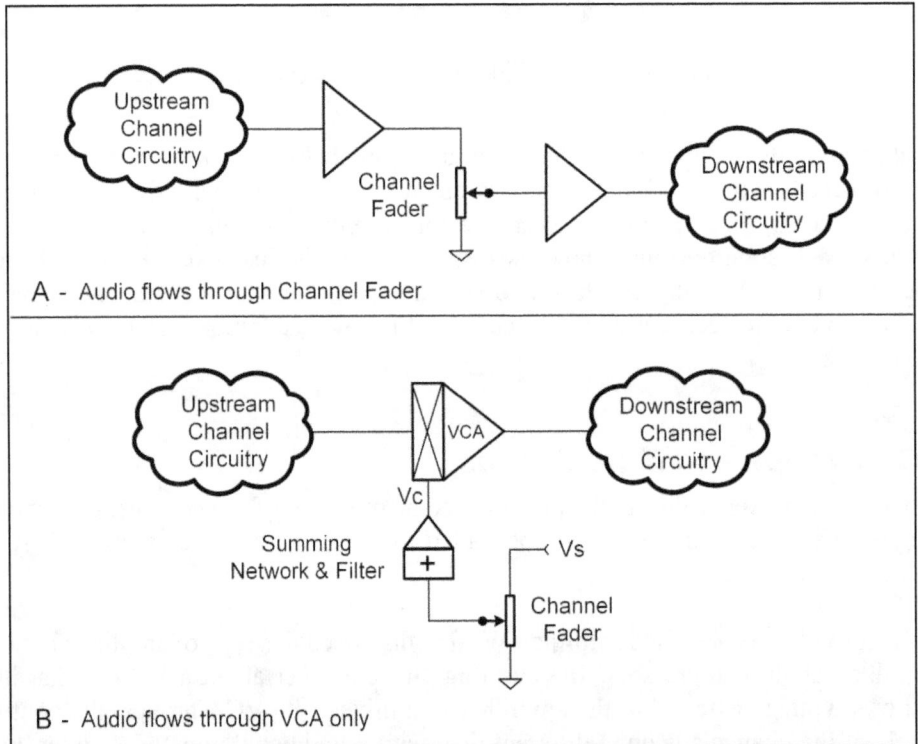

Figure 9-8 Regular Fader and VCA Fader Comparison

On a mixer with VCA groups, any input channel can be assigned to a VCA group, and a channel can be assigned to multiple VCA groups—just as with traditional subgroups. However, the similarity ends there because a VCA group is strictly a control structure. When we route a channel to a subgroup, we actually place its audio signal onto a physical bus, as Figure 9-9(A) shows, using the example of routing to subgroups 1 and 2. This audio is then summed with all the audio from the other channels that are similarly routed, and passed through the subgroup fader to the subgroup output. A subgroup has a physical output that can be used to feed an amplifier or effects processor, etc.

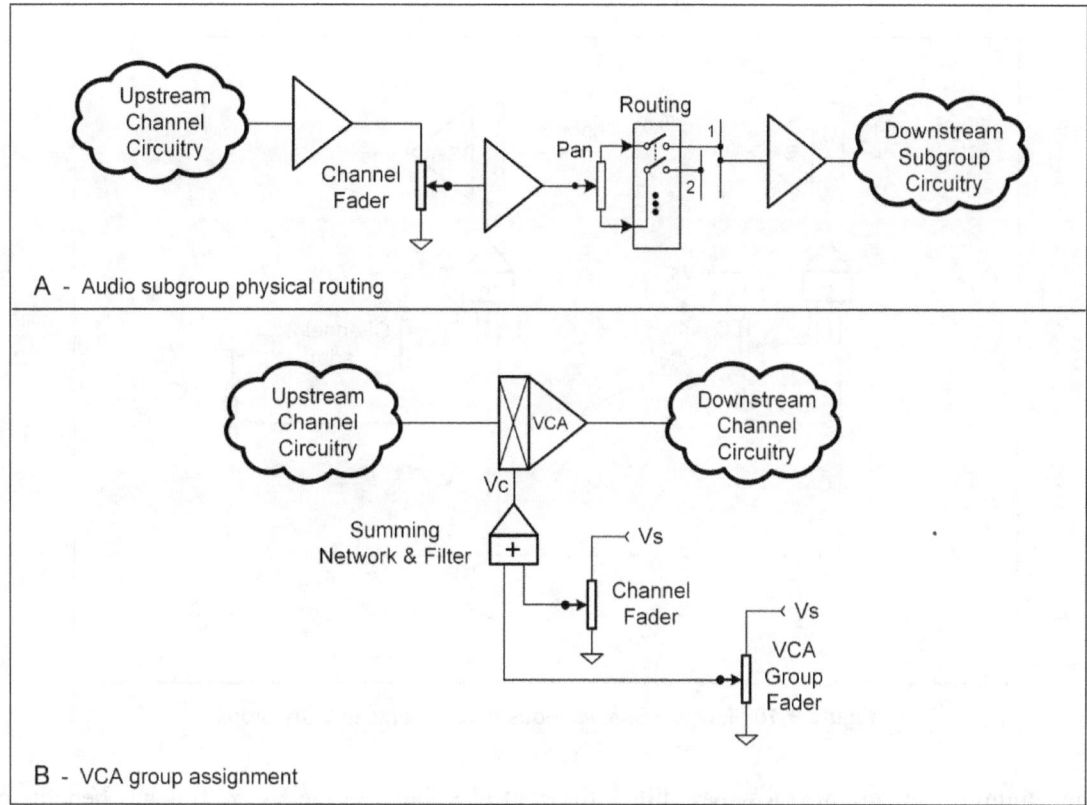

Figure 9-9 Audio Subgroup and VCA Group Comparison

In contrast, when we assign or add a channel to a VCA group, we are not actually sending its audio anywhere; we are not actually touching the audio stream at all. What we are doing is linking that channel's VCA control port to the VCA group fader. Figure 9-9(B) shows us that we now have two ways to control the gain of the channel's VCA—via the channel fader itself, and via the VCA group fader. Basically, the voltage from the channel fader and the voltage from the VCA group fader are combined in a summing network and fed to the channel's VCA control port. Note that since a VCA group fader controls the channel level directly, any feed to a post fader aux send will be directly affected by movement of the VCA group master.

When we assign a channel to multiple VCA groups, we are simply linking the output voltage from all of these group faders into the channel's summing network. Figure 9-10 shows us the linkages when we assign Channel 2 to both VCA Group 1 and VCA Group 2. In this case, we now have three ways to control the VCA of Channel 2. Figure 9-10 also shows that a single VCA group may be associated with multiple channels—here we see that both Channel 1 and Channel 2 are assigned to VCA Group 2. Again, this just means that the output voltage of VCA Group 2 fader is linked to the summing network of each individual channel VCA.

The VCA typically has a large gain control range of around 130 to 140 dB, and the control voltage from the summing network is usually scaled and offset to allow the working gain to vary from around −110 dB to +15 dB or so. Thus, it does not matter how may faders are linked to the summing network, even if all of them were to be pushed up to their maximum, we would still get only the maximum gain allowed by the summing network. Going the other direction, if any single fader in the assigned set is pulled all the way down, the gain of the VCA would be set to its minimum, as determined by the summing network.

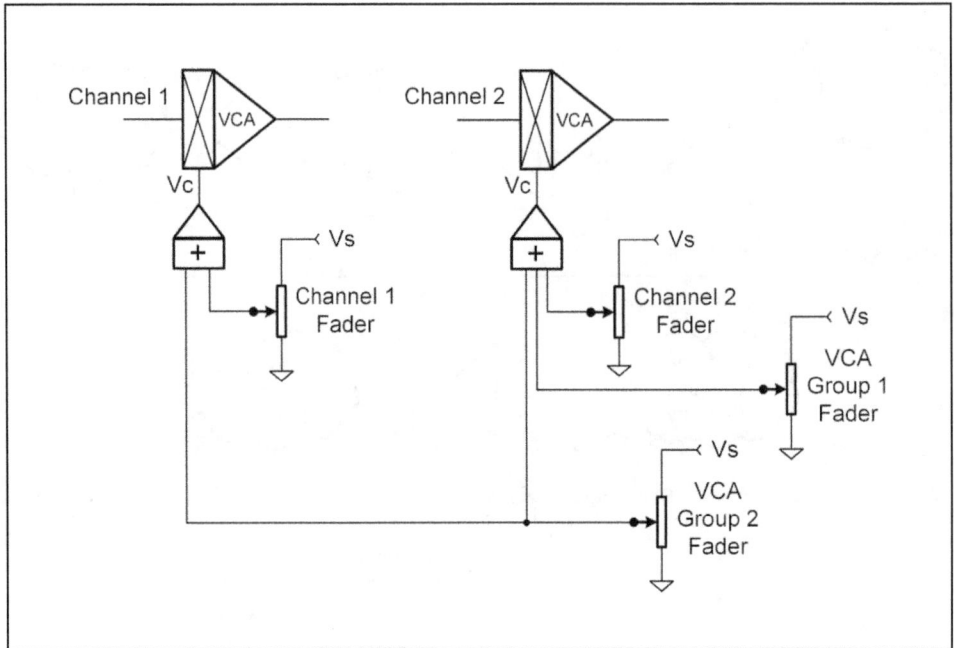

Figure 9-10 Multiple Simultaneous Assignments to VCA Group

The summing network also low-pass filters the control voltage to the VCA. This has benefits of removing any glitchy noise that may be generated in the fader itself, e.g., when a fader is worn (in regular non-VCA mixers, we hear this as scratchy audio when the fader is moved). Filtering is also useful in removing noise in systems that have digital fader control. In this case, the linearly encoded output of the fader is converted to a usable voltage by a small word length (e.g., 8 bit) digital to analog converter (DAC). That voltage is then added in the summing network. You will remember from Section 7.12.2 that the output of a DAC looks like a series of stair-steps, which is similar to an undesirable high-frequency signal (noise) superimposed on the desirable low-frequency signal (the level control voltage). The filter in the summing network removes this jaggedness and makes the changing of levels a smooth operation.

As Figure 9-7 shows, each VCA group master has associated SOLO and MUTE buttons that remotely control these functions for all the channels that are assigned to the group. By remote control, we mean that these functions solo or mute each channel individually at the channel level, typically using the channel's own solo or mute logic.

Consoles that implement VCA groups also usually implement additional and separate *mute group* functionality. You may also find mute groups on some higher end consoles that do not have VCA groups. The concept is similar to the VCA group feature in that it is purely a control structure wherein a set of channels is assigned to a particular mute group for the purpose of remote control. Now when the master MUTE button is engaged, all the channels assigned to that group are muted, again at the individual channel level, using the channel's own mute logic.

We can readily envisage a use for mute groups. It is not hard to imagine, for example, a theatrical production where different sets of actors are on stage during different scenes. We can then assign the microphone channels to appropriate mute groups so that with the press of just one MUTE button, only the microphones of the actors on stage are open, while all others are muted. As we move from one scene to a next with a different set of actors, another mute group would be activated. A VCA group, though, has no audio output and can't be routed anywhere—so what is its use? The main advantage of VCA groups is sophisticated grouping control. As a bonus, VCA groups give us this sophisticated control without introducing any further noise-inducing electronics into the audio stream. Anytime we need to set up a group-within-group or group-across-group scenario, VCA groups are the only trivial way to accomplish this.

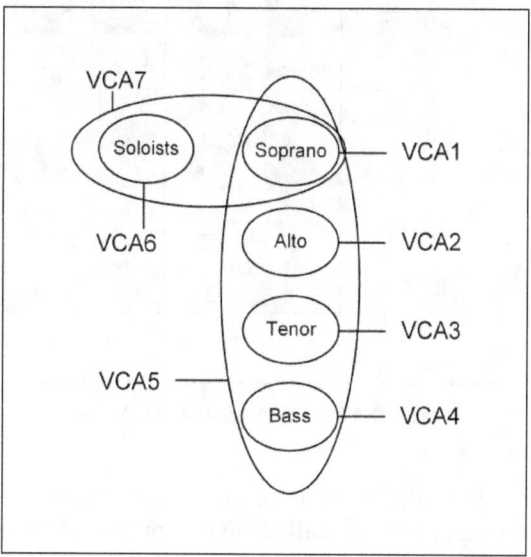

Figure 9-11   Example of VCA Grouping

Consider, for example, a large multipart choir that has to be dynamically balanced within itself as well as with one or more soloist and the band. The multiple microphones of each choir part (e.g., soprano, alto, etc.) may be assigned to a corresponding VCA group. These VCA group masters may then be used to balance the choir within itself. However, all the choir microphones may also be assigned to a single VCA group. This master may then be used to raise and lower the level of the choir as a whole and balance it against the band.

Now imagine further that the arranger has a scene where the sopranos have to be brought up together with the soloists, and these two groups balanced with the whole. Figure 9 11 shows us that now we have a subset of the choir microphones linked with extra-choir microphones and controlled as one (VCA7) while still maintaining autonomy of control of each choir part and the choir as a whole from the previous setup (VCA1 to VCA5). This type of grouping control could not be accomplished with regular subgroups without patching subgroup outputs into input channels, thus increasing physical wiring requirements and consuming scarce input channels.

9.8 Mix Matrix

On some larger mixers, the subgroup mixer-within-mixer concept is taken a step further to include a matrix section. Figure 9-12 shows what this may look like on a four-bus mixer.

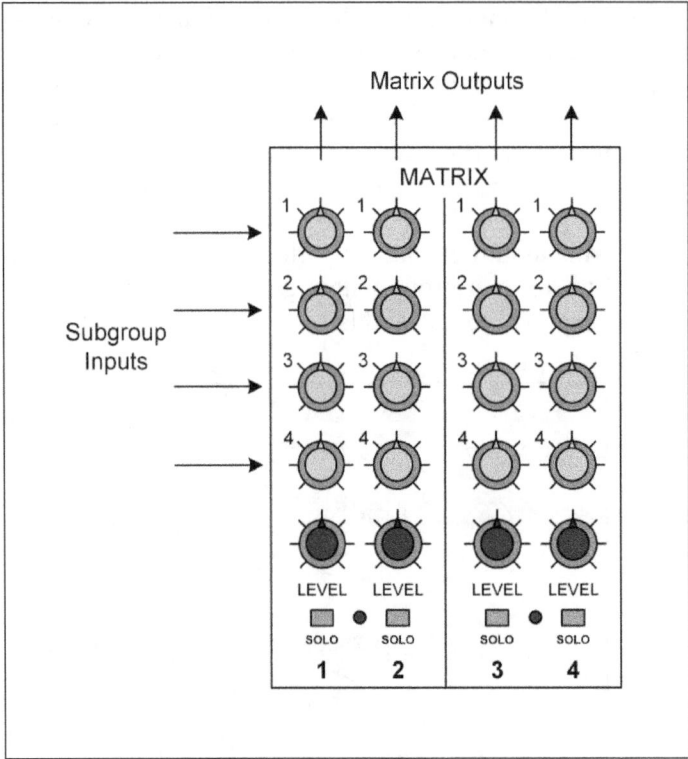

Figure 9-12 A Simple 4x4 Mix Matrix

Conceptually, a matrix is the intersection of a number of input busses—in this case, the subgroup busses—with a number of output busses, called matrix outputs. The intersection points allow a variable amount of the input signal to be placed on the matrix output bus. Thus, if we had a four-bus mixer and we wanted to have a matrix output for each subgroup, we would have at minimum, a 4x4 matrix (since the main L and R output busses sometimes also serve as inputs to the matrix, we could in fact have a 6x4 matrix). With a matrix arrangement, each subgroup can be mixed in any proportion with any and all other subgroups and the resulting signal made available at the matrix output jack. These signals can be used to feed multitrack recorders, provide refined monitor mixes, feed additional house amplifiers, provide a broadcast feed, etc.

9.9 Aux Returns

The aux returns are sometimes referred to as *effects returns*. This section is where we control the routing and levels of the signals coming back from the off-console effects processors. Better mixers have at least as many aux returns as aux send channels.

Figure 9-13 shows the controls for four aux return channels. Each aux return is a stereo input channel. Unless operating as a mono signal (usually using only the Left input jack), the left signal can be routed only to the main left or odd busses, while the right signal can be sent only to the main right or even busses. In some cases, the only routing option available may be the main left and right channels. Some aux return channels may have a balance control which sets the relative amount of the left and right signals that feed to the busses, just as for regular stereo input channels. In Section 9.13, we will see how the aux returns are used in conjunction with the aux sends to connect effects processors.

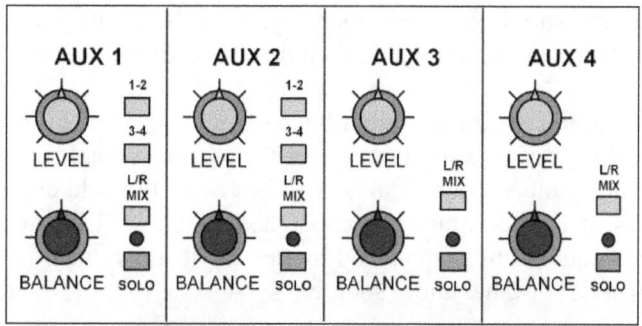

Figure 9-13 Aux Return Controls

9.10 Monitoring

The monitor section is a very important part of the mixer. It allows the engineer to listen to any set of signals flowing through the mixer. This section is the one area that can vary widely from mixer to mixer. Figure 9-14 shows the controls that may be found on a mid-sized mixer. All mixers will at least allow the engineer to perform three basic tasks required in monitoring, *viz.* select the source signal to be monitored, set the level of the signal being monitored, and select the destination where the monitored signal will be heard. Common destinations found on many mid-sized mixers are headphones, control room, and studio. In live sound reinforcement, there is no control room or studio, but these signals may be used to feed engineer's wedges all the same.

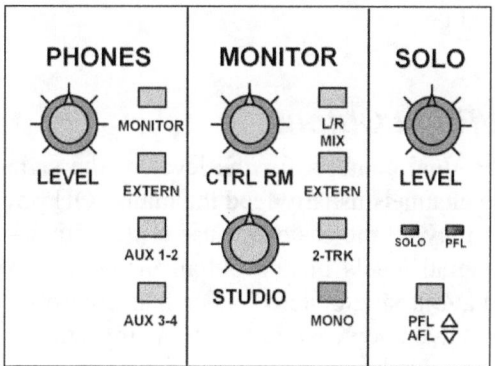

Figure 9-14 Monitor Section of Mixer

One very common monitoring task is soloing a signal. Engaging the SOLO function does not disturb the main mix output in any way, so it can be done at anytime, anywhere on the console. The solo function allows the engineer to monitor a specific signal or group of signals in isolation from the rest of the mix, while sourcing these signals from either before or after the channel faders (or in the case of aux returns, before or after the level controls). To support this, the monitor section should contain a level control for the SOLO monitoring function and a means of selecting between pre-fader listen (PFL) and after-fader listen (AFL) modes.

The solo function is engaged by first depressing the SOLO button at the source, i.e., the input channel, aux return channel, or subgroup. This action places the signal of interest on the solo left, solo right, and PFL busses and makes it available to the monitor section. PFL mode is used to get an accurate indication of the level of the signal on a channel. The PFL bus is thus usually a mono bus. This is the mode that should be used when adjusting the channel levels for optimum gain structure (i.e., gain vs. fader level). We will look at this in Section 9.16.

The AFL mode uses the solo left and solo right busses instead of the PFL bus. In contrast to PFL, AFL is a stereo mode that preserves the imaging of the channel signal, i.e., the solo left and solo right busses receive the same proportional signal levels as the odd/even subgroup and main left/right routed busses. It is thus sometimes called solo in place. This mode should be used for checking the presence or quality of a signal and to see how it sits in the mix. This is also the mode to use when cueing up a CD or other audio source.

The level meters are a very important part of the monitoring process. At a minimum, metering should be provided for each subgroup and the main L/R busses. These busses are where multiple signals are summed together, and the possibility for overload is high. The meters give a readily discernible indication of possible distortion on a bus, especially where in a noisy live environment it may not be easy to detect small amounts of distortion in the audio output.

On consoles without a dedicated solo level meter, one thing to remember is that the main L/R meters are wired so that they always display the level of the signals being monitored. When monitoring a regular mix (i.e., not solo), the zero mark on the meter is referenced to +4 dBu (1.23 V). When any SOLO button on the mixer is engaged, the main output L/R meters are used to display the internal level of the signal being soloed. In this mode, the zero mark on the L/R meters is referenced to 0 dBu (0.775 V). This makes the meter more sensitive in the lower region of its scale, and thus it becomes easier to make adjustments to the channel settings. This also means that it is common to see the level displayed on the meters jump higher when a SOLO button is pressed, even though the actual signal level did not change.

9.11 Main Mix – Left/Right Output

The main mix faders are the final controls for the level of the signal leaving the left and right channels of the mixer. These channels usually feed the main FOH power amplifiers and two-track tape outputs. These controls may be the easiest to use on the mixer—push up for more and pull down for less! Watch the signal levels on the output meter, use your ears, and set the levels accordingly. Some mixers have a single stereo fader that controls both left and right channels simultaneously, while others have two separate faders, as in Figure 9-15. Mixers with two faders allow easy adjustment of the final balance between the output signals, something that would otherwise have to be done either at the subgroups or at the power amplifier.

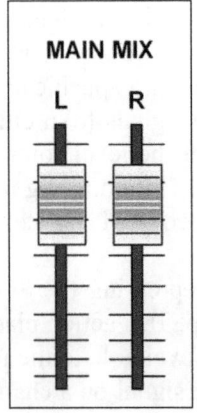

Figure 9-15 Main Mix Faders

9.12 Connecting Effects Processors via Inserts

There are two ways that effects processors are usually incorporated into the signal path. In the first, we want the entire signal to be processed by the unit. This is the case where we are connecting processors such as equalizers, compressors, or de-essers. The mixer accommodates this by providing insert jacks on each mono channel. Insert jacks are also found on each subgroup and each main mix output channel.

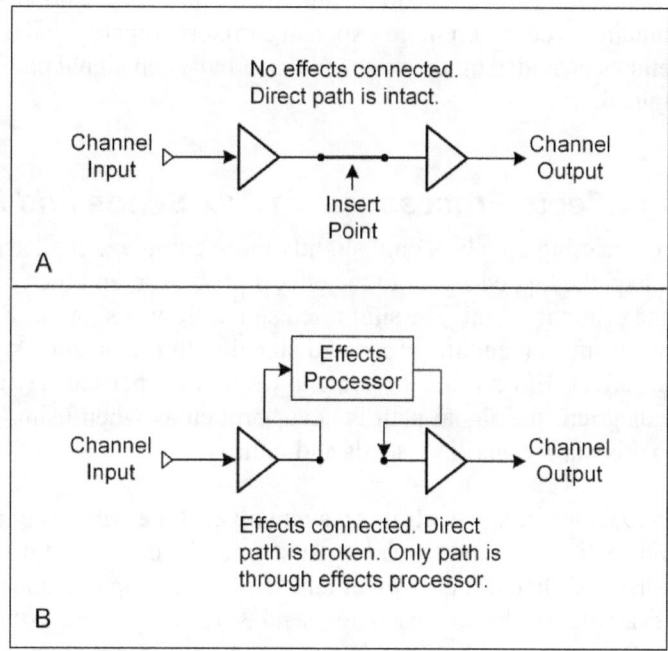

Figure 9-16   Connecting Effects via Normalized Insert Point

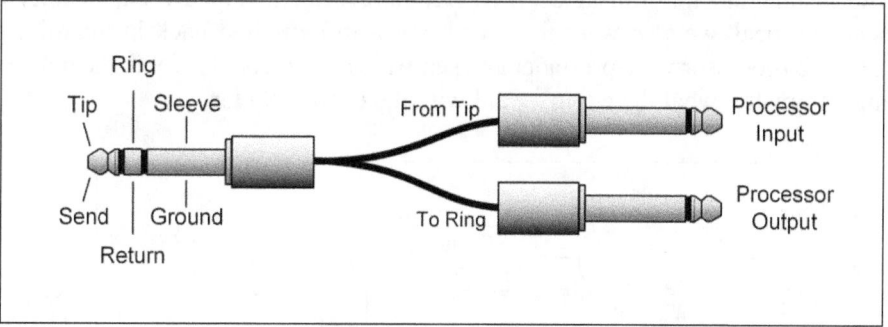

Figure 9-17   Insert Cable

The insert jack is normalized. This means that when nothing is plugged into the insert jack, it maintains the continuity of the circuit, as shown in Figure 9-16(A). However, when a plug is inserted, it breaks the path, and the only way to complete the circuit is through whatever the plug connects to, e.g., an effects processor as in Figure 9-16(B).

The insert jack accepts a 1/4-inch TRS plug, with the convention that the tip takes the signal from the mixer to the external processor, the ring connector brings the signal from the external processor back into the mixer, and the sleeve is a common ground between the mixer and the external processor. A special insert cable, like the one shown in Figure 9-17, is used to make the connection between the mixer and the external processor.

Because the signal path is broken when the TRS plug is inserted, the entire input signal is sent to any device that is plugged into the insert point. It can also be seen that with this arrangement of TRS connector, the signal is sent in an unbalanced configuration to the external device. However, since the signal is at line level and the distance between the effects processors and the mixer tends to be short, noise pickup along this connection would be minimal and a good signal-to-noise ratio would be easy to maintain. Even so, on more expensive mixers, separate balanced insert send and return jacks are sometimes provided on the mixer so that a balanced signal path between connected devices can be maintained.

9.13 Connecting Effects Processors via Aux Sends and Returns

The other method of connecting effects is only slightly more complex. The basic principle is to tap the signal of interest, feed the tapped signal to the effects processor, and then mix the output of the processor back into the original signal. The simplified signal flow is show in Figure 9-18. We take this approach when we want to blend the processed signal with the original signal, so as to finely control the level of the effect. Effects like Reverb and Chorus are particularly used in this way. As we can see from the diagram, the signal path is never broken as when using an insert point. The mixer allows us to do this via the auxiliary sends and returns.

As shown in Figure 9-19, connecting effects this way involves three sections of the mixer—an input channel strip aux send section, the corresponding aux send channel in the master section, and an aux return channel. First, we choose a post-fader aux send on the input channel strip as the source of the signal. In the example, we have chosen aux send 3. Here we control the level of the signal going to the corresponding aux send bus, and hence to the corresponding aux send channel in the master section of the mixer. We choose a post-fader aux send channel because we want the level of the signal coming back from the effects processor to track what the engineer is doing with the house signal via the channel fader. For instance, if we have a reverb effect on a vocal channel and we pull back the channel vocal, we also want the reverb vocal to be pulled back in the mix. If we were feeding the reverb processor via a pre-fader aux send, the reverb vocal level in the house mix would stay the same, no matter what the engineer did with the channel fader.

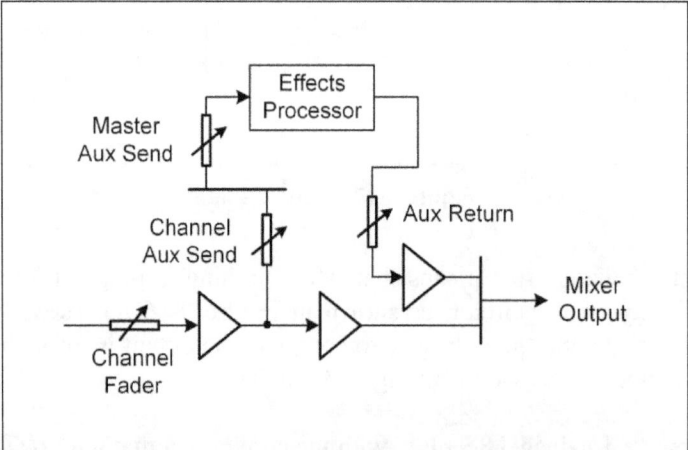

Figure 9-18 Connecting Effects via Aux Send & Return

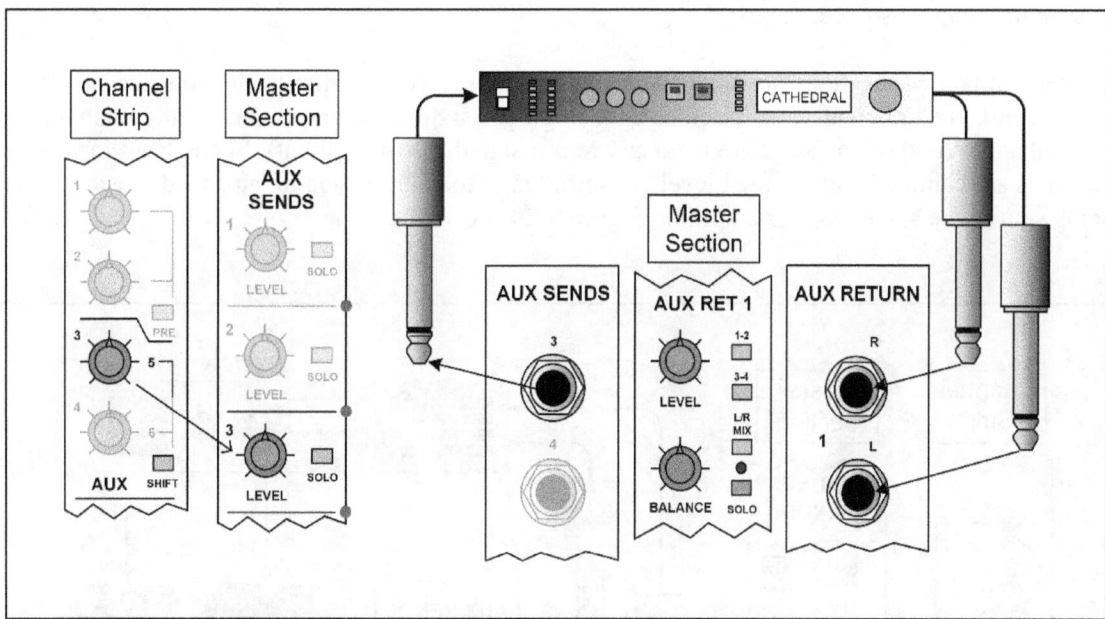

Figure 9-19   Mixer Connections for Effects via Aux Sends and Returns

In the master section of the mixer, the level of the signal taken from the aux send bus is controlled before sending it to the input of the effects processor. Unlike the case with using an insert point on an input channel, any number of input channels can send a signal to the connected effects processor simply by sending a signal to aux send bus 3. As these signals are summed on the bus, care must be taken not to overload the aux send summing amplifier. However, we can see that this is an excellent way to share a processor across many channels.

The mixer also provides a number of dedicated stereo input channels to accept the output from the effects processor and get the signal back on board. These are called aux returns or effects returns. Each pair of aux return channels has its own set of routing, level, and balance controls in the master section of the mixer. Here we decide where to send the returned signal. Depending on the sophistication of the mixer, we may have the option of routing the aux return signal anywhere in the mixer, or we may be restricted to sending it only to the main mix. This is also where we control how much of the processed signal we want to blend into the original signal. One thing to notice in this example is that we are taking the signal back into the mixer via aux return channel 1, even though we sent the signal to the processor via aux send channel 3. In this type of setup, there need not be any correspondence between the aux send and aux return channels used. However, to maintain simplicity and reduce confusion, it is better to keep the send and return channel numbers the same.

9.14 Providing a Stage Monitor Mix via Aux Sends

Without a dedicated monitor mixer, the pre-fader aux sends of the FOH mixer are usually used to provide a monitor mix for the performers. Unfortunately, unless the mixer is very large, this approach is quite limiting, with typically only between two to four pre-fader aux send channels available. We can have as many different stage monitor mixes as we have pre-fader aux send channels. We know that the audio engineer often times adjusts the channel faders during a performance to tweak the FOH mix. As we discussed in the previous section, pre-fader aux signals are immune from any kind of movement of the channel faders—the signal being tapped from *before* the fader after all. The pre-fader signal thus remains stable despite the actions of the engineer and so is ideal as a monitor signal source.

As far as mixer connections are concerned, the situation is very similar to the connections for an effects send. The only real difference is in choosing a pre-fader aux send channel rather than a post-fader channel (and of course, there is no aux return signal to worry about). Sends from individual channels are summed and the final level is controlled at the corresponding aux send master before being sent to the stage monitor amplifier. Figure 9-20 shows the setup.

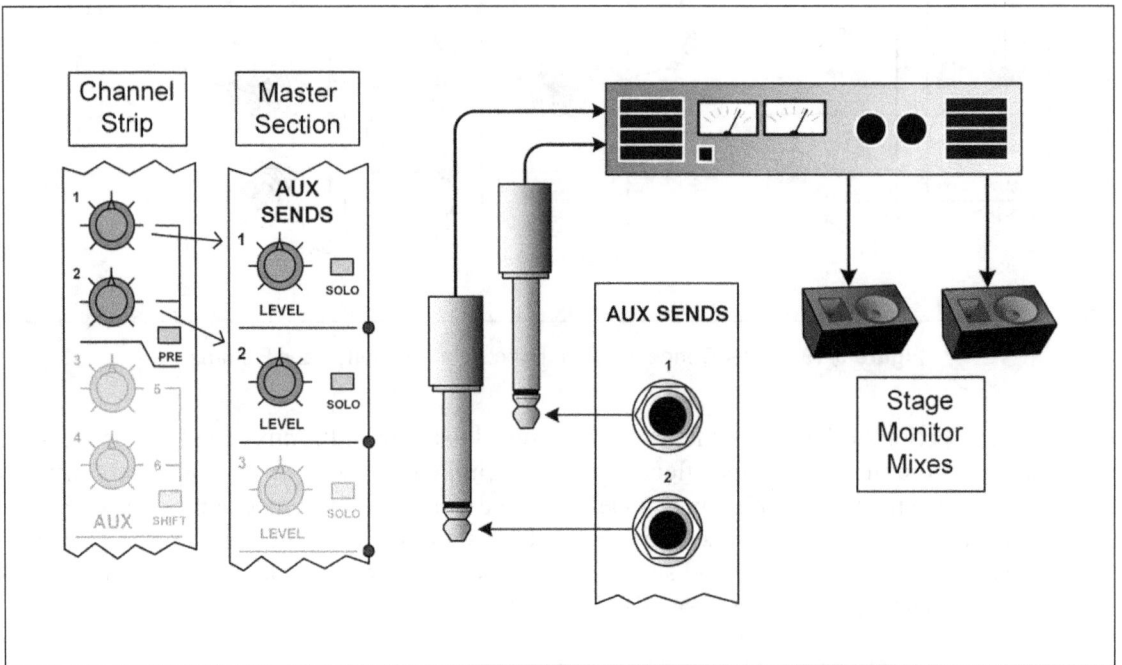

Figure 9-20 Stage Monitor Mix Using Pre-Fader Aux Sends

If we desperately need more stage monitor mixes than we have pre-fader aux send channels, we can use post-fader aux send channels that are not being used for effects sends. In a pinch we can also use subgroup outputs. Where the mixer has a Mix Matrix, these outputs can also be used. The main drawback with all of these approaches is that they are affected by the movement of the channel fader. In the end, the only real comprehensive solution to providing more than a few effective stage monitor mixes is to utilize a dedicated monitor mixer.

9.15 Mixer Gain Structure

Gain structure in general refers to the gain of the various stages of processing that a signal passes through, from initial input to final output of the system. Thus, it is sometimes referred to as gain staging. In general, the aim is to drive each stage optimally. This means that the input signal to each stage is well above the noise floor, insuring a good signal-to-noise ratio (SNR), but not so high as to cause overload distortion (clipping), while leaving enough headroom to handle the peaks of the signal. There are many ways of representing and working through the gain structure of the mixer, but we will build on the intuitive approach of Buick[1] to introduce the concept.

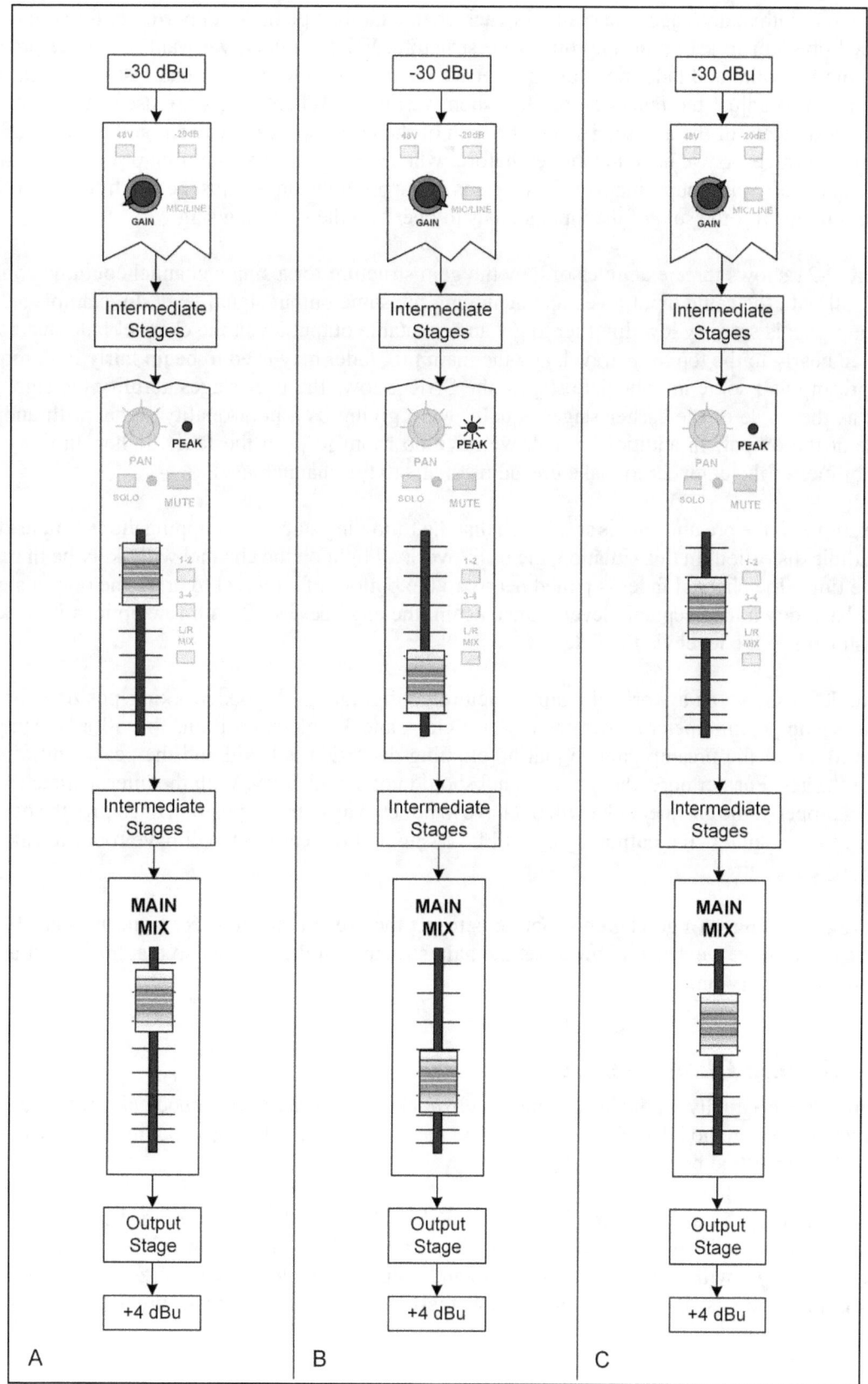

Figure 9-21   Low SNR, Clipping Distortion, and Good Gain Structure Settings

The mixer, with many stages internally for each channel, offers plenty of opportunity for an inexperienced operator to set up an improper gain structure. For the mixer, we want to set the preamp gain, the EQ gain, and fader positions so that we have high SNR, no overload distortion, and still have room to adjust the faders on the fly when warranted. When effects are used, we must also consider the gain of the aux send stage, the gain of the effects processor itself, and the gain of the aux return stage. As we have mentioned before, whenever we have a summing point in the mixer (e.g., aux sends, subgroups, main mix), the gains of the previous stages must be carefully controlled, since it is easy for the sum of the input signals to overload the summing stage.

Figure 9-21 shows three examples of how the gain structure for a single channel could be configured, all with the same input level, and achieving the same output signal level. In example *A*, the preamp gain is set very low. In order to get an acceptable output level, the channel fader has to be pushed nearly to the top of its travel, and the main mix fader may need to be up fairly high too. In the preamp and some intermediate stages, the SNR is low. The later stages amplify the signal as well as the noise of the earlier stages equally well, giving us a poor-quality signal with audible noise at the output. In addition to that, we have no room to push the fader higher, limiting the capabilities of the engineer to make fine adjustments to the channel level.

In example *B*, the preamp gain is set so high that the following stages are clipping the signal, leading to audible distortion. In this situation, the peak/overload light on the channel will likely be lit much of the time. The channel fader is pulled nearly to the bottom of its travel to bring the output signal level back down to acceptable levels. Once again, the engineer is left with few options in making adjustments to the level via the fader.

Example *C* shows us a workable gain structure configuration. A good working position for the preamp gain potentiometer is between the 11 o'clock and 3 o'clock position. This affords plenty of room to adjust the preamp gain so that no clipping distortion is heard and the peak light almost never flashes. Furthermore, the output signal should not sound noisy. With the intermediate stages driven properly, the channel fader will not have to be set in an extreme position to control the overall level of the channel. The output signal will be clean, and the engineer will have room to ride the fader when required.

The key to setting up a good gain structure is to get the preamp gain correct from the start. In the following section, we discuss how to set the gain structure of the channel so that the output audio sounds right every time.

9.16 Setting Channel Levels

So how do we actually set the levels for each channel to ensure we have a good gain structure? You should devise a method that works and use it all the time, so that it becomes habitual. We will walk through one method below.

Setting the proper gain structure for a channel should ideally be done during the sound check. However, it can still be done after the show has started. If we are setting up the mixer during a sound check, we will set all the group faders and output faders to unity gain (zero mark on fader scale). If the show is already in progress, leave these faders alone, and make sure the channel is not routed anywhere.

For each channel, start at the top and work your way to the bottom. Set the preamp gain to minimum (the knob fully counterclockwise). If this is a vocal microphone channel, the low-cut filter should also be engaged. Next, set all EQ bands to unity gain, i.e., no boost or cut, and then set the channel fader to unity gain (zero mark on the scale). Now depress the SOLO button on the channel. On some mixers, this may be called the PFL button. Now move to the master section of the mixer and set the solo mode to pre-fader listen, or PFL. The channel is now initialized to allow proper level setting.

Once the channel is initialized as above, we can proceed to set the level. We will use the main output meters, which double as solo meters on most mid-sized mixers, as an aid in setting accurate internal signal levels. Remember, when the SOLO/PFL button on any channel is depressed, the output meters are tied to the solo busses and are internally rescaled. The 0 mark on the meters is now referenced to 0 dBu instead of the usual +4 dBu. Remember too, in PFL mode the signal is mono, and on some mixers only one meter will show a signal. On others, the two meters are tied together.

Next, feed a typical signal into the channel. For a vocal microphone channel, have a performer speak or sing at the level expected to be used during the performance. For an instrument channel, the instrument should be operated at its expected output level. If at this stage the channel is already overloading, then either turn down the output level of the instrument, or engage the pad on the mixer's input if one is available. Now watch the meters and increase the preamp gain until the continuous signal is showing 0 dB and the peaks are between +2 and +6 dB. Do not forget to use your ears as well and listen for any distortion and noise. This is an audio exercise after all, and what we hear is more important than what we see.

Note that changes to the channel EQ gain will affect the overall gain structure of the channel. Therefore, if any adjustments are made to the EQ gain, the above procedure should be repeated. Similarly, any effects processor that is inserted into a channel will have an effect on the gain structure. Gain setting should be done with the processor in place and switched on.

Once you are done with a channel, disengage the channel's SOLO/PFL button and move on to the next channel.

9.17 Auditioning a Signal

Auditioning a signal involves procedures that are similar to level setting, but the goal is actually quite different. The goal of auditioning is to check the presence, state, quality, level, and imaging of a signal or group of signals in the final mix. Imaging refers to a signal's left/right position in the stereo field. This means that we need a stereo signal, after the fader, with appropriate effects returns.

Again, as with level setting, this procedure can be used during the sound check, or after the show is already in progress. It is frequently used during the performance to confirm that a signal is where we want it to be. In another example, we may need to cue up an audio source while a performer is in action. In this case, we just need to make sure that the channel is not routed anywhere and that the stage monitor aux sends are turned all the way down for that particular channel.

As in level setting, the SOLO button on the channel being auditioned channel is depressed, allowing us to hear the signal in isolation. However, instead of PFL, we select the after-fader listen (AFL) mode in the master section. This mode is sometimes called solo in place. This mode uses a solo stereo bus that receives signals in the same proportions as the group and main mix busses, and so preserves the stereo imaging of the channel in the final mix. Any number of channels can be soloed together. Thus, it is possible to audition a channel and its effects return, or two related channels to see how they sit together in the final mix.

9.18 Monitor Mixers

In Section 9.14, we discussed how we could provide a stage monitor mix by using the pre-fader aux sends of the FOH mixer. We also noted that since on most mid-sized mixers the number of aux sends is limited, and the available aux sends must be shared between effects processors and provision of monitor mixes, we may be left with only two easily controllable and stable monitor mixes. Thus, this approach will be unsatisfactory in all but the smallest of live performances. The real solution to this problem lies in the use of a dedicated monitor mixer to provide the necessary stage monitor mixes.

The concept of the monitor mixer is really quite simple. It takes the idea of the mix matrix that we discussed in Section 9.8 and expands that to be the structure of the entire mixer. From Figure 9-22, we can see that the entire aux send section of each input channel is replaced by monitor sends. These form the matrix inputs, so to speak. Figure 9-23 shows us that each monitor send is tied to a corresponding monitor bus. There are no routing assignments to subgroups or main L/R mixes, as these busses typically do not exist in a mid-sized monitor mixer. The subgroups and L/R outputs are replaced by the monitor outputs, which, as Figure 9-24 shows, get their signal directly from the monitor busses. These form the matrix outputs. There is one monitor output for each monitor bus. One thing to notice is that each and every output has an insert point. A dedicated EQ is normally inserted here to tailor the sonic characteristics of each individual monitor mix and help control feedback.

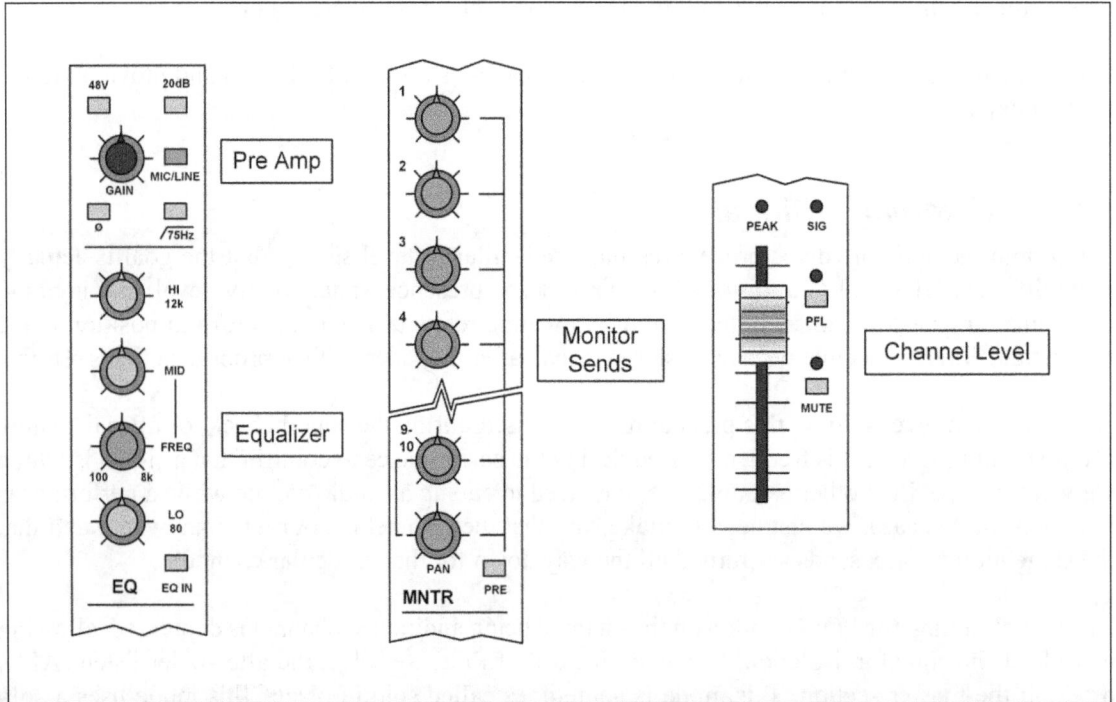

Figure 9-22   Monitor Mixer Channel Strip (In Three Sections)

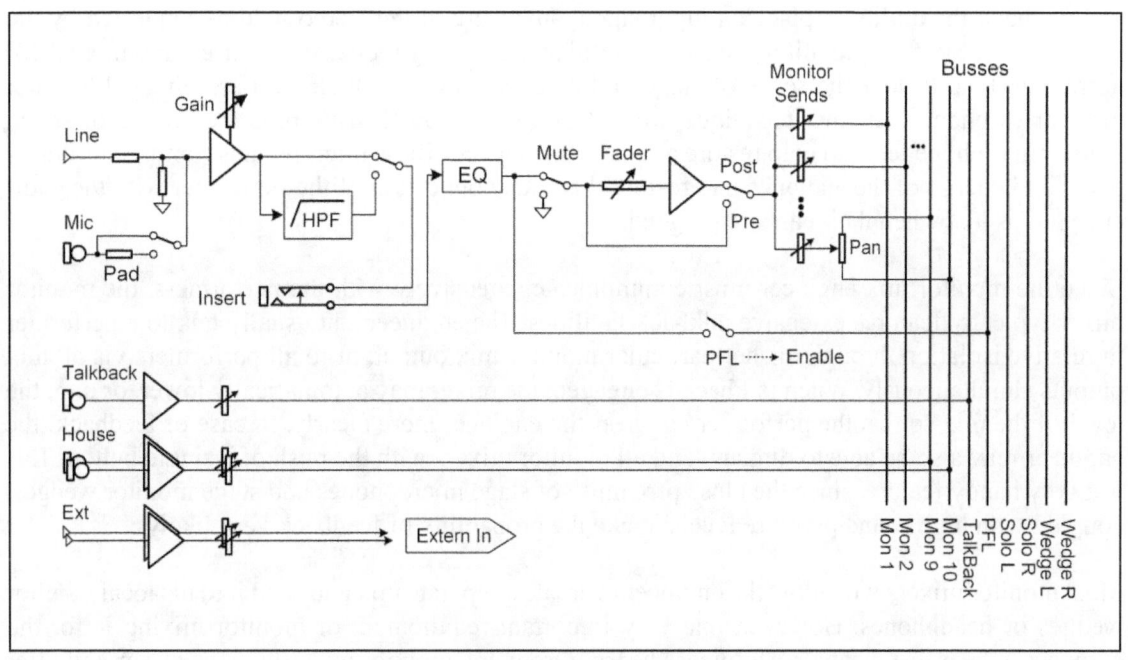

Figure 9-23 Monitor Mixer Input Side

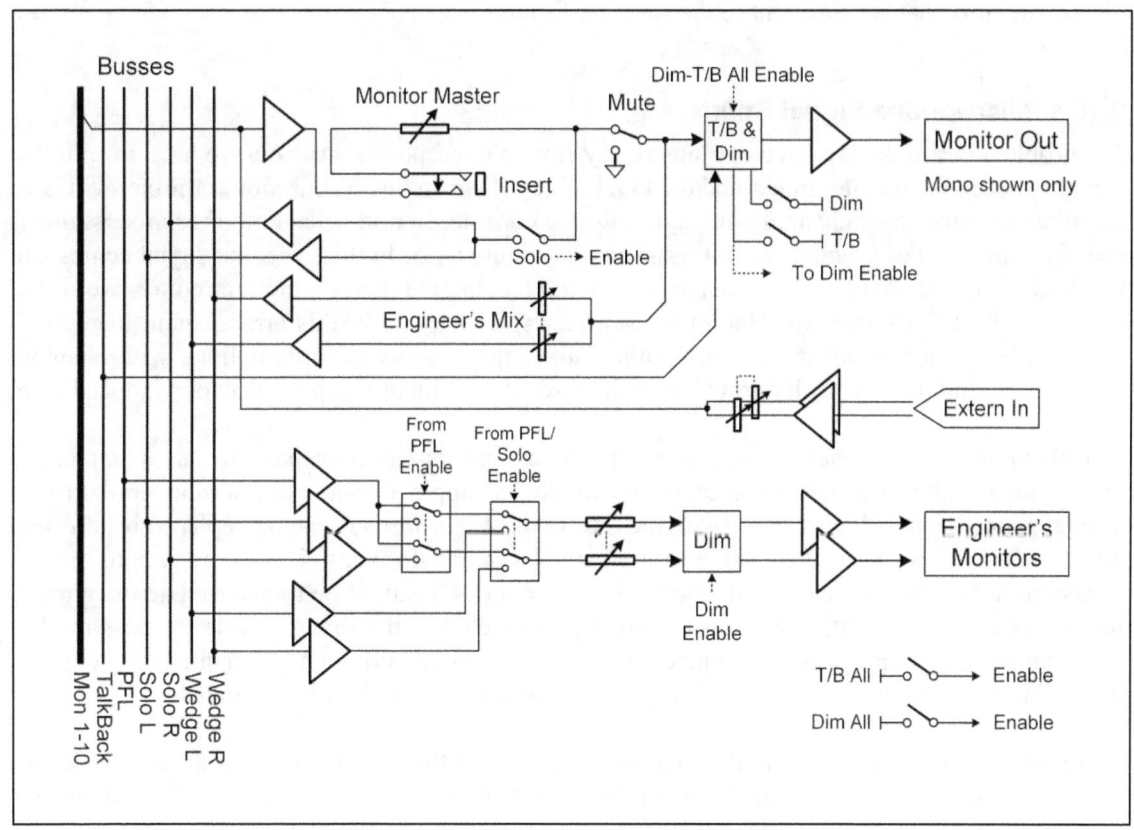

Figure 9-24 Monitor Mixer Output Side

As can be seen in these figures, some monitor busses (9 and 10 in the example) are stereo bus pairs, and we have the ability to place an input signal anywhere in the stereo field as requested by the performer. These stereo monitor mixes are useful and generally requested by in-ear monitor (IEM) users who tend to lose the sense of stage and ambience because their ears are plugged by their monitor earphones. Another technique utilized extensively by IEM users is the mixing of stereo house microphone feeds to help restore a sense of ambience. These microphones capture the atmosphere and energy of the audience and restore that vital connection of the performer with the audience that is lost when their ears are plugged.

Since the monitor mix engineer must communicate extensively with the performers, the monitor mixer typically features extensive talkback facilities. The engineer can usually talk to a performer in relative isolation through his/her particular monitor mix output, or to all performers via all mix outputs simultaneously. When talkback is engaged, the mixer may automatically lower, or dim, the level of the mix so that the performer can hear the engineer more clearly. In case of feedback, the engineer may also be able to dim any and all monitor mixers with the push of a single button. This is a very handy feature, since the close proximity of stage microphones and stage monitor wedges, coupled with high sound pressure levels, make the probability of feedback very likely.

Most monitor mixers will allow the engineer to create a separate mix and send it to his local monitor wedges or headphones. However, one very important requirement of monitor mixing is for the engineer to hear just what a performer is hearing from any monitor mix. Thus, the mixer will offer the ability to PFL any input channel, or solo any individual monitor mix. These functions will typically override any personal mix that the engineer has created, but as with the FOH mixer, will not disturb any mix that is being sent to the stage performers.

9.18.1 Microphone Signal Splitters

The problem is simple. We have one signal, say from a microphone, but now we have to split that signal between the monitor mixer and the FOH mixer. If we are lucky, our monitor mixer will have a splitter for each input channel built right into the console. In mid-priced monitor mixers, this is usually comprised of a simple Y-split, as shown in Figure 9-25. In this case, the signal comes into the mixer and goes right back out again, with a tap feeding the mixer's input circuitry. We would then feed the FOH mixer from the monitor mixer split outputs. In this arrangement there is no isolation between the two mixers, and either mixer may supply the microphone with phantom power. A ground lift is typically provided on the downstream link to help control loop induced hum.

Not all monitor mixers have split outputs, however. In this case, we have to use a standalone microphone splitter. The very cheapest ones will have a simple Y-split configuration similar to the arrangement shown in Figure 9-25 (with the feed to the *Mixer Input Circuitry* replaced by a second XLR output). More expensive passive splitters will use a transformer to isolate the various outputs as shown in Figure 9-26. Note that since direct current (DC) cannot be transferred across a transformer, one of the outputs must still be directly connected to the input in order to pass the DC phantom power. The mixer that is connected to the direct output must supply phantom power to the microphone. In this case, though, the connected mixers are isolated for each other.

If we need to split the microphone signal more than three times, it is better to go with an active splitter. Remember that as more mixer inputs are fed from a single microphone, the load on the microphone output increases. This causes the microphone output voltage to fall, thus decreasing the signal-to-noise ratio. Basically, the microphone signal is degraded, becoming noisier. Active splitters are typically used when we have to feed several mixers simultaneously, such as the FOH, monitor, live recording, and live broadcast mixers.

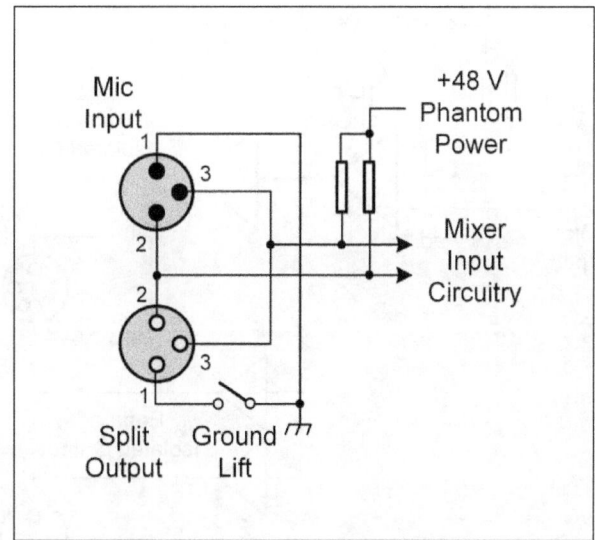

Figure 9-25   Microphone Signal Splitter Built Into Mixer

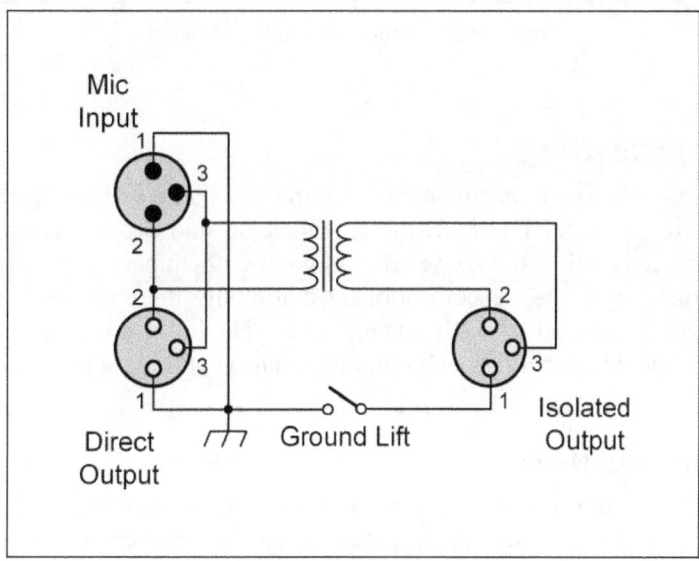

Figure 9-26   Passive Microphone Splitter

Figure 9-27 shows the block diagram of the active splitter. In this case, no output necessarily needs to be directly connected to the input, as phantom power can be sent to the microphone from the splitter's own power supply. The load on the microphone output is restricted to the splitter's single input preamp, no matter how many outputs the unit is feeding, so the signal quality of the microphone is maintained. All outputs are isolated from each other. As we can see, this arrangement is quite flexible—some outputs may be electronically buffered, while others may be transformer isolated from the mixer.

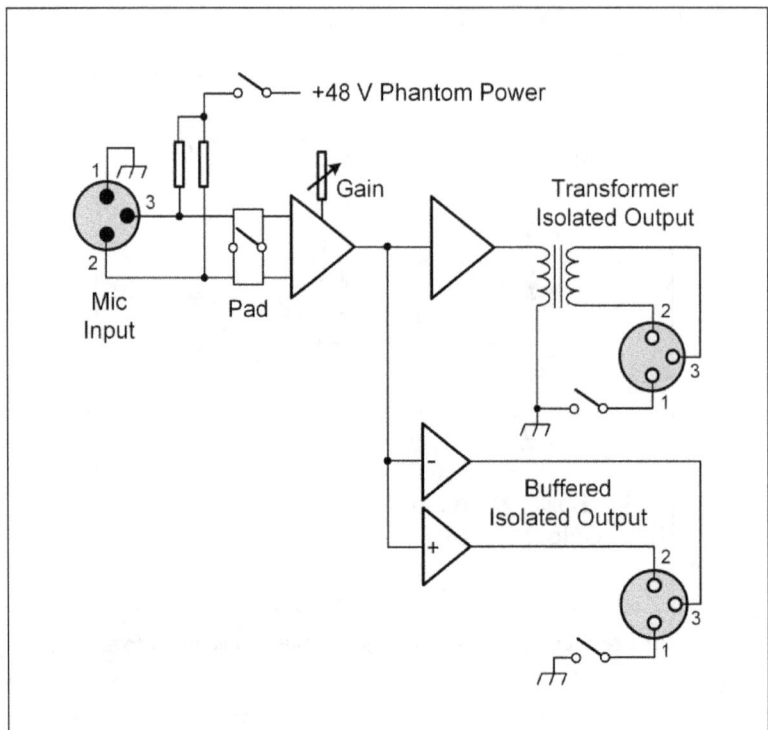

Figure 9-27 Active Microphone Splitter

9.19 Mixer Specifications

A mixer shares many of the same specifications of amplifiers that we discussed in Chapter 5. These include Frequency Response, THD, CMRR, Residual Noise, and Crosstalk. Their interpretation is the same as we discussed earlier, and so we will not go over them here. Refer back to Chapter 5 to refresh your understanding on these specifications. Additionally, there will be a list of specifications describing the equalizer section of each channel strip. Having covered equalizers in depth in Chapter 8, these too should pose no problem in interpretation at this stage.

9.19.1 Equivalent Input Noise

One specification that we did not cover before is Equivalent Input Noise, or EIN. EIN is a specification that is associated with microphone preamps, and since every microphone channel on a mixer contains a microphone preamp, we will see this specification associated with the mixer. EIN attempts to tell us the noise that we will see at the output of the preamp due to the internal noise producing components of the preamplifier itself. But if it is output noise that EIN is measuring, why is it called input noise, and what is it equivalent to? The answers to these questions lie in how the noise is modeled. Consider that the self-noise of the preamp is produced by many discrete components that are distributed throughout the amplifier circuitry. As a simplification, all of these discrete noise sources can be lumped into a single equivalent noise source, connected to the input of the amplifier, while the preamplifier itself is considered as noiseless. This single noise source is called the equivalent input noise source, and the noise thus modeled is called equivalent input noise. With the noise modeled this way, it is easy to see that the actual noise appearing at the output of the preamp is highly dependent on the gain setting of the preamp. This self-noise will be amplified equally as well as the microphone signal that we are interested in. As you can imagine, with several microphone channels on our mixer all being summed to the same output, it pays to have very quiet preamplifiers.

EIN is measured by first setting the preamp gain to a high value, typically maximum. Then a 150 Ω resistor is strapped across the input of the preamp. An RMS voltmeter, fitted with a band-pass filter and perhaps a weighting filter, is then used to measure the noise at the output of the preamplifier. The gain of the preamp is finally subtracted from the noise reading to derive the EIN number. A filter is used to band-limit the noise because the greater the bandwidth of the measurement, the greater the noise reading will be, and we are only interested in measuring the noise in the range of human hearing.

You should see all of the conditions of the test measurement listed with this specification in order to make full sense of it. Thus, a manufacturer should publish a statement like:

EIN: −129.0 dBu, 150 Ω source, 22 Hz to 22 kHz, A-Weighted

The resistance used across the input is important because this resistance is itself a source of noise. Recall from Section 5.9.1 that even passive components generate noise due to the random motion of their electrons. This noise can actually be calculated and depends on the resistance of the component, its temperature, and the bandwidth of the measurement. At 17 °C, a 150 Ω resistor generates about 0.219 µVrms of noise over 20 Hz to 20 kHz. This is equivalent to −131 dBu. Thus, we would normally never expect to see an EIN specification better than −131 dBu. However, manufacturers may do several things to improve this specification, such as use a narrower band-width for the measurement, or use a weighting filter. As we saw in Section 5.9.1, the A-weighting filter rolls off the high-end noise by 3 dB and low-end noise by over 6 dB. The manufacturer may also short the input instead of using a 150 Ω resistor and eliminate this source noise altogether.

Note that mixers destined for markets other than North America may use ITU-R 468 methodology when measuring the noise of the preamp. This recommendation specifies the use of a quasi-peak measuring device rather than RMS. Additionally, the ITU-R 468 weighting filter which emphasizes midrange noise in the region around 6 kHz (shown in Section 5.9.1) is also used. You will know that you are looking at European test results because noise levels measured according to ITU-R recommendation are presented as dBqps (some manufacturers present this as dBqp) rather than dBu. Furthermore, the European tests typically use a 200 Ω resistor on the input rather than 150 Ω. A 200 Ω resistor produces 1.2 dB more noise than a 150 Ω resistor. These three factors combine to give an EIN figure that appears to be 10 to 12 dB worse when compared to the EIN figure prepared for the North American market. We must therefore be careful to note the dB reference and test conditions when making comparisons of EIN to make sure we do not come to the wrong conclusion when reviewing a mixer's microphone preamp noise specification.

9.19.2 Other Noise Specifications

Noise is a big deal in a mixer. A mixer replicates functional blocks of preamps, equalizers, groups, etc. many times over. We can view each of these blocks as a source of noise. In the end, everything is summed together and presented as one output. Therefore, in addition to the EIN, there may be several noise specifications given for various sections of the mixer—channel, subgroup, main mix, etc. These are the real noise figures that must be contemplated when looking at mixer noise because, no matter how good an EIN specification the mixer has, it is at one of these points (i.e., channel, subgroup, main mix) that you will get the audio out of the mixer. The key things to look for here is what the noise is referenced to (e.g., +4 dBu) for SNR measurements, the gain setting and fader level used during the test, and number of channels assigned or muted. A noise specification produced with all the channels muted or not routed, minimum gain setting, and faders at -∞ may be interesting, but not particularly useful, as this is not the way we use the mixer. Look for noise figures specified with all the channels assigned and not muted, with reasonable (e.g., unity) gain set, and reasonable (e.g., 0 dB) fader position. A figure of −95 to −100 dBu for output residual noise is typical for good mixers.

9.20 Choosing a Mixer

When it comes to choosing a mixer, we will be looking at far more than the few specifications discussed above. Outside of budget, there are a few key things that must be considered. One of the first questions to be answered is the role of the mixer—FOH or monitor. While some consoles may function just as well in either role, most are suitable for only one.

The number of input channels available on the mixer is a consideration that must be dealt with early on. You always want to get more channels than you think you will need. In a dynamic situation you will always need more channels than you initially thought, and it is always a good idea to have a few spare channels in case a channel in use fails.

If the mixer is to be moved from show to show, then size and weight become a consideration. This, of course, depends on the number of input channels. As we will see in the next chapter, some digital mixers offer modularization that distributes the weight over a few pieces instead of concentrating it all in the console. Even further savings in size and weight may be had if the console uses layering.

Depending on the complexity of the mixing session, features such as the number of subgroups and availability of matrix outputs or direct outputs may become important. The ability to store console settings also becomes important when the shows become large and the number and speed of change of acts become high. Again, you may be looking at a digital mixer for some answers and relief.

When it finally comes down to it, though, sound quality and personal taste will also play a part in the decision. Sit in front of a few consoles and see how you like the feel and layout of the surface. Do the controls feel solid and does the layout facilitate an efficient workflow? Push the mixer with both low-level and high-level signals and listen for noise and distortion. The mixer is the one piece of equipment in the sound system where you will be spending the most time, and its character will have the greatest effect on the sound you produce. Take a little time up front to get it right and get the right one. The investment in time will pay big dividends in the long run.

Review Questions

1. Explain the functions of a mixer.
2. Discuss the pros and cons of a powered mixer.
3. What are the outputs found on a typical mid-sized mixer and from which busses are they fed?
4. Explain whether a pre-fader or post-fader send would be used to feed a stage monitor mix.
5. Discuss the use of the subgroups in a typical midsize mixer.
6. What is a VCA group? How is it different from a regular subgroup and how is it used?
7. What is a mix matrix and what are its uses?
8. Describe how you would use the SOLO and PFL/AFL functions of the mixer to monitor a channel's signal and hear how it sounded in the final mix.

9. Explain the calibration of, and the indications on, shared output/solo meters when switching from monitoring the main output to SOLOing a signal.

10. Explain how the normalized insert path works to connect an off-console effects processor.

11. Explain how the aux sends may be used to connect an off-console effects processor.

12. Describe the step-by-step process used to set the proper gain structure for a channel.

13. Discuss the main differences between a monitor mixer and a front of house mixer.

14. What are the various types of microphone signal splitters and what are the advantages and disadvantages of each?

15. What does the EIN specification of a mixer tell us?

Chapter 10

Digital Mixers

10.1 System Overview

As in most things electronic, mixers have also moved into the digital realm. While initially expensive, mid-sized digital mixers are now competitively priced with good analog mixers. As uptake increases and they become more popular, we can expect to see digital mixers moving into the lower end of the mixer scale, just as happened within the effects processor space (practically every effects processor available today is digital). A good example of an affordable digital mixer with an extensive feature set is the Behringer X-32 shown in Figure 10-1.

Figure 10-1   Behringer X-32 Digital Mixer

At a very high level, the digital mixer is very similar to a digital effects processor in its conceptual layout, with a number of extra things thrown in. On the input side, Figure 10-2 shows us that the analog audio is first converted to the digital domain by the analog to digital converters (ADC). There is an ADC for each analog input port. The mixer will also have a digital interface receiver (DIR) for each digital input port it has. Where the audio stream goes after entering the port is determined by the input routing module, under the control of the user. On the output side we have the reverse of the input. The output routing module feeds audio streams to the physical output ports, which can be either analog or digital. The analog ports are fed from digital to analog converters (DAC), while the digital ports are fed from digital interface transmitters (DIT) and possibly sample rate and format converters. There is a DAC or DIT for every physical output port provided on the mixer.

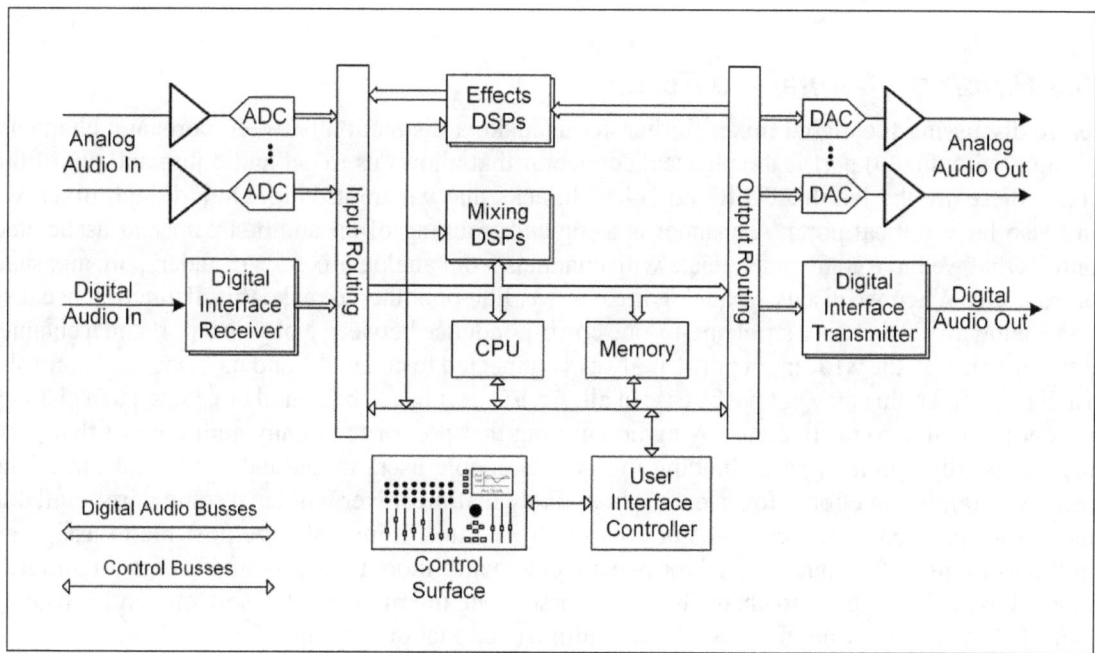

Figure 10-2 Digital Mixer Overview

Once the digital audio stream gets into the mixer, the user can send it practically anywhere via the input and output routing modules. The user only indirectly interacts with these modules through the mixer's user interface (discussed below). The routing tasks consist of familiar functions, such as aux sends and returns, matrix sends, and output sends. Within the mixer, channels can also be assigned to virtual groups and their signals controlled by a single fader, just like in the analog world. These groups are called digitally controlled amplifier (DCA) groups in the digital mixer terminology, borrowing from VCA groups in the analog domain. DCA groups conceptually operate the same as VCA groups discussed in Section 9.7.

Because of the large amount of data flow and complexity of operations being performed, the mixer will likely have a dedicated central processing unit (CPU) to direct much of the operation of the many modules in the mixer. There is thus a conceptual separation between control data and audio stream data and the busses used to carry that data within the mixer. For example, data that is generated as a result of a user setting a parameter via the control surface is treated quite differently from data that is generated due to sending the audio stream from the ADC to a routing module. The mixer will also have a certain amount of memory that is used in data manipulation and memory that is used to store user settings for later recall.

The actual mixing of the signals is done by the mixing digital signal processors (DSP) under direction from the CPU. The mix busses are conceptually contained in, and shared among, the mixing DSPs. The effects DSPs are used to provide a virtual effects rack built right into the mixer, as well as multiple sets of EQs. It is not uncommon to have up to eight virtual effects processors and twelve 31-band graphical EQs available in some of the better digital mixers. The effects DSPs may take their input signals from the output routing module and send their output to the input routing module, or they may have dedicated input and output routing modules. These effects and EQ processors are available to be patched into any bus in a way that is conceptually similar to the procedures used with analog mixers. The effects DSPs also provide in-line filter, EQ, delay, and dynamics processing for the various channels of the mixer.

10.2 Binding Channels to Ports

Before discussing the digital mixer further, let us make a distinction between ports and channels. An input/output (I/O) port is the physical connector that allows us to get audio into and out of the mixer. These are the familiar XLR and 1/4-inch jacks that we are used to. On a digital mixer, we may also have optical ports. A channel is a logical grouping of an audio stream and associated controls that we have come to associate with channels from analog mixers, *viz.* fader, pan, aux send controls, etc. When we discussed analog mixers, we did not bother to make this distinction because in an analog mixer there is a strict one-to-one correspondence between a physical port and a channel on the mixer, e.g., the XLR input port 1 is always connected to channel 1 and its associated controls. In a digital mixer, this may not be the case at all. An I/O port has to be bound or patched to a channel in order to be able to manipulate any audio entering that port or to get any audio out of that port. There is usually a default patch binding in place to enable users to get audio into and out of the mixer with minimum effort. However, it is also likely that the mixer will have some ports available that are not patched to any channels or controls by default. It may also be possible to assign an input port to more than one channel, or patch a channel to more than one output port. In a digital mixer, all patching of ports to channels is, of course, done internally via the software and is usually controlled by the user from the mixer's large multifunctional display interface.

10.3 The Control Surface

The human-machine interface of a digital mixer is called the control surface. This consists of all the knobs, sliders, buttons, etc., and perhaps a multifunctional display (MFD) that the operator uses to route the audio, adjust levels, assign and adjust effects, etc. The MFD may have touch-screen capabilities, presenting a familiar operational model to any tablet-computer user. The term control surface is appropriate because on a fully digital mixer no audio stream flows through any control device on the control surface. For instance, on most mid-sized analog mixers, the audio actually flows through the channel fader potentiometer, whereas on a digital mixer, the channel fader's position is read by the user interface controller and sent to the DSP which uses it to perform the appropriate level setting calculation on the digital audio stream. Similarly, rotary potentiometers of the analog mixers are replaced by rotary encoders on the digital mixers, and simply provide parameter level increase or decrease indications to the processor.

On a typical mid-sized digital mixer, the control surface may look rather sparse when compared to a similarly sized analog mixer. Unlike an analog mixer, there may not be a dedicated input fader for every input channel. There surely will not be a dedicated physical control (i.e., slider, rotary encoder, or switch) for every function (e.g., level setting, EQ adjustment, routing) on every channel. For example, a 32-channel mixer may have only 16 physical channel strips on the control surface. As Figure 10-3 shows, each of these channel strips may consist only of a multifunction rotary encoder, a select switch, function button, solo and mute buttons, and a linear encoder for the fader. There will, however, be an area on the control surface which has a single complete set of controls for all of the functions we are used to seeing on a typical channel strip, *viz.* channel gain, high-pass filter (HPF), and comprehensive parametric EQ. In addition, there may also be dedicated controls for the dynamics processors that are available to be inserted on every channel, but on some mixers, these controls may be the same ones used to adjust the EQ and HPF. These controls can be associated with any channel simply by pressing the SELect button of that channel.

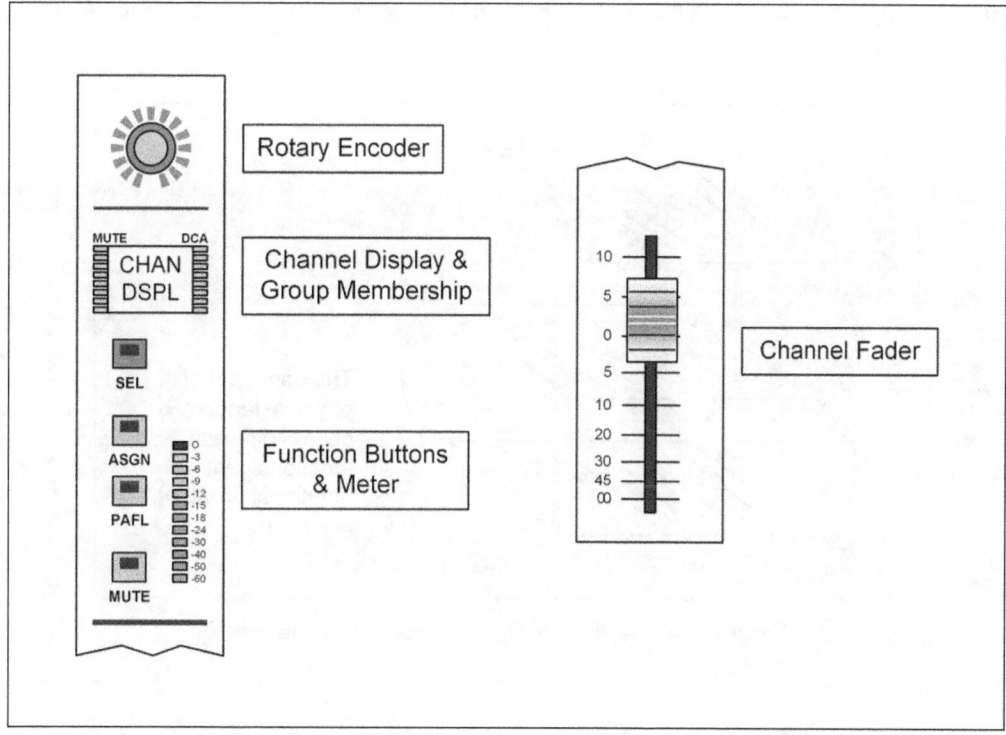

Figure 10-3   Digital Mixer Channel Strip (In Two Sections)

10.4 Layering

The concept of sharing physical controls across multiple functions is called layering and is shown in Figure 10-4. This approach saves a lot of space on the control surface. Different layers are accessed by pressing the appropriate layer button on the control surface. For example, the faders of the control surface may be grouped into banks of eight, and there may be four layers associated with each fader bank. These faders are typically freely assignable, so we could have a configuration of a bank where layer 1 is the stereo input channel faders, layer 2 is the effects returns, layer 3 is the DCA groups, and layer 4 is not configured. Since the positions of the faders are likely not the same in each layer, the faders must move automatically to the correct positions when a new layer is selected. Therefore, you will find that the faders of a digital mixer are motorized. It is also likely impossible that anyone could remember which functions and channels are assigned in each layer. Digital mixers therefore have means of allowing the user to record the channel assignments in the mixer, either via small LCD/LED displays associated with each channel, shown as CHAN DSPL in Figure 10-3, or the large MFD. As mentioned above, the controls used for EQ and dynamics may also be layered, but these layer assignments may not be changeable by the user. For example, the manufacturer may set layer 1 to be channel gain, filter, and EQ; layer 2 to be Delay, Compressor, and Gate; and layer 3 to be internal effects adjustment.

It is not difficult to see that too much layering in a digital mixer can be a hindrance to workflow. Manufacturers are constantly balancing the size of the control surface with the amount of layering. A good compromise in digital mixer control surface layout is one where the input channels are layered in their own banks, and the auxiliary functions (e.g., aux sends, DCA groups) are layered in their own banks, separate from the input channels, or not layered at all. The EQ and dynamics block should not be layered either. The main mix (L/C/R) faders should be separate from all others and should not be layered. This approach requires more control surface real estate, but can save a lot of hassle and time paging through layers in the heat of a show to tweak a single parameter just right.

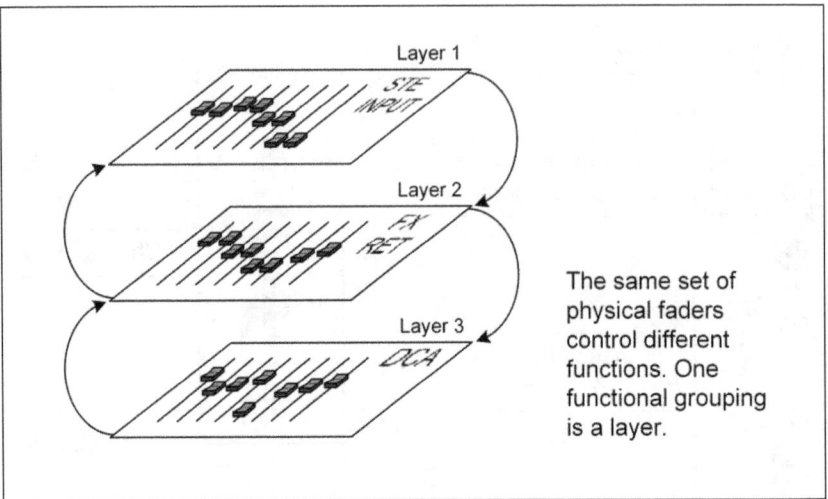

Figure 10-4 Layering in Digital Mixer Control Surface

10.5 Total Recall

The control surface will also have a comprehensive set of controls for saving and recalling the mixer settings, as well as copying and assigning channel settings among channels. The state and settings of the mixer at any instant is called a scene. When we copy and save the mixer state, this is called a snapshot. The ability to save and recall the settings of the entire mixer is one of the big advantages of a digital mixer. Imagine engineering a show where multiple bands will be performing. During the sound check, we can store the mixer settings for each band, or even each song. Now during the show, as each new band takes the stage, we can press the appropriate scene memory buttons and have instant recall of the starting mixer settings from the sound check, including not just the fader and encoder positions, but also the channel names, the internal patching, the assigned effects and their settings, etc.

10.6 Mixer Modularity

One of the good things about digital mixers is their modularity. In many small digital mixers, the DSP and CPU modules may be housed in the same unit as the control surface. However, in larger mixers, they may be housed separately. For example, Figure 10-5 shows that the mix and effects DSPs may be housed in a DSP module, separate from the control surface. The DSP hardware may be on a card that can be plugged into a slot in the DSP module. The ADCs and DACs may also be on plug-in cards and housed in an I/O module. It is not uncommon to see the I/O, DSP, and CPU modules hosed in one relatively small cabinet. This modular approach makes repair or upgrading very easy, as defective or outmoded cards can simply be swapped out for new ones.

Modularizing the functional blocks of the mixer like this also has other operational advantages. For example, we can keep the I/O and DSP modules on or near the stage, thus eliminating the need for long analog cable runs. We can then connect the on-stage module to the control surface via a small, light cable, which carries only digital information. In a single-system configuration with all the modules on stage, this cable would be required to carry the digital audio from the stage to the control surface for monitoring purposes only. The only audio stream we really need to send from the control surface to the on-stage module would be talkback or test audio. The cable would also carry the relatively low bandwidth control data stream which flows in both directions.

Another manifestation of modularization is the virtualization of the control surface. In Section 10.3 we mentioned that some control functions are implemented via the mixer's MFD. However, with large (i.e., bigger than 60 cm, 24 inches) touchscreen monitors now being affordably priced, we have the option of controlling all functions of the mixer via a touchscreen interface and computer with dedicated control software. Here, all elements of control (e.g., faders, push buttons) and information (e.g., meters, labels) are drawn on the screen, and a paging or layering function is used to manage clutter and access all possible operations. This fully software-controlled surface can replace the physical control surface at a lower cost, giving engineers who are comfortable working with touchscreen computers the option of mixing on a pane of glass.

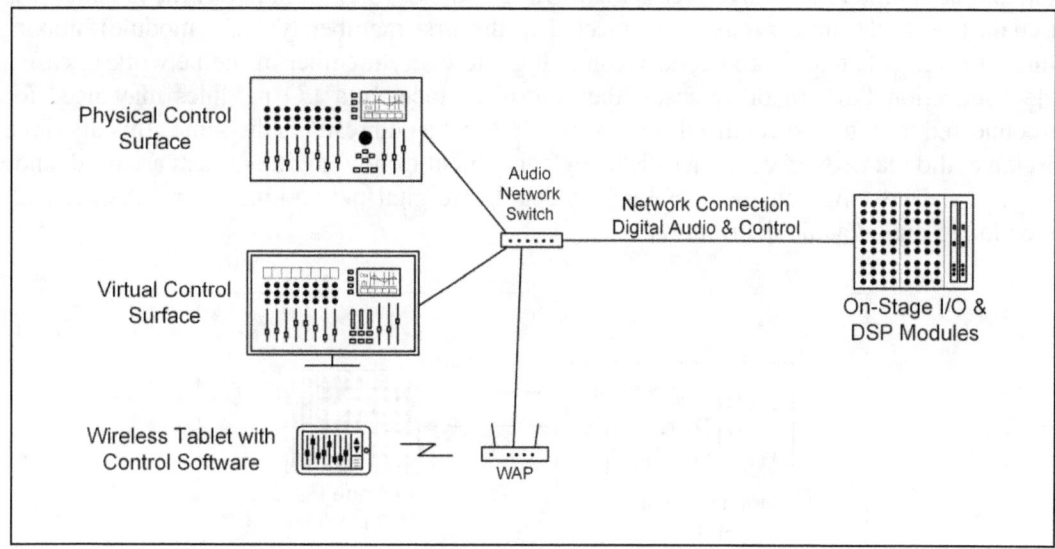

Figure 10-5   Components of Digital Mixer can be Distributed

Figure 10-6   Ruggedized EtherCon Connector

10.7 Component Interconnection and Networking

Many manufacturers have adopted an Ethernet-based interconnection between the control surface and other remote modules. The connection may be based on an open standard such as AVB or AES50, though in many cases a proprietary protocol is used, e.g., Audinate's Dante or Digigram's EtherSound ES-100. The physical link may use inexpensive Cat 5e cable with ruggedized EtherCon connectors, as in Figure 10-6, if using the lower speed 100BaseT Ethernet. Higher speed links may utilize Cat 6 or fiber optic cable and gigabit Ethernet. This saves on the cost and effort of handling bulky, expensive multicore snakes that are common in the analog world.

Interconnecting the various components mentioned above allows us to build a network over which we can share the digital audio streams amongst all the components. A mixer can be set up to take all audio streams entering on port *A* say, and send them out port *B*, or even back out port *A*. Thus, Figure 10-7 shows that with a digital FOH mixer and a digital monitor mixer, we can connect the on-stage I/O module to the monitor mixer control surface, and connect the FOH mixer control surface to the monitor mixer in a daisy chain, and eliminate the need for expensive microphone splitters on stage. The FOH mixer will get all audio streams (for monitoring) from the monitor mixer, and both mixers will have access to all DSPs and I/O ports. In practice, the last member of this chain (the FOH mixer) may be connected to the first member (the I/O module), making a redundant ring so that audio and control can still get to every member of the network in case any single connection fails. In other cases, the control surfaces and I/O modules may need to be interconnected through a specialized central switch, but the end result is the same. This distribution of digital audio via networked devices is a key component of how digital mixers are used, and we will spend the next two chapters looking in detail at digital networking over Ethernet and its application to digital audio distribution.

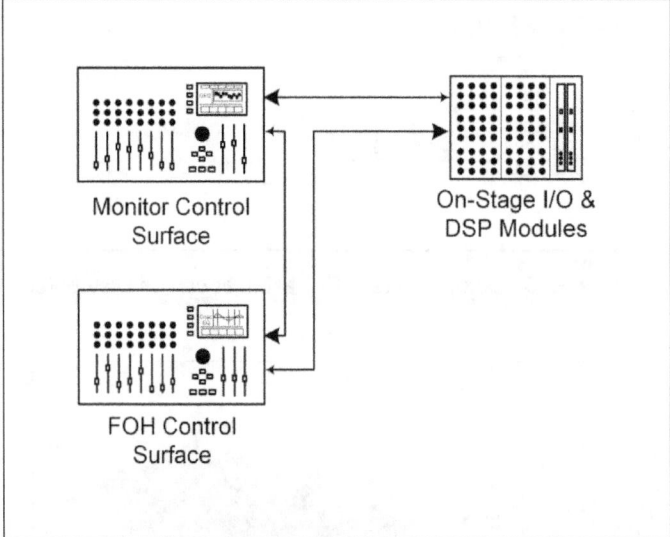

Monitor Control Surface

On-Stage I/O & DSP Modules

FOH Control Surface

Figure 10-7   Digital Monitor and FOH Mixers Share Same I/O

10.8 The Digital Advantage

Digital mixers offer other substantial advantages as standard. As we saw earlier, the built-in, comprehensive dynamics suite available on every input channel, as well as the virtual effects rack with multiple effects engines and EQ, eliminate the need for an external effects and EQ rack when using a digital mixer. All patching is done internally, again eliminating the need for external error prone wiring and cables that tend to break and fail at just the wrong time. Even if we did provide external effects processing, the bulk of the equipment required to match the kind of dynamics processing internally available to every channel would be fairly large.

Due to the underlying software nature of digital mixers, you will find that most digital mixers offer offline editing whereby the console can be set up or configured on a PC and then later, that configuration can be uploaded to the mixer. This may save some time at the performance venue, where things tend to get hectic during the pre-show setup. Digital mixers can also be easily upgraded simply by installing new firmware. This allows the manufacturer to fix bugs or add new functionality to the mixer with nothing more than a USB or Ethernet connection and a PC.

One interesting thing about digital mixers is the possibility of remote control, e.g., via a wireless tablet and a dedicated wireless access point (WAP) that is linked to one of the mixer's communications ports, as shown in Figure 10-5. In some mixers, this access point may be built right into the mixer, eliminating the need for an external device. This kind of arrangement may be very useful in a setting where the traditional location of the FOH engineer is not allowed, either for esthetic or other reasons, and the control surface has to be in a suboptimal location for mixing. This is not an uncommon situation in many places of worship, for example. In this case, an assistant with a wireless tablet running remote control software may unobtrusively take any position in the audience area and adjust the mix or tweak effects as necessary.

The modern digital mixer has many things in its favor. They may be a little intimidating at first, but once a user gets used to working with them, switching back to an analog console may not be a cherished idea. Indeed, if one is fortunate enough to have 'grown up' on a digital mixer, moving to an analog mixer will surely be viewed as a step backwards, especially during setup, or the sound check for a multi-act performance.

Review Questions

1.   What is the purpose of the DIR and DIT in a digital mixer?
2.   Why are the mixing controls of a digital mixer often called a control surface?
3.   Compare and contrast the routing of signals in a digital mixer with that of an analog mixer.
4.   What is a DCA group and how does it compare with a VCA group?
5.   What are the implications of having a digital effects rack integrated into the mixer?
6.   What is meant by layering in a digital mixer? What are the pros and cons of layering?
7.   What is a scene and how is the ability to store it useful?
8.   What are the advantages to performing the analog-to-digital and digital-to-analog conversions on stage and transmitting the digital data stream to the mixer (control surface) at the remote mixing location?
9.   What are some of the standards that digital mixers use to transport the digital data between external modules?
10. Explain how digital FOH and monitor mixers can share input signals.
11. How does a digital mixer facilitate remote wireless control?
12. Discuss the advantages and disadvantages of using a digital mixer as compared to an analog mixer.
13. Discuss the high-level organization of a digital mixer, explaining data flow, its various interfaces, and how the mixing DSP and effects DSPs interact with the CPU, memory, and control surface.

Chapter 11

Digital Data Networks

11.1 Networking Basics

In the analog world, we interconnect our analog devices so that we can share signals between them. For example, we are all familiar with connecting the analog output from our microphone to the analog input of a mixer, and then connecting the output of the mixer to the input of a power amplifier. With digital devices, we can also share signals the same way, e.g., we connect the analog output on the digital effects processor to the analog input of our digital mixer, and so on. In this case, the digital signal which the digital processor is manipulating is converted into an analog signal before sending it to the next device in the chain. However, it is possible to share information between digital devices without first converting it to the usual analog audio signals that we are used to. Indeed, with the proliferation of digital devices in the audio realm, it is now common to have a complete network of digital audio devices which share digital data amongst themselves in very intelligent ways. In this chapter, we will examine how Ethernet based digital networks are created and how devices share data across the network. In a subsequent chapter, we will look at how a digital audio distribution system is built on top of such networks.

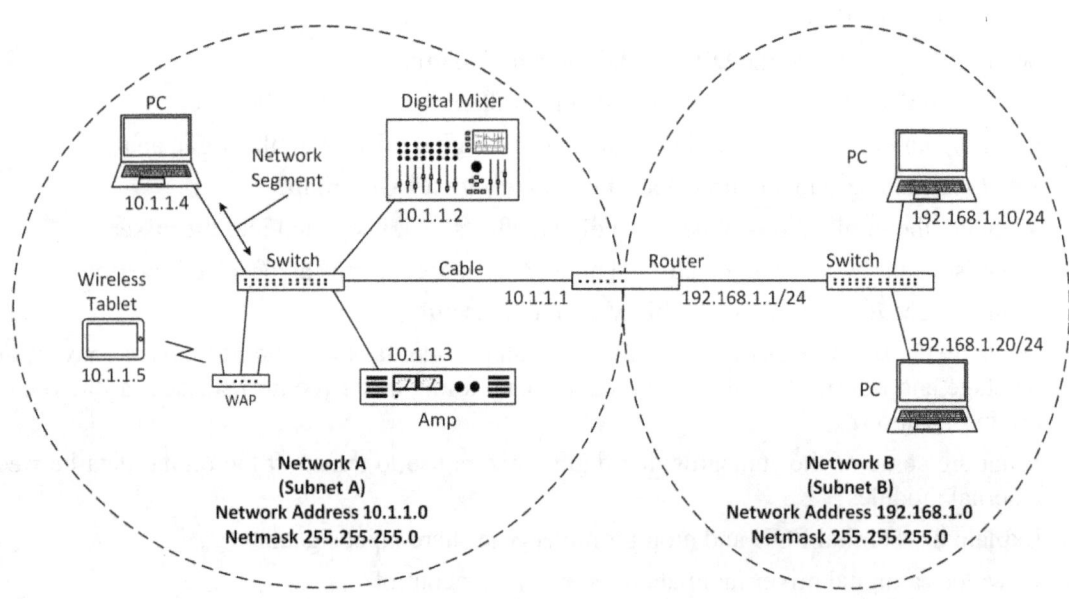

Figure 11-1 Basic Digital Network

11.1.1 Digital Networks – The Principal Elements

There are many players that are required to make a digital network function properly. Generally, there are the network devices, and then there are the physical media that interconnect them. These are highlighted in Figure 11-1. We will take a brief look at the network devices now and elaborate on them further below.

In a network, you will find three broad classes of devices. First, there are the devices that produce and consume data. These are typically end user devices, such as a PC or tablet, a digital mixer, a digitally controlled amplifier, and so on. These devices are usually the first or last points in a digital communications link, and so we will sometimes call these the *endpoints* of a communications link. You will also hear such endpoints being referred to as hosts. In any network, a host is identified by a set of unique addresses which allows a data stream to be directed specifically and only to it. In a typical Ethernet network based on the IEEE 802.3 standard,[1] these addresses are the Media Access Control (MAC) address and the Internet Protocol (IP) address. We will discuss these in detail later in this chapter.

The next common device is a network switch. The switch consists of a number of input/output (I/O) ports to which other network devices are connected. The primary role of a switch is to forward data from one endpoint to another within the same network, while keeping that data stream isolated to the links that connect the endpoints to the switch. In other words, most of the time, the switch only sends data destined for Host A over the link (network segment) on which Host A is connected. There are some cases where the switch will send data destined for Host A to most other devices in the network, but this typically only happens when the switch is learning where Host A is located (i.e., the port where it is connected to the switch), or if the switch's forwarding mechanism is forced to operate in an abnormal mode. You will sometimes hear the function of a switch referred to as bridging, and a switch may sometimes be called a bridge.

A wireless access point (WAP) is a device in the same class as a switch, in that its forwarding decisions are based on the same class of addresses that switches use. The role of a WAP is to provide a bridge between the wireless domain and the wired network, i.e., transform data that is packaged and transported wirelessly into the form that is required by the wired network and then deliver such data onto the wired network medium, and vice versa. Wireless networks based on the IEEE802.11 standard[2] have become extremely common, and the term *Wi-Fi* is commonly applied to such networks.

The third class of network device is the router. Routers are used to pass data between different networks, i.e., across a network boundary. Their name comes from the fact that this device determines a route to the destination of the data and then forwards that data along that route.

The notion of the *same network* is important to the concepts of switching and routing. What constitutes the same network is determined by its extent, the range of addresses that the network architect reserves for hosts on the network, and the policies that are applied to how data can be moved through the network. The term *subnet* is often used to describe a network that is partitioned based on a subset of IP addresses and in which all the hosts share a common network address. We will discuss this in more detail below.

Of course, in order to pass data amongst themselves, network devices must be interconnected somehow. Above, we used the term *physical medium* to refer to this interconnection. In a wired network, a cable is used to connect the I/O port of one device to that of another device. In an Ethernet network, this is typically a multi-conductor copper cable as shown in Figure 6-15, but fiber optic cables may also be used. In the wireless network segment, the medium is an electromagnetic wave which allows over-the-air[a] connections.

11.1.2 Network Classification

As alluded to previously, two of the design considerations of a digital network are its extent and administrative control. By extent, we are considering primarily the geographic reach of the network, i.e., its physical size, which can range from very small with as few as two devices, to worldwide with perhaps thousands of devices.

A local area network, or LAN, is the smallest type of network that we will encounter. LANs are owned and controlled by a single entity. The geographic scope of a LAN typically will range from a single room, to an entire building, and even a small campus. A LAN may consist of one or several subnets, thus may or may not have a router in its core, but will typically have one or more switches.[b] The interconnecting physical links in a LAN are usually Cat 5e or better copper cable, though some larger LANs may utilize fiber optic cables for some connections. Wireless links have also become common in LANs as the popularity of Wi-Fi devices has grown in the last several years.

Metropolitan area networks (MAN) encompass very large campuses, several city blocks, or even entire cities. MANs typically interconnect LANs. Thus, if an administrator of one LAN wanted to have a reliable data connection to devices in another LAN on the far side of a city, one option would be to tap into a MAN. These network interconnections are usually owned and administered by small telecommunication providers, and typically consist of copper or fiber optic cables. In some cases, wireless microwave links may also be used.

The last type of network that we will consider is the wide area network (WAN). WANs have large country-wide, continental, and even intercontinental geographic scope. An entity with wide geographic distribution, e.g., a large multinational corporation, may have need of a WAN to interconnect its widely dispersed offices. The interconnecting links of a WAN are usually owned by one or more third parties, typically large telecommunications providers. These links may consist of many kilometers of copper or fiber optic cables, as well as long range wireless microwave links.

One other consideration for the network is its administration and the policies that govern the flow of data through the network. A closed LAN is inherently private. By default, the only data allowed on the LAN is that of its owner and administrator. On the other hand, the links in MANs and WANs may be shared by several entities. Entities requiring privacy for their data in such a network would need special policies to be implemented to ensure such privacy. If it is acceptable for an entity to have its data mixed with that of other entities, then it is possible to send data over any geographic scope using the public Internet, which has worldwide reach.

Since live sound events typically occupy relatively small geographic areas, e.g., an arena or a park, the type of network that we will construct for live sound distribution will typically be a LAN. With this in mind, we already will have some predetermined design considerations for the network in terms of the core technology (e.g., switching vs. routing, wired vs. wireless, copper vs. fiber optic)

---

[a] Note that even though the term *over-the-air* is often used to describe wireless communication, actual wireless transmission does not use or require air to work, as it utilizes electromagnetic radio waves which require no intervening matter to transport a signal.

[b] Remember that a subnet defines a network boundary, and we need a router to forward data *across* a network boundary, whereas switches forward data *within* a network.

and management of data on the network. We will see how these design elements come into play as we consider the requirements for a distribution network in the next chapter.

11.1.3 Data Transmission Modes

When two connected devices send data to each other, that communication must be ordered, i.e., there are some rules around when a device can transmit its data. There are three modes of operation that a pair of communicating devices can assume. The first mode is called simplex. In this case, the data is sent in one direction only along the communications link, say from Device A to Device B. You will not likely encounter this type of communication in your typical digital network, but this may be the mode used between a dumb sensor and a computer, for example, where data is sent only from a remote temperature sensor to the computer, but not from the computer to the sensor.

The second mode is half-duplex. In this case, data can be sent in both directions along the link, but not at the same time. This is analogous to the push-to-talk walkie-talkie type communications. Common Wi-Fi standard communications between a wireless device, such as a tablet, and an access point is half-duplex.

Finally, we can have full-duplex communication on a link. In this case, data can be sent in both directions simultaneously. This is analogous to two people carrying on a conversation over the telephone, where they can both speak and be heard at the same time. Clearly, each communicating device must be able to transmit and receive at the same time. This is the normal mode of operation in an Ethernet network over wired connections. This mode of operation can be used, for example, to allow a PC to download a large file from a server and still send (i.e., upload) an email at the same time.

11.2 Ports, Interfaces, and Communicating Entities

When speaking casually about communicating devices in the network, we may say that one device communicates with another device, and do not bother to make a distinction between the physical endpoint, e.g., a PC, and the software running on the endpoint. In general, this is perfectly acceptable. However, to better discuss and understand the communications process, it is useful to distinguish between the major elements of the communications chain internal to the network device with which the user may interact. For our purposes, we will restrict these to the I/O port, the interface, and the communicating software entity.

The I/O port is where the user makes the physical connection to the device. In an Ethernet network using copper physical medium such as Cat 6 cabling, the physical port will be an 8-position 8-contact (8P8C) modular socket. If using fiber optic media, then small form-factor pluggable (SFP) ports are common. Figure 11-2(A) shows these two types of ports (SFP optical on top and electrical at the bottom) as may be found on an Ethernet switch. Figure 11-2(B) shows the 8P8C (commonly called RJ45) plug that would terminate the Cat 6 cable, while (C) shows the optical duplex connector used for the SFP port.

The network interface is a notion that encompasses the port and can be thought of as all the functions and hardware that allow a device to connect to a network. The communicating software entities send and receive data through the network interface. The interface is a construct that is tightly bound to the physical port, and sometimes the term is used uncritically to refer to the port itself.

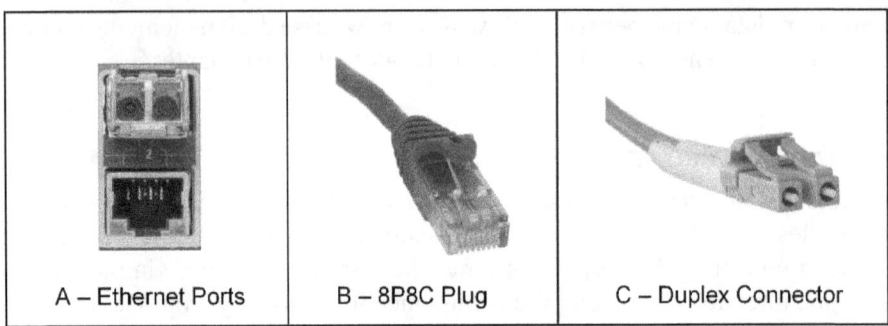

A – Ethernet Ports   B – 8P8C Plug   C – Duplex Connector

Figure 11-2 Electrical and Optical Ethernet Sockets and Plugs

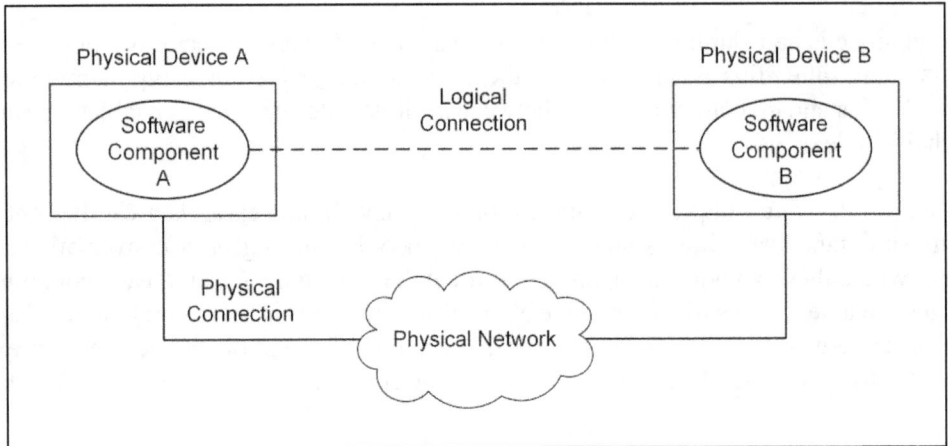

Figure 11-3 Physical vs. Logical Connection

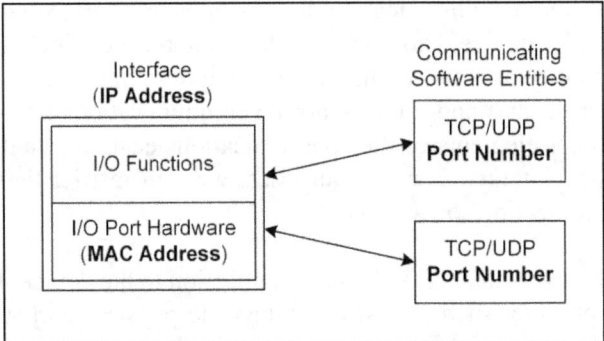

Figure 11-4 Relationship Between Port, Interface, and Software Entity

Typically, it is a software component running on the endpoint that really needs to communicate with another software application in another device. For example, a Web Browser (software application) running on a PC may need to communicate across the network with a Web Server (software application) running on a different computer. In our discussion below, we will refer to these communicating software components as communicating entities, or peer entities. Conceptually, communicating peer entities logically form a connection between themselves, as illustrated in Figure 11-3. Thus, we can consider that the Web Browser in the PC forms a direct logical connection with the remote Web Server without regard to that fact that the connection goes over a physical cable and perhaps several switches and routers. On any given physical endpoint, there may be several software entities that need to communicate with peers in remote devices. The port, interface, and communicating entity within a single network device may be thought of as having a relationship as depicted in Figure 11-4.

11.3 Network Device Addressing

When one endpoint wants to send data to another endpoint on the network, it must be able to uniquely identify that endpoint on the network. In an Ethernet network, the addresses commonly used are the MAC address (also called the Ethernet address), the IP address, and the TCP port number. For our purposes, we can consider that the MAC address is associated with the physical I/O port and the IP address is associated with the interface on the device. Thus, if a device has more than one I/O port, it will have more than one MAC address and may have more than one IP address. This is typically the case with devices such as routers that have multiple ports, and computers that have both wireless and wired interfaces. Many basic switches, though they have multiple ports, do not have addresses associated with their ports because there is no need to be able to directly address these ports. The TCP port number can be thought of as being associated with the actual software entity that is communicating with its peer in another device. Figure 11-4 shows how these addresses are associated with the elements in the communications chain.

MAC stands for Media Access Control, which is a function in the device that controls when the device takes control of the physical medium, i.e., transmits its data. The MAC address associated with a port is permanently programmed into the device by the manufacturer, and is globally unique. Thus, even though it is possible to temporarily override the MAC address of a device, it is expected that no two devices in the world will have the same MAC address. The MAC address is made up of six bytes. Recall from Chapter 7 that a byte is a binary number consisting of 8 bits. Thus, a byte can have a decimal value ranging from 0 to 255. When presented in human-readable form, it is common practice to write the separate numbers of the MAC address in hexadecimal[c] format, separated by colons. Thus, you may see a MAC address written like 01:23:45:67:89:AB, and this would uniquely represent the port that is associated with that number, much like a street address such as "123 ABC Street, Some-Town, Some-Country" could uniquely represent an apartment complex somewhere in the world.

The other address that is used extensively by most devices in the network in processing data is the IP address. IP is the Internet Protocol, which we will discuss further below. There are currently two versions of IP in use—these are IP version 4 (IPv4) and IP version 6 (IPv6). However, IPv4 is far more popular, and we will discuss only this version. Thus, when we speak of IP, we implicitly mean IPv4.

An IP address consists of 4 bytes. In human-readable form, it is presented as four decimal numbers separated by periods (dots), and is referred to as dotted decimal format. You may therefore see an IP address like 10.11.12.13 associated with the interface (and thus port) on your device. Since each byte can have a value ranging from 0 to 255, the biggest IP address that you would expect to see is 255.255.255.255. An IP address may be likened to the number associated with a specific apartment within an apartment complex, e.g., "Apartment 1." Just like there can be more than one apartment within a complex, we can have more than one IP address associated with a single port (though this is not the usual arrangement).

---

[c] Hexadecimal is a number system of base 16, where each digit can be one of {0, 1, 2, 3, 4, 5, 6, 7, 8, 0, a, b, c, d, e, f} and have respective decimal values of 0 to 15.

Although the majority of assignable IP addresses are expected to be globally unique, there is also a large set of addresses that are not globally unique. This latter set is reserved for use in private LANs, and includes addresses in the following three ranges: 10.0.0.0 to 10.255.255.255, 172.16.0.0 to 172.31.255.255, and 192.168.0.0 to 192.168.255.255. Since anyone can use them in a LAN anywhere in the world, they are not globally unique, and data destined to an address in this set cannot be directly routed over the Internet. These addresses may be used by anyone to identify endpoints that are reachable on their LAN. Therefore, when we are configuring the devices (mixers, amplifiers, PC, etc.) on our distribution network, we will typically use addresses from these ranges.

In any given LAN, the IP address of an interface should be unique. Unlike MAC addresses, however, IP addresses are not permanently programmed into a device by the manufacturer. Indeed, a manufacturer may not program any IP address on a device. Thus, it is up to the user to determine what IP address to assign to a device. We will look at this in Section 11.3.2.

Each *instance* of a communicating application uses a third address to identify itself.[d] This address is confusingly called a TCP/UDP port number, but has nothing to do with the physical I/O port on the device. TCP (Transmission Control Protocol) and UDP (User Datagram Protocol) are protocols which we will discuss later, and their port numbers range from 1 to 65536. If we continue with our addressing analogy of apartment dwellings, we can liken the port number to the names of individual people living in an apartment, as they are all instances of humans. Except for servers (e.g., a Web Server) which must be reliably reachable, the port number assigned to each instance of an application is usually temporary. When that instance of the application is no longer running, its TCP/UDP port number can be reassigned to a different application.

Some devices, like switches and wireless access points, use only the destination MAC address embedded in the data to direct that data to the intended recipient, while others such as routers, computers, and digital mixers use both the MAC address and IP address to send data. If the endpoint is a computer, then when data arrives, the operating system further looks at the TCP/UDP port number to correctly direct the data to the intended software application instance.

11.3.1 IP Addresses and Subnets

In the last section, we discussed how each endpoint in the LAN needs to have a unique host IP address. This address is used by other devices to send data to that particular endpoint. If we look back at Figure 11-1, we see that there are two subnets (**A** and **B**) connected by a single router in the middle. The host IP address of each device is shown in the diagram. For example, the digital mixer is shown as having a host address of 10.1.1.2. Any device in the LAN that wants to send data to this mixer will use that address as the destination of their data. Generally, though, a network device needs to know three other pieces of IP information to function properly. These are its own host IP address, the size of the subnet to which it belongs, and the IP address of the router that it uses to send data outside of the subnet. We will now take a closer look at these last two bits of information.

As mentioned earlier, a router is needed to pass data across a network boundary. Thus, a router must have connections in at least two different networks. That is why we see that the router in Figure 11-1 has two IP addresses, one in each subnet. The router is the gateway for data that must travel between these two subnets. When an endpoint in one subnet wants to send data to an endpoint in another subnet, that first endpoint will send the data to the router that is the gateway to the other subnet and let that router forward it on to the correct next device. This takes the burden of having

---

[d] It is important to note that the address is associated with each instance of the application. If we had 2 browser windows opened on your PC, these would be considered as 2 separate instances of the application.

to know how to reach every possible device away from the host endpoints and places it on the gateway router. Thus, we typically will configure every host in the subnet with the IP address of this gateway router, called the *default gateway*, so that they can reach devices in other subnets.

The above discussion raises a question: how does a host know if a given IP address is outside its subnet? In the subnet, there is the concept of the *network address*. Every host IP address in the network can be thought of as being composed of two parts, namely, the network part and the host part. Furthermore, every endpoint in the same subnet has the same network IP address part. Looking again at Figure 11-1, we see that all the hosts in subnet **A** have IP addresses that start with 10.1.1, while those in subnet **B** all have addresses that start with 192.168.1. The subnet address is determined by the number of bits that are reserved for the network address, and this in turn determines the value of what is called the subnet mask. Let us look at an example of how this works.

Remember that the IPv4 address is composed of 4 bytes. Since each byte is 8 bits long, that means that the address is 32 bits in length. A configuration where we reserve 24 bits for the network address is very common, so we will look at that scenario. In Figure 11-5(A), we show a generic IP address laid out in binary form, with each bit represented by n. Counting from the left, it shows 24 bits (or 3 bytes) reserved for the network address part. It also shows that there are 8 bits left over for the host part of the IP address. Remember that each bit can take only one of two values, *viz.* 0 or 1. To figure out the network mask, we set each bit that is reserved as part of the network address to 1, while those that are left for the host address part are set to 0, as in Figure 11-5(B). Thus, for the case of 24 bits reserved, we get a mask of 11111111.11111111.11111111.00000000, which when written in decimal form is 255.255.255.0. Because 24 bits are reserved for the network address, you will sometimes see "/24" appended to the written IP address to indicate this. For example, we see in Figure 11-1 that all the hosts in subnet **B** have this form, e.g., the router's IP address is shown as 192.168.1.1/24.

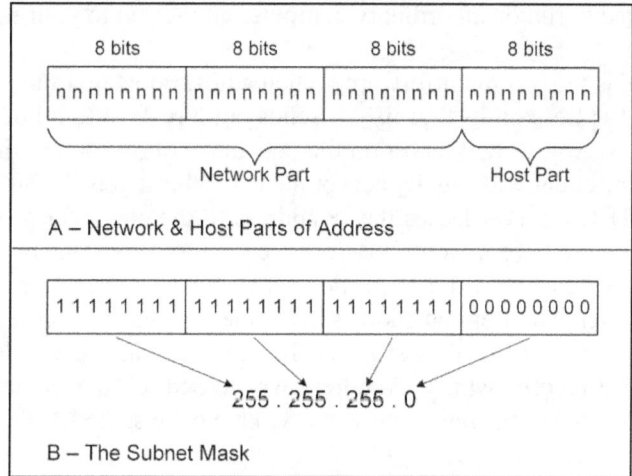

Figure 11-5   Deriving the 24-bit Subnet Mask

The number of bits reserved for the network address directly determines the size of the subnet, i.e., number of hosts that can be part of the subnet. If 24 bits are reserved for the network address, then only 8 bits are left for the host part of the address. Now we ask the question: how many unique numbers can we represent with 8 bits? We know from Chapter 7 that the answer to that is 2^8 unique numbers, or 256, in the range 0 to 255. However, we still cannot have 256 hosts in our example subnet because in every subnet, the first number in the range is reserved as the network address, and the last number in the range is reserved as the broadcast address. Thus, in our example of subnet **B** with 24 bits reserved for the network address, the network address is 192.168.1.0 (the first address in the range) and the broadcast address is 192.168.1.255 (the last address in the range). The broadcast address is used as the destination address when sending data to all hosts in the subnet. In this case, then, we could have a maximum of 254 different hosts in the subnet.

Thus, setting the network mask on a device allows it to figure out the range of host addresses that are in its own subnet. The device wants to know this range because the way it attempts to communicate with a peer differs depending on whether that peer is in the same subnet or a different subnet. If the peer is in the same subnet, the device will attempt to contact that peer directly. If it is in a different subnet, the device will simply send data destined for that peer to the default gateway and let the gateway figure out how to get the data to the peer.

11.3.2 Assigning IP Addresses in Subnets

Of the three types of addresses (MAC, IP, TCP port) discussed above, the IP address is the only one that a user will typically have to worry about setting. Within your working subnet, the host IP address of a device should be unique. There are three common ways that a device can get an IP address. The first is automatically, assigned by a special DHCP server; the second is that the host can assign itself an address if assignment via DHCP fails; and the third is statically, assigned by the user.

DHCP is the Dynamic Host Configuration Protocol. Its purpose is to facilitate the assignment of unique IP host addresses to clients in a given subnet. DHCP is thus implemented in two parts. The first part runs on the client, i.e., the device that wants to get an IP address. Such devices would typically be our PCs, wireless tablets, digital mixer, etc. The second piece runs on the DHCP server. In small LANs, this server software typically runs on a router in the network, but in fact a DHCP server can be configured to run on an ordinary computer connected to your network.

The working of DHCP is fairly straightforward. When a host boots up (and periodically thereafter if it does not get an IP address on its first try), it sends out a request to all devices on the network asking for an IP address. Any DHCP server on the network will respond to the client with an offer of an IP address, and the client will usually accept the first offer it gets. If the client indicates that it accepts the offer, the DHCP server leases the IP address to the client for a limited period of time (typically several minutes to several hours). Before the lease runs out, the client will ask the server to renew its lease of the address. If the client does not renew its lease (e.g., if it has been turned off), then the DHCP server can offer that address to another client. One of the pieces of information that the client sends along with its request for an IP address is its MAC address, and the DHCP server keeps a table that records which IP address was leased to a particular MAC address. That way, it ensures that no two devices on the network will have the same IP address.

What happens in the above scenario, where a host boots up and is configured to get an IP address via DHCP, but there is no DHCP server to assign an address? In this case, the host will try to get an address via DHCP, but will fail. If it is designed to use zero configuration networking (zeroconf), the host will then randomly choose an address from the range 169.254.1.0 to 169.254.254.0[e] as a possible address for itself. It will then use the Address Resolution Protocol (ARP) to determine if any other device on the local network has already reserved that address for itself. ARP is typically used to learn the MAC address of a host when we already know its IP address, but in this case, the

---

[e] The address space actually goes from 169.254.0.0 to 169.254.255.255 but the first 256 and last 256 addresses in the range are not used by hosts when selecting an IP address.

host just wants to see if any other device responds to its query. As an example, suppose the host chooses 169.254.1.100 as a candidate address. The host will then transmit an ARP probe, essentially asking the question: who has IP address 169.254.1.100? If any device responds to the ARP probe, then the host randomly chooses a different address from the same range and repeats the process. When no device responds to its ARP probe, the host knows that the chosen address is not used on the local network and assigns it to itself. Addresses assigned this way are called link-local addresses because no router will route any packets to any destination in this range, and these packets are limited to the switched local network (the switched network being considered a logical link). It should be noted that there is no guarantee that any host will assign itself a link-local address—it is dependent on the implementation of the network stack on the device. However, if your network is switched and consists of hosts that do assign link-local addresses, and if your devices advertise themselves, and your digital network management software auto-discovers your devices, then this is the simplest way to manage the addresses on your network.

Finally, a user can choose to manually assign an IP address to a device. In this case, the address is said to be static. When configuring addresses statically, the user must be careful not to assign the same IP address to multiple devices. Configuration will usually involve assigning the three basic addresses (host IP, network mask, and default gateway) on the device. When the number of devices in the network is small (e.g., less than 10), static assignment is quite feasible, but this approach to managing addresses can get tedious and error prone when the number of network devices is large. Figure 11-6 shows an example of the network configuration screen on a digital mixer where the three addresses are being configured statically. You can also see that the MAC address of the mixer is displayed at the lower right of the table, and the option to select the use of DHCP is shown in the lower left of the screen.

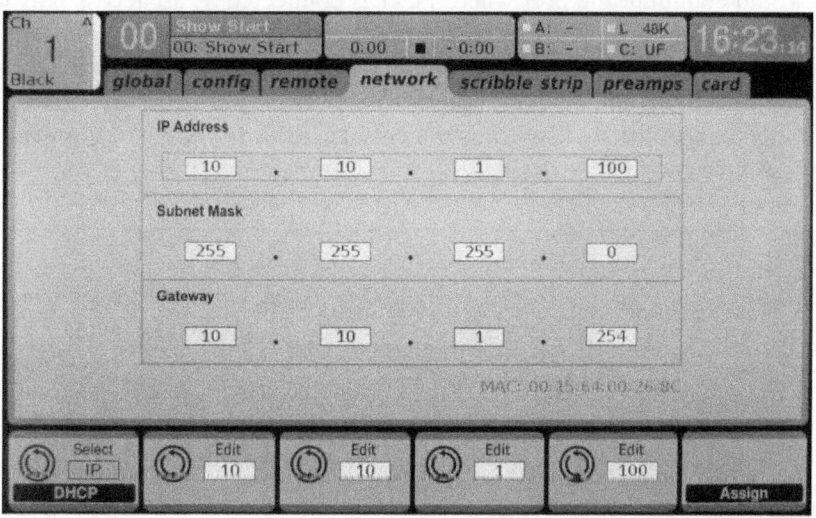

Figure 11-6   Configuring Network Addresses on Digital Mixer

In a network, it is also possible to have some devices that get their IP addresses via DHCP, while others are assigned statically. We would do this if we want some devices to always have the same IP address. For example, we may want our mixer to always have IP address 10.10.10.10 and our main amplifier to be always at 10.10.10.20 so that we are confident about where to reach them. In contrast, we may not really care what address our wireless tablet has because we are not trying to initiate a connection to it from some other device. Note that if we use both static and DHCP assignments, then we must ensure that the addresses that we assign statically are outside the range that can be assigned by the DHCP server, otherwise we may end up with two devices on the network having the same IP address. Note, too, that it is also possible to configure a DHCP server to always assign the same IP address to a particular device based on that device's MAC address. Essentially, we reserve the IP address for that device on the DHCP server, and it will be assigned to no other device. This approach is good, as we have to configure only one device (the server), and it achieves the same goal as having addresses statically assigned.

11.4 Virtual LANs

In the last section, we discussed how a large network can be broken into smaller subnets by use of an IP addressing scheme and the subnet mask. Subnets are useful in partitioning a network and isolating some devices from others. A router is required to pass data between different subnets, and some types of data that would typically be flooded over an entire subnet, e.g., broadcast data, will never traverse the network boundary.

If we have a single subnet where the endpoint devices are connected only via Ethernet switches, there may also be times when we want to partition this subnet to isolate some subset of the devices from others without resorting to using a router and IP subnetting. In this case, we can create virtual local area networks, or VLANs. Basic Ethernet switches (also referred to as unmanaged switches) typically do not support the creation of VLANs, but more sophisticated managed switches do.

VLANs are identified by a VLAN ID, which in practical use is an integer that ranges from 1 to 4094. Typically, the VLAN configuration is done only on the switch itself. For example, suppose we have a network like the one shown on Figure 11-7. We want to isolate the data that flows between the digital mixer, amplifier, and PC-2 from the data flowing between PC-1 and PC-3. On Switch-1, we would configure the port where PC-1 is connected as an access port for VLAN 10. Similarly, we would configure the ports where the digital mixer and amplifier are connected as access ports for VLAN 20. On Switch-2, we would configure the port where PC-2 is connected as an access port for VLAN 20, and the port where PC-3 is connected as an access port for VLAN 10. The ports that interconnect the two switches are configured as trunk ports, allowing them to pass data for both VLAN 10 and 20.

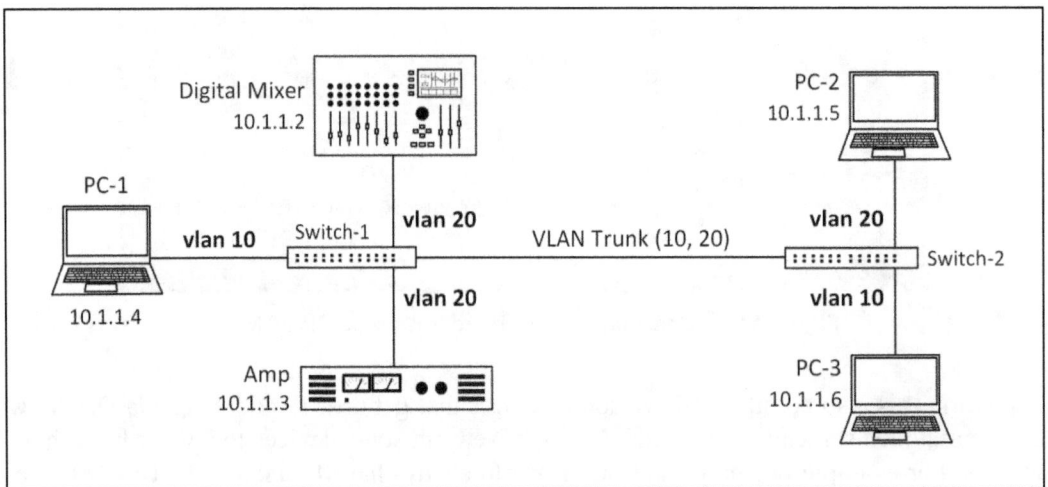

Figure 11-7 Switched Network Segmented Using VLANs

Access ports add a special VLAN tag to the data coming from the connected device, e.g., PC-1, before sending it onward into the network. Amongst other things, this tag contains the VLAN ID and the priority treatment level that switches should apply to this data as it traverses the network. Access ports also strip the VLAN tag from data destined for a connected device before sending the data to that device. Trunk ports do not strip the VLAN tag from the data before sending it to the next connected device, and so allow VLAN tagged data to be relayed through larger multi-switch networks.

Access ports act to isolate data between devices because the switch will not send data to an attached device if that data does not have the VLAN ID that matches the VLAN configured on the port. Thus, in Figure 11-7, PC-1 would not be able to access the digital mixer because its data would be tagged with VLAN 10 on entering the switch. However, the only data that the switch will send out to the mixer would have to be tagged with VLAN ID 20. The switch will simply discard any data that would otherwise be sent out this port, but is tagged with the wrong VLAN ID. Thus, VLANs can be used to help prevent unauthorized access to certain devices by placing restricted devices on a separate VLAN and only allowing authorized devices on that VLAN.

VLANs can also be used to give priority treatment to specific data streams. As mentioned above, the VLAN tag contains an indication of priority for the tagged data. This priority is represented as an integer in the range 0 to 7, with 7 being the highest priority and 1 being the lowest (0 is the default *best effort* priority, a little bit higher than 1, which is classed as *background*). In a busy network with many types of data, we would typically tag audio stream data at 5 or higher to ensure it is treated with high priority. A network switch will forward higher priority data onward towards its destination before handling lower priority data. Thus, higher priority data will experience lower network delay and loss, all at the expense of lower priority data.

11.5 Data Distribution and Network Topology

There are many different forms that a network of audio devices can take in order to share data amongst themselves. The layout of the network is referred to as its topology, and this has an impact on its reliability and the efficiency of how data may be distributed throughout the network. When we consider the network, we make a distinction between physical and logical connections. The physical connections are what we create with the physical wires and devices, but we also create logical connections when we think about the communicating endpoint (e.g., the software entity running on a mixer) forming a connection with its peer in another device. Logical connections are built on top of physical connections. As we will see later, sometimes the logical network may not resemble the physical network which it represents. We will first take a look at forms of data distribution and then consider the topologies over which we can send such data.

11.5.1 Unicast, Broadcast, and Multicast Distribution

As far as an endpoint sending data to a peer is concerned, the transmission may be unicast, broadcast, or multicast. In all three cases, the type of transmission (i.e., unicast, broadcast, multicast) is determined by the destination IP and MAC addresses of the data being sent. In a unicast transmission, the peer sends data to a single other endpoint which may be located in either the local network or in a remote network. In this case, the destination IP address set in the data packet is that of the intended recipient. The destination MAC address is that of the recipient if it is in the same subnet as the sender, or it will be set to the sender's default gateway if the recipient is in a different network. As shown in Figure 11-8(A), if an endpoint wanted to send data to three different peers using unicast transmission, it would have to create and send three different data flows to each of them, even if the packets contained the exact same data. For a large number of such flows, this could become inefficient and put a burden on the sender.

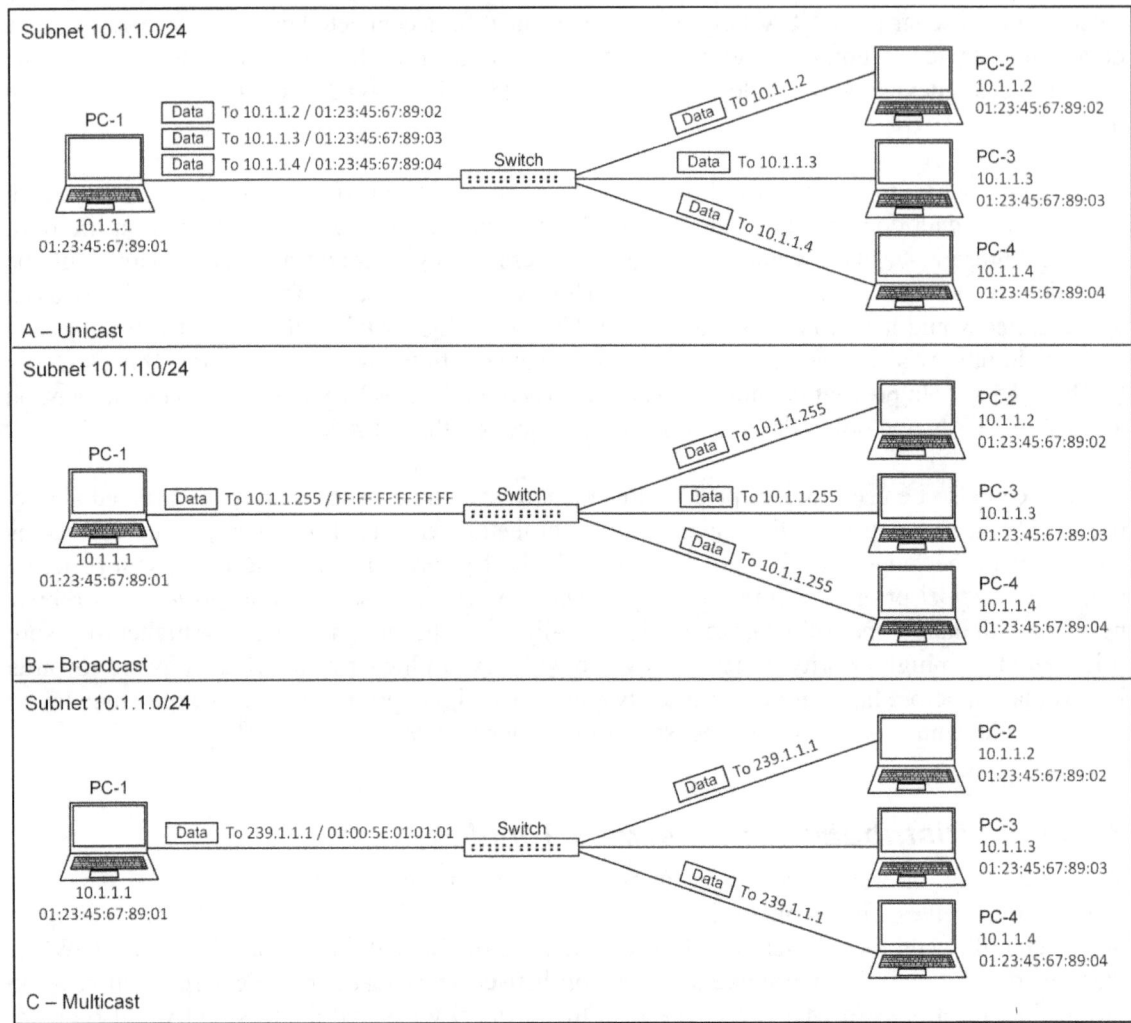

Figure 11-8 Unicast, Broadcast, and Multicast Data Flows

In a broadcast transmission, the endpoint sends data to all other endpoints on the local network. However, it does not create a separate data flow for each separate peer endpoint. In fact, the sender does not even need to know how many other endpoints there are in the network. The sender creates a single data flow and sets the destination IP address to the broadcast address of the subnet. This is the last address in the subnet address space, and in the case of Figure 11-8(B), is 10.1.1.255 (given the subnet of 10.1.1.0/24). The sender also sets the destination MAC address to the broadcast address of FF:FF:FF:FF:FF:FF. While an endpoint will normally examine the MAC address of every received Ethernet frame and discard any frame whose destination MAC does not match its own, all endpoints will accept a frame whose destination MAC is the broadcast address FF:FF:FF:FF:FF:FF. When the broadcast packet enters the switch, it will forward it out every other port. Thus, the burden of distributing the broadcast data shifts from the sender to the switch, which is better suited to doing this function. Note that broadcast data does not cross network boundaries. Recall that a router is required to forward data from one network to another, and a router will not pass broadcast data between networks.

Finally, in a multicast transmission, an endpoint will send its data to a group of endpoints, members of which may be located locally or in a remote network. Multicasting is another form of data distribution that takes the burden of individual data streams off the sender and places it on the network switch or router. In networks with multiple hops, it can lead to more efficient use of limited network bandwidth. Once again, the sender creates only a single data stream, and sets the destination IP address to a specific multicast IP address. The destination MAC address set by the sender is derived from the multicast IP address. In a LAN, we can freely use multicast IP addresses in the range from 239.0.0.0 to 239.255.255.255. These are treated the same as private IP addresses and are not globally unique (and thus not routable over the Internet). Typically, unless properly configured, a switch will treat multicast data just like broadcast data and send it out every port except the one where the data entered the switch. However, when configured to snoop the traffic using the Internet Group Management Protocol (IGMP), the switch will listen for requests from connected devices and only send the multicast data out the ports where requests for that specific multicast stream arrive on the switch. Thus, in Figure 11-8(C), we see the case where PC-2 and PC-4 have requested the multicast data, while PC-3 has not.

11.5.2 Logical and Physical Topologies

The most basic network topology is the bus topology, depicted in Figure 11-9(A), where a common shared medium carries data for all communicating entities. We can think of this as being like a single cable that is tapped along its length and connected to all the devices that need to communicate. In the past, this bus topology was implemented using network hubs, but you will not find a wired hub in any modern network. This is because a bus topology is inefficient in that data collisions are common on the bus when many devices are connected. A data collision happens when two endpoints send data at the same time on the shared medium. Since collisions cause data corruption and loss, a network with a large number of collisions will suffer from degradation of network throughput.[f] An example of a bus topology still in use is the common wireless Wi-Fi network, where the electromagnetic spectrum in the shared medium.

Figure 11-9(B) shows the star topology. This topology is built around a central switch to which communicating endpoints are connected. This is sometimes called a hub-and-spoke topology and is the basis for the most common wired topology currently deployed. Switched networks are more efficient than bus topologies because once the switch learns where an endpoint is connected, it sends data destined for that endpoint only out the port where that device is connected. This eliminates almost all collisions, as the link between the device and the switch is essentially a point-to-point link. In this topology, the switch becomes a single point of failure for the whole network. If it fails, no device will be able to communicate with any other device. On the other hand, a single link failure would affect only the device connected to that link.

Though less common, a third topology which may be encountered is the ring, shown in Figure 11-9(C). In this case, we have several devices connected in a daisy chain fashion, with the first device connected to the last, forming a closed loop. It is especially important in this topology to configure devices properly to make sure that data does not circulate around the loop indefinitely. Assuming that data can flow bidirectionally around the ring, this topology offers the advantage that if any single link or device fails, all remaining operational devices can still reach each other.

Other topologies exist, such as mesh and tree, but the above are the ones you are most likely to encounter.

---

[f] Throughput is a measure of network performance that tells us how much error-free data we can send over a connection in a given time, usually measured in bits per second.

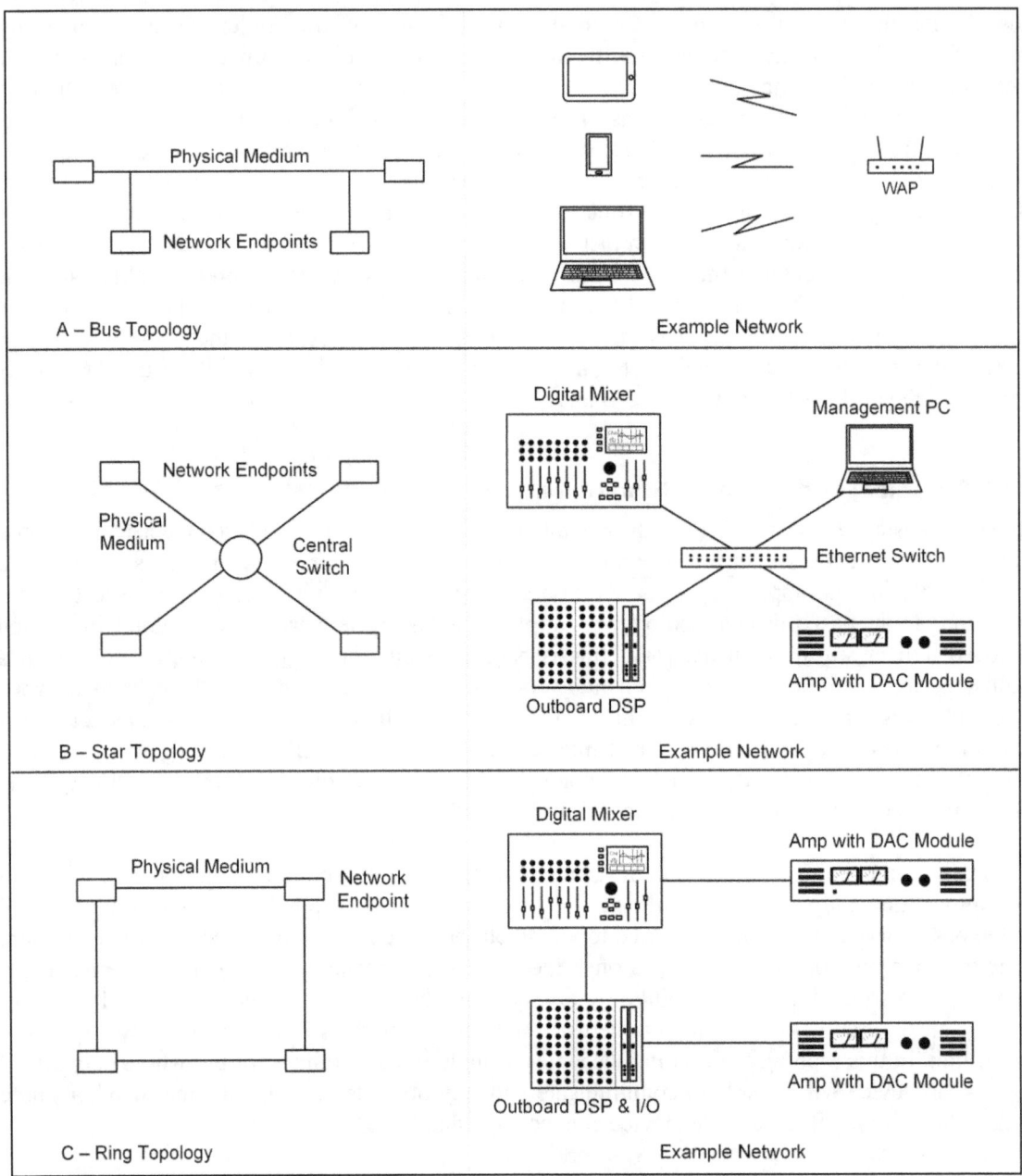

Figure 11-9 Bus, Star, and Ring Network Topologies

11.6 Digital Transmission Encoding and Bandwidth

An important concept that you will encounter when designing and implementing a digital network is that of the bandwidth of a communications link. We have already talked about bandwidth when we discussed band-pass filters in Chapter 4. For a digital communications link, the concept is related, but described differently. When we talk about the bandwidth of a digital connection, we are usually concerned with the amount of data that the link can correctly transfer (i.e., error free) in terms of bits per second. However, that concept is directly related to our usual (i.e., analog) understanding of bandwidth in terms of the upper and lower frequency that the link is able to pass unimpeded. To see how this is so, let us look at a simple example.

Recall from Chapter 7 that digital data is represented simply as a string of ones and zeros. Thus, if we wanted to transmit digital data across a link, we could simply use two voltages, say -1 V and +1 V, to represent binary 0 and binary 1, respectively. To transmit a string like 1010 1010 we would transmit a signal like the one shown in Figure 11-10 on the link. As you can see, that signal looks like a square wave, with one cycle, shown in bold in the figure, representing the binary string 10. Thus, we can transmit two bits in one cycle. If we wanted to transmit 1000 bits per second our square wave would have to have a fundamental frequency of 500 Hz. The analog bandwidth required in this case would go from 0 Hz up to the maximum required to transmit a 500 Hz square wave. To transmit a perfect square wave would require infinite analog bandwidth (because the square wave is made up of an infinite number of odd harmonics of the fundamental), so in reality, transmitted pulses are low-pass filtered to remove high-frequency components and retain enough low-frequency harmonics to allow the receiver to correctly extract the data from the waveform. This method of transmitting data where we send pulses (square waves) directly on the link is called baseband transmission. The other way we can send data over a link is to use a high-frequency carrier wave and modulate that carrier with the data that needs to be transmitted. Common Ethernet networks use a form of baseband transmission to send data over wired links, while wireless Wi-Fi networks send data by modulating a high-frequency carrier in the 2.4 GHz or 5 GHz range.

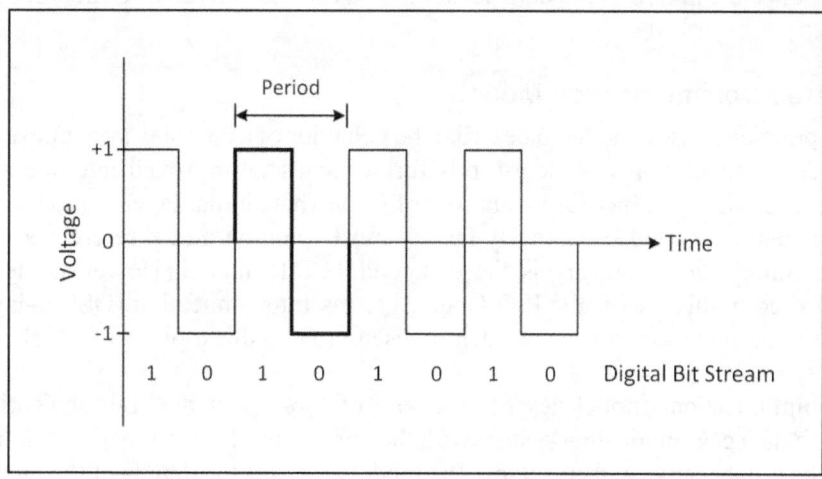

Figure 11-10   Binary Data Transmission Using PAM

The way we represent digital ones and zeros (e.g., with different voltage levels) on the physical medium is called a coding scheme, and a signal is said to be *encoded* when transformed into the representation determined by the coding scheme. Further, the use of pulses of a specific amplitude to represent specific binary values is called pulse-amplitude modulation (PAM). In our example, we had only two levels of amplitude, representing binary 0 and 1, but more levels can be used to transmit more data in the same time. For example, if we had four amplitude levels, we could represent binary 00 with one level, 01 with the second level, 10 with the third, and 11 with the fourth. Thus, you see that each pulse would then transmit twice as much data as in our example. The example we gave above is very basic, and in reality, the coding schemes used to transmit data in typical Ethernet networks are much more complex and utilize a form of PAM with more levels of amplitude. However, the basic concept remains the same.

Thinking of bandwidth in terms of coding schemes and transmission signals is not very practical for everyday use. It is more practical and much simpler to describe the capacity or bandwidth of a digital link in terms of bits per second that can be transmitted over the link. This gives a good idea of how much real or useful data we can correctly send from one device to another. For example, if we had 10 uncompressed audio streams where each stream was digitized at 24-bit, 48 kHz sample rate, and we wanted to send all 10 streams over a link simultaneously, we would know immediately that we would need at least 11.52 Mb/s of bandwidth.[g] We would actually need more bandwidth than the sum of all the streams because, as we will see later, the communicating devices need to transmit more information than just our raw data. Still, we see that considering bandwidth in this way allows us to quickly gauge the capacity requirements of the link.

Currently, typical Ethernet devices that we will likely use in building our networks for audio distribution offer link speeds of 10 Mb/s, 100 Mb/s, and 1 Gb/s. Manufacturers of equipment used in audio networking increasingly recommend the use of 1 Gb/s links, as that offers more than enough bandwidth to adequately meet the requirements of current audio distribution protocols.

11.7 How Devices Communicate in the Digital Domain

Earlier we mentioned that it is the communicating software peers, or peer entities, that share data between themselves, and in a sense, these communicating peers can be considered to carry on a conversation. In this section, we will look at ways of describing the communication between these peers and introduce the notion of the data packet that is used to convey information between them.

11.7.1 Layered Communication Models

It is common practice to use a model to describe the behavior of a complex system in simpler terms. If such a model is adopted as a standard, this further leads to improved interoperability among various vendors' equipment. There are a number of somewhat similar layered models that are used for describing communicating systems. By far, the most common-model referenced in day-to-day use in the communications industry is the extended TCP/IP model. However, the only formal standard published in this area is the ISO Open Systems Interconnection (OSI) 7-layer reference model,[3] and it is common to introduce academic discussion on this topic with a look at this model.

A layered communications model describes a way of looking at the communication functions (processes) of a data communication system with the aim of simplifying the discussion of how such systems can share data between themselves. The model abstracts the details of the underlying functionality and presents them as a set of layers. The key concept of a layered model is that a particular layer will provide a service to the layer above it so that the upper layer can carry out its tasks, while at the same time utilizing the service of the layer below it. Because the layers are shown stacked one atop the other, the software implementing this type of model is often called a communications stack or protocol stack.

As an example, consider a manufacturer needing to create a configuration application which will run on a PC and upload a configuration to a digital mixer. At the highest level, this application on the PC communicates with an application on the mixer that is able to accept this configuration and apply it to the mixer. By using a layered approach to the design of the communication function of the application, the application designer does not have to worry about how to actually send the data to the mixer. The application simply assembles the data in a form that is required by the remote mixer application, sets the destination to the mixer's address, and passes the data to the next lower layer. The software at each layer manipulates the data units so as to help ensure their delivery at the correct destination device, and finally at the lowest layer the data is transformed into an electrical

---

[g] Each sample is 24 bits long, and there are 48000 of them every second. Thus, we need 24 x 48000, or 1152000 b/s for each stream. For 10 streams we need to pass 1152000 x 10, or 11.52 Mb/s.

or optical signal that can be sent to the remote device over the communications link. The implementation of the software and hardware in the lower layers are of no concern to the application designer, and so this modularity shields the designer from the details of how to actually get the data across the network.

11.7.2 The OSI Reference Model

The OSI reference model consists of seven layers. It is shown in the left panel of Figure 11-11. The topmost layer—layer 7—is the application layer. This is the layer that captures the user's interaction with the application process and provides the user interface to network-wide services.

OSI Model		Internet Model		TCP/IP Communications Role
7	Application	5	Application	Convert user data to the format required and present it to the communications API for transport to peer endpoint.
6	Presentation			
5	Session			
4	Transport	4	Transport	Establish the end-to-end connection between me and remote endpoint.
3	Network	3	Network	Figure out next addressable network device in the path to remote endpoint.
2	Data Link	2	Data Link	Send data to the next device physically connected to me.
1	Physical	1	Physical	Convert data into physical signal and send it over the connecting medium.

Figure 11-11   Layered Communication Models

Layer 6 is the presentation layer. This layer is mostly concerned with data format translation (encoding), e.g., from application-specific to application-independent transfer syntax, data encryption and decryption, and data compression.

The session layer allows applications to set up a connection to manage their data exchange across the network, i.e., establish a session. This session is a dialog which may be full-duplex or half-duplex. This layer's functionality is concerned with synchronization, which consists of setting checkpoints so that in case of communication failure, the dialog may be resumed at a predetermined point rather than from the beginning. Another function is exception reporting to handle non-recoverable communication failure. Yet another of this layer's responsibility is token management. Sometimes it is essential that certain actions not be attempted by both sides of the session at the same time, e.g., accessing a shared database object. In this case, a token may be handed back and forth between sessions and only the session entity with the token may perform some action on the shared object.

In the OSI model, layer 4—the transport layer—provides an error-free data delivery service to the session layer. Its functions consist of repackaging upper layer data into manageable sized chunks for transmission on the network, and in the reverse direction, making sure that data is delivered to the session layer in the order it was sent by the remote peer. Other tasks include detecting and solving for loss data, and detecting and solving for duplicate data. This layer also provides metering, i.e., a flow control function of data to the network.

Layer 3 is the network layer, whose overall task is to provide a transparent network-wide path between two transport layer entities. Two very basic roles of the network layer entity are to determine the next hop, or next device, along the route to the remote endpoint, while presenting a scalable network addressing scheme that can be used to identify these devices.

The data link layer, or layer 2, presents the physical layer as a reliable transmission service to the network layer. This layer reassembles physical data into meaningful datagrams to present to the network layer, while performing error detection on that data. This layer may also provide some error correction. A key function of this layer is providing orderly access to the physical medium, i.e., deciding when to transmit data onto the physical link.

At the bottom of the stack, the physical layer is concerned with the physical transmission of 1's and 0's across the communications link. This layer deals with issues such as electrical and optical cable sizes, electrical signal voltage levels and impedance, laser light intensity and wavelength, and timing of the transmission waveform.

11.7.3 The Internet Model

While the OSI reference model is a good introduction to layered view of communicating endpoints, in the day-to-day discussions in the data communications industry, a simpler model is generally referenced. The Internet model, as it is sometimes called,[4] is a five-layer model, itself based on the TCP/IP model with an explicitly shown physical layer. This is shown in the middle panel of Figure 11-11. Note that the Internet model is not a formal standard, but has rather evolved into a current de facto standard in the data communications industry.

In the Internet model, the application layer—layer 5—provides user application access to network services, and combines the functionality of top three layers of OSI reference model, while the bottom three layers (1 to 3) are identical in function to their OSI counterparts.

The transport layer, layer 4, now provides an end-to-end data delivery service to applications and may be reliable or not (TCP vs. UDP). As a reliable service, its functions are similar to those of the OSI model. As an unreliable service, it is focused more on low-latency delivery of data.

In the sections that follow, our discussion of the communication process will be with reference to the Internet model.

11.7.4 Data Packets

When sharing a signal (e.g., audio from a microphone) in the analog world, we simply send a continuous electrical signal along the connecting cable from the sending device to the receiving device. However, in the digital domain, data is sent from one device to another in well-defined blocks. In the layered communications model, these blocks of data are passed from layer to layer, and are called protocol data units, or PDUs. These data units are further called frames, packets, or segments, depending on which layer in the communications stack we are considering. However, we will use the term *packet* when speaking generically about these data units. Generically, a packet has the format shown in Figure 11-12. Specific frames or packets will have a different structure, but typically all the functional fields shown in the figure, with the possible exception of the start of packet delimiter, will be present. Note, too, that even though we show the packet as being composed of well-delimited sections (referred to as fields), it really is just a string of ones and zeroes, or bits, in the data stream.

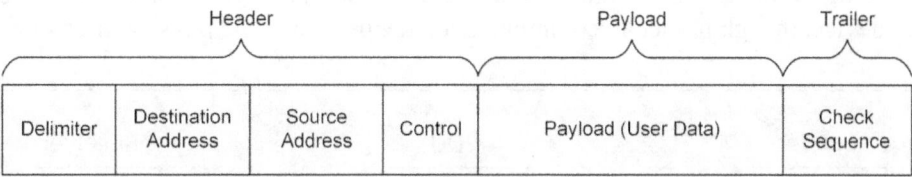

Figure 11-12   Structure of Generic Data Packet

As can be seen from Figure 11-12, a packet may have three broad distinct areas. These are the header, the payload, and the trailer. The header information is used by the devices in the communications path to properly forward the packet along the way from the sender to the receiver, and may contain other control information. The payload area is where the actual user data, e.g., a number of audio samples, are encoded, while the trailer contains information to allow the receiver to determine if the data in the packet was corrupted or otherwise altered while it was being transported from sender to receiver. From the end user's perspective, only the payload bits are useful for transporting their data. The header and trailer bits are termed *overhead* as they are necessary, but take valuable link capacity from transporting actual user data.

Looking more closely at the header in Figure 11-12, we see that it is structured into several fields. The delimiter is a special bit sequence that is used to allow the receiver and other intermediate devices to determine when the packet starts. As mentioned above, this may not be present in some types of packets. The destination address field holds a number that uniquely identifies the intended receiver or receivers of the packet (recall that a packet may be sent to a single device, a select group of devices, or to all devices in a network). Similarly, the source address uniquely identifies the sender of the packet. It is added to the packet because most times, the receiver needs to know who sent it the data it is receiving. The control field may carry various pieces of information to allow proper handling of the packet by the receiver and other network devices. It may hold such information as the version of the application that created the packet, the total length of the packet, the type of data being transported in the payload, the priority treatment that should be given to the packet, and much more.

The trailer may not be present in all types of packets, but it usually contains a check sequence in the form of a cyclic redundancy check, or CRC. In some cases, the CRC may be included as part of the control bits in the header instead. The CRC included in the packet is a calculation performed by the sender on the other contents of the packet. When the receiver gets the packet, it performs a similar calculation, and if the results differ, then the receiver knows that the data has been corrupted while in transit. The receiver may then decide to discard that packet, and depending on whether the communication was intended to be reliable or not, may cause the sender to transmit that packet again.

11.7.5 Data Encapsulation

As noted above, the payload field is where the actual application data is carried in the packet. When data is sent between peers, the PDU travels down the stack, from the application layer, to the transport layer, then network layer, then the data link layer. At each layer, the communicating entity adds its own header, and possibly trailer, encapsulating the upper layer PDU. The payload for each intermediate layer includes everything it got from the upper layer, including the upper layer's header. Thus, you can see that as the PDU descends the stack, more and more of its content consists of overhead, and the actual data transport efficiency drops. This can be seen in Figure 11-13, where the PDU at each layer is shown increasing in size as it travels down the stack. The horizontal dotted arrows in the figure indicate conceptually that the entity at a layer communicates with its peer in the remote device, though in fact all communication happens over the physical medium.

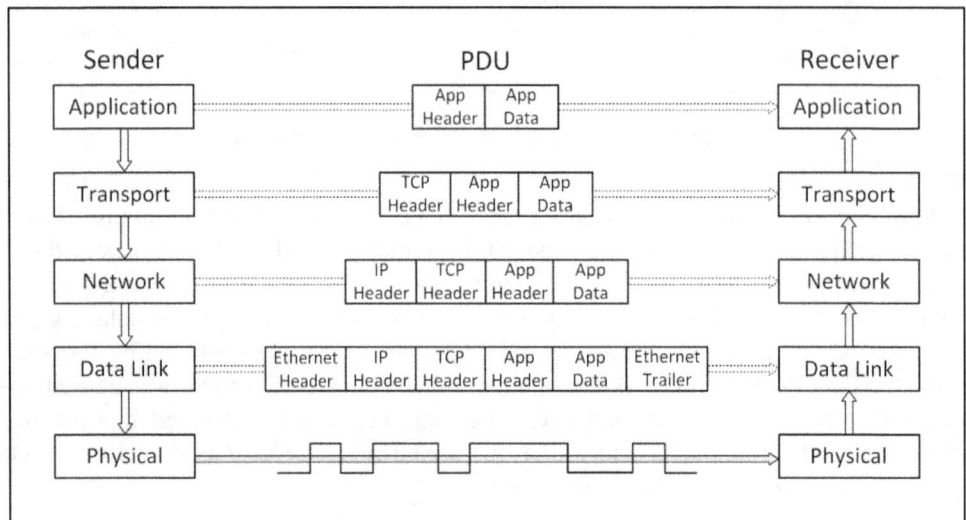

Figure 11-13 PDU Encapsulation when Sending Data between Peers

In the physical layer, the device transforms the data into an electrical or optical signal representing the bits that make up the packet and sends it over the connecting medium—copper or optical cable, or wirelessly over the air. When the signal reaches the destination device, the physical layer takes the signal and transforms it into a series of bits and hands it to the data link layer. The PDU then travels up the stack, with each layer looking only at its appropriate header (and possibly trailer) in order to determine what to do with the packet. Before sending the PDU up to the next layer, the software at that layer will strip its header and trailer from the packet. Eventually, the PDU will reach the application layer, where the application software will extract the original application data and act on it accordingly. For example, if the data is a set of audio samples, the application software may assemble them into a stream and send them to a digital to analog converter.

11.8 Communications Protocols

The communication between two end devices can be thought of as proceeding through several stages. Different protocols are used at different stages of the communication process.

In Figure 11-13, we see that even though communication between remote devices happens over the physical medium, a peer in a given layer on one device conceptually talks to its peer in the same layer in the remote device. In order for this conversation to happen properly, these peers must speak the same language. In other words, they must use the same packet format; the fields of the packet must have the same meaning, and they must calculate error-checks the same way. At the physical layer, the devices must use the same timing between bit transitions to properly interpret the signals. The combination of these attributes—format, order, meaning, timing, etc., imposed on the data is called a *protocol*. The protocol thus describes the language and set of rules that govern the communication between peers. Generally, communicating peers must use the same protocol to talk to each other. However, it is possible that communicating peers use different local protocols, but in such a case they must have some kind of protocol translator between them.

A defined protocol is a form of intellectual property, and so protocols can either be proprietary or non-proprietary. With proprietary protocols, the intellectual property rights are owned by a single person or organization, and anyone who wants to implement that protocol in a product generally has to pay a licensing or royalty fee to the owner of the protocol. The owner also retains control over the improvements that may be made to the protocol. A non-proprietary protocol, on the other hand, may be freely used in any product without any licensing restrictions. Non-proprietary protocols are also *open*, i.e., the specification and workings of the protocol are freely available to the general public. Proprietary protocols tend to be *closed*; the actual working specification of the protocol may not be made public. However, it is also possible that a proprietary protocol be open— its specification is publicly available to anyone who wants to implement it, but the creators still retain the intellectual property and licensing rights to the protocol, and implementers must still pay a fee for using the protocol in any devices produced.

Proprietary protocols tend to be created by private individuals or organizations seeking to solve a particular problem, fill a gap in some market segment, or take advantage of some opportunity in a market. If a company can create a protocol and bring it to market fast enough to achieve significant market penetration, that protocol may be adopted as a *de facto* standard. Though so adopted, these protocols are not formal standards. Non-proprietary formal standards for protocols are generally created by world-recognized standardization bodies, such as the Audio Engineering Society (AES), the Institute of Electrical and Electronics Engineers (IEEE), the Internet Engineering Task Force (IETF), and the International Telecommunication Union (ITU).

Unfortunately, the process of creating and ratifying a formal standard can take several years, which for many manufacturers wanting to take advantage of market opportunities, is too long a time to wait before introducing a product to that market. Thus, some manufacturers will create implementations based on incomplete or draft standards, or implement a proprietary protocol in order to be able to get a product to market before their competitors. This can lead to incompatibility between products that, on the surface, should work together. One great advantage of basing a product on a formal standard protocol is that that product should be interoperable with any other vendor's product that implements the same standard.

If we consider the TCP/IP-based layered communications model shown earlier, it is typical that many manufacturers' devices use formal standard protocols to implement the bottom three layers, while the manufacturer may use proprietary protocols at the upper two layers. This has advantages in that the manufacturer may use any off-the-shelf router or switch to build an audio distribution network, while their endpoint devices may use any specialized upper-layer implementation to suit the needs of their application.

In the sections that follow, we will take a closer look at some of the protocols used at various layers of the communications stack.

11.8.1 Physical Layer Protocols

The bottom layer (layer 1) of the communications model is the physical layer. In a typical Ethernet network used for audio distribution, the protocol specification for this layer may be 1000BASE-T, which is the Ethernet physical layer specification for 1 Gb/s transmitted over Cat 5e copper cable. This implementation allows the use of standard 8P8C connectors and a maximum cable length of

100 m between devices. We may also deploy 1000BASE-SX, which also gives us 1 Gb/s and uses short range fiber optic cables with SFP connectors, and allows for cable lengths of up to 550 m. With the cost of physical layer implementations dropping, you may also see devices with 10GBASE-T used here, which is the specification of 10 Gb/s transmitted over Cat 6 or Cat 6A copper cable. This also allows the use of common 8P8C connectors, and with Cat 6A cable, a run of 100 m between devices. The Ethernet physical layer takes the data frame provided to it by the layer 2 protocol and transforms that into an electrical or optical signal that it then sends on the physical cable. In the receiver, it performs the reverse operation, taking an electrical or optical signal and creating a data frame that it passes to the layer 2 protocol.

11.8.2 Data Link Layer Protocols

The second layer of the model is the data link layer. In a typical Ethernet network, the protocol commonly used here is Ethernet version 2 (Ethernet II), with modern versions specified in the IEEE 802.3 standard[1] and supported by every Ethernet switch on the market. The data link layer is concerned with framing the data packets, timing of data transmission, and error detection.

7 octets	1 octet	6 octets	6 octets	2 octets	46 to 1,500 octets	4 octets
Preamble	Start of Frame Delimiter	Destination Address	Source Address	Type	Payload (User Data)	Frame Check Sequence

Figure 11-14 Ethernet II Frame Structure

The Ethernet II frame structure is very simple, as shown in Figure 11-14. The field sizes are given in octets.[h] The Preamble and Start of Frame Delimiter are used to synchronize the receiver to the start of the frame. The Destination Address is the MAC address of the next endpoint device that must process the frame (note that that device may not be the final recipient of the data, which will be the case when that final recipient is in a different subnet). The Source Address is the MAC address of device sending the frame. The Type field identifies the type of data that is carried in the Payload, while the Payload itself carries all the data that was passed from the upper layer. The Frame Check Sequence (FCS) field contains the CRC value used to check the integrity of the contents of the frame once it arrives at the destination device.

Part of the job of the data link layer is to determine when to transmit data, i.e., send data to the physical layer so that it can be sent on the physical link. As originally developed, Ethernet was intended to be used in a broadcast network. That meant that every device in the network could hear all transmissions of every other device in the network all the time, and the probability of data corruption due to collisions (two devices transmitting at the same time) was high. Thus, when a device had data to transmit, it had to first listen on the physical medium to make sure that no other device was transmitting, and if the medium was clear, then it could transmit its data. In a modern Ethernet network, a switch is typically used to interconnect the devices, and since switches usually only transmit data destined for a given device out the port on which that device is connected, the links in a modern Ethernet network behave like point-to-point links most of the time. Still, the behavior of the Ethernet layer 2 protocol has not changed in this regard, and the devices still operate in a listen-before-talking mode. Furthermore, to make sure that its transmission does not get corrupted by the transmission from another device, the transmitting device continues listening to the physical medium while it is transmitting. If a device hears that its transmission was corrupted by

---

[h] The term *octet* is used to specify exactly 8 bits, which in modern computer hardware is equivalent to a byte.

another device's transmission, it stops its own transmission, waits a random amount of time, and then starts its listen-before-talking transmission process again. In the Ethernet operation, this scheme is called carrier sense multiple access with collision detection (CSMA/CD).

The third important role of the data link layer is error detection. In the case of Ethernet, the FCS field of the frame is 32 bits in length, and holds the result of a specific type of calculation called a cyclic redundancy check (CRC) that is calculated from the other contents of the frame. The Ethernet protocol is implemented in hardware at the data link layer, and this hardware very efficiently and quickly calculates and checks the CRC value. If the calculated CRC does not match the value stored in the frame, the data link layer will simply discard the frame without any notification to any other entity. This saves the upper layers from having to repeat this type of calculation on a corrupted packet, which would be done using software, and thus be much slower than when done by the Ethernet hardware.

11.8.3 Network Layer Protocols

Layer 3 is the network layer. As mentioned earlier, the most common protocol used at layer 3 is Internet Protocol version 4 (IPv4). Like the protocols in other layers, the job of IP is to simplify the process of communicating for entities in the layer above it. In this case, it presents a transparent network-wide path between upper layer entities. This network-wide path may be confined to a single subnet, or it may span multiple subnets. The important point is that the transport layer entity does not have to worry about such detail—it simply presents its data to the network layer and IP will figure out where to send it.

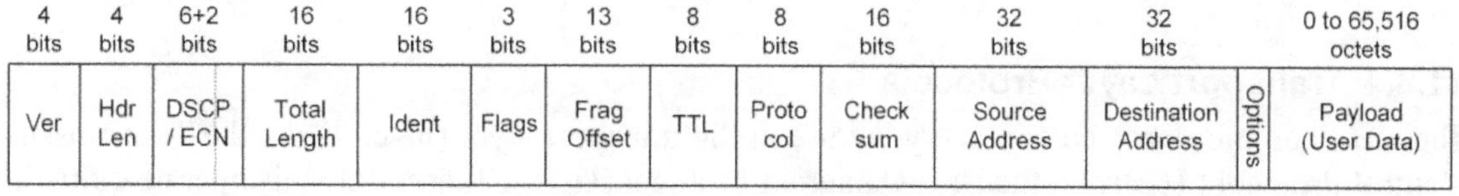

4 bits	4 bits	6+2 bits	16 bits	16 bits	3 bits	13 bits	8 bits	8 bits	16 bits	32 bits	32 bits		0 to 65,516 octets
Ver	Hdr Len	DSCP / ECN	Total Length	Ident	Flags	Frag Offset	TTL	Proto col	Check sum	Source Address	Destination Address	Options	Payload (User Data)

Figure 11-15   IPv4 Packet Structure

As can be seen from Figure 11-15, the header of an IP packet is fairly complicated. There are 13 mandatory fields and one optional filed, which carries options. The Options field is not typically used when transporting user data. There are a few interesting fields that we can look at to get the essence of how this protocol operates.

Two important fields in the header are the destination and source IP addresses. Similar to the data link protocols, IP does its forwarding by looking at the destination address of the data. In order to determine how best to reach the destination, IP keeps a table of *next-hops* that allows it to choose the best interface through which the data should be forwarded. This table, called a routing table, may be updated periodically by the system itself as it learns where other devices are located, or it may be entirely static if entries are all made by a human operator. The source address is sent along with the IP packet so that the receiver can know who sent it the data, and to allow it to figure out a path back to the sender.

The DSCP field is important in setting the priority that should be given to the packet by the next router that this packet may pass through on its way to its final destination. DSCP stands for *differentiated services code point*. This is simply a number that indicates the desired level of service, or *quality of service* (QoS), for a data stream. With audio streams, DSCP should be set to give the highest priority, providing for low latency and low loss. This field would have to be set by the vendor's endpoint device because the network router does not typically set this field.

The TTL field stores a *time to live* value. This value is set by the initial source of the packet and decremented by every router as soon as it starts processing the packet at the network layer. If the value reaches zero, the router will discard the packet. This mechanism prevents packets from being forwarded forever around a network, as can happen if the routing table becomes corrupted or is improperly configured.

Notice that IPv4 also has a Checksum field in its header. This checksum works like others we have mentioned before, but is calculated over the header only. As in other cases, when the receiver gets an IP packet, it recalculates the checksum, and if it does not match that of the header, it will discard the packet.

Without the Options field, the header of the IPv4 packet is 20 bytes long. However, the entire packet can be up to 65535 bytes long, so you can see that a single IP packet can transport a large amount of data. However, IP packets are typically not that large in practice because the lower Ethernet layer restricts the user payload size (commonly referred to as the *maximum transmission unit*, or MTU) to 1500 bytes, and the higher layer protocols usually package their data to fit within that size restriction. If the upper layer payload is such that the total IP packet is bigger than the MTU, then IP must break the payload into fragments before passing it on to the Ethernet layer. In this case, the Identification, Flags, and the Fragment Offset fields are used to track these fragments for reassembly at the receiver. However, IP fragmentation leads to inefficiencies in data transport, and so upper layers try to avoid this.

11.8.4 Transport Layer Protocols

The most common protocols in everyday use at the transport layer (layer 4) are the Transmission Control Protocol (TCP) and the User Datagram Protocol (UDP). These protocols operate quite differently and are used for different purposes.

TCP is one of the most complicated protocols in the communications stack, and its header, shown in Figure 11-16, hints at this complexity. TCP is used when we need reliable delivery of data across the network. It has built-in mechanisms that ensure that user data is delivered to the remote endpoint application exactly as it was sent by the sending application. These mechanisms include acknowledgement of received data, detection of corrupted data, and a way to ensure correct sequencing of delivered data. We can now take a look at some of the key fields that help TCP provide this functionality.

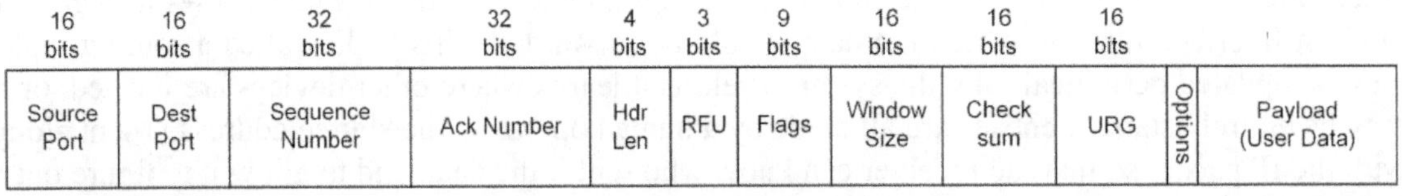

16 bits	16 bits	32 bits	32 bits	4 bits	3 bits	9 bits	16 bits	16 bits	16 bits		
Source Port	Dest Port	Sequence Number	Ack Number	Hdr Len	RFU	Flags	Window Size	Check sum	URG	Options	Payload (User Data)

Figure 11-16 TCP Packet Structure

The Source Port represents the address of the software entity on the local system that is transmitting the data, while the Destination Port is the address of the intended recipient of the data on the remote device. This recipient would be a software application that is already running on the remote system and actively listening for incoming data on its port number. Usable TCP/UDP port numbers can range from 1 to 65536. Port numbers below 1024 are the so-called well-known port numbers that are assigned to server applications that provide common services (such as a web server, whose well-known port is 80). The port numbers from 1024 to 49151 can be registered to a particular application by the Internet Assigned Numbers Authority (e.g., port 19000 is registered to the JACK Sound Server). Port numbers above 49151 are called ephemeral port numbers and are temporarily assigned to applications by the operating system of a computer when an application instance wants to transmit data.

The Ack Number field allows the receiver to acknowledge all the data it receives. This means that the receiver sends a confirmation to the sender that it has received the data correctly. This acknowledgement typically covers more than one packet, but the actual number acknowledged will vary as both the sender and receiver can adjust certain parameters of the TCP connection on the fly. If the sender does not receive an acknowledgement for a particular block of transmitted data, it will resend it. Thus, data that is lost or corrupted during transmission will eventually be resent.

The block of data that is transmitted in the TCP payload is called a segment. These segments are numbered sequentially and tracked via the Sequence Number field of the header. At the receiver, the sequence numbers are used to put the segments in the correct order before passing the data up to the application layer. However, the sequence numbers also allow the receiver to detect lost segments (when the next expected sequence does not arrive), as well as duplicate segments (when the receiver gets multiple segments with the same number).

The Checksum field in the header is used to detect corrupted data. TCP's checksum works similarly to the mechanism described previously, but is calculated on the fields in the TCP header itself as well as the included payload, and a subset of fields from the network layer (i.e., IP) header. If the receiver concludes that the received data segment is corrupted, it will discard it. Furthermore, the receiver's TCP will not acknowledge corrupted data, so the sender will eventually resend the segment that was discarded.

The above mechanisms work in concert to allow TCP to provide a reliable transport mechanism for data. This is very good for certain types of data transfer, such as when downloading a file from a remote server. However, it also makes TCP slow and inappropriate for some types of data streams, such as low latency audio or real-time video. In the latter case, a protocol such as UDP can be used.

In comparison to TCP, UDP is a fairly simple protocol. UDP does not signal the sender to retransmit lost or corrupted data. It does not offer any resequencing of out of order datagrams it receives, and even though there is a checksum field incorporated in the header, in many cases it is not used (i.e., no checks are performed). As can be seen in Figure 11-17, its header is only 8 bytes long, and consists only of a source port, destination port, length, and checksum fields. Thus, UDP has very low overhead, and along with its minimal processing, can move data up or down the stack much faster than TCP. This makes UDP better suited to transporting data for applications that have a requirement of low latency (such as streaming audio).

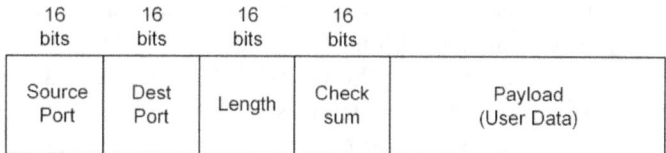

Figure 11-17 UDP Packet Structure

Because UDP offers no error correction or sequencing, it is up to the application using UDP to deal with these issues itself. In many cases, the application will just ignore the corrupted data, but missing or out of sequence data may manifest as pops and clicks in an audio stream or glitches in a video stream.

11.8.5 Application Layer Protocols

The application layer provides the highest level of services that allow end-to-end communication between an application and its peer. Here, the term application refers to the software entity that an end user may directly interact with, e.g., an audio streaming server or a web browser. The protocols at the application layer are used to facilitate and manage the communication for an application, and so may be application specific, i.e., may be designed for use by a single application. Some applications, though, may integrate more general protocols in their implementation in order to provide services. One such protocol that is used for transporting digital audio streams and which is popularly integrated into many streaming applications is the Real-time Transport Protocol[5] (RTP).

RTP is popular because it is simple to implement, is flexible, and allows a manufacturer to overcome the major deficiencies of UDP while still keeping its low latency, low overhead benefits. RTP typically uses UDP for the transmission of data, but adds a sequence number and timestamp to each packet before passing it on to UDP. This extra information allows the receiver to deliver the packets in correct sequence, and to compensate for jitter[i] in the delivered data. With RTP, there is no fixed way to package the audio data in the RTP packet. While there are many defined ways—called profiles—to package and interpret the included data, users are able to define any new way, i.e., profile, they want to package the included data. Thus, we could define a profile that specifies the data as four channels of uncompressed 24-bit 48 kHz audio.

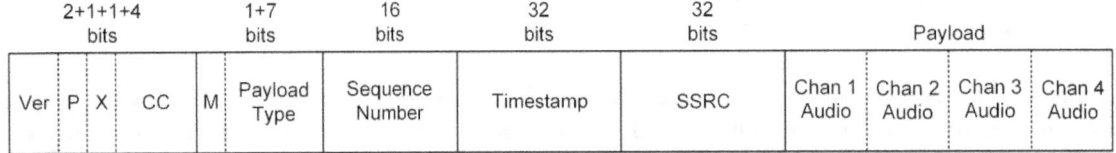

Figure 11-18   An Example RTP Packet

Figure 11-18 shows the header fields of an RTP packet and one way that four channels of digital audio may be packaged. Some of the key fields in the header are the Payload Type field, which specifies the profile for the packaged data; the Sequence Number field, which stores the sequence number of the packet; the Timestamp field, which gives the instant the first byte of the packaged audio was sampled; and the SSRC (synchronization source) field, which identifies the source of the audio stream.

While it is typical that a single protocol is used at each of the lower layers to effect communication, multiple protocols may be used simultaneously at the application layer by the communicating peer. For example, an application that uses RTP for packaging and transporting its data will use the RTP Control Protocol (RTCP) to help manage various QoS statistics of the session. Before sending any audio data, it may use the Session Description Protocol (SDP) to tell the remote peer how to actually interpret the encoded audio (e.g., how may channels, sample rate, bit depth, etc.), and may use yet another protocol for the actual set up and tear down of connections to peers. Similarly, applications

---

[i] Jitter is the variation in the interarrival time of packets at the receiver due to inconsistent delay in traversing the network.

may rely on the Precision Time Protocol (PTP) to accurately synchronize certain actions, such as when to play out an audio sample from a buffer.

11.9 Application to Live Sound

In this chapter, we took a brief look at the basic concepts of design of Ethernet digital data networks and communication between digital endpoints. In the next chapter, we will see how these concepts are applied specifically to creating networks for the distribution of digital audio in the context of live sound reinforcement.

Review Questions

1. In networking terms, explain what is meant by a subnet.
2. Discuss the use of switches vs. routers in a network, and when you would need to use one over the other.
3. Explain the network classification of a LAN, and why this would or would not be the appropriate network type for a medium sized live sound event.
4. What is the difference between half-duplex and full-duplex data transmission? Why is one more suitable than the other as a data transfer mechanism in an audio distribution network?
5. In an Ethernet network, what does a MAC address identify?
6. Explain what is meant by a globally unique IP address, and whether IP addresses used in a LAN have to be globally unique or not.
7. Explain three ways that an endpoint device (host) in a network can get an IP address.
8. What are VLANs used for in a digital network?
9. What is the role of the network layer in the ISO OSI reference model?
10. What is a communications protocol, and why is it important to have standard protocols for networking applications?
11. Explain the difference between closed proprietary and open non-proprietary communications protocols.
12. What is a data packet and how is it used in data communications?
13. What is the DSCP field in an IP packet used for?
14. When sending an audio stream between two devices, which of the two transport protocols, TCP or UDP, should be used? Why?
15. RTP is a protocol that is often used to send audio stream data between devices. RTP adds a sequence number to its packets before sending. Why does it do this?

Chapter 12

Digital Audio Distribution Networks

12.1 Requirements for a Digital Audio Network

Having looked at the fundamentals of data networking in general, we now will look at how this technology can be deployed to facilitate distribution of digital audio over a geographic area. There are many digital audio transportation protocols that use Ethernet cabling, but not standard Ethernet switching, e.g., AES-50, EtherSound, and SuperMAC. These types of protocols are not the focus of this chapter. Rather, we will be considering deployment solutions that run on top of standard Ethernet protocols and equipment at the core.

We will start our investigation with a look at what aspects of networking are required for acceptable distribution of audio over a digital network. Requirements can be examined from various angles. When determining the requirements to examine, we start with the end goal of digital audio distribution and work backwards from that. So, let us state our goal: the delivery of high-quality audio from source to loudspeaker, with the ability to easily troubleshoot and solve connectivity problems, within reasonable budgetary constraints. The key terms in this goal that will drive the requirements are "high quality," "easily," and "reasonable." Although these terms are all qualitative and too imprecise for developing actual requirements for a specific real network, they are adequate as the basis for an introductory discussion on the requirements of a digital audio distribution network. Based on such a goal, we can arbitrarily break our requirements into three broad categories: technical—which pertains to the capabilities of the network nodes; operational—which pertains to those aspects of the system that affect user interaction and operation; and financial—which pertains to the cost of deploying the system.

12.2 Technical Requirements

Suppose we consider high quality audio to be as good as what we would get from a high-end pro-audio analog system. The analog transducers on the edge of the network, i.e., our microphones and loudspeakers, are the same whether analog or digital distribution is used, so we are not concerned with these devices in this discussion. Instead, we ask the question: what aspects of digital distribution can degrade the quality of an audio stream. The answer is: bit depth (i.e., word size) and sample rate of the data stream, latency, jitter, and interference immunity. Bit depth and sample rate have further implications for the bandwidth of the network links and the processing capabilities of network nodes. Latency and jitter have further implications for process synchronization, quality of service implementation, and buffering on the network nodes.

244

12.2.1 Bandwidth and Throughput

Currently, analog to digital converters (ADC) produce output with a wide range of word sizes and sample rates. A common combination is 24-bit 48 kHz samples. This combination is very acceptable for live audio distribution, with a resolution better than a typical compact disc (which uses 16-bit samples at 44.1 kHz sampling rate). If we take this as a baseline for uncompressed audio streams in the network, how much bandwidth would it take to transport one such stream using RTP, assuming that our RTP packet transports 1 ms of audio for 8 different channels? Figure 12-1 shows the overhead associated with such a packet. In all, there are 62 bytes of header and 4 bytes of trailer, for a total of 66 bytes of overhead. At 48 kHz, 1 ms worth of 24-bit data would contain 144 bytes[a] for a single channel, or 1152 bytes for 8 channels. The total Ethernet packet size would thus be 1218 bytes. There would be 1000 such packets transmitted every second, giving a bandwidth requirement of 1218 × 1000 × 8 bits/s, or just over 9.7 Mb/s. If our scenario required, say, 24 channels of audio to be transmitted (thus requiring 3 streams), we would need just over 29.2 Mb/s of bandwidth. Since RTP also uses RTCP for out of band data associated with audio streams, and it is recommended that 5% of the actual audio bandwidth requirement be further allocated for RTCP,[1] we would need another 1.5 Mb/s, giving a total of approximately 30.7 Mb/s just for the audio. On top of this, we have to consider all the other data that must be sent over the network for management and control of the system. The quantity of this data is typically small compared to the digital audio streams. From the foregoing discussion, it is clear that 100 Mb/s network links would be adequate to meet our requirements. Note that this discussion does not consider how the audio data is actually encoded, which could add additional bytes to the audio payload.

Ethernet Header + Preamble & SFD	IP Header	UDP Header	RTP Header	Audio Payload	Ethernet Trailer
22 bytes	20 bytes	8 bytes	12 bytes	1440 bytes	4 bytes

Figure 12-1   Encapsulation of 10 ms of 24-bit 48 kHz Audio Payload

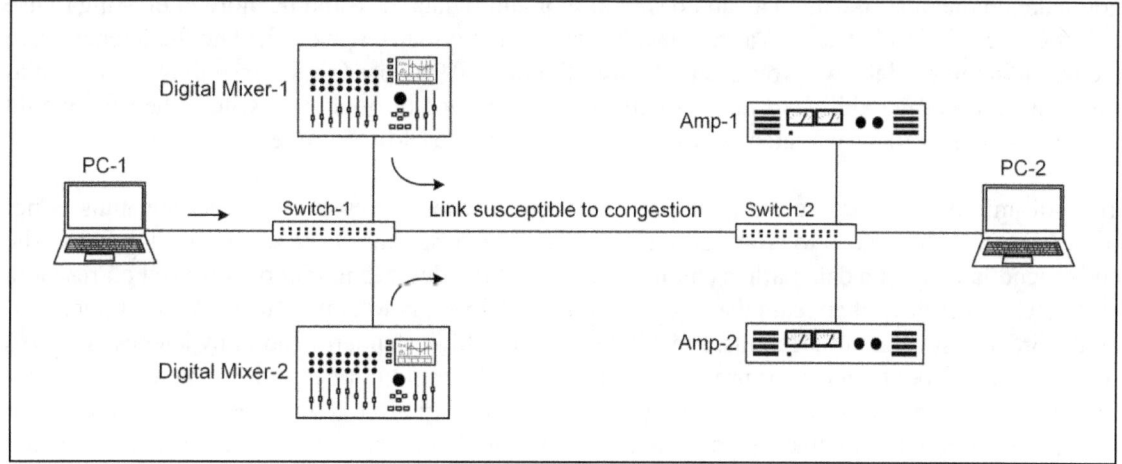

Figure 12-2   Interswitch Link Congestion

However, consider the scenario in Figure 12-2, where we have multiple sources of digital streams in a multi-switch network. We can immediately see that the inter-switch link could potentially be

---

[a] 1 ms of 24-bit data sampled at 48 kHz contains (48000 × 1/1000) × 24/8 = 144 bytes (where 1 byte = 8 bits).

subjected to a traffic load that could congest a 100 Mb/s link. Thus, to mitigate the risk of this happening, we would specify 1 Gb/s as a minimum bandwidth for such links. With the cost of gigabit Ethernet switches reasonably low, we could specify gigabit Ethernet for all links of the network.

Of course, there is no point for a switch to have all the Ethernet ports needed, but not the switching capacity, i.e., the processing power, to actually move the data through the switch in a timely manner. This is typically not a problem in a modern switch, where we find that an average small switch will have twice the switching capacity of the aggregate bandwidth of all the ports (i.e., an 8-port gigabit Ethernet switch would be expected to have at least 16 Gb/s full duplex switching capacity), and midsize switches (24 ports) may have 40 Gb/s full duplex switching capacity. However, we should determine what our total aggregate peak data rate is likely to be and make sure that the total switching capacity is easily able to handle that amount of data. We will likely find that we stand a greater chance of congesting an uplink port than the switching backplane.

12.2.2 Latency

Network switches typically operate in a store-and-forward fashion, where a frame that arrives at the switch is placed in a queue (a block of memory) for processing before being forwarded out the switch. Amongst other things, this processing consists of checking the frame for errors, checking its destination address, and then forwarding it out the appropriate egress port on its way to its destination. The time it takes for the switch to process the frame in this way is called the latency of the switch. Every digital device has some latency which depends on the speed of its memory operations, its processing power, and the nature of the processes that must be applied to the data.

The latency specification for switches can vary widely. Small switches with less capable processors tend to have greater latency than larger switches, which usually have more powerful processors. For example, the latency specification for one typical midsize 24-port switch shows 95.2 Mfps (million frames per second) for 64-byte frames. Note that 64 bytes is the minimum valid frame size, typically used by manufacturers for specifying attributes like latency, but this size of frame would not make up the bulk of data in an audio stream. For audio data, it would be more interesting to use a frame size of 1518 bytes (i.e., a frame with maximum payload). If we calculate the latency from the manufacturer's data, we would get a figure of about 0.01 μs per frame. Assuming a linear relationship between the ability to move small vs. large frames through the switch, then we would expect this switch to have a latency of about 0.25 μs for a 1518-byte frame.[b]

This sub-microsecond latency is typical of midsize switches currently available, and thus is not something we would need to worry about unless our network consists of several switch hops. The end-to-end latency of a data path is cumulative over all the devices in that path. Based on research by Lester and Boley,[2] it appears that musicians playing live can tolerate a wide range of latencies, on the order of several milliseconds, and still perform to a high standard. Thus, any latency concerns we might have about devices or processes in the network would not be on the switches, but rather on more software centric nodes and processes, such as digital effects processors, digital mixer plugins, and the buffering that needs to happen at the end device to ensure smooth playout of audio (we will discuss this further in the following sections).

---

[b] The listed latency specification for switches varies widely from a high of 15 μs for a small 8-port gigabit Ethernet switch (108T) to 0.1 μs in midsize switch (S3300).

12.2.3 Jitter

Jitter is the variation in interarrival time of data packets at an endpoint. This variation depends on the devices that the packets go through on their way to the destination. In our audio distribution network, these devices would be the digital mixer, any digital effects processors, and the Ethernet switches. For an audio distribution network, jitter matters to the last device in the chain, the one that does the final digital to analog conversion. These devices use buffering, i.e., storing a small number of packets in their internal memory and then playing them out at precisely the required rate, to compensate for jitter in the arriving packets. However, excessive jitter can cause buffer overflow (running out of buffer memory and having to throw away newly arrived packets, or worse, writing packets to memory where they are not supposed to) or underflow (where the buffer does not contain enough data for the expected number of samples). Both of these scenarios could lead to audio dropouts, depending on their severity and the quality of the playout algorithm. Given that jitter, like latency, is cumulative over the entire network path, it pays to have network devices with enough computing power to minimize this quantity at the endpoint device. The current generation of network switches available to us is very capable in this regard. From our discussion above, we see that these switches have sub-microsecond forwarding rates, and this will help us to achieve this requirement. However, even such capable switches can exhibit jitter of 1 ms under load.[3] To help manage jitter, some approaches, such as with Dante, allow the user to configure a buffer (latency) setting on the endpoint, while others, such as AVB, specify a maximum number of allowed switch hops to the endpoint.

12.2.4 Layer 2 and Layer 3 QoS

In an ideal world, a digital audio distribution network would transport only audio stream data. There would be no data from applications like web browsers, file transfers, or database queries also vying for bandwidth on the network. In the real world, however, this is not always the case. A campus distribution network, for example, is typically a shared network with all kinds of data flowing through the network devices. Even in a tightly controlled audio distribution network, there will still be multiple kinds of traffic, such as timing, control, and monitoring data, besides the actual audio data itself. Given that audio stream data is very susceptible to latency and jitter induced problems, we would want to make sure that non-audio traffic on the network does not consume so much bandwidth as to degrade the quality of the audio played out by the endpoints. One solution to this problem is to utilize network switches and routers that have extensive quality of service (QoS) capabilities. Of course, if the network had enough bandwidth to handle all the data that needed to be transported without prioritization, there would be no need for a QoS implementation.

Switches and routers prioritize forwarding by placing the data to be forwarded in one of a few queues. A queue is simply a small amount of memory where data is stored until the switch is able to send it out an egress port. The switch's queue works just like a human queue—the first packet placed in the queue is the first one to be taken from the queue. A midsized switch may have four queues, with priorities like 3 = very high, 2 = high, 1 = medium, 0 = low. It may then utilize sophisticated algorithms to determine when to service a particular queue, i.e., when to forward data from a particular queue, but generally, data in a higher priority queue will be sent before data from a lower priority queue. Some algorithms try to be fair; they will send a large amount of data from a higher priority queue and then a small amount of data from a low priority queue before returning to service the higher priority queue (e.g., weighted round robin). If there is audio data in the higher priority queue, you can see that this approach would introduce some jitter to the audio stream. Some simpler algorithms may service higher priority queues exhaustively at the expense of low priority queues (e.g., strict priority queuing). This approach may starve the low priority queue to the point where it becomes full, and the switch would have to throw away data that would otherwise be placed in that queue.

QoS can be implemented at both layer 2 (data link) and layer 3 (network). Recall that layer 2 QoS requires adding a VLAN tag to the Ethernet frames. The switches in the network must therefore recognize VLAN tagged data and honor the embedded priorities of any received frames. Recall, too, that at the network layer, an audio application can set the priority of audio packets via the DSCP field. Even though this is a layer 3 attribute, many switches can read this information and prioritize the forwarding of packets based on it. The creators of digital audio distribution protocols may assign any VLAN priority or DSCP marking they want to the packets that constitute the protocol's data. For example, Dante (which we will look at further below) marks its high priority packets with DSCP value 56 (CS7), medium priority with 46 (EF), and low priority with 8 (CS1).[4] Our network switches must therefore allow us to arbitrarily map specific DSCP and VLAN priorities to its queues so that we can prioritize critical timing and audio stream data above other types of traffic. Figure 12-3 shows an example configuration screen for a switch, where we configure the mapping of DSCP priorities to queues for the values mentioned above.

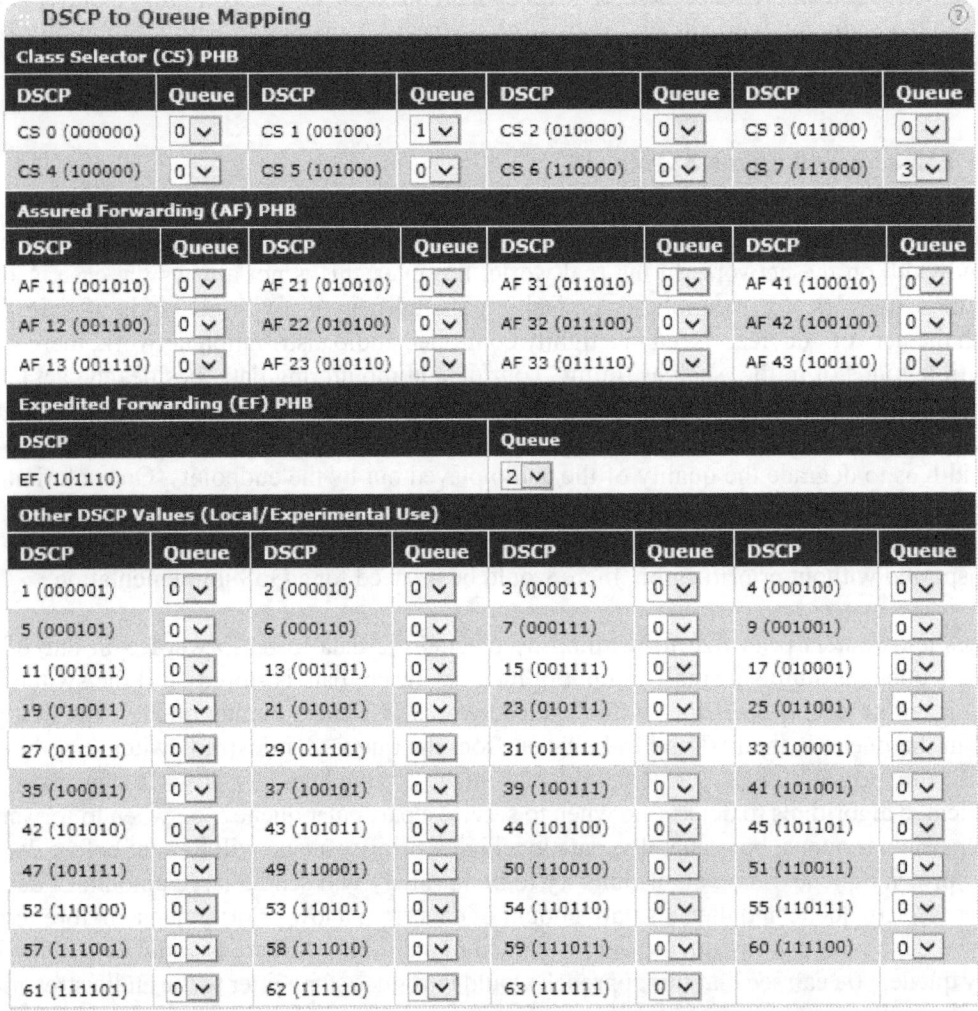

Figure 12-3 DSCP to Queue Mapping for QoS on Network Switch

12.2.5 IGMP Snooping

As discussed earlier, if a source device has to send the same audio stream to several endpoint devices, it can take one of two approaches. In the first case, it can send separate unicast streams to each endpoint. This requires a lot of computational work on the part of the source and uses bandwidth inefficiently on the egress link from the source. Alternatively, the source can send a single multicast stream to a single multicast destination, and each endpoint device can subscribe to the stream. This significantly reduces the workload on the source and uses the egress link bandwidth much more efficiently, but shifts the distribution workload to the network switch, as shown in Figure 11-8(C).

Normally, a switch will handle multicast data just like it would broadcast data, i.e., it will forward a packet with a multicast destination MAC address out every port of the switch except the port through which it entered the switch. If there is a lot of multicast data flowing through the switch, this can be very inefficient, and floods most network segments with data that may not be needed by any device in those segments. However, with IGMP snooping, the switch can examine the IGMP requests to join a group and learn the ports on which the requesting devices are located. It will then send the multicast data out only those ports, and thus operate more efficiently and avoid flooding the network with unnecessary data.

12.2.6 PTP Support

The precision time protocol (PTP) is used by many digital audio distribution protocols to disseminate a precise timing reference throughout the network. This timing reference is used to synchronize the playout of audio samples at multiple endpoint devices. PTP works on a master-slave model, where a single device—the master—is the source of the timing reference, and all other endpoint devices that need to be synchronized are slaves of this master, i.e., they use the reference disseminated by this source to adjust their clocks. The amount by which a slave adjusts its clock, called the offset, is calculated by averaging the time it takes for timing packets to travel from the master to the slave and from the slave to the master. As we mentioned in Section 12.2.4, when a packet enters a switch, its priority treatment is determined, and it is stored for a brief time in a queue as it waits to be forwarded. This store-and-forward method of operation leads to some packet delay variation, or jitter, in the arrival times at the destination endpoints. This in turn leads to some variation in the offset calculated by different endpoints, i.e., degradation in the time synchronization amongst the endpoints.

To overcome the above degradation in synchronization, the IEEE-1588 PTP standard[5] (and its derivative IEEE 802.1AS[6]) describes two approaches to handling PTP synchronization packets by switches. In both cases, the switches must have more advanced functionality beyond a basic Ethernet switch. In the first approach, the switch operates in a boundary clock mode. Here, the switch operates normally as a slave of the master clock that supplies the sync packets, but then becomes the master clock for all attached devices that request synchronization. The single device thus operates as both a slave and a master. In the second approach, the switch operates in transparent clock mode. In this case, the switch measures the time that each timing packet spends in the switch and adds that time to the correction field of the packet. Each slave then uses this correction to adjust its calculated offset. Switches that can operate in either of these modes are said to be IEEE-1588 compliant.

The above modes of operation are useful in maintaining tight synchronization amongst endpoints that are several switch hops away from the master clock. In a live sound distribution network with a small maximum number of switches (e.g., 3 or less) where switching latency is on the order of a microsecond, we can expect that endpoints would be synchronized to within a few (i.e., less than 10) microseconds of each other, even without IEEE-1588 boundary or transparent clock functionality. Synchronization differences of several microseconds between the playout devices would not produce any more adverse sonic artifacts than what would naturally occur due to differences in distance from the various loudspeakers to the listener. Thus, we can conclude that in a small network with low latency switches, compliance of the switches with IEEE-1588 is not a necessary requirement to preserve acceptable synchronization amongst endpoints. (Note that some applications, e.g., AVB, may force this requirement via a protocol check, and fail to function without it.)

12.2.7 RSTP Support

When a distribution network consists of one or two switches, it is easy to correctly cable all the devices together so that there is only one path between the network switches. However, as the number of switches in the network grows, the probability of inadvertently wiring the network in a loop increases. The creation of a loop in the switching network causes the switches to operate inefficiently in the best case, as they continually update their forwarding tables. In the worst case, broadcast frames can circle the loop forever, degrading performance to the point that the network becomes unusable for delivering the audio stream data to the endpoints. To guard against this, the network switches should run the Rapid Spanning Tree Protocol (RSTP).[9] RSTP automatically detects loops in the switching network and disables the appropriate switch ports so that there is only one active path between any two switches at any time.

However, not all loops in the switching network are inadvertent. As we will see below, we may deliberately create a loop in the network when building redundant paths between critical switches. In this case, RSTP must be considered from the start as a key element of the network deployment.

12.2.8 Interference Immunity

Depending on the environment, a live sound deployment may very well be an electrically noisy environment. Ethernet cable consists of four twisted pairs, and since Ethernet lines are driven in a balanced manner, a deployment with unshielded Cat 5e cable should be able to reject most electrical noise typically encountered. However, as insurance against EMI and ESI when using gigabit Ethernet in a very noisy environment, a move to shielded Cat 6 cable, or even fiber optic cable, may be warranted.

Interference and noise, though, will only typically be an issue for Wi-Fi (wireless) links. Though it may be adequate over short ranges, the Wi-Fi signal degrades quickly with distance, and in the presence of interference may fall back to a modulation scheme that is too low to properly transport multiple audio streams. As such, it is currently not advisable to send our critical aggregate audio streams over a Wi-Fi link. However, even when there is interference, we may still be able to use a Wi-Fi link to control some of our devices. Wi-Fi devices operate in the 2.4 GHz and 5 GHz bands. The 2.4 GHz band is generally more crowded than the 5 GHz band, and suffers more interference from non-Wi-Fi equipment. If noise poses a problem in such a case, it is wise to move all control to wired endpoint applications.

12.3 Operational Requirements

Moving to a digital distribution network can have many advantages. In order to realize these advantages and make this move worthwhile, there are many operational requirements that must hold true. Let us take a look at some of the key ones now.

12.3.1 Setup Flexibility

As far as the distribution network itself is concerned, the physical setup is very flexible. Connections are made between endpoints and switches or routers. When the network consists of a single subnet without any VLANs, we can connect an endpoint (or another switch) to any free port on the Ethernet switch. When VLANs are configured, we must connect the endpoint or uplink switch to the correct port, but we have total control over which ports we choose for the VLAN configuration. Similarly, if the network consists of multiple subnets and a router is used between them, we have to make the connections on the router via the correct ports. Again, though, we have full control over which ports on the router we configure as being in each subnet.

12.3.2 Ease of Configuration

With the move to digital audio distribution, one major issue that we will face is the identification of endpoints as sources and destinations of audio streams. A further issue is that of actually setting up the routing of streams from particular sources to destinations, i.e., binding a particular endpoint to an audio stream. These tasks would actually be quite tedious to configure by hand. Thus, we need a software application that allows us to accomplish these tasks in a manageable way. Fortunately, protocols have been defined, and software have been written, to allow automatic discovery of endpoints and make it easy to establish such connections.

12.3.3 Network Security

The switches and endpoints in our distribution network should have the same level of physical security as any other piece of audio equipment in a typical audio deployment. In other words, no unauthorized person should be able to physically access our switch and connect an Ethernet cable to it.

Of more concern is the security of the wireless part of our network. As mentioned previously, we may use wireless devices on the Wi-Fi portion of the network to control and manage certain aspects of the system. Wi-Fi is a totally uncontained broadcast technology. This means that anyone within range of the wireless signal can intercept it, and moreover, can maliciously target our wireless access point. There are a few measures that can be taken to increase the security of this part of the network. First, the network identifier (technically called the SSID, or service set identifier) should not be broadcasted. This does not vastly increase the security of the network, but will prevent the casual snoop from attempting to associate with the access point. Secondly, the access point should be capable of maintaining an access control list so that only devices whose MAC addresses are in the list are allowed to access the network. Thirdly, the network security should be set to utilize WPA3 or WPA2 (encryption should be set so that Advanced Encryption Standard is used when WPA2 is used). Finally, make sure that a strong passphrase is used with the above security settings, and it is changed on a reasonable schedule, e.g., every live show, or at least once a month.

Depending on our network deployment, firewalls may also be a consideration. Firewalls are mostly used to keep certain unwanted traffic out of a network, and they come in all shapes and sizes. Firewalls must be very carefully configured. They can easily be misconfigured to lock us out of our own network, or block legitimate data that should be allowed through the network. If the distribution network is attached to a larger network, or perhaps to the Internet, a hardware firewall may be deployed at the edge of the distribution network. If our network is an isolated LAN, then there is really no need for a discrete hardware firewall to be used. If we are using a wireless access point in our network, this device will likely have a built-in software firewall that can be configured to block unwanted traffic.

Security vulnerabilities are constantly discovered in network devices, such as managed switches, computer operating systems, and audio endpoints. These vulnerabilities may allow an attacker to gain unauthorized access to the network or otherwise corrupt the network data. Once discovered, manufacturers typically address these vulnerabilities by issuing updated firmware and patches for the software that run on these devices. In many cases of network breach, access is gained via network devices that have known vulnerabilities and that are left unpatched. Therefore, it is important to make sure that, as far as possible,[c] all the devices in the network have the latest firmware updates and security patches installed.

12.3.4 Robustness

The physical robustness of digital network equipment should be on par with what we expect for equipment that must meet the demands and rigors of a live sound system, including harsh handling during transportation and setup, many power cycles, and hundreds (perhaps thousands) of connector mating cycles. Ruggedized connectors, as in Figure 10-6, should be used where possible.

Considering the network as a whole, there is another aspect to robustness which deals with the ability to recover from network misconfiguration and network component failure. Misconfiguration has already been discussed in the requirement for RSTP, and we will discuss component failure in the following sections.

12.3.5 Network and Data Stream Monitoring

In the live sound context, the distribution network is critical infrastructure. Failure of a switch port could potentially prevent all audio streams from reaching the destination endpoints and result in silence at the loudspeakers. Being able to monitor the health of the network nodes is an important part of being able to effectively operate the network. The simple network management protocol (SNMP)[7] is commonly used to allow the health of critical network components, such as switches and routers, to be constantly monitored. When used in this role, SNMP requires management software to be run on a PC in order to receive notifications from the monitored devices. When a significant event (e.g., port failure) occurs on a switch, a specific alarm is raised on the monitoring PC, allowing the engineer to quickly pinpoint the problem and decide on the best course of action to solve or work around the problem.

Many audio data distribution applications use RTP as the core streaming protocol. This has certain benefits, as RTP is paired with RTCP, which provides the application with statistics regarding bandwidth usage, packet loss, latency, and jitter. The application can monitor these statistics and raise an alarm if any one crosses some predetermined threshold indicative of a serious degradation in streaming performance.

12.3.6 Network Redundancy

Redundancy is an important part of any critical infrastructure. Network redundancy is the incorporation additional or duplicate components in the network such that if a failure occurs in a single active component, we still have an intact and functional path from source to destination. Redundancy may be implemented in areas such as power supplies, ports on switches, or entire switches

---

[c] In some cases, updating the firmware or patching a device may cause its behavior to change to the extent that it no longer interoperates with other network devices. If the device is critical to the operation of the network, patching or updating the device may have to be postponed until the manufacturer can fix the issue.

and routers. Figure 12-4 shows a number of ways that redundancy may be implemented in the network.

Some manufacturers build redundancy into their endpoint devices by incorporating two ports which are always connected to the distribution network. There are multiple approaches to how this redundancy may be implemented. For example, one approach transmits the same data on both ports, but only one of the receive ports is active (the active path), while the other is in a hot standby mode. Two completely separate networks (one primary and the other secondary) connect the transmit ports to the receive ports. If a failure occurs along the active path, the standby port takes over in a fraction of a second. Another approach is based on the Parallel Redundancy Protocol (PRP),[8] which operates similarly to above, except that both receive ports are always active, and the receiver simply keeps the first packet of the replicate pair that arrives and discards the second. In this case, there really is no switchover time if one path fails. A network that supports these approaches is shown in Figure 12-4(A). This is typically the best approach, as it gives full end-to-end path redundancy, and the manufacturer is able to implement an optimal algorithm that gives the fastest switchover time.

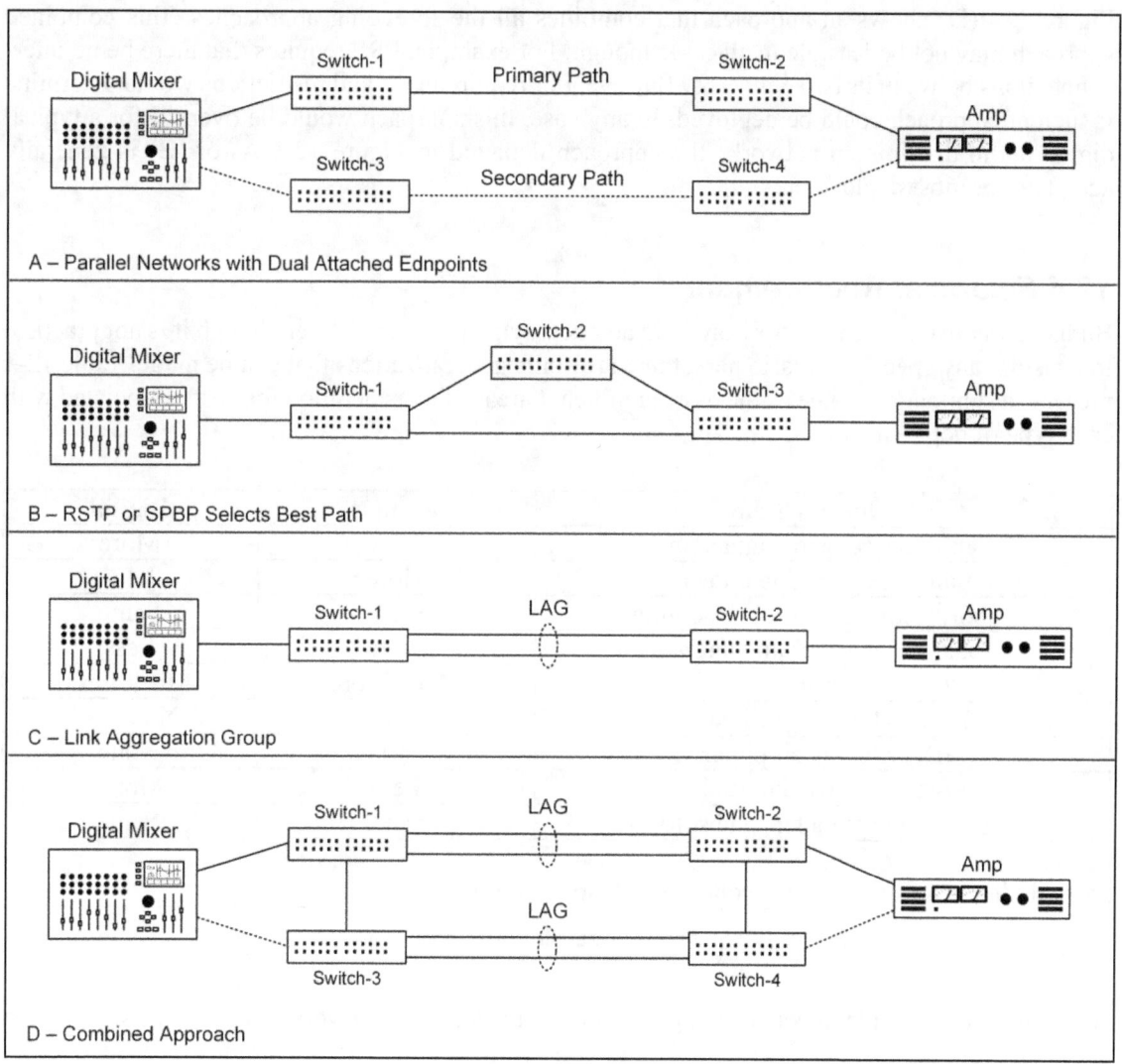

Figure 12-4 Approaches to Redundancy in the Distribution Network

Figure 12-4(B) and (C) show how we can get partial path redundancy by using other standard Ethernet protocols to create multiple paths in the core of the network. In (B), we see that there are two paths from Switch-1 to Switch-3 (Switch-2 has been added to illustrate that the core of the network may be arbitrarily complex). When we run the Rapid Spanning Tree Protocol (RSTP) on these three switches, it selects the best single path from Switch-1 to Switch-3. If the currently in-use path fails, the protocol selects the next best intact path. A default switchover time of 6 seconds is typical, though this can be configured to be shorter. If the switches are capable of running the Shortest Path Bridging (SPB)[10] protocol, then all links can be used to forward data simultaneously in a load sharing fashion, and path recovery times are typically under 100 ms.

Figure 12-4(C) shows an approach to redundancy using a Link Aggregation Group (LAG).[11] A LAG combines two or more physical ports and uses them as one logical port. Data may be load balanced across all the ports in the LAG. If one of the ports fail, data is redistributed on the remaining operational ports. The SPB protocol mentioned above combines aspects of RSTP and LAG, and can be used in place of both of them.

Figure 12-4(D) shows an approach that combines all the foregoing approaches. This combined approach may not be feasible in all cases though. For example, PRP requires that there be no inter-connections between its two networks. Thus, vendor requirements and restrictions would determine if such an approach could be deployed. In any case, this approach would be overkill for a typical digital audio distribution network. The approach depicted in Figure 12-4(A) or (B) is generally adequate for most deployments.

12.4 Financial Requirements

Budgetary considerations factor heavily in any deployment scenario. Even though it is not practical to consider any specific scenario here, there are some generalizations that can be made. Table 12-1 presents a summary of some of the more significant areas of consideration for costs associated with each type of deployment.

Budgetary Item	Analog	Digital
Labor: design and engineering	Less	More
Labor: setup and teardown	More	Less
Labor: testing and troubleshooting	Same	Same
Labor: maintenance and upgrades	More	Less
Cabling	Much More	Much Less
Nodes (Ethernet switches)	None	Nominal
Distribution amplifier	Nominal	None
Stage box (snake head)	Less	More[†]
Management computers and software	Same	Same
Notes		
† Includes ADC and DAC for connected analog equipment		

Table 12-1 Comparison of Costs for Analog vs. Digital Deployment

In any non-trivial deployment, a large part of the cost is labor. At the very start, the skillset required to specify, design, and engineer a stable digital network is quite different from that needed for an analog deployment, and for a digital deployment, this stage may be more costly. System testing and troubleshooting connectivity issues in a digital network again requires a similarly different skillset. However, monitoring and control software help to significantly reduce the cost of these activities.

The setup cost for a digital deployment may be low compared to an analog deployment, as storage and handling of Ethernet or fiber optic cable, and interconnecting switches and audio endpoints with Ethernet cable is fairly straightforward and nowhere near as labor intensive as when using multicore snakes in an analog system. In permanent installations, the labor costs associated with reconfiguring the system can be high for an analog deployment compared to digital. This is especially true if new cables need to be pulled in the analog deployment, as opposed to merely making a new connection via management software, as would be the case with a digital deployment.

With the exception of cabling, the core equipment (mixer, amplifiers, loudspeakers, uninterruptable power supply for sensitive equipment) costs in either an analog or digital deployment are likely to be about the same. The cabling costs of an analog deployment can easily be an order of magnitude more expensive than a comparable digital deployment, as we are essentially replacing bulky multicore snakes with thin and lightweight Ethernet cables. The same Ethernet cable also carries all the audio, monitoring, and control signals.

Ultimately, a reasonable requirement would be that in the long term, the digital distribution network should be no more expensive to deploy than a similarly specified analog system. Current cost trends support the achievement of this requirement, and even point to digital deployment as being more economically attractive than analog deployment in the long term.

12.5 Technical and Operational Requirements Summary

We discussed several diverse requirements for our digital audio distribution network in the foregoing sections, and at this stage, it may be hard to keep track of all of them. Table 12-2 summarizes these in one place for easy comparison and tracking.

Requirement	Needed	Nice to Have	Not Needed
Switch port bandwidth	1 Gb/s		
Latency	< 1 ms per hop		
Jitter	< 1 ms per hop		
VLAN Capable	✓		
Layer 2 QoS	✓		
Layer 3 QoS	✓		
IGMP Snooping	✓		
Switch PTP Compliance			✓†
RSTP Compliance		✓‡	
Wired Link	Cat 5e	Cat 6	
Wireless Link Security	WPA2-CCMP	WPA3	
Endpoint Management	Software Configured		
SNMP Capable	✓		
Robustness		Ruggedized	
Redundancy		Parallel	
Notes			
† Unless required by specific protocol implementation, e.g., AVB.			
‡ Only if network size and complexity warrants.			

Table 12-2 Summary of Audio Distribution Network Requirements

12.6 Example Digital Audio Distribution Systems

In this section, we will take a brief look at two currently used digital audio distribution systems that run on top of standard Ethernet technology, and which can be deployed in the type of networks that we have been discussing up to now. In fact, from a topological and design point of view, we can use the same network, shown in Figure 12-5, to highlight the basic operation of both systems. Note that since this is a standard Ethernet network, non-audio related devices and their data can coexist and function normally along with all the digital audio related devices on this network.

There are many issues that need to be addressed by every digital audio distribution system. One convenient list, as referenced in the AES67 standard[12] (with regards to what it covers and excludes), is given here. We will see how they are handled by the example systems that are detailed below.

- **Network-wide clock synchronization** – All participating endpoints in the network need to know what time it is with reference to a master clock (so-called wall clock). This is needed so that, for example, an endpoint knows *when* to play a sample of audio.
- **Media clock derivation** – Endpoints need to be able to derive the clock that determines *how fast* to play out audio samples. The media clock is synchronized to the network-wide wall clock.
- **Audio encoding** – There needs to be a defined way of packaging the audio samples in the packets that are sent between endpoints. Besides the sample packaging itself, this determines things such as sample rate, bit depth, and number of samples per packet.
- **Stream (session) description** – Some agreed format to describe the contents of a stream so that the connection management endpoint application can understand what it needs to do to accept and process the stream. This includes a description of the encoding and source of the stream.
- **Transportation** – Once encoded, there must be defined rules for sending the data across the network. This includes aspects such as network addressing, transport protocol (e.g., RTP over UDP), unicast vs. multicast, and QoS.
- **Endpoint discovery and enumeration** – How are participating endpoints identified and their capabilities listed?
- **Connection management** – A way must be available to setup and tear down streams of audio between endpoints.
- **Endpoint control** – For a complete system implementation, we usually need to interact with remote endpoints and be able to change parameters such as channel gain.

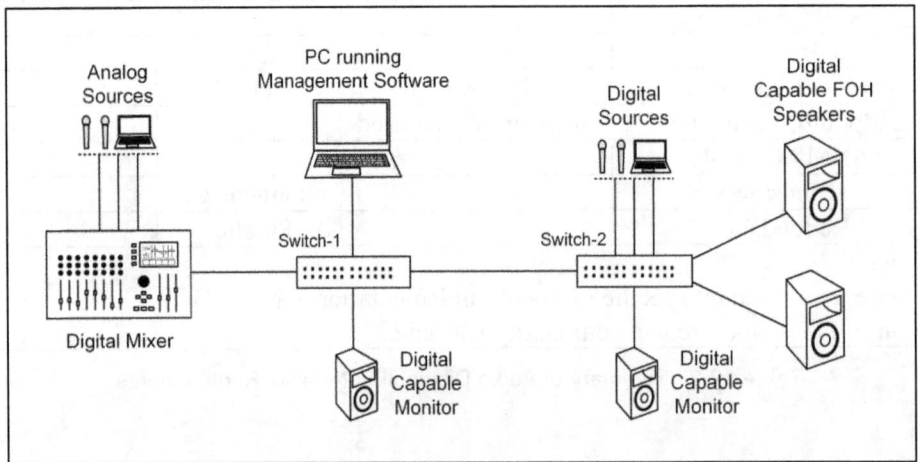

Figure 12-5   Basic Digital Audio Distribution Network

12.6.1 Audio Video Bridging

Audio Video Bridging (AVB) is a suite of protocols for implementing digital audio transport and distribution. AVB is built on a set of open, non-proprietary standards developed and maintained by the Time-Sensitive Networking (TSN) Task Group within the IEEE. These include IEEE 802.1BA, *Audio Video Bridging Systems*, for determining the acceptable members of an AVB network; IEEE 802.1AS, *Timing and Synchronization for Time-Sensitive Applications*, an application of PTP for AVB (generalized PTP, or gPTP); and two extensions to IEEE 802.1Q, namely, IEEE 802.1Qat, *The Stream Reservation Protocol (SRP)*; and IEEE 802.1Qav, *Forwarding and Queuing for Time-Sensitive Streams (FQTSS)*. IEEE 1722-1, *Audio/Video Device Discovery, Enumeration, Connection Management and Control (AVDECC)*, is an extension of IEEE 1722, and can be used by a management computer to discover and control AVB capable endpoints, e.g., amplifiers, effects processors, and audio sources in the network.

Currently, an AVB network is a switched network with endpoints communicating over some chosen VLAN. Therefore, AVB endpoints must be in the same subnet, and AVB does not work across a network boundary. This may change in the future as the IEEE TSN group works to expand the scope of AVB.

In AVB, the sources of digital audio data are called talkers, while the endpoints that receive the audio data are called listeners. The connection from talker to listener may be one-to-one (a talker sends data to a single listener), one-to-many (many listeners simultaneously receive the same audio data transmitted by a single talker), or even many-to-one (e.g., a mixer receiving several audio streams from several digital microphones). AVB uses the concept of a stream as a conduit to transport data from talkers to listeners. A single stream can bundle many separate channels of audio. The actual number of streams and channels that a single network link can accommodate is determined by the bandwidth available on that link, and the bit depth and sample rate of the audio being transported.

The communication between talkers and listeners is isochronous. This means that the talkers generate audio packets at a specified set rate, and the listeners expect to receive and playout the packets at the same rate. AVB defines two classes of operation, *viz.* Class A and Class B. For Class A, the talker generates packets every 125 μs and end-to-end latency must be held to a maximum of 2 ms. For Class B, the talker generates packets every 250 μs and end-to-end latency must be held to 50 ms.[13] Maintaining this timing relationship between talkers and listeners requires a network with low latency, low jitter, and a precise timing distribution mechanism.

Low latency and low jitter require network switches that are powerful enough to move data through the switch with minimal delay. Moving data quickly is not a problem with today's switches when the switch is lightly loaded. However, other factors come into play in the presence of non-audio stream data or when the volume of audio data is such that the switch may become congested. AVB uses a multi-pronged approach to optimize audio transport. First, AVB uses the SRP to guarantee that the required bandwidth for a particular stream is available along its entire path, i.e., from talker to listener and through every intermediate switch. On every switch, SRP will reserve bandwidth for streams, but only up to a maximum of 75% of the bandwidth of the ports used. This ensures that some bandwidth is available for data other than audio streams (e.g., control data). If a switch cannot guarantee the required bandwidth for the stream, then the path will not be established through that switch.

The other tact that AVB uses to reduce latency and jitter is employing FQTSS to implement the priority treatment of audio packets flowing through the switch. In addition to reserving the bandwidth for a stream, SRP also causes the switch to automatically update its QoS related tables.[14] The forwarding and queuing implementation of FQTSS is then used to give priority treatment to the AVB stream on its requested VLAN. This guarantees delivery of stream data from talker to listener with a maximum end-to-end latency for the chosen class of operation (e.g., 2 ms for Class A[d]).

Thirdly, participants in the AVB network must run the generalized PTP (gPTP). This is the protocol that distributes a precise timing reference, allowing talkers, listeners, and switches to synchronize their clocks to within 1 μs. In an AVB capable switch, each port must be capable of running the gPTP. Furthermore, an AVB compliant switch will not forward AVB advertisements out a port if the device that is connected to that port does not also run the gPTP.

For managing an AVB network, a controller application implementing the AVDECC protocol can be run on any host computer. This allows automatic discovery and enumeration of AVB capable devices. It can also be used to set up and tear down streams between endpoints and modify certain playback parameters, such as output level.[13] The core of the controller GUI is typically a matrix type display, such as in Figure 12-6, that shows all discovered devices, and makes it extremely easy to set up or teardown a stream simply by clicking on the intersection point between the two devices. For example, the highlighted intersection in Figure 12-6 shows that we are selecting a stream from a device called STB01 to a device call FOH01, while the panel in the upper left of the figure gives details about the stream and its status.

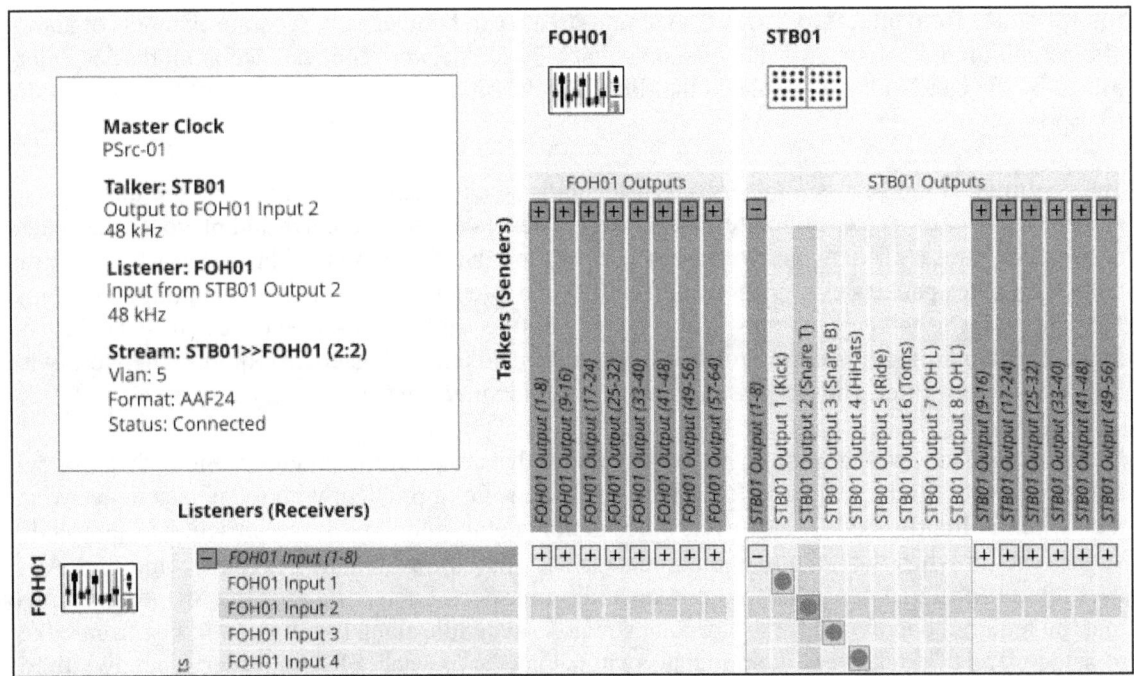

Figure 12-6 Matrix View of Network Controller

To walk through an example with reference to the network in Figure 12-5, a user could set up a stream from the mixer to the FOH loudspeakers. The switches propagate the advertisements only

---

[d] In order to achieve this upper bound of 2 ms, the AVB standard also specifies a maximum of seven switch hops in a Class A network.

through those ports that meet both SRP and gPTP requirements. Thus, Switch-1 would not send the advertisement to the management PC, but to the monitor and to Switch-2. Switch-2 would send the advertisement to all its attached devices. At each hop, the advertisement accumulates the latency of that network segment, and carries that quantity along to the next device it reaches. When it reaches an endpoint (listener), that listener will know the value of the end-to-end latency along the entire path from the talker to itself. In our example, since only the FOH loudspeakers have been programmed by AVDECC to receive the stream, only they respond to the advertisement. Switch-2 combines their separate responses and forwards them back to the mixer (the talker) along the same path that the original advertisement took. When a switch processes a response, it reserves the bandwidth necessary, and updates its forwarding and queuing table (based on the VLAN ID and priority of the stream), to guarantee that the stream data is forwarded with the high QoS needed for low-latency audio. As noted above, a switch will only allow up to 75% of a link bandwidth to be reserved for AVB streams. When the response reaches the mixer and is processed by it, the stream set up will be complete, and the mixer can start sending audio data on the stream. When this stream is torn down, a reverse operation takes place, where the bandwidth on the switch is freed and the QoS entry for the stream is flushed from the forwarding table. Thus, we see that bandwidth and QoS management are entirely automatic in AVB.

All switches in the network path linking AVB endpoints must implement the SRP and gPTP. Currently, most basic Ethernet switches on the market are not SRP and gPTP capable, as this requires an upgrade to the hardware or firmware of the switch, and the few switches that are SRP capable are more expensive than commonly available Ethernet switches. The lack of widely available, low-cost switches is likely one of the hurdles in the widespread adoption of AVB.

As can be seen from the above discussion, AVB is built around several different open standards that must work together to create a usable audio distribution system. Because no single private entity controls the production of hardware implementing these standards, and several interpretations of the standards are possible, it is possible that different manufacturers can create separately AVB compliant systems that do not interoperate. This may be another reason that AVB has been slow to take off in the pro-audio space. To solve this problem and encourage the adoption of AVB, the Avnu Alliance has created the Milan protocol,[15] which is really a set of rules and restrictions, or profiles, regarding the mandatory support for certain media stream formats, media clocking, device discovery, connection management, and control. It also provides guidance for providing network redundancy. Thus, any device that is certified by the Avnu Alliance as Milan compliant should be able to interoperate with any other Milan compliant device.

12.6.2 Dante

Dante is a popular digital audio distribution technology that has seen good adoption rates.[16] Even though Dante utilizes several open standard protocols (e.g., PTP, UDP) in its operation, its core design and operation are based on proprietary and closed technology developed by Audinate Pty. From a manufacturing perspective, this means that there are licensing costs associated with using Audinate's intellectual property in any devices that implement Dante, and the low-level operational aspects of the technology are not publicly disclosed. From the user's perspective, products implementing Dante may be a bit more expensive than products based on non-proprietary technology, as licensing costs are typically passed on the end user. However, as a private entity, Audinate Pty can move much more quickly in bringing new features and capabilities to market, which ultimately benefit the end users.

With regards to the network topology, a network based on Dante would look no different from one based on AVB, and there are also many operational similarities between the two. Similar to AVB,

Dante uses PTP to distribute an accurate timing reference to all endpoints participating in the distribution network. While its PTP requirements are less stringent than that of AVB (e.g., AVB requires gPTP and switches that implement boundary clock, while Dante does not.), it is still completely capable of maintaining microsecond level synchronization between endpoints. Timing distribution and master-slave operation are as previously described, and so will not be repeated here.

Dante uses the concept of a flow to bundle several channels of digital audio for transportation between endpoints. This is similar to a stream in AVB. Dante uses UDP to transport the audio, but the actual encoding is proprietary. By default, a unicast flow in Dante bundles four channels of audio. Remember, unicast means that the data is going from one sender to one receiver, so this type of flow would be useful for sending four different channels of monitor mixes from a mixer to a single amplifier, for example. To send the same audio to several different endpoints, a multicast flow can be used instead. A multicast flow can bundle up to 8 different channels of audio. When using multicast flows in a larger network, it is beneficial to enable IGMP snooping on the switches to prevent the audio data from flooding the entire network. Without this, a switch will simply broadcast the multicast traffic out every port on the switch, unnecessarily consuming bandwidth. Dante endpoints support IGMP, and so will send the proper join requests for a particular multicast flow. The switch will then send the multicast traffic only to those endpoints that actually request it.

Unlike AVB, Dante does not need any specialized networking equipment beyond what is generally available today in order to work. It will run over any Ethernet switch that has enough throughput. However, in a Dante network, there is no automatic reservation of bandwidth on the network switch for an audio flow as there is with AVB. Therefore, bandwidth must be manually and carefully managed when using Dante. Audinate recommends using no more than 70% of network link bandwidth for audio traffic,[17] but manually calculating actual bandwidth usage is tedious. As a rule of thumb, Audinate suggests that a typical unicast flow with four channels consumes about 6 Mb/s of bandwidth. Thus, a single 100 Mb/s link could easily handle 11 unicast flows, or 44 channels, within this bandwidth restriction.

In a network with limited or shared bandwidth, priority treatment is required for the various different types of Dante specific data if good results are to be had. This means that the user must configure the QoS settings of the switch manually to ensure reliable transmission of audio through the network. In a Dante network, there are four types of data that a user must be concerned about. These are PTP timing data; the flows carrying the actual audio data; Dante monitoring and control data; and everything else. Dante uses IP packets to transport its data and utilizes the DSCP field in the IP header to signal the priority of that data. Dante marks its high priority packets with DSCP value 56 (CS7), medium priority with 46 (EF), and low priority with 8 (CS1).[17] Audinate recommends that the critical PTP data be given the highest priority, the audio flows given the next highest, and the monitoring and control data the third highest priority. Figure 12-3 shows an example of how the QoS table of a switch may be configured to accomplish this.

Dante implements dual attached primary and secondary interfaces. This allows network path redundancy of the type shown in Figure 12-4(A). In case of a loss of communications on the primary path, communications will seamlessly switch to the secondary path. Not all Dante devices have this feature. However, it can be very useful to have this type of redundancy between endpoints on high value, high risk paths, e.g., between the FOH mixer and amplifiers.

For managing a Dante network, Audinate provides the management software. This software will automatically discover and enumerate all Dante endpoints in the local network using the standard DNS-SD and mDNS protocols. Once the endpoints have been identified, the software allows a user to make interconnections of flows between them. Dante also uses specific monitor data to transport health and performance statistics about each endpoint to the management software. Since Dante does not automatically keep track of the latency of each hop along a network path, another key function of the management software is to configure an approximate latency value at each playout endpoint. The latency value is based on the number of hops between the source and the destination endpoint and the network link speeds in the path. Audinate provides guidance in its management software for the values to set for any given network topology.

12.6.3 AES67 and SMPTE ST 2110-30

As digital audio networks become popular, users will want some choice in deciding which vendor's equipment they want to use for implementing their network. For example, a user may want to use an amplifier from vendor A which implements AVB and a mixer from vendor B which implements Dante. Normally, these devices would not be able to talk to each other. However, by adopting some common standards for key elements of digital networking, different vendors' equipment may be able to interoperate.

AES67[12] is not a complete digital audio distribution solution, but an interoperability framework. Of the list of elements itemized at the start of this section, AES67 specifies only the network-wide synchronization, media clock, transportation, audio encoding, stream description, and connection management aspects. It does not specify endpoint discovery and enumeration or endpoint control, which are required for a fully functional digital audio distribution system. However, with the elements that it does specify, if vendors implement the protocol specific functional areas according to AES67, then equipment from different vendors running different base distribution protocols should be able to successfully exchange digital audio. Thus, for example, equipment based on RAVENNA,[e] which is compliant with AES67, can easily work with equipment implementing Dante when the Dante equipment is configured to operate in AES67 mode. However, because there is no common endpoint management specification, the endpoints would still be managed using different vendor-specific software, rather than one common application.

In its most basic form, the SMPTE ST 2110-30 standard[18] may be viewed as an adaptation of AES67 with a number of restrictions. The underlying mechanisms are the same in both standards, but ST 2110-30 specifies the use of a specific PTP profile, the option to force a device to operate in PTP slave-only mode, and zero offset between the device media clock and the RTP stream clock. It also mandates support for IGMPv3. It therefore constrains an AES67 compliant device to operate in a more restricted fashion. Thus, a strictly ST 2110-30 compliant device may not be able to interoperate in a fully conformant AES67 environment, but a properly configured AES67 device should be able to interoperate in a fully conformant ST 2110-30 network.

12.6.4 AES70

As mentioned above, AES67 stops short of specifying certain elements needed to provide a unified end-to-end networked audio solution. Essentially, on the road to providing this complete solution, AES70 picks up where AES67 leaves off. The Open Control Architecture[19] (OCA) was developed by the OCA Alliance trade association to specify the control and monitoring of professional media networks. AES70[20] is the formal open standard that codifies OCA.

---

[e] RAVENNA is another digital audio distribution system that operates in IP over Ethernet networks. It is primarily targeted at the broadcast industry.

Like AES76, AES70 is not a single protocol, but a framework that provides for the unified management of a digital audio distribution network made up of multivendor endpoints. Conformance to AES70 thus allows for the creation of a solution where two very important things can be achieved.

First, devices that conform to AES70 will make themselves discoverable and manageable by adhering to the specified format of the protocol packets that they transmit, and accepting and acting appropriately upon the protocol packets that they receive.

Secondly, we can now create a software application, i.e., the controller, that can discover all the devices on the network that can be controlled and monitored. Furthermore, the controller can then be used to set up and tear down the media streams between the endpoints; for configuring the devices by changing allowable parameters of the devices, such as gain, EQ, and soft-switch positions; for monitoring the health of the devices by receiving and interpreting informational data from the devices pertaining to such things as network interface status, device temperature, stream statistics; and even upgrade the firmware or operating software of the device. All of the above functions can be performed in a secure manner over a wired or wireless network connection.

Note that since AES70 is strictly concerned with control and monitoring of endpoints, it could be implemented in a network quite separately from AES67. However, for a deployment where full multivendor interoperability and unified management is a desired goal, it makes sense that both AES67 and AES70 be implemented on all the network endpoints.

12.7 Advantages of Digital Audio Distribution

The trend towards digital audio distribution continues to advance at a rapid pace. While there are advantages and disadvantages to adopting any new technology, the real advantages that accompany a move to digital audio networks outweigh the disadvantages. Here we will summarize the greatest of these advantages.

- **Cost** – The cost of deploying a digital audio network can be viewed from two angles—labor and infrastructure. Labor cost is directly related to time cost. The time it takes to set up and tear down a digital audio distribution network is a fraction of an equivalent analog system. The second element is infrastructure cost. In a digital network, heavy multicore snakes are replaced with far cheaper Ethernet compatible cabling, and relatively inexpensive switches are used to interconnect the equipment.
- **Reduced connection errors** – There is an ease of deployment in that we can plug our network cable into any unused port on the Ethernet switch. On the mixer, there is a single primary connection port. This leads to fewer interconnection errors and less time troubleshooting routing problems.
- **Ease of configuration and control** – A digital audio distribution system is managed from a central application running on a computer. Here the endpoints are configured and the interconnections between sources and destinations are made, and these interconnections can be reconfigured at any time with minimal effort.
- **Maintenance and troubleshooting** – A misconfigured endpoint can easily be identified and pinpointed in the control application. If a used network port fails, the fix is a simple matter of plugging the connecting cable into another unused port.
- **Redundancy** – In many deployments, path redundancy can easily be built into the network. If the active path fails, audio can still flow from source to destination, with the switchover to the standby path occurring automatically.

- **Realtime monitoring** – Every endpoint can report its health status and other metrics (e.g., packets processed, latency) on an ongoing basis to the management application. An alarm can be raised to alert an operator when some monitored parameter on an endpoint gets out of tolerance.
- **Sound quality** – With strict timing distribution and proper bandwidth management, there is no perceivable degradation in sound quality when digital audio is transported across a digital network. Furthermore, issues like hum and buzz do not typically plague this type of deployment.
- **Market opportunities** – The distribution requirements and complexity of deployment scenarios are increasing such that the traditional analog approach may no longer be feasible in some cases. However, these cases are readily accommodated with a digital network solution, giving the operator of digital distribution networks a competitive advantage over analog operators.

However, there are some costs associated with digital audio distribution compared to traditional analog deployment. One of them is endpoint cost. Every participating endpoint in the digital audio network must include hardware and firmware that support and implement the protocols for the deployed system, as well as the associated ADC and DAC components. This can add significantly to the cost of each endpoint.

Another cost, potentially quite high, is the training needed to acquire the knowledge and skills to properly design and engineer a digital audio distribution network. However, once these skills have been acquired, they can be reused and marketed, so this may be better thought of as an investment with potentially good returns, rather than a simple expenditure.

In summary, there are compelling advantages to moving to a digital audio distribution system. The readily available equipment and protocols have been hardened through many years of use in the IT industry, and the professional audio industry can confidently build on this foundation. The shift towards all digital audio distribution is well underway, and we can expect that this trend will continue well into the future.

Review Questions

1. Discuss three advantages of moving to a digital audio distribution system and how these compare to providing the same services in an analog deployment.
2. What is the primary purpose of running the Rapid Spanning Tree Protocol in the distribution network?
3. Discuss the sources and cause of latency in a digital network and the possible effects of latency on audio playback at the endpoints.
4. What is network jitter? Why is it important that an endpoint device that plays out an audio stream be able to compensate for jitter?
5. Why is network redundancy important in a digital audio distribution network? Describe one form of this type of redundancy.
6. What is the role of IGMP snooping in a network switch?
7. Explain what is meant by the term isochronous when applied to a digital audio stream.
8. For a pro-audio digital distribution network based on AVB, which class of operation (i.e., Class A or Class B) would be used in the network? Explain why.

9. Briefly explain the roles of AES67 and AES70 in the area of digital audio distribution and internetworking.

10. Discuss the need (i.e., necessary or not) for sub-millisecond timing synchronization in playout endpoints in a small (less than 5 hops from source to destination) digital audio distribution deployment.

11. Discuss three issues that need to be addressed by every digital audio distribution system.

12. Explain how advertisements are propagated and used in an AVB network to assist in stream establishment.

13. What are the likely negative effects of not properly configuring QoS on the switches in the distribution network?

14. Compared to AVB, Dante utilizes some closed and proprietary technology at its core and is owned and marketed by a private enterprise. What are the potential pros and cons of this?

15. Compare and contrast the costs of deploying a digital audio distribution system to that of an analog system.

Chapter 13

Designing the Sound System

13.1 Design Considerations

At this stage, we have learned the fundamentals of sound: wave propagation, intensity falloff, interaction with objects in the sound space, etc. We have also looked at the various components that make up a sound system, at their internal design and application. We have learnt how to read and understand their specifications and have learnt some things about cables and equipment interconnection. We are now thus in a position to apply all these separate pieces of knowledge, put it all together, and build the sound system.

Whether we are putting together two loudspeakers, an amplifier and a single microphone, or a multi-everything system targeted for a large space, there are a number of fundamental factors that we must consider. We will take a look at some of these factors, which include primary system use, venue size, audience distribution, as well as human traffic considerations at the venue, the permanence of the sound system setup, safety concerns, and of course, the working budget.

This chapter does not intend to make you a sound system design professional—there are entire textbooks dedicated to that subject alone. The goal here is to provide enough fundamental knowledge so that you can ask the right questions and make estimates that are on the correct side of the line in order to satisfy the requirements of the sound reinforcement system that you may provide.

13.2 Primary System Use

The first issue we consider is the primary use of the sound system—speech, music, or combination of both. Knowing this up front will assist us in making the right decisions regarding key design and budgeting issues. A sound system designed primarily for speech reinforcement will likely be a poor performer for music. A system designed primarily for music may not meet critical requirements for speech.

13.2.1 Speech System Characteristics

The most important requirement of a sound system designed primarily for speech reinforcement is intelligibility. If this is not met, the system is useless. There are several measures of speech intelligibility defined. Thus, unsurprisingly, intelligibility can be measured by several means, either by electronic equipment (e.g., STI method IEC 60268-16) or by trained listeners listening to trained talkers in the target area (e.g., ANSI/ASA S3.2).

Speech is composed of a combination of vowel and consonant sounds. Intelligibility is conveyed primarily by the consonant sounds in speech, which have a level about 27 dB below the vowel sounds and a duration of about one-third that of the vowels.[1] The meaning of the phrase "separate the wheat from the chaff" is quite different from "separate the wheel from the shaft." You can discern that the difference in meaning is all in the consonant sounds. With such critical information contained in lower level short duration sound, you can see that anything that causes this information to be lost will easily affect the usefulness of the sound system.

One of the biggest factors affecting speech intelligibility is room reverberation. Long reverberation times in rooms can muddle the consonant sounds and make understanding difficult. Articulation loss of consonants (Alcons) is just one example of a measure of speech intelligibility that is formulated around this muddling of consonants. Alcons is a function of reverberation time and direct-to-reverberant sound ratio. It only takes a small loss in Alcons—around 10 to 15%—before speech intelligibility becomes unacceptable. For a room to have good speech intelligibility, its reverberation time should be 1.5 seconds or less. There are several inexpensive software applications on the market that can be used to measure the reverberation time of a room. In rooms with long reverberation times, acoustical treatment may have to be used to help control the quality of the audio. This may include adding absorption panels to the walls and ceiling, or adding diffuser panels to certain locations to reduce specular reflections. Additionally, the loudspeaker system may have to be more distributed in order to bring the direct sound sources closer to the listeners (thus increasing the direct-to-reverberant sound ratio). Further, close miking techniques may have to be used more extensively to get more direct sound into the audio chain.

Background noise also plays a large role in the intelligibility of speech in any given room. Some studies have shown that the speech level at the audience location should be about 65 to 75 dB SPL for optimum intelligibility, and that the background noise from all sources should be at least 10 dB below the speech level.[2] In such a case, the background noise level should not exceed 55 dB SPL. We will see later how to determine the amplifier and loudspeaker combination to achieve any required dB SPL level at the audience location. One particular type of background noise that is particularly effective at reducing intelligibility is the so-called distractor voices. These are other competing, typically lower level, speech sounds occurring in the same space as the target speech. Rooms with large reverberation times allow the original speech to bounce around the room multiple times and act as multiple distractor voices.

The frequency response of the sound system should be tuned to match its role as a speech reinforcement system. The average fundamental frequency of human speech is about 150 Hz (lower for men, higher for women). Vowel sounds are created at harmonics of these fundamental frequencies, while consonant sounds occur in the range of 2 kHz to 9 kHz. Low-frequency noise is very effective at degrading intelligibility when its level is comparable to the speech level. Therefore, all speech circuits should have all frequencies below 80 Hz severely cut to minimize masking from low-frequency noise. At the other end of the spectrum, the response of the sound system should extend up to at least 10 kHz. As a further precaution, in low power systems (where feedback is less likely an issue), omnidirectional microphones may be used instead of cardioids because the proximity effect of cardioids can cause the speech signal to mask itself when the user gets close to the microphone.

A speech reinforcement system typically has to deal with few input sources—maybe only one or two microphones at a time. This implies a small mixer, with likely a small set of auxiliary effects to help shape and control the sound. The bulk of effort in the design of such a system will be in the choice and placement of loudspeaker systems and the control of the acoustic properties of the sound space.

13.2.2 Music System Characteristics

Though not critical, speech intelligibility is still very important for a music reinforcement system. In fact, one way to design a music system is to design it as for a speech system and then extend it to handle the requirements of a music system, while maintaining the good intelligibility characteristics of the speech system. Since speech intelligibility is not critical for a music system, larger reverberation times can be tolerated, up to 2.5 seconds for large spaces (e.g., bigger than 25000 cubic meters).

A music system requires extended low and high ends to be able to reproduce the entire range of musical sounds. Whereas we cut the energy below 80 Hz in a speech system, we want to be able to reproduce cleanly all the sound down to below 30 Hz in the music system. Moreover, to reproduce those tinkling bells and cymbals, we want to get at least up to 18 kHz. Thus, the loudspeaker system will need to include precision high-frequency units, and for the low end, subwoofers may be necessary. The sense of music sounding 'just right' is highly subjective. This usually creates the need to have a whole slew of effects processors to tweak the audio one way or another to the satisfaction of the engineer, producer, or performers. We may find, for example, that we have to shape the tone of a singer's voice more for balance and fit in the total mix than for some objective attribute such as intelligibility.

There is no doubt that a typical music system is run at higher SPL than a comparable speech system. Bigger amplifiers and bigger and more loudspeakers may be required to get the required sound level out of the system. However, the system will also be required to deal with high-level transients that are typically part of any modern-day pop music, necessitating a further consideration of the power handling requirements of the system.

One major difference between a speech system and a music system is the sheer number of input sound sources to be handled and managed at any one time. In a music system, you may have well over 10 microphones, plus multiple signals from keyboards, samplers, sequencers, guitars, etc., all needing to be mixed. Many of these signals will need some form of effects processing. Thus, in addition to the larger number of effects processors required, a bigger, more complex mixer will be called for. Digital mixers, with multiple on-board effects processing options may provide some advantage here over the separate external effects rack and mixer combination. Furthermore, for anything but simple live stage shows, the performers may require separate monitor mix handling from the FOH, necessitating yet another mixer. To get all of these various signals between the stage and the mixing locations, some form of multichannel signal distribution will be needed, either in digital form as discussed in the previous chapters, or analog in the form of multicore snakes. Thus, the system is overall quite a bit more complex than a speech system, and its budget is comparably bigger.

One thing that is often forgotten early on in the planning process for a music reinforcement system is its electrical power requirement. To feed and run all of this equipment, adequate power sources are required. Thus, you should check that the venue has the required power outlets needed to safely and adequately power all of the equipment.

13.2.3 Hybrid System

A sound system that is used for both speech reinforcement and music reinforcement must have the good characteristics of both systems as described above. The best approach in this case is to design the system as for speech reinforcement and make sure that intelligibility requirements are met.

When this is done, the system may be expanded with any additional loudspeakers and processors to get the needed range and shaping capabilities required to deliver pleasing sounding music.

If it is possible, it is best to reserve certain mixer channels only for speech. These channels will be EQed appropriately, with little to no effects processing. These channels may also be amplified separately and fed to specific sets of loudspeakers that may exclude certain types of drivers, e.g., subwoofers. When channels are shared by both purely speech and music, the major concern becomes adjusting the EQ and effects processing between changing roles of the channel. This can truly be a hassle with an analog mixer, to the point that the operator will eventually do nothing when roles change, and system performance will suffer as a result. In such a case, a digital mixer with recallable parameter settings would eliminate this shortcoming.

13.3 Indoor or Outdoor

A sound reinforcement system that is operated outdoors will behave quite differently than one operated indoors. This difference arises because the walls, ceiling, and floor of the indoor space are usually quite good at reflecting and containing the sound energy from the sound system. In an outdoor space, the only surface of note is the ground, and this may be relatively absorptive compared to the typical indoor floor. Thus, in an indoor space, reverberation must be taken into consideration. It may work in our favor, or it may work against us. In an indoor space, reverberation will cause the sound system to sound louder than in an outdoor space, other things being equal. However, we also know that too much reverberation will cause intelligibility problems. In an outdoor space, there will be little to no natural reverberation. In fact, reverb processing may have to be added to some channels to make the sound more desirable. Additionally, you will find that large vertical structures, such as the sides of a building, may produce very strong and annoying echoes that must be factored into the system layout.

13.3.1 Critical Distance

Say we have a sound system operating outdoors, with a single loudspeaker delivering the sound to the audience. In this space where there is practically no reflection of the sound energy, the inverse square law, discussed in Section 1.7, holds. This is because with no reverberation to speak of, we are pretty much always hearing the direct sound from the loudspeaker. Thus, as we move away from the loudspeaker, the sound energy decreases rapidly, and the SPL is reduced by approximately 6 dB every time we double the distance from the loudspeaker.

Now let us move the sound system indoors, where the walls, ceiling, and floor reflect much of the sound energy, creating a reverberant space. This is illustrated in Figure 13-1. Initially, as we move away from the loudspeaker, the direct sound will be stronger than the reflected sound. We will find this to be true in every direction that we take from the loudspeaker. As the delivery of sound from the loudspeaker is quite directional in nature, this gives rise to a rather distorted spherical space, called the direct sound field, around the loudspeaker. In this space, the inverse square law can be observed to roughly hold. For example, in Figure 13-1 position B is twice as far from the loudspeaker as position A, and C is double the distance of B. The SPL at position A is 80 dB, while at B and C it is 74 and 69 dB, respectively. We can see that as we double the distance from the loudspeaker, the SPL is reduced by approximately 6 dB.

As we move farther away from the loudspeaker, we will eventually enter the reverberant sound field. Here, the reflected sound has higher intensity that the direct sound. The sound waves in this area have bounced around thousands of times, and the sound is very diffuse. The SPL in this space is relatively constant. This is the case at positions **X**, **Y**, and **Z**, where the SPL is 62, 63, and 61 dB, respectively.

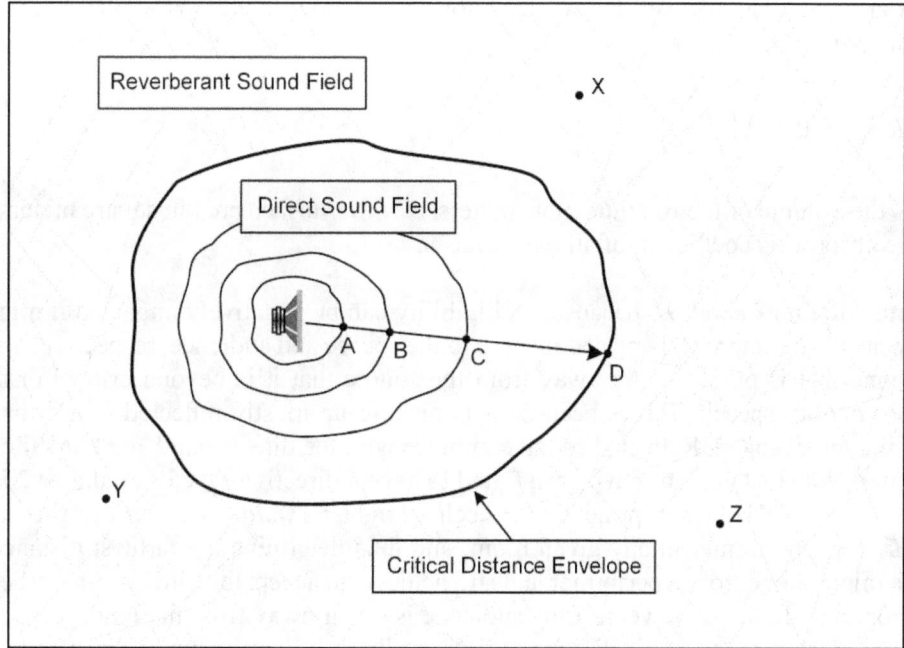

Figure 13-1   Critical Distance in Indoor Space

There is a transitional boundary between the direct sound field and the reverberant field. In this boundary region, the inverse square law ceases to hold. The center of this boundary region is the critical distance (D_c). At the critical distance, the direct and reverberant sounds have equal intensity. The SPL at D_c is 3 dB higher than the reverberant field alone, and is also 3 dB higher than the direct sound field would be in a non-reflective environment (e.g., in an open-air space). In Figure 13-1, the SPL at **D** may be seen to be around 65 dB. Note that critical distance is a concept that applies to indoor spaces, where we have walls, a floor, and a ceiling that can reflect sound.

D_c varies with the directivity of the sound source. With a directional source, such as a typical loudspeaker, D_c will change as we move around the source. This gives us the notion of a critical distance envelope. D_c also depends on the reflectivity of the walls of the room. More reflective walls will tend to reduce D_c.

Critical distance can be determined by using a pink noise source and an A-weighted SPL meter. Start at the source and double the distance from the source while noting the reduction in SPL. The SPL will initially fall by 5 to 6 dB for each doubling of distance. At some point, the SPL will be reduced only by 1 to 2 dB. This is the transition area through D_c. The critical distance can also be found by calculation. If we restrict ourselves to the longitudinal axis of the sound source, then critical distance (in meters) may be approximately calculated using the following equation, which takes the directivity factor of the source into account:[3]

$$D_c = 0.057 \sqrt{\frac{Q \cdot V}{RT_{60}}}$$

where V is the volume of the room in cubic meters, Q is the directivity factor of the sound source, and RT_{60} is the reverberation time of the room in seconds. The reverberation time of a room can most easily be found by using one of many available software/hardware packages, but it can also be estimated using Sabine's well-known equation if we know the absorption coefficients of the walls, floor, and ceiling:

$$RT_{60} = 0.161 \left(\frac{V}{S \cdot a} \right)$$

where V is the volume of the room in cubic meters, S is the surface area in square meters, and a is the average absorption coefficient of all the surfaces.

It is important to know about D_c because intelligibility can be negatively impacted if microphones and loudspeakers are improperly placed relative to the source and audience, respectively, and to D_c. A microphone that is placed so far away from the source that it is beyond critical distance will deliver poor quality speech. This is because it is picking up mostly reflected sound that is over-whelming the direct sound. Reflected sound combines with the direct sound and blurs the words of the performer. Note that humans have been found to have a directivity factor of almost 2 in the mid frequencies (1 to 8 kHz) when projecting speech straight forward.[4] We can use this number to calculate D_c for a performer in any given room, and thus determine the farthest distance that we can place a microphone from a performer and still achieve an acceptable direct-to-reverberant level at the microphone. In the same vein, if the audience is so far away from the loudspeaker that they hear mostly reflected sound, we will have speech intelligibility problems as discussed in Section 13.2.1.

13.4 Design Model and Layout Plan

If you are going to design something, it is good practice to have a model on which the design is based. We already have a model of a sound system that was discussed in Section 1.2. Our model gives a simplified, big-picture look at the system components and how they interact with each other. The model shows microphone and line-level sources feeding into a central mixer, where routing, levels, and effects are coordinated, and then fed to the house and monitor amplifiers. Our final implementation will follow this basic structure.

Models can be used to produce a more detailed design. If the sound system is large and complicated, producing a detailed design is a good idea. It is much easier and cheaper to find and fix mistakes at the design stage than after the cables are laid down and components interconnected. Even with simpler systems, a basic layout plan like the one shown in Figure 13-2 can be a useful tool in uncovering issues with power sources, traffic flow, bylaw conformity, audience safety, obstructions, equipment requirements and security, etc. For example, even the basic plan of Figure 13-2 can give us a rough idea of the length and routing of the cables needed. It shows us that we will need some protection for the snake from being trampled. We also see that the FOH area is wide open to the general audience and will need some measure of security. In most cases, a proper layout plan can only be created after consultation with the performers' representatives and facilities management. A site survey is often necessary to help determine things like cable routing paths, mixing locations, loudspeaker positions, and the acoustic characteristics of the sound space. A reasonable layout will help ensure both safety of equipment and minimization of hazards to the performers, personnel, and the audience.

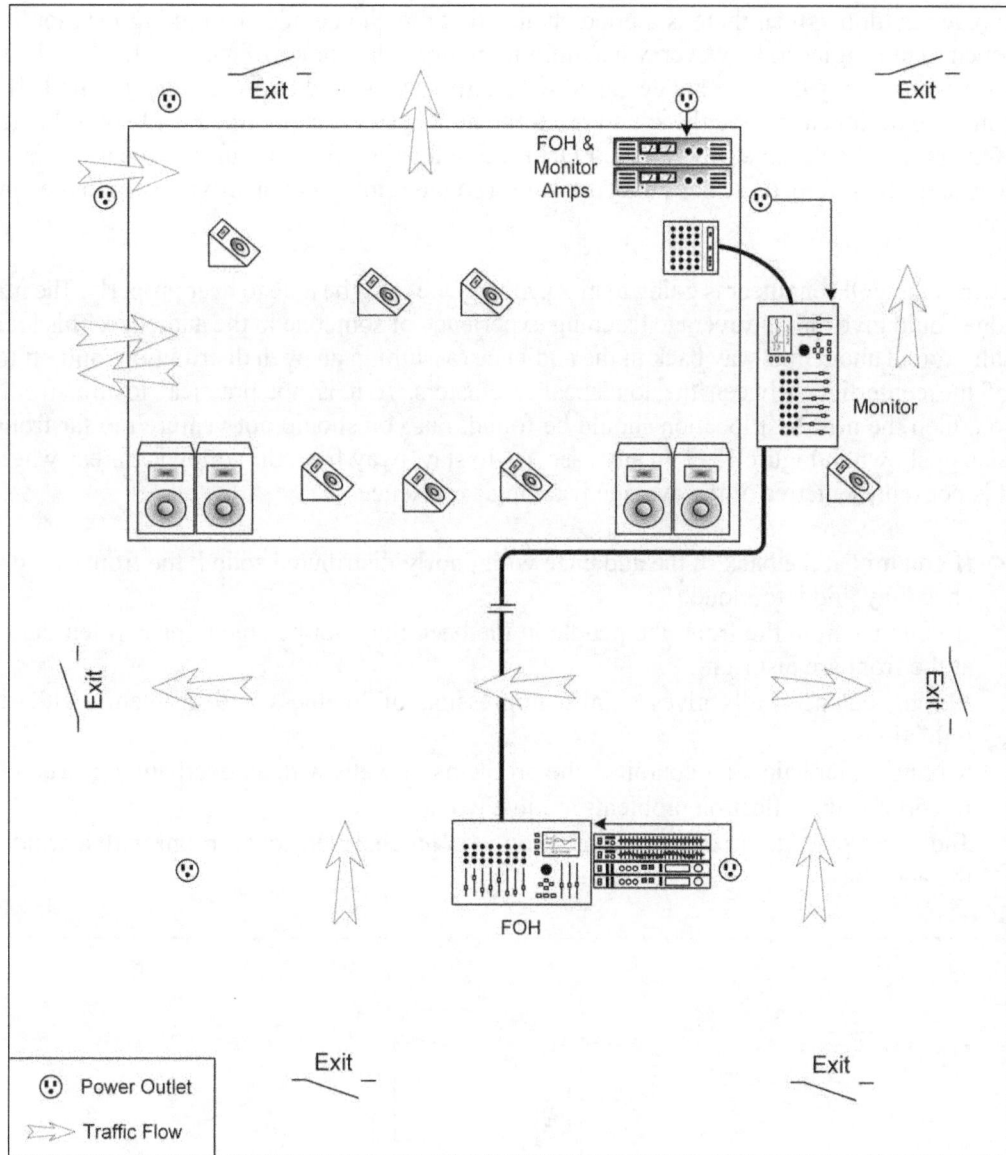

Figure 13-2   A Basic Layout Plan Helps to Resolve Conflicts

13.5 Mixing Locations

When determining the setup for the sound system, the locations of the monitor and FOH mixers should be one of the first things to be decided. Two things are important to choosing these locations, *viz.* sight and sound. The audio engineers need to be able to see what is happening on stage so that their actions can be properly coordinated. The monitor engineer must have a very strong feel for the sound *on stage* in order to control its overall quality and level, and manage feedback. The monitor engineer must work so closely with the performers that he is usually located right next to the stage, or if the stage is big enough, sometimes even on stage, where eye contact with the performers is possible and communication may be unobstructed.

On the other hand, the FOH engineer needs to hear the average sound in the audience space. This means that he should ideally be located somewhere out in the audience area. If the performance is taking place outdoors, then there is a good chance that the choice of FOH mixing location is also wide open to the engineer. However, when mixing indoors, the choice of location tends to be much more restricted. Many halls are not designed with active live sound reinforcement in mind. That is, the rooms are designed to have the sound reinforcement system set up, fixed, and not to be fiddled with. Others may be designed so that the FOH mixing location does not take space away from the audience area. Therefore, these halls may not have a sound reinforcement mixing area in the optimal location.

However, if the FOH engineer is going to mix properly, he must be able to hear properly. The mixing location should give him the average listening experience of someone in the audience. This location is usually found about mid-way back in the audience (assuming an even distribution) and off to one side of the centerline between the loudspeaker clusters. If it is not practical to mix from that location, then the next best location should be found, but you should not venture too far from this central area shown in Figure 13-3. In any case, try to stay away from the following areas where the sound is not representative of the average listening experience.

- If you mix at the back of the audience with poorly distributed sound, the front of the audience may find it too loud.
- If you mix from the front, the people in the back may not be able to hear when the levels at the front are just right.
- Mixing against walls gives a false impression of loudness and problems with strong reflections.
- Mixing under balconies combines the problems of walls with an overhanging structure.
- In corners, the reflection problems are just worse.
- The worst place to mix in a live situation is in an enclosed sound room with a window to the audience.

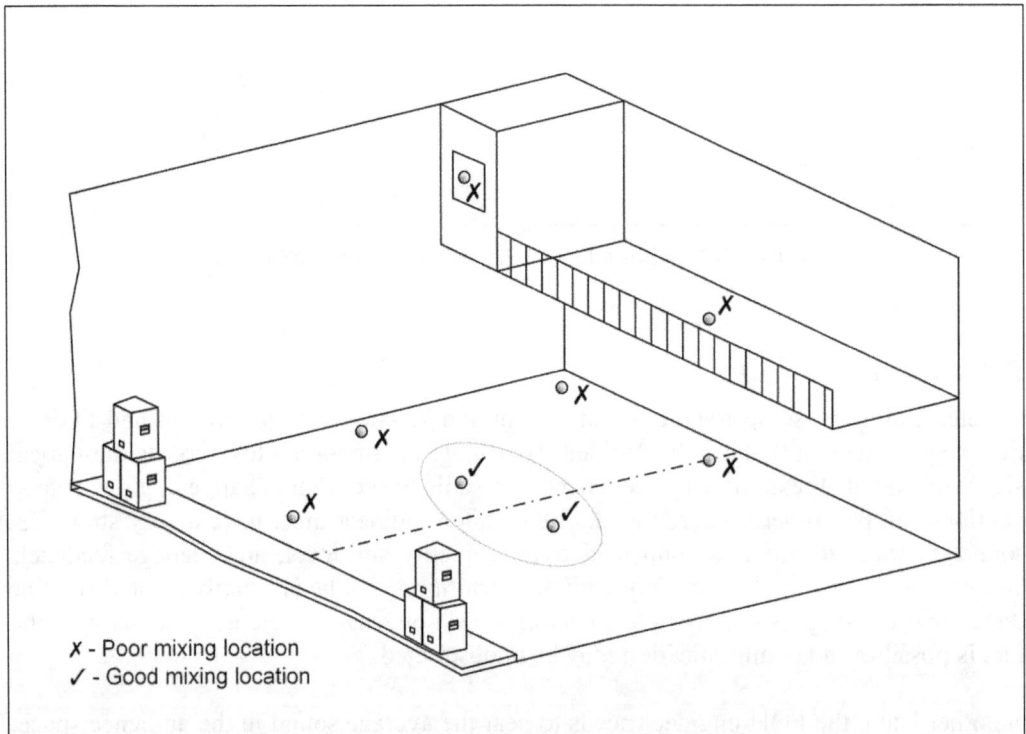

Figure 13-3 Good and Poor Choices for FOH Mixing Location

13.6 Front of House Loudspeaker Placement

Let us consider for a moment where to put the FOH loudspeaker system. One may think that this should be a fairly simple issue to resolve—just put the speakers on each side of the stage and point them at the audience. In fact, in many cases with smaller systems, it is just that simple. However, what we want to do here is look a little deeper into some of the issues related to positioning loudspeaker systems, given the size of the venue and the audience distribution, with the aim of optimizing the loudspeaker coverage.

13.6.1 Loudspeaker Coverage vs. Height

For illustration purposes, we will start with the simplest case—a small rectangular space that can be covered by just one loudspeaker positioned at the front and centered left to right. We will ignore the influences of the room and audience for now and look at just the area of coverage of the speaker system.

One of the first things that we must determine about the loudspeaker system being used is its directivity characteristics. Figure 13-4 shows a loudspeaker system with the longitudinal axis and the vertical and horizontal planes highlighted. The longitudinal axis of the speaker system is an imaginary line running from the back of the cabinet and extending through the front of the cabinet at some point determined by the manufacturer. It is usually centered left to right on the front of the cabinet. The directivity characteristics, typically published as a set of dispersion plots, are referenced to this axis and these planes. There is usually one set of plots for the vertical spread, or dispersion, and one set for the horizontal dispersion. Figure 13-5 shows a typical set of plots for the vertical dispersion plane of a loudspeaker system. Note that plots are given for a select set of frequencies, showing how the sound level decreases as we move above or below the longitudinal axis of the loudspeaker. The plots for the horizontal plane will look rather similar, but show how the sound level varies as we move left or right of the longitudinal axis.

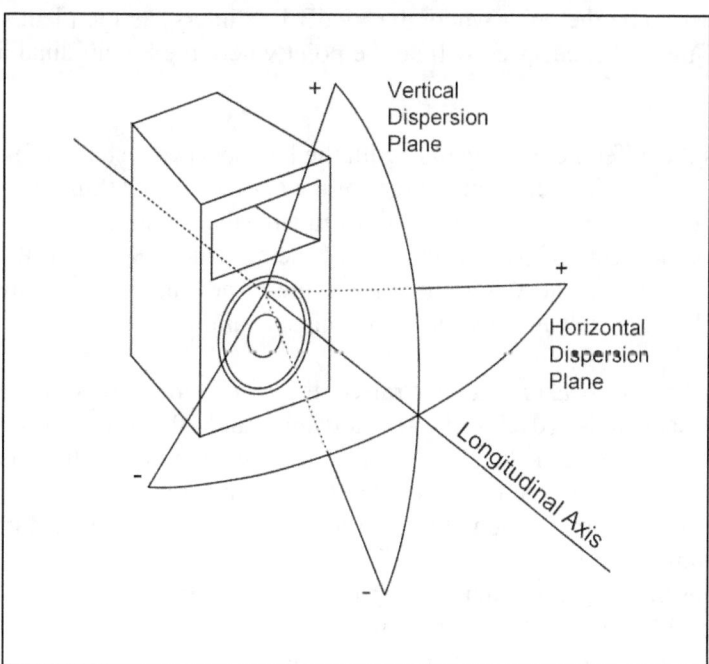

Figure 13-4   Reference Axis and Planes of Dispersion Plots

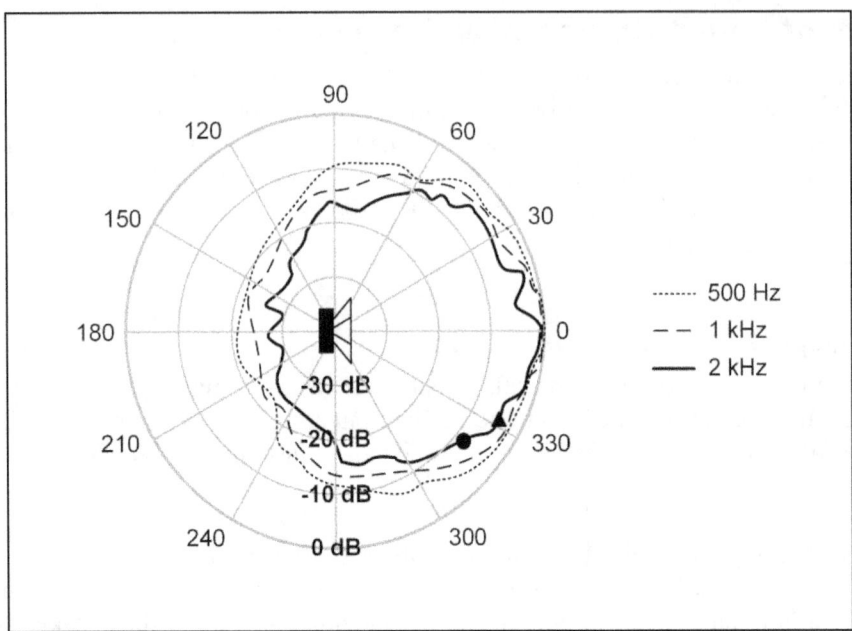

Figure 13-5 Vertical Dispersion Plots Showing Three Frequencies on One Graph

To examine some of the issues of coverage that depend on the height of the loudspeaker, let us only consider a single frequency and a fixed coverage angle. This simplified approach will allow us to easily compare differences in coverage as we move the speaker around. To that end, we will use the information derived from the 2 kHz vertical dispersion plot and the angle where the sound level falls by 9 dB to determine the limit of coverage towards the front of the audience. This angle is shown by the round dot on the 2 kHz plot in Figure 13-5 and highlighted directly in Figure 13-6. It is about 40° below the longitudinal axis in this particular plot. So now we can determine the forward limit of the audience to be the point where the −9 dB line intersects the plane of the average ear height. The rear of the audience space will be the point where the longitudinal axis intersects this same plane.

Figure 13-6 shows the difference in coverage with the loudspeaker system at four different heights while keeping it the same distance from the rear of the audience. One thing that we may conclude from Figure 13-6 is that there is no advantage to raising the loudspeaker system higher than is necessary to have a clear line of sight to the rear of the audience. However, that would be a premature and incorrect conclusion. While there are some disadvantages to raising the loudspeaker system, there are also some advantages, and so we must examine the situation a bit more closely.

We can see that as the loudspeaker system is raised, the −9 dB line moves further and further back into the audience space, thus reducing the forward limit and effective area of coverage. In fact, Figure 13-7 allows us to conclude that there is a disadvantage to raising the speaker system more than necessary because we can see that in order to maintain the same area of coverage when we raise the speaker system from position *A* to position *B*, we have to move it farther back over the stage, thus increasing the distance to the audience. You may also recall from the discussion in Section 1.7 that the SPL decreases not in proportion to the distance, but with the square of the distance. Thus, the SPL in the audience area may drop quite a bit if we raise the speaker too high and try to maintain the same area of coverage. (Of course, if we needed to raise the loudspeaker and maintain the same area of coverage without having to move the loudspeaker deeper into the stage area, we could simply change to a speaker system with larger vertical dispersion.)

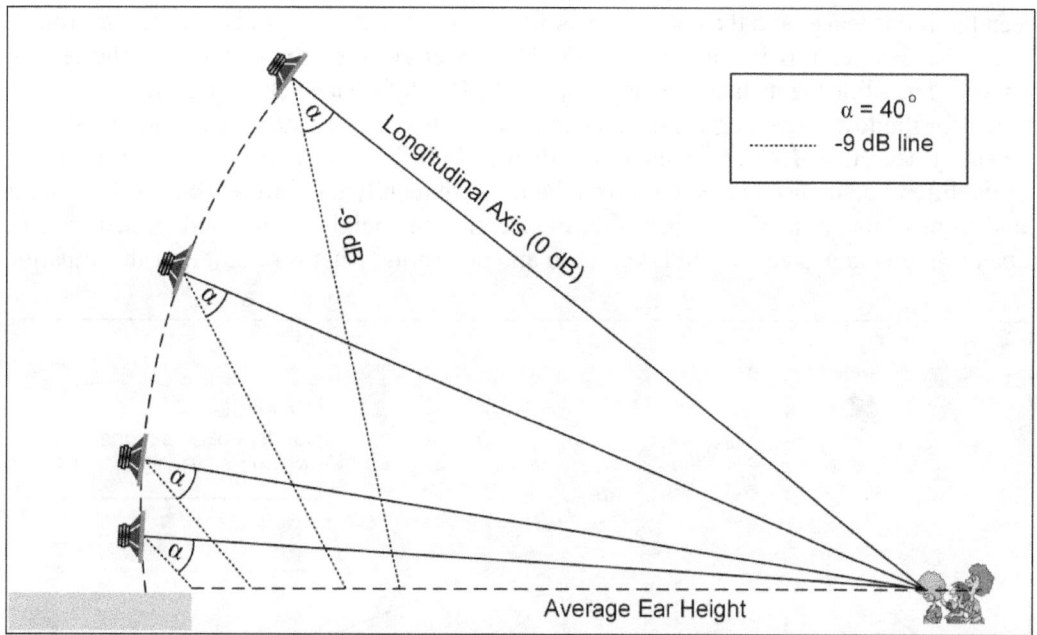

Figure 13-6   Coverage Area Varies with Loudspeaker Height

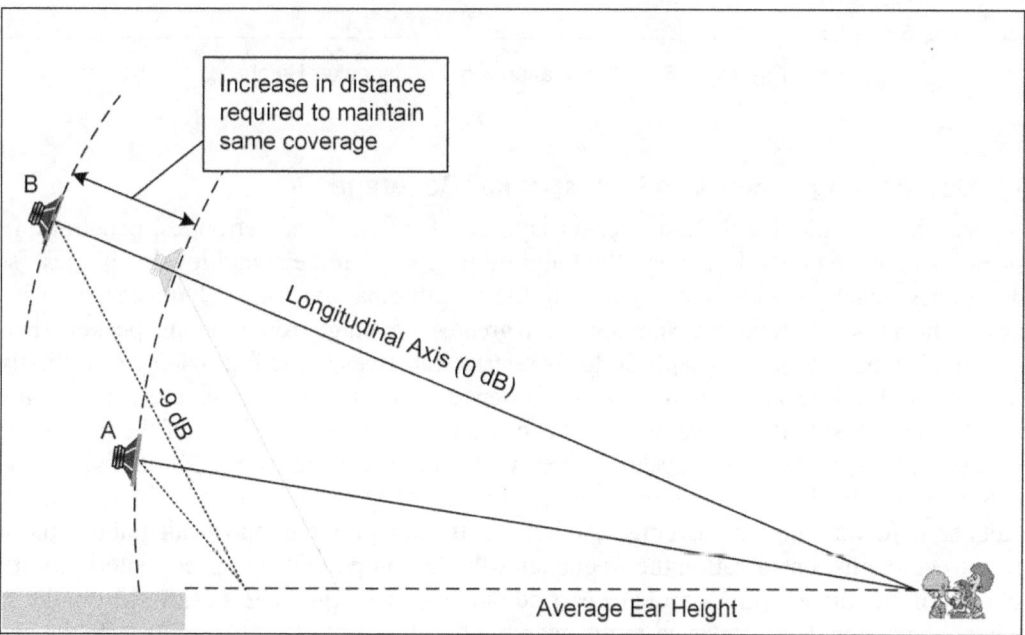

Figure 13-7   Raising Loudspeaker While Maintaining Same Area of Coverage

On the other hand, placing the loudspeaker system just high enough to have a clear the line of sight to the rear of the audience also has some disadvantages. Firstly, we are aiming more of the high-energy output of the system at the back wall. This means that we will in turn get higher intensity reflections from the wall back into the audience area. Additionally, when the loudspeakers are kept low, they are closer to the microphones on stage, and so the likelihood of feedback is increased. The biggest problem, though, with having the loudspeaker system down low is the large difference in SPL between the front and rear of the audience. This results from the large difference in distance between the loudspeaker and the rear of the audience as compared to its distance to the front of the audience. We can see this in Figure 13-8. For the upper speaker, the distance to the rear of the audience is AB, while the distance to the front is AC. The difference between these two is $AB-AC$. Similarly, for the lower speaker, the difference in these distances is $DB-DC$. From the diagram, it can be easily seen that $AB-AC$ is much less than $DB-DC$. Thus, when the speaker system is up high, this difference in distance is minimized, and consequently the difference in SPL between the rear and front is also minimized. Therefore, in selecting the height of the loudspeaker system, we must balance coverage area with SPL variation and determine what is optimal for the situation.

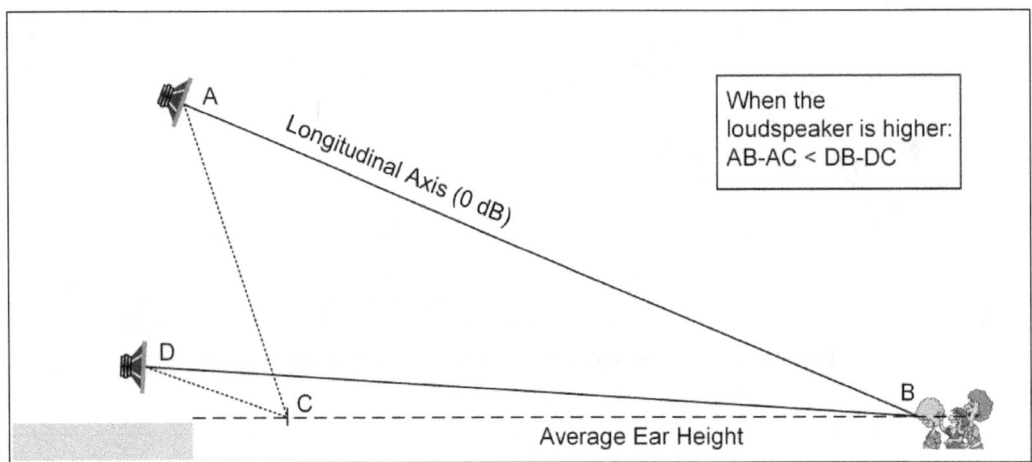

Figure 13-8 SPL Variation with Loudspeaker Height

13.6.2 Determining Acceptable Loudspeaker Coverage

One approach to positioning the loudspeaker is to aim for farthest and strongest penetration from the system as our first goal. It is along the longitudinal axis that the sound level is highest, so we get the longest reach or penetration by aiming the longitudinal axis of the loudspeaker system at the rear of the audience. Next, we want to get the greatest coverage from the loudspeaker. There are many variables that we can play with, so let us restrict ourselves in the following ways: Firstly, we will not fly the loudspeaker system over the audience area—as a matter of safety, many municipalities disallow this. Furthermore, we will keep the system generally ahead of the performers on stage, which means that the loudspeaker system will generally be above the forward section of the stage. Secondly, we will use the information derived from the vertical dispersion plots to determine the acceptable forward limit of coverage and help set the height of the loudspeaker above the stage. In order to take into consideration the frequencies that are important for speech intelligibility, the calculation of the forward limit should be carried out by considering levels derived from the 2 kHz to 10 kHz dispersion data. However, in the example that follows, we will demonstrate the process by using the data for the 2 kHz plot. The process is the same for any frequency chosen.

Now suppose that the front and rear physical limits of the audience space are predetermined (as is the case with any already constructed enclosure), and that when we position the loudspeaker at some height h and aim its longitudinal axis at the rear limit, the forward limit of the audience space falls 40° below the longitudinal axis, as shown in Figure 13-9. We now must determine, using the loudspeaker's dispersion data, whether the SPL at the front of the space is within acceptable limits relative to some reference level established at the rear. Ideally, we would like the SPL to be the same at all points from the front to the rear of the audience space.

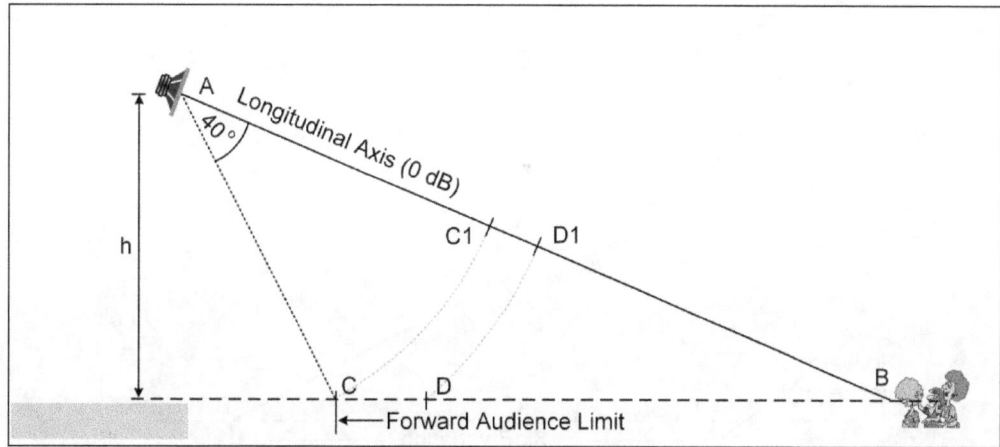

Figure 13-9   Determining Acceptable Loudspeaker Coverage

The arrangement in Figure 13-9 allows us to use the inverse square law to our advantage. Recall that the inverse square law tells us that every time we reduce the distance to the sound source by half, the SPL increases by 6 dB. As noted above, we would like that the SPL remain the same as we move from the rear to the front of the audience space. It is unlikely that this is exactly achievable, but using the combination of inverse square law and speaker directivity, we should be able to come close to this goal. For example, let's say that the SPL at the rear of the audience is 100 dB. Points **D** and **D1** are on the arc that is half the distance from the rear of the audience to the loudspeaker. Point **D** falls on a line that is about 28° below the longitudinal axis of the speaker system. From Figure 13-5, we see that such a line gives us about a 5 dB reduction in SPL (shown by the triangle mark on the 2 kHz plot), but with the +6 dB SPL gain due to the inverse square law, the change in SPL at point **D** should actually be about +1 dB, giving an SPL of 101 dB.

Point **C**, the forward limit of the audience space, is on an arc that intersects the longitudinal axis at **C1**, which is roughly 0.4 of the distance from the loudspeaker to the rear of the audience. Previously, we noted that point **C** is 40° below the longitudinal axis. Again, from Figure 13-5, we can determine that we have a 9 dB reduction in level due to being 40° off-center. At the same time, though, the inverse square law gives us an increase of $20\ log(1/0.4)$ dB, or about 8 dB. Hence, the actual SPL at point **C** is around 99 dB. Thus, we see that with the loudspeaker mounted at a reasonable height, we can keep the variation in SPL from the front to the rear of the audience acceptably small.

If we relax the requirement of *no variation* in SPL from the rear to the front and allow the SPL to vary by ±3 dB (which is common industry practice), then we can increase our coverage area by tilting the longitudinal axis of the loudspeaker down further, as shown in Figure 13-10. That is, the axis is no longer pointed at the rear of the audience, but at a point forward of that, so that the SPL at the rear of the audience is 3 dB less than at the point where the longitudinal axis intersects the plane of average ear height. By doing this, the forward limit of coverage moves closer to the stage. Viewed another way, we can leave the longitudinal axis of the loudspeaker pointed as above, and the rear limit of the coverage area will extend farther from the stage.

One thing of note is that with this type of loudspeaker setup, about half of the speaker's output is impinging on the rear wall. If the wall is highly reflective, a lot of sound will be reflected back into the audience and may lead to problems with intelligibility that we discussed earlier. If this type of arrangement were used in a permanent setup, one solution would be to place absorptive panels on the wall.

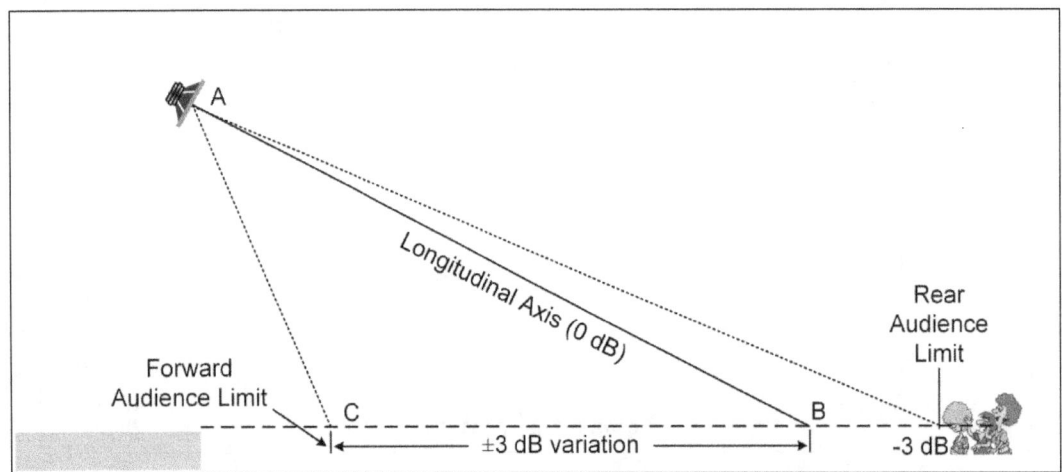

Figure 13-10   Loudspeaker Coverage with ±3 dB Variation

Doing this type of analysis by hand is clearly tedious, especially when we consider that this must be repeated for various frequencies and loudspeaker heights. Even a simple spreadsheet with the appropriate formulae would help a great deal, but the dispersion data for each loudspeaker would still have to be entered by hand. There are, however, software packages available that make the above analysis easy to perform, and manufacturers often publish data sets for their installation series of loudspeakers that can be imported into some of these software packages, eliminating even the need to manually enter the dispersion data.

13.6.3 Increasing Horizontal Coverage with a Single Cluster

Up to now, we have been discussing the use of just one speaker system to achieve horizontal coverage of the audience space. Now suppose we want to increase the horizontal coverage of the system while still using just a single cluster of loudspeakers centered left to right. We can do this by placing the loudspeakers side by side and angling the cabinets so that the edges of their horizontal nominal coverage areas just overlap, as shown in Figure 13-11. This nominal coverage area is typically published by the manufacturer as the horizontal nominal dispersion specification, which is usually the −6 dB included angle averaged over a set of midrange frequencies (e.g., 1 kHz to 4 kHz). This arrangement should give us good results while almost doubling the coverage of the loudspeaker system. In the example shown in Figure 13-11, the −9 dB combined coverage is around 200°. Assuming the loudspeakers are driven from the same source, then the sound waves along the centerline of the *cluster* will be correlated, and we should get a boost of around 6 dB (see Section 1.14) along the centerline due to the addition of waves from the two sources. Thus, the SPL drop along the original centerline should be limited, and the level should not be far off compared to the original level when only one loudspeaker system was used.

Note that the above discussion applies to mid and high-frequency drivers only. For low-frequency drivers, the coverage pattern is basically omnidirectional, and thus the horizontal extremes of the audience are already likely to be well covered. However, in placing the loudspeaker cabinets next to each other to achieve the above coverage, we necessarily place the woofers close to each other. We know that if two low-frequency drivers are placed next to each other, an increase of up to 6 dB SPL will be had from the usual doubling of sound power along the centerline of the pair when driven from the same source. However, mutual coupling occurs when large drivers are placed next to each other and the two drivers behave as one larger driver. This mutual coupling increases the

SPL in any direction for very low frequencies. Thus, with this arrangement, some EQing of the low end will likely be necessary to restore tonal balance of the overall sound.

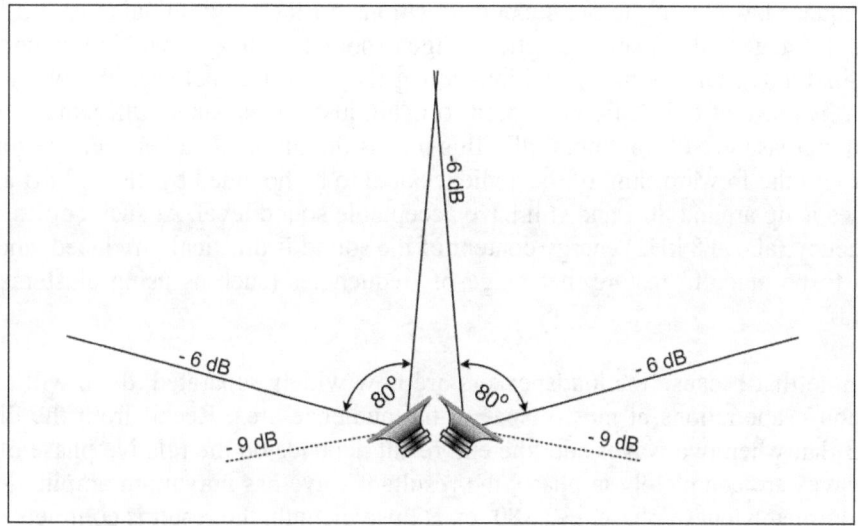

Figure 13-11   Splaying Loudspeaker Systems to Increase Coverage

13.6.4 Increasing Horizontal Coverage with Spaced Sources

Unless the loudspeakers are being flown, we do not usually have a cluster of speakers in the middle of the stage. The usual arrangement is to have the cabinets at opposite sides of the stage, as in Figure 13-12.

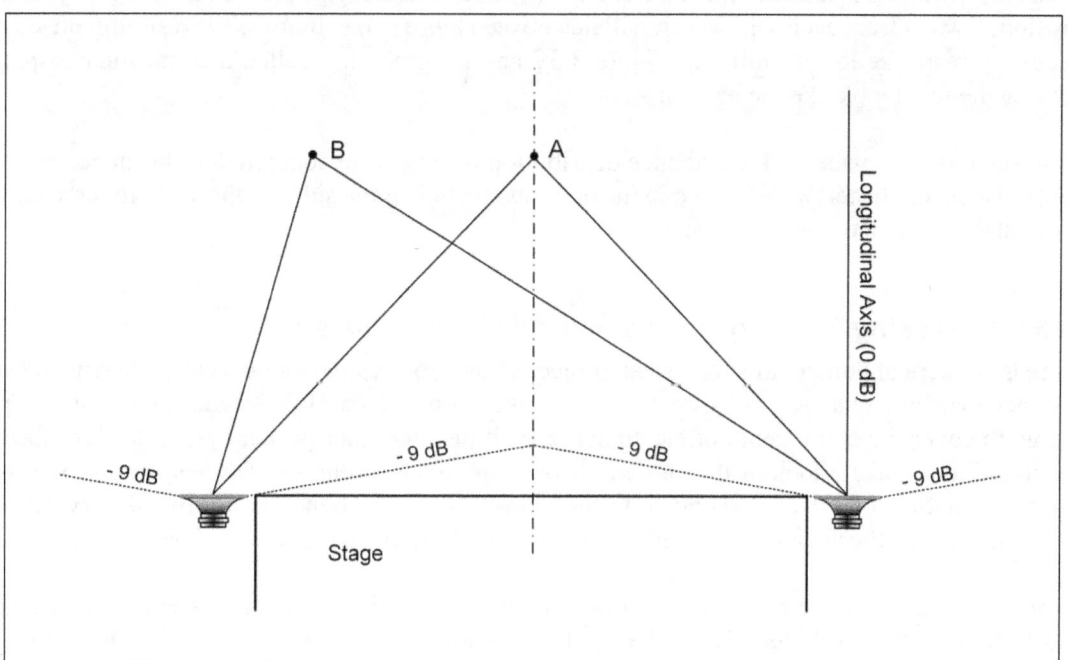

Figure 13-12   Coverage with Loudspeakers at Edge of Stage

The horizontal dispersion of FOH loudspeakers tends to be quite generous, with the −9 dB angle being up around 70°. By placing the loudspeakers far apart, we will be able to cover most or all of the horizontal extent of the audience space. Again, similar to the discussion in Section 13.6.1, with the inverse square law coming to our support, positions off the longitudinal axis of a loudspeaker but closer to the stage will not suffer drastic change in sound level (compared to other positions off the longitudinal axis). Furthermore, positions along the centerline between the two loudspeakers will have an increase of 6 dB SPL compared to using just one speaker. Off-center positions will also have an increase in SPL, but not 6 dB. This means that in the area between the loudspeakers, we could extend the forward limit of the audience area to be bounded by the −12 dB angle, which in many cases is up around 80°, and still have acceptable sound level. At such angles, though, the higher frequency (above 8 kHz) energy content of the sound is drastically reduced, and provisions would need to be made to restore that range of frequencies (such as using clusters or front fill loudspeakers).

We should note that because the loudspeakers are now widely separated, there will be some frequency response aberrations at most places in the audience area. Recall from the discussion in Section 1.13 that when two waves add, the end result depends on the relative phase of the waves. When the waves are completely in phase, the resultant wave has maximum amplitude. When the waves are completely out of phase, i.e., 180° or 1/2 wavelength, the result is complete cancellation of many components of the waves. There are an infinite number of phase combinations between these two extremes.

At positions such as *A* along the centerline between the two loudspeakers, all waves from the two speakers arrive at the same time and are in phase. Therefore, assuming no other external influences, an increase of 6 dB results for all frequencies and the frequency response should be relatively smooth. At positions off centerline, such as *B*, the *difference* in distance from *B* to each loudspeaker will exactly equal 1/2 wavelength of some wave. When this happens, waves of that particular frequency will be canceled. If the difference is a whole wavelength, the waves will add just like position *A*. Most frequencies, however, will lie between these two extremes. The resulting off-center frequency response looks similar to Figure 8-37 and is commonly called a comb filter response because the nulls look like the teeth of a comb.

If the stage is very wide, or the audience distribution is very wide compared to the stage, we may use loudspeaker clusters as discussed in the previous section at the sides of the stage to increase the horizontal coverage of the speakers.

13.6.5 Increasing Reach in Large Venues by Using Clusters

There is a practical limit to how far we can project sound from a single loudspeaker system and still have adequate and relatively consistent coverage over a given area. If the venue is large, we will be unable to cover the entire depth of the audience with just one loudspeaker system as described in Section 13.6.1. Take a look at the vertical dispersion plot for your speaker system and you will likely see that beyond the −9 dB range to the loudspeaker, the sound level falls off very rapidly. Thus, the front of the audience will suffer if we are pointing the loudspeaker as described.

To restore coverage at the front, one solution is to utilize a vertical array or cluster of loudspeakers in a fashion similar to Figure 13-13. Here, speakers lower in the array are angled downward to provide coverage towards the front. The loudspeakers may be driven and equalized separately to balance their output level and maintain a fairly consistent SPL along the depth of the audience. In such a case, the loudspeakers covering the front will be driven at a lower level that those covering the rear, and the rear coverage speakers may need the high-frequency end boosted since their output has to travel much farther through the air than that from the other speakers (recall that higher frequencies are absorbed more readily by air and the audience than low frequencies).

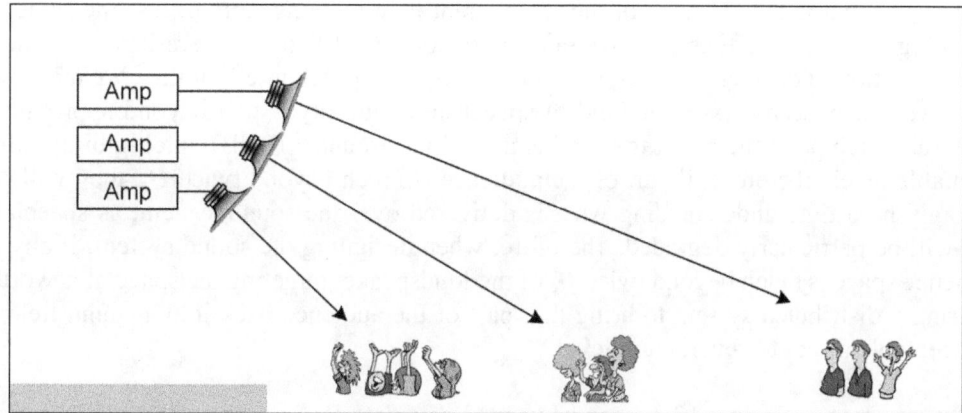

Figure 13-13   Increasing Coverage by Using a Loudspeaker Cluster

13.6.6 Reverberation and the Limitations of Long Throw Loudspeakers

Even if we use a cluster as described above, still there are practical limitations to how far we can project sound from a loudspeaker system. You may have had the experience of being in a large gymnasium with hardwood floor, painted concrete walls, and a hard ceiling—all good reflectors of sound—and recall having the hardest time trying to understand what was being said over the loudspeaker system. The problem, of course, stems from the long reverberation time of the room and being situated beyond critical distance (D_c) of the loudspeakers being used.

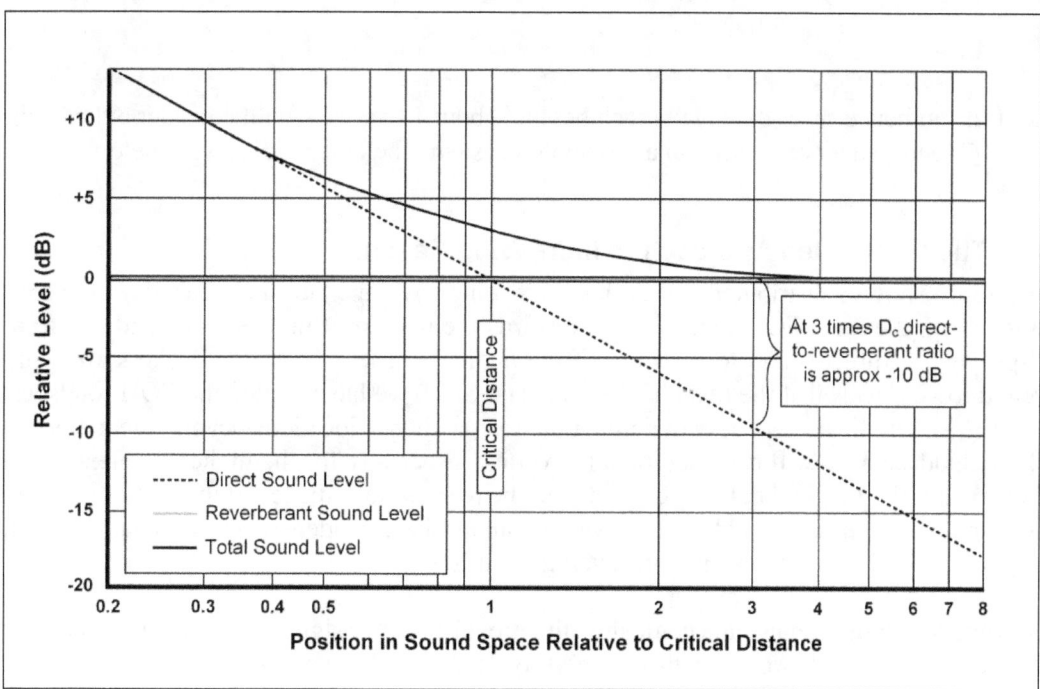

Figure 13-14   Relationship between Direct Sound and Reverberant Sound in a Room

In Section 13.3.1, we discussed the notion of critical distance and how it is affected by the reverberation time of the room. Recall that at critical distance, the direct sound level produced by the source is the same as that of the reverberant sound field. Beyond critical distance, the direct sound continues to decrease, but the level of the reverberant field remains more or less the same. This is shown in Figure 13-14. Furthermore, we said in Section 13.2.1 that long reverberation times facilitate the formation of distractor voices that work to degrade speech intelligibility. We may conclude from the work of researchers in the field of speech intelligibility[5, 6] that beyond approximately 3 times D_c (at which point the direct to reverberant ratio is around -10 dB), intelligibility falls to an unacceptable level. Therefore, if part of the audience is much beyond twice D_c, they will have an increasingly hard time understanding what is delivered over the sound system, as speech intelligibility will be particularly degraded. Therefore, when designing the sound system, if any part of the audience space is much beyond twice D_c of the loudspeaker targeting the space, it is worthwhile considering a distributed system to bring that part of the audience back into a sound field with a more acceptable direct-to-reverberant level.

The equation given in Section 13.3.1 can be used to calculate D_c for a given loudspeaker and room combination. For example, suppose we have a room whose volume is 2500 cubic meters and a reverberation time of 1 second. Suppose, too, that we are using a loudspeaker system with a directivity factor of 10 in this room. The D_c for this combination would be:

$$
\begin{aligned}
D_c &= 0.057 \times \sqrt{\frac{10 \times 2500}{1}} \\[2mm]
&= 0.057 \times \sqrt{25000} \\[2mm]
&= 0.057 \times 158.1 \\[2mm]
&= 9.0 \text{ m}
\end{aligned}
$$

Thus, if the audience space extended much beyond about 18 m, we would consider either utilizing a higher Q loudspeaker or moving to a distributed system. The latter is discussed below.

13.6.7 The Distributed Approach to Increasing Reach

Figure 13-15 shows a distributed approach to extending coverage the audience space. The solution here is to aim the main FOH loudspeakers towards the middle of the audience and use front fill loudspeakers to direct sound to the front of the audience. These front fill speakers are typically driven at lower levels that the main house speakers. Since we have angled the FOH loudspeakers so that the longitudinal axis does not point at the rear, audience locations beyond the major axis of the FOH loudspeakers will need additional coverage, especially for the higher frequency sounds. In fact, even if the longitudinal axis is pointed at the rear of the audience, if the distance to the rear is greater than 30 m or so, additional coverage should be considered. Again, smaller equalized loudspeakers can be used to extend the coverage in this area.

To prevent doubling of sound and confusing directional cues, the signal to the rear fill loudspeakers will need to be delayed. We know that sound travels about 340 m/s. To determine the total delay, we need to figure out the difference in distance from the rear of the audience to the main FOH loudspeaker and from the rear of the audience to the rear fill loudspeaker and delay the sound by 1 ms for every 0.34 m. That sounds confusing, so let's look at an example. Suppose the distance from the rear of the audience to the main FOH speaker system is 45 m, and the distance from the rear fill to the audience is 10 m. Then the difference in distance is 45 m minus 10 m, or 35 m. Therefore, our delay for this difference in distance would be 35/0.34 ms, i.e., 103 ms. To maintain precedence (that is the sense that the sound is coming from the direction of the main FOH loudspeaker), the rear fill signal should be delayed even further so that the sound from the main FOH loudspeaker arrives at the listening position slightly before the sound from the rear fill. A value of around 10 ms has been found to work well for this. Thus, in our example, the total delay to the rear fill loudspeaker system would be 113 ms.

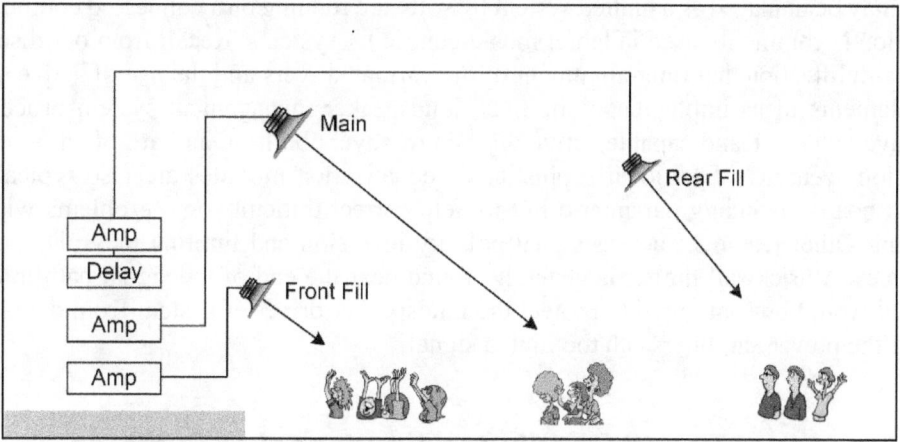

Figure 13-15   Increasing Coverage by Using Distributed Loudspeakers

13.6.8 Loudspeaker Management Systems

As we discussed in the previous section, we will need to delay the signal going to some loudspeaker units when deploying a distributed system. We may also need to delay the signal to some drivers when using certain cluster arrangements if we are trying to time-align their output. If we only have to do this for a single driver or speaker unit, this will be easily manageable with a simple delay processor. However, if we have to do this for several drivers or speaker systems, then this task becomes unwieldy to manage with separate delay processors. This is where a specialized loud-speaker management system processor comes in handy. An example of such a device is shown in Figure 13-16. These processors typically have 2 to 3 input ports and 6 to 8 output ports, with very flexible routing between inputs and outputs, allowing an input signal (or sum of the input signals) to be routed to many different combinations of outputs. Key to the operation and use of this kind of processor is the ability to apply different delays to any input or output separately, typically with compensation for ambient air temperature (recall from Section 2.2 that the speed of sound—and hence required delay—is affected by air temperature). To allow greater utility, it is usually possible to daisy-chain several of these processors together, thus allowing the unified control and alignment of practically any combination of drivers or loudspeaker systems.

To help with the alignment process, manufacturers offer a level of automation in this type of processor. First, the user connects a microphone to one of the inputs of the device, and the main output of the system is specified. Next, the processor cycles through the outputs feeding the various drivers or loudspeaker units while generating a test signal (typically a noise signal). It measures the difference in arrival time of the test signal at the microphone's position for all outputs compared to the main output and automatically calculates the delay necessary to align the audio delivered at the microphone's location.

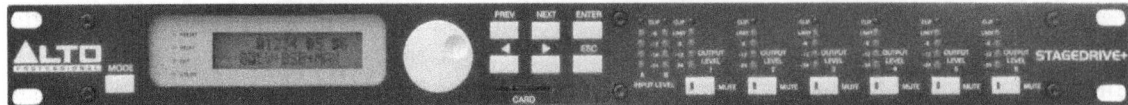

Figure 13-16 Loudspeaker Management System Processor

Sometimes, you will hear these processors referred to as PA management systems, and that is because they typically offer much more than just simple time alignment of audio signals. As mentioned above, multiple processors may be combined to meet the deployment requirements, and the whole set may be managed as a unified system by software running on a connected computer. Multi-amplification is commonly used in larger music-centric PA systems. Recall from our discussion on bi and tri-amplification that time-alignment of the various drivers and the use of active crossovers are key elements in its implementation. Thus, loudspeaker management system processors will usually have extensive and capable active digital crossovers built in, and are often used in multi-amplification scenarios. Additional sophisticated equalization modules are also typically implemented on board, including parametric EQ to help correct difficult sonic problems with a given deployment. Other features, such as signal polarity inversion and limiting may also be found in these devices. A brickwall limiter is generally found near the end of the signal path in these processors and would be configured to protect the loudspeaker driver or system from damage caused by driving the power amplifier with too high a signal.

13.7 Transformer-Based Distributed Loudspeaker Systems

There is another form of distributed loudspeaker system that is quite different from the one described above. You will not find this as the main FOH system at a concert event, but you may see it in a large conference room, or perhaps even to provide distributed audio to remote locations such as the foyer and cry room in a church. Unlike in the previous section where one amplifier channel was used to drive one loudspeaker, in this arrangement one amplifier channel drives several discrete loudspeakers.

The problem with trying to drive several loudspeakers conventionally via a single channel is that the load impedance becomes very small, thus potentially overloading the amplifier output stage and causing it to shut down or fail. As we see from Figure 13-17, the more resistors we connect in parallel, the smaller the final impedance becomes. For example, suppose we had eight speakers wired in parallel, each with a nominal impedance of 8 Ω. Based on this formula, the total impedance of the group would be 1 Ω, which would tax even most high current amplifiers.

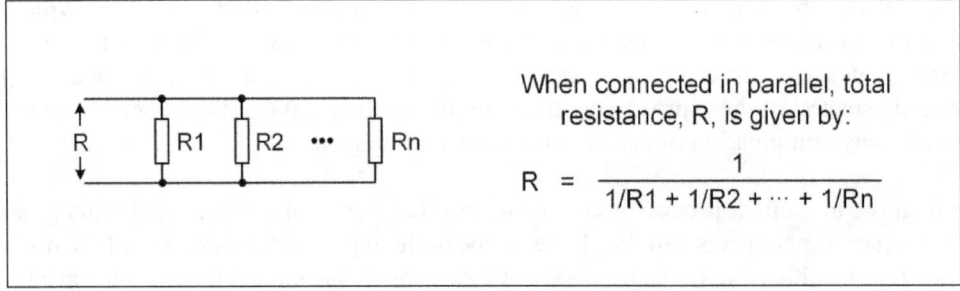

When connected in parallel, total resistance, R, is given by:

$$R = \frac{1}{1/R1 + 1/R2 + \cdots + 1/Rn}$$

Figure 13-17 Connecting Resistors in Parallel

To overcome this limitation, we can connect a transformer to the output of the amplifier, as shown in Figure 13-18. Under normal operating conditions, the transformer's primary winding presents the amplifier with a safe working impedance even as several loudspeakers are connected to the secondary winding. In practical implementations of this system in North America, the output transformer is a step-up transformer with a high impedance (i.e., several tens of ohms) secondary winding that provides 70.7 Vrms when the amplifier is driven to it maximum output by a sine wave signal. This system is thus commonly called a 70-volt system. Clearly, the transformer must be matched to the amplifier to be able to produce the proper voltage on its secondary winding. Some manufacturers sell amplifiers with the matched transformer already built into the amplifier. In some municipalities in North America, the secondary voltage is restricted to 25 Vrms, while in Europe, systems like this are configured to produce 100 Vrms on the secondary.

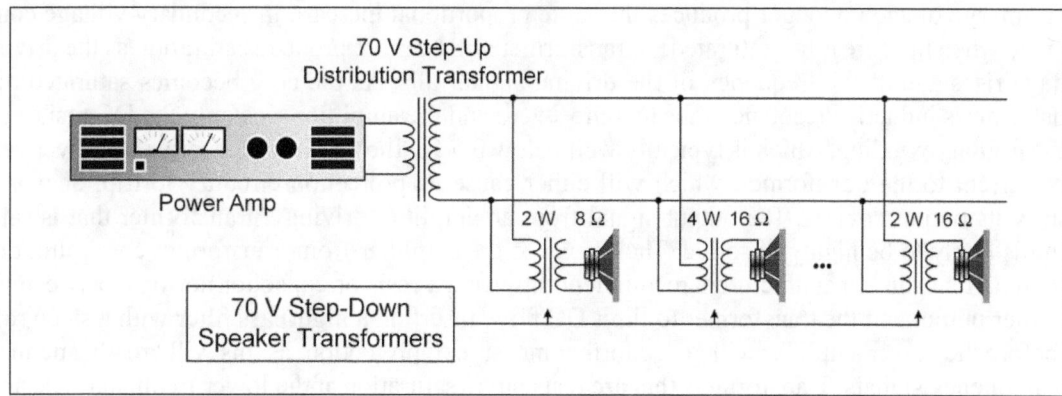

Figure 13-18   Transformer-Based Distributed Loudspeaker System

As shown in Figure 13-18, the individual loudspeakers are connected in parallel with the secondary via a local step-down transformer. These transformers are purpose-built for this application and make it easy for the system to be assembled. The speaker transformer has multiple taps on the primary side that allow selecting the maximum power that the system designer wants to deliver to the loudspeaker. On the secondary side, there are also a number of taps that allow matching the transformer to the speaker's nominal impedance.

This system will support a large number of loudspeakers being driven from a single channel, so long as the impedance of the entire set of loudspeakers is not lower than the minimum safe imped-ance required on the secondary side of the distribution transformer. This impedance requirement can be restated to say that total power drawn by all the loudspeakers combined must not exceed the maximum that the amplifier can supply. Fortunately, it is easy to determine the total power that will be drawn. It is simply a matter of adding up the power drawn by each individual speaker, which is determined by which tap is used on its local transformer. This is a much simpler affair than trying to figure out an equivalent scheme when using a set of directly (i.e., non-transformer) connected loudspeakers in series and parallel.

There are other advantages to using a transformer distributed system. Remember that power is the product of current and voltage. If power remains the same, as voltage goes up, current must go down. Because the voltage on the distribution wires is so high, that means that the current flowing in the wires is proportionately lower. However, power loss due to current flow (heating up the wires) is proportional to the square of the current. Therefore, it pays to make the voltage high and the current low. When that is done, smaller gauge (i.e., thinner and cheaper) wire can be used to distribute the audio signal to the loudspeakers. This can make a big difference in cost when the cable runs to the loudspeakers are long (which is typically the case in this type of system).

Of course, there are some disadvantages to this scheme as well. The power delivered by the amplifier is transferred across the windings of the transformers. If the transformer had no internal losses, all the power delivered at the primary side would be transferred to the secondary side. However, transformers have small insertion losses; they do not transfer all of the input power across their windings. Thus, the more transformers in the system, the more power is wasted due to insertion loss. Another problem is sound quality. Transformers have leakage inductance that attenuates the high-frequency end of the audio signal.[7]

The biggest problem with this type of system, though, is saturation of the transformer's core. When a transformer's core goes into saturation, its magnetization approaches its limit and an increase in the primary voltage no longer produces the same proportional increase in secondary voltage compared to when the core is not saturated. A transformer will be more prone to saturation as the driving voltage rises and as the frequency of the driving signal falls. As the core becomes saturated, the transformer's inductive reactance falls to almost zero and the amplifier sees only the DC resistance of the primary winding, which is typically well below 0.5 Ω. The amplifier thus tries to deliver very high current to the transformer, which will either cause its protection circuitry to trip, or worse, destroy its output devices. The output signal from an amplifier driving a transformer that is saturating is likely to be highly distorted. The danger to the amplifier from transformer core saturation is so great that some manufacturers recommend inserting a resistor-capacitor network between the amplifier output and the transformer to limit DC flow. Inserting a high-pass filter with a steep roll-off before the power amplifier will give a further measure of protection, as this will greatly attenuate low-frequency signals. Transformers that are resistant to saturation at the lower frequencies desired for good audio fidelity will tend to be larger and heavier, which translates into higher cost.

13.8 Monitor Loudspeaker Placement and Microphone Interaction

Now that we know how to set up the FOH loudspeakers, let us move on to the stage monitors and take a look at where and how to place them on stage. The primary use of stage monitors is to give the performers enough information in the midst of a noisy stage environment so that they can stay in key and in time with each other. These are key elements in being able to play tightly in synch with each other and in delivering a good performance. To fully achieve this, they must be able to hear themselves clearly as well as other key players in the group.

Since everyone wants to hear his own monitor distinctly, the stage can get very loud as each performer asks for more volume from his monitor. The best solution may be to give everyone a personal in-ear-monitor, but since we are talking about loudspeaker placement, we will discuss that later. The next best solution is to give everyone lower levels to start with, and ask the acoustic producers—the drums, bass, brass, etc.—to play softer. This may not work for very long, though, as musicians get excited and expressive. However, if the stage volume can be kept down, everyone will benefit. There will be less problems with feedback, and there will be better sound on stage and in the house. Loud stage monitors often spill sound into the front of the house and muddy the house sound as it mixes with the FOH loudspeaker output.

The usual solution to satisfying each performer's need for foldback is to have several monitors, one or two for each key area of the band. The monitor should be placed close to the performer. It is better to have a few more precisely aimed monitors at lower volumes than too few monitors at high volumes trying to cover everyone. Having many close-in monitors will allow for better monitoring, but it also means that the monitors are closer to the microphones, which can lead to higher incidence of feedback. Therefore, the placement and control of monitor loudspeakers are closely tied to the placement and type of microphones in use.

13.8.1 Feedback

We want to pay attention to the microphone and monitor interaction not just to optimize the fold-back to the performer, but also to reduce or eliminate feedback. Feedback is one of the sure problems that we will face in the stage monitoring environment sooner or later. Figure 13-19 shows that feedback occurs when the sound emitted from the loudspeaker enters the microphone, progresses through the electronic processors, and finally out the loudspeaker again. The overall gain around this loop must be at least 1 in order for feedback to be sustained, i.e., the sound from the loudspeaker must arrive at the microphone with a level at least equal to that of the original sound. When the gain is less than 1, the system will be stable and there will be no sustained feedback. When the gain is more than 1, the feedback howl will get progressively louder (to the limit determined by the gain of the amplifier or until the loudspeaker fails).

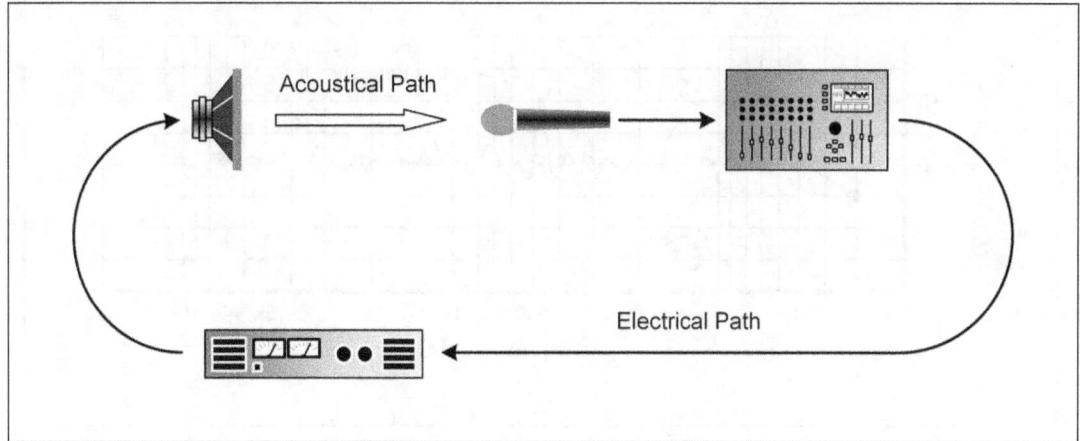

Figure 13-19   The Feedback Loop

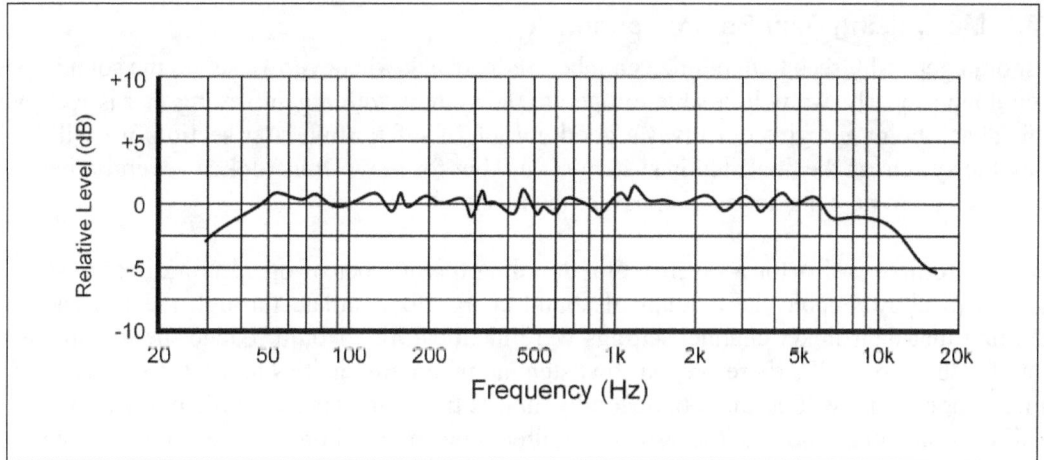

Figure 13-20   Stable System Frequency Response

Let's say we have the sound system set up in an indoor space. We can move around the room and sample the frequency response at various points. Suppose the frequency response at a particular location is as in Figure 13-20. As the gain of the system is increased, the frequency response at any given location will likely change. That is because the amount of sound energy reflected from surfaces in the area increases and room resonances become excited, causing the results of interaction with direct sound to change. There are many frequencies at which resonance will occur due to the reflection of sound from opposite surfaces or adjacent sides of a room. The reflections from opposite surfaces of the room are the strongest and cause the most problems, especially below about 300 Hz. These reflections give rise to standing waves whose fundamental frequency is given by $c/(2L)$, where c is the speed of sound, and L is the distance between the opposite surfaces.[a] Resonance also occurs at whole number multiples of these frequencies. In a system that is on the verge of feedback, large peaks appear in the frequency response at frequencies that are being excited by the output of the system or where some tuning of the system (e.g., EQ setting) favors a particular resonance, as shown in Figure 13-21. Any increase in gain at this point will cause the system to ring at the frequency of the highest peak. This is the so-called maximum gain before feedback point for this system.

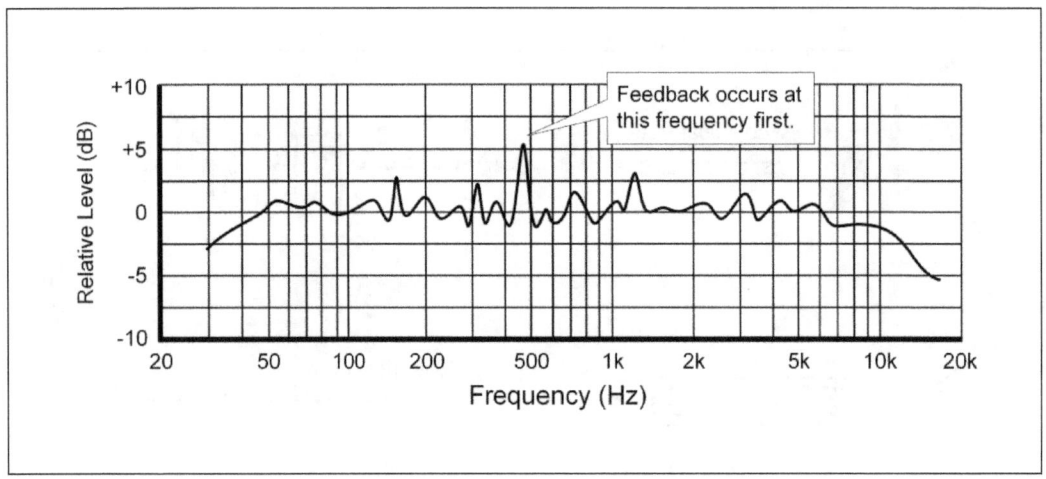

Figure 13-21 Unstable System Frequency Response

13.8.2 Maximizing Gain Before Feedback

We should generally seek to maximize gain before feedback whenever we set up the sound system. When done properly, we will be able to operate the system with a gain setting that is well below the feedback point and give ourselves a good margin of safety. In a later section, we will look at tuning the system as the final step in system set up, but for now, let us look at several other things that we can do to help maximize gain before feedback.

One thing to note is that for a given setup, doubling the number of open omnidirectional microphones will allow double the amount of sound energy to circulate through the feedback path. Assuming that their mixer channel settings were identical, this would reduce the maximum gain before feedback by 3 dB. Therefore, our first step in maximizing gain is to use the smallest number of microphones that will do the job. Secondly, unless trying to capture ambient sound, we should use directional microphones. The use of unidirectional microphones to reject unwanted sound coming from the direction of the loudspeakers will allow an increase in gain before feedback by about 4 to 6 dB. This is a large increase over what can be achieved with omnidirectional microphones.

---

[a] A standing wave is formed when a wave traveling across a room is reflected from the far wall, and the reflected wave, traveling in the opposite direction to the original wave, combines with the original wave, forming an interference pattern where the points of total cancellation (nodes) and the points of maximum addition (antinodes) are fixed with respect to the reflecting surfaces. The frequency of the wave has to be just right in order for the position of the nodes and antinodes to remain fixed in space.

Next, we should use the inverse square law to our advantage. Remember that in the direct sound field, SPL falls off quickly with increasing distance from the source. Thus, we should be aiming to put as much distance between the loudspeakers and microphone as we can. In a similar vein, the distance between the microphone and whatever we are miking should be as short as practical. Note we say practical rather than possible. It may be possible to place the microphone very close to the sound source, but if the source is physically large, the microphone may pick up sound from only a small section of the source, giving an unnatural color to the sound. For example, if we are miking a piano and place the microphone 2 cm from the treble strings, we will get a thin sounding piano with little body and bass, and with lots of hammer percussion. Additionally, you will recall that cardioid microphones exhibit the proximity effect of boosting the bass sounds when the source is very close to the microphone, and this must be compensated for accordingly, especially if the system is prone to feedback in the lower frequencies.

Some engineers attempt to combat feedback by wiring pairs of stage monitors as in Figure 13-22, where one monitor is wired with opposite polarity to the other. Depending on the exact location of the microphone, this will lead to cancellation of sound at several frequencies. One of those frequencies just may be a troublesome feedback sustaining frequency. This technique may be effective at canceling lower frequency sounds, but requires lots of experimentation, and may not produce adequate results when time is limited (as it usually is when preparing for a show).

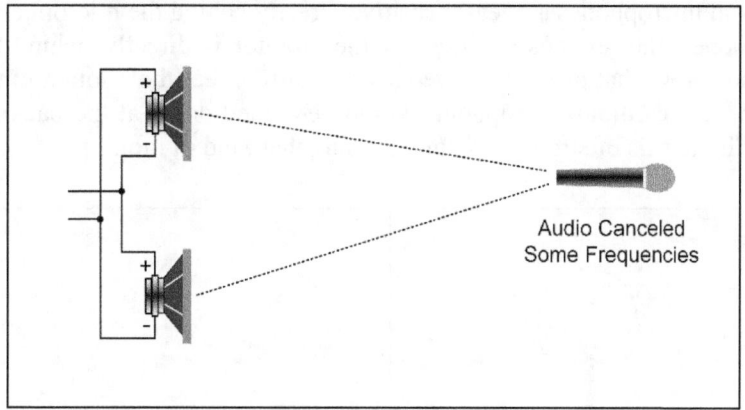

Figure 13-22   Monitors Wired with Opposite Polarity

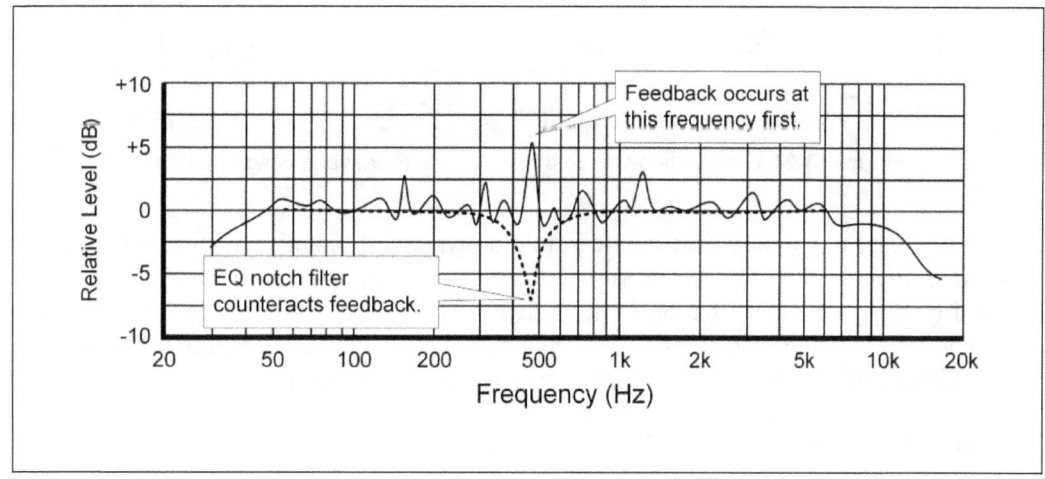

Figure 13-23   EQ Cut at Troublesome Frequency Suppresses Feedback

Another effective tool in the fight against feedback is the 31-band equalizer. A parametric equalizer, when used properly, can be even more effective. We mentioned earlier that when feedback occurs, it initially starts at a single frequency. The frequency response of the system at that time will show a spike at that frequency where feedback is occurring. If the feedback frequency can be determined, an equalizer with narrow bands can be used to cut the gain of the system at that frequency. As shown in Figure 13-23, we essentially apply a notch filter centered at the problem frequency. Conceptually, this is a job best suited to a parametric equalizer where the center frequency of the filter can be arbitrarily selected and the bandwidth of the filter can be set very narrow indeed, but 31-band equalizers do a reasonably good job here as well. The end result of this manipulation is that the frequency response of the system is flattened. The large spike at the feedback frequency is removed and the system is stabilized. This approach works so well that some manufacturers package parametric equalizers with a feedback detector to create an automatic feedback suppressor.

The way we position the stage monitors relative to the microphones also has an impact on the maximum gain before feedback that we can achieve. The general rule with monitors is to position them in the least sensitive area with respect to the microphone. Obviously, if a performer is roaming the stage with a microphone, it will not be possible to maintain the ideal relative positions of microphones and monitors. However, performers should be made aware of the interaction of the microphone and monitor so that they can work with engineers as a team in reducing feedback.

Recall that cardioid microphones are least sensitive directly behind the microphone. Therefore, for cardioid microphones, the obvious position for the monitor is directly behind the microphone. Figure 13-24 also shows that since floor wedges normally direct their sound output up from the floor at about 45°, the cardioid microphone should be angled down at the back to align its least sensitive area with the axis of strongest sound, if using that kind of monitor.

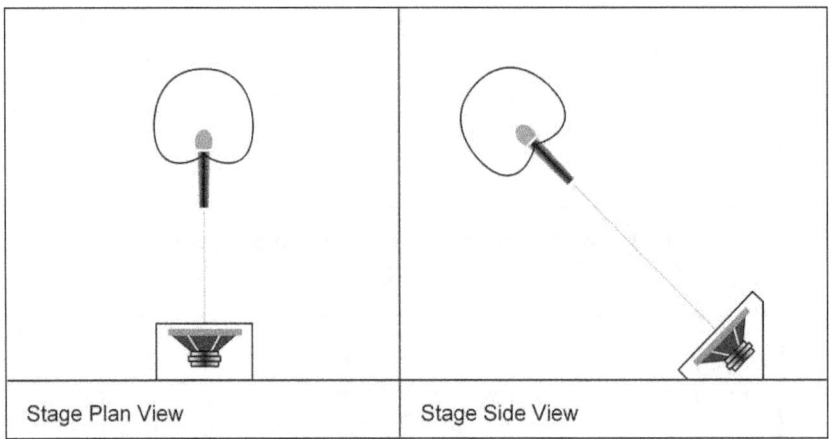

| Stage Plan View | Stage Side View |

Figure 13-24 Positioning Cardioid Microphone Relative to Stage Monitor

Looking at the polar plot of a typical supercardioid or hypercardioid microphone, we see that there is a rear lobe, making this kind of microphone least sensitive at about 45° off axis to the rear. Therefore, Figure 13-25 shows that on stage, instead of placing the monitor directly to the rear of the microphone, we would place it off to the side at about 45°. These microphones work well with the corner fill technique, where the monitors are shooting back from the front corners of the stage. Since floor wedges already project their sound up at about 45°, it makes sense not to align the axis of the supercardioid microphone with the major axis of the loudspeaker, but to keep the microphone more or less level with the stage floor.

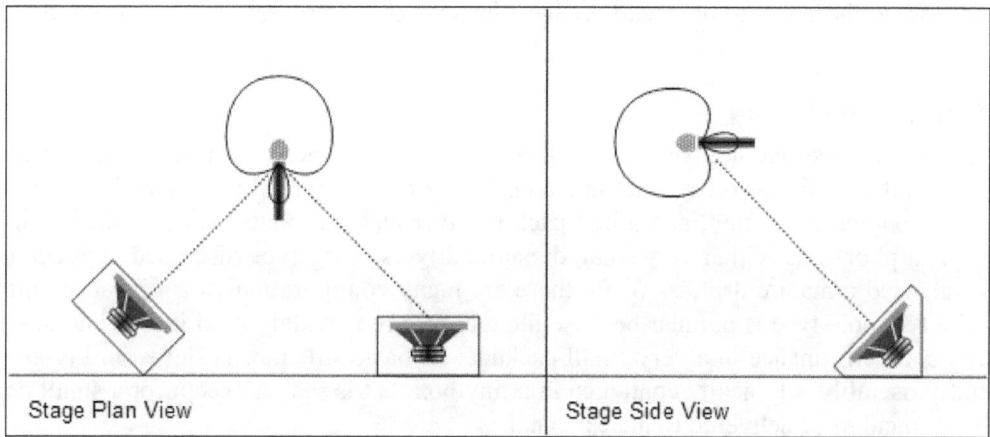

Figure 13-25   Positioning Supercardioid Microphone Relative to Stage Monitor

13.8.3 Side Fill Monitor Placement

Side fill monitoring, also called cross stage monitoring, or reverse stereo cross stage mix, is really useful when the stage is large and the group is spread out, or the performing group is large, and there is a need for musicians on one side of the stage to hear the musicians from the other side of the stage. In this monitoring setup, the audio from the performers at the right of the stage is fed to the left stage monitor, while sound from the left stage is fed to the right monitor—hence the term *cross stage*. You will also notice that in this case the monitors are raised on stands to bring them up to ear level. That way, they can project farther with lower sound level. Some engineers will simply reproduce the FOH mix in the side fill monitors, but clearly this is not the best approach to realizing the benefits of this technique, and may only serve to degrade the FOH sound as the FOH mix bleeds into the stage microphones and makes its way through the system multiple times (like reverb).

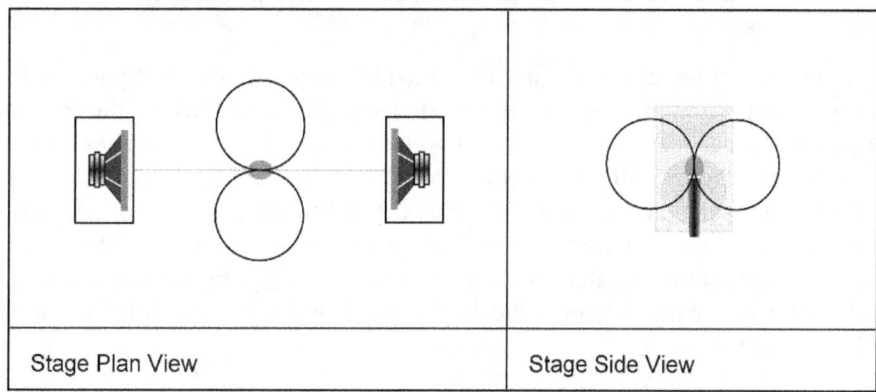

Figure 13-26   Using a Figure-8 Microphone with Side Fill Monitors

While not particularly common, it is possible to use a figure-8 microphone to good effect when using side fill monitors. The deep null in the pickup pattern of the microphone is aligned with the loudspeaker axes, as shown in Figure 13-26, giving good rejection of the monitors. The forward pickup sphere is directed towards the performer, while the rear pickup sphere is obviously pointed away from the performer. Depending on the microphone angle, this may pick up a large amount of ambient sound from the audience area. In such a case, a nearby ambience microphone may be added (out of phase) to the signal to drastically reduce the level of the processed ambient sound.

13.8.4 In-Ear Monitoring

If you ever get tired struggling with feedback and monitor placement, get personal in-ear monitors (IEM) for all the musicians on stage. These consist of a transmitter, which typically occupies rack space at the engineer's station, and a belt pack receiver and earphones worn by the musician on stage. The earphones use either very small dynamic drivers of the type discussed in Section 4.4.1, or tiny balanced armature drivers. While there are many configurations for a balanced armature driver, the two-pole type is popular because the technology is widely used in hearing aids, and it delivers good performance in a very small package. Its basic structure is shown in Figure 13-27. This entire assembly is typically contained in a tiny box that is sealed except for a small port that allows the sound to be delivered to the ear canal.

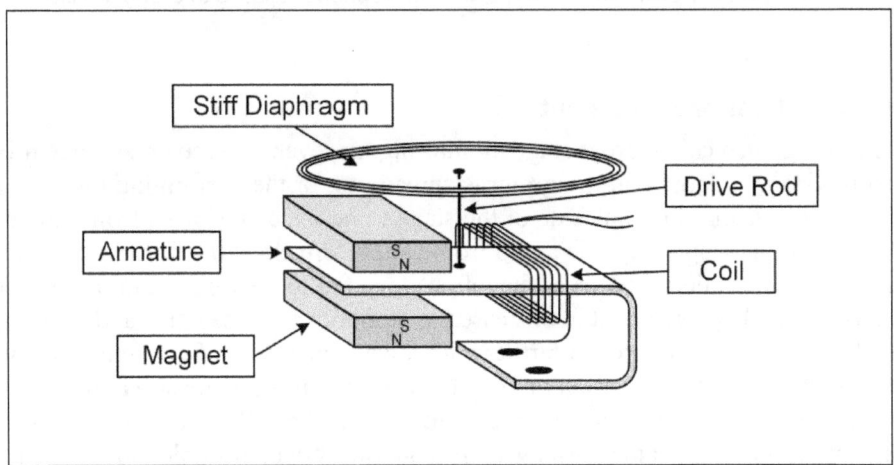

Figure 13-27   Balanced Armature Driver Internal Structure

In this design, one end of the armature sits suspended between the opposite poles of the magnets. A coil also surrounds the armature as shown. When no current flows through the coil, the armature remains still, balanced between the pull of the two magnets. When the current of an audio signal flows through the coil, the resulting magnetic flux induced in the armature causes the free end to move away from one magnet and towards the other, depending on the direction of the current flow. When the current is removed, the torsional stiffness in the armature's arm restores it to its neutral position. The movement of the armature is transferred to a stiff diaphragm via a connecting rod. The sound thus created from the movement of the diaphragm is delivered deep into the ear canal via the port coupled to a tuned duct.

IEMs typically form a tight seal between the ear drum and the outer ear, isolating the user from ambient noise, and thus improving their efficiency by allowing the audio to be delivered at a much lower level. Because of this seal, the diaphragm needs to be quite stiff in order to overcome the high acoustic impedance (i.e., the resistance to its movement from pushing against the trapped air) without deforming and consequently distorting the sound. The diaphragm used in dynamic drivers tends to be soft or compliant in comparison and does not perform as well in such a high acoustic impedance environment. Thus, monitors with dynamic drivers may vent the drivers to the outside to maintain their performance, but this somewhat defeats the tight seal needed in the IEM.

Some manufacturers package two or even three balanced armature drivers in a single monitor and use a passive crossover, just like in a large loudspeaker system, to channel the low, mid, and high-frequency signals to the right driver to enhance the efficiency and overall sonic characteristics of the unit. Other units may use a hybrid design, with a balanced armature driver for the high end of the spectrum and a dynamic driver for the low end.

There are many advantages to having an IEM system, not the least of which is drastically reduced sound level on stage and the virtual elimination of feedback. Bleed into the FOH system is reduced, as well as spill into the forward section of the house space, leading to a cleaner FOH sound. Assuming that the monitor mixer has enough sends, IEM allows each musician to have a totally personal monitor mix, completely separate from all other musicians. Of course, it also means a few less bulky pieces of equipment to move around from show to show.

There are a few disadvantages to IEM, though. While wireless monitors allow total freedom of movement, they suffer from the same problems as wireless microphones in terms of dropouts and interference. Therefore, for those musicians who tend to remain relatively stationary, such as the drummer, keyboard player, and horn section, wired monitors are the better and more secure way to go. The ear pieces of the monitors must fit and seal very well in the ear canal in order for the user to get the required signal level to use them effectively on stage. The best way to get this kind of fit is with a personalized molded earpiece that is usually made by an audiologist. This adds significantly to the cost of acquiring and using this technology. This seal also effectively isolates and disconnects the musician from the audience and the stage. The solution to this has been to set up ambience microphones and blend them into the monitor mix. Furthermore, the mixes provided to the musicians now have to be stereo, and reverb has to be added to some channels (typically vocals and drums) to restore that sense of space and presence. This also means that the monitor mixer and the monitor equipment must be capable of providing and transmitting a stereo mix, which adds to the cost of the system. Additional equipment may also be needed. For example, while the receivers typically incorporate limiters to protect the hearing of the users, compressors may be needed on some channels (e.g., kick, snare, bass) to prevent the mix levels from fluctuating too widely and overly distracting the musicians. There must also be a unique transmitter/receiver pair for each musician, though some more expensive units allow multiple simultaneous mixes over several channels on a single transmitter. Finally, having IEMs may not eliminate all wedges from the stage. As the low-end response of the earphones tend to be somewhat lacking, the drummer and bassist especially may request supplemental augmentation of the low end via subwoofers or other means.

13.9 Power Amplifier Requirements

One sure question that we will come face to face with when designing the sound system is this: what size of power amplifier is adequate to drive the loudspeakers and provide the signal level desired in the audience space? There are many common sense guidelines and rules of thumb that can be used as a starting point. Clearly, a bigger venue will require more amplifier power than a small venue. A rough rule of thumb for a starting guess is 1 watt per person that the venue is capable of holding. Naturally, the efficiency of the loudspeakers plays a role in the final outcome. More efficient loudspeakers allow the use of smaller amplifiers. Indoor venues usually require less power than outdoor venues because the walls of the structure reflect some of the sound energy and create a reverberant field that keeps sound levels higher on average. Then there is the old adage instructing us that we should have more amplifier power than we need. These are all good things to know, but what we want to do here is step through a simple analysis that will give us a more concrete answer to our question.

13.9.1 Basic Analysis

We want to calculate the power required to drive the loudspeaker system. Before we can do that, though, we must know the sensitivity specification of our speaker and the sound level we want to achieve at our target location in the audience space. Figure 13-28 shows the basic layout for the system. For this example, let us assume that the loudspeaker has an on axis sensitivity of 97 dB SPL at 1 W at 1 m, and is pointed at the rear of the audience, which is 20 m distant. Let us also say that we want to achieve a sound level of 95 dB SPL at the rear of the audience. We will ignore room and audience effects of gain and absorption for now. Furthermore, let us assume that due to physical constraints, we will need 20 m of cable to connect the amplifier to the loudspeaker, and that the loudspeaker's nominal impedance is 8 Ω.

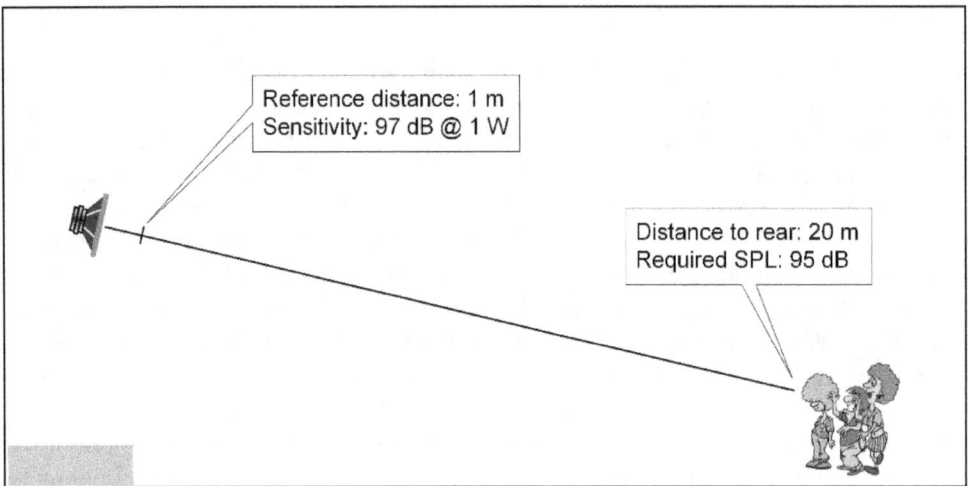

Figure 13-28   Basic Layout Information for Finding Amplifier Power Required

Now that we have the basic facts about the setup, we can systematically step through the process of determining the size of the amplifier needed to give us the required level. Knowing the sensitivity specification of the loudspeaker, we can use the inverse square law to help us determine how much power we need to produce at the loudspeaker itself.

Step 1: First of all, let us find the decrease in sound level (in dB) from the loudspeaker reference distance, d_{ref}, to the rear of the audience. While any distance can be chosen as the reference distance, we must choose one where we know what sound pressure level the loudspeaker can produce at this distance for a given input power. We are given exactly that information in the sensitivity specification of the loudspeaker. Thus, in our case the reference distance is taken to be 1 m. The distance to the rear of the audience is 20 m and represented by **d**. Now we can use the inverse square law to determine the decrease in sound level. Recall that the sound power varies with the square of the distance from the sound source. We can therefore express the change in dB SPL as *10 log (d^2/d_{ref}^2)*, which is the same as *20 log (d/d_{ref})*. Thus:

$$dB\ SPL\ drop \quad = \quad 20\log\left(\frac{d}{d_{ref}}\right)$$

$$= \quad 20\log\left(\frac{20}{1}\right)$$

$$= \quad 26\ dB\ SPL$$

Step 2: Now we will figure out the sound level required at d_{ref} (1 m from the loudspeaker) in order to get the required SPL (95 dB) at the rear of the audience location. We now know that the sound level drop is 26 dB at the rear audience location. This is true no matter what power the speaker is putting out. So, it is a simple addition to figure out the level at d_{ref}.

$$dB\ SPL\ at\ d_{ref} = 95 + (dB\ SPL\ drop\ from\ Step\ 1)$$
$$= 95 + 26$$
$$= 121\ dB\ SPL\ at\ 1\ m$$

Thus, the loudspeaker system must be producing 121 dB SPL at 1 m in order to deliver a level of 95 dB SPL at the rear of the audience.

Step 3: We have calculated the continuous dB requirements at the loudspeaker reference distance. However, our system must also be able to handle the peaks in the music program material. We should have at least 3 dB dynamic headroom, and some practitioners even suggest as high as 10 dB as a minimum. Headroom of 6 dB is adequate for most applications, so let us add this to our required SPL at the loudspeaker reference distance.

$$dB\ SPL\ at\ d_{ref} = 121 + headroom$$
$$= 121 + 6$$
$$= 127\ dB\ SPL\ at\ 1\ m$$

Step 4: Now that we know the highest SPL that we need to produce at the loudspeaker reference distance, let us find out how much this exceeds the reference level given in the speaker specification for 1 W at 1 m.

$$dB\ SPL\ excess = (dB\ SPL\ required) - (Speaker\ sensitivity)$$
$$= 127 - 97$$
$$= 30\ dB\ SPL$$

The loudspeaker system must therefore produce 30 dB SPL more than the reference SPL at the reference distance of 1 m.

Step 5: At this stage, we should consider the power loss in the loudspeaker cable. This loss can be quite significant if the cable is too thin and the distance from the amplifier to the loudspeaker is large. Instead of being used to drive the loudspeaker and produce sound, the electrical energy flowing in the cable is converted to heat and dissipated. Power loss is due mostly to the resistance of the cable and is proportional to the square of the current flowing through the cable.

The situation is detailed in Figure 13-29. This shows the output stage of the power amplifier, producing an output voltage of V_o, with an output impedance of Z_o. The loudspeaker is shown with a coil resistance of R_s. The cable is shown in two legs, each with a resistance of R_c. Thus, the total series resistance due to the cable alone, R_{ctotal}, is $2R_c$. If we ignore the output impedance of the amplifier in this case (it is small compared to the other resistances, on the order of 0.01 Ω for a typical power amplifier), we can derive that the signal loss due to the cable resistance is:

$$dB\ loss\ in\ cable \quad = \quad 20\log\left(\frac{R_s}{R_s + R_{ctotal}}\right)$$

Now, suppose the installation uses relatively small 16 AWG cable between the amplifier and the loudspeaker. Since the distance between the amplifier and the loudspeaker is 20 m, that means the round trip is 40 m. Various suppliers list the resistance of 16 AWG stranded wire to be about 4.018 Ω per 1000 feet, which works out to be 0.0132 Ω per meter. Thus, our cable in this example has a total resistance of 0.528 Ω. We can thus calculate the power loss due to the cable to be:

$$dB\ loss\ in\ cable \quad = \quad 20\log\left(\frac{6}{6 + 0.528}\right)$$

$$= \quad 20\log\left(\frac{6}{6.528}\right)$$

$$= \quad 20\log(0.919)$$

$$= \quad -0.73\ dB$$

So, we will have a loss of 0.73 dB in signal due to the cable alone. That may not seem like much, but it actually represents about 15% power loss. Thus, we need to produce an additional 0.73 dB from the amplifier to overcome the power loss in the cable.

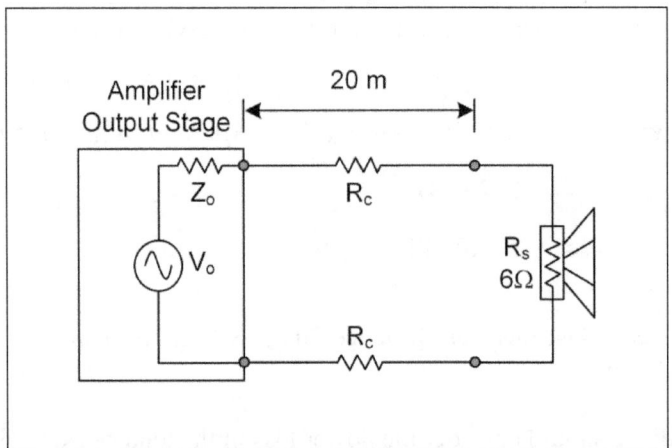

Figure 13-29   Power Loss Due to Loudspeaker Cable Resistance

Step 6: What we really want to do now is figure out the amplifier power, *P*, required to produce an increase of 30 dB, plus an additional 0.73 dB to compensate for cable loss, when the reference power, P_{ref}, is 1 W. Recall that the dB relationship for power is ***dB change = 10 log(P/P_{ref})***. We can thus write what we know:

$$30.73\,dB = 10\log\left(\frac{P}{P_{ref}}\right)$$

$$= 10\log\left(\frac{P}{1}\right)$$

$$= 10\log(P)$$

We want to find **P** in this equation, so we will just rewrite it so that **P** is on the left-hand side. Doing that does not change anything. It is merely a convention that makes it easy for us to manipulate the equation.

$$10\log(P) = 30.73$$

We divide each side of the equation by 10 and find the antilog of both sides to get rid of the log. This gives us the value of **P**, the required power.

$$\log(P) = \frac{30.73}{10}$$

$$= 3.073$$

And thus:

$$P = \text{antilog}(3.073)$$

$$= 10^{3.073}\,\text{W}$$

$$= 1183\,\text{W}$$

So, we need our amplifier to produce 1183 W. Rounding that up to a nice even number, we can conclude that a 1200 W amplifier will provide the amount of power we need to continuously produce 95 dB at 20 m and adequately handle the transient peaks in the music program.

13.9.2 Multiple Loudspeaker Units

The above analysis was performed around a single loudspeaker unit. However, we will typically use at least two loudspeaker units when designing and implementing the sound system. Recall from Section 1.14 that when the loudspeakers are driven equally with the same source material, their outputs are correlated, and there will be a 6 dB increase in SPL along the centerline of the pair in the forward direction. If the feed to the loudspeakers is not identical, we can expect at most a 3 dB increase in SPL. One approach, then, to calculating the amplifier requirements with two loudspeaker units is to reduce the required SPL at the audience reference location (in our case, the rear of the audience) by 3 to 6 dB (depending on whether the loudspeakers are driven by identical signals or not), knowing that we will get that much increase in SPL when both loudspeakers are driven with equal power.

For example, let us repeat the above analysis, assuming that we will get a 3 dB increase at the audience reference location by employing an additional loudspeaker unit. This means that instead of 95 dB SPL as our target, we use 92 dB SPL as the requirement at the rear of the audience. When the analysis is completed, we arrive at a figure of 593 W for the amplifier power requirement to drive one loudspeaker. Therefore, we will need a two-channel amplifier, where *each* channel delivers 593 W to separately drive both loudspeakers. The combined output of the two loudspeaker units will then give us the required 95 dB SPL at the rear of the audience. Rounding up, we see that a 600 W per channel amplifier would satisfy our requirements.

13.10 Ground Loops and Induced Noise

Hum and buzz—every engineer will come face to face with this menace at some point in their career. There are many ways that unwanted noise can contaminate the signal of your program. For example, we know that all active components generate a certain amount of self-noise, which is cumulative in the signal chain. This white noise is heard as hiss in our output signal. However, by far the most severe and annoying source of noise is external to the system, and by far the most annoying and common interference noise is hum at the power line frequency and its harmonics that infiltrates the wiring and finds its way into the signal chain.

13.10.1 The Anatomy of a Ground Loop

How does hum get into the system in the first place? Well, consider that in a venue's power distribution wiring, an outlet on a particular branch circuit may be quite far away from the distribution panel, perhaps with a wire length of 50 m. Another outlet on another branch circuit may be equally far away from the panel. In a 3-wire system, the neutral and safety ground conductors of these outlets are bonded together at the distribution panel and then connected to the installation earth ground, as shown in Figure 13-30. Over the length of the wire between these two outlets, the resistance of the ground wire is almost 1 Ω. With the large currents flowing in the load wires of the venue and inducing small currents in the safety ground wires, it would not be uncommon to have a voltage difference of 1 to 2 volts between the ground contacts of these two outlets.

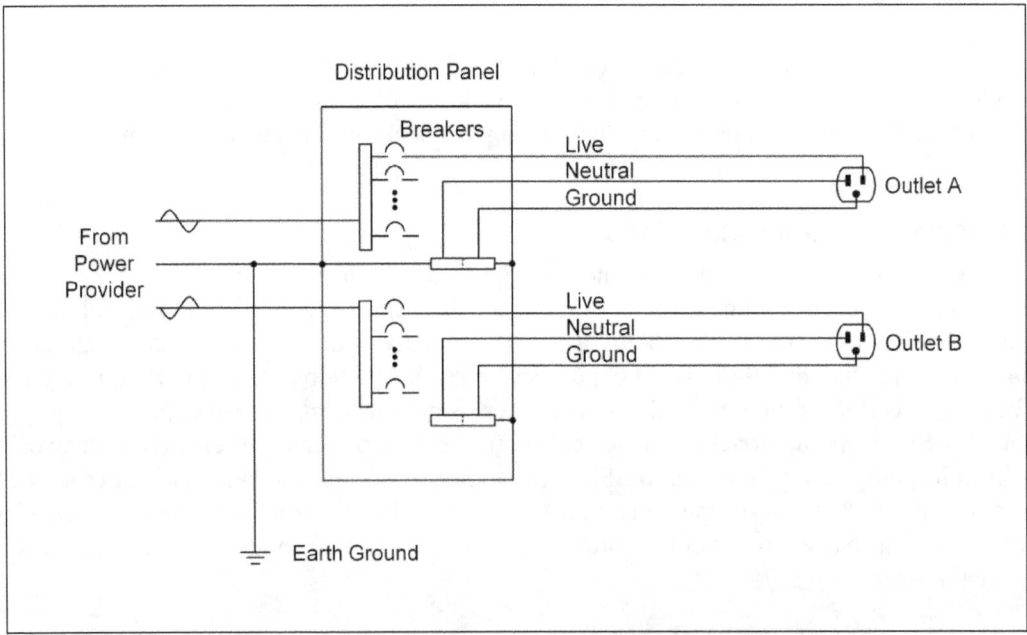

Figure 13-30 Electrical Power Distribution of Building

Now let's say that outlet *A* is onstage where we connect a signal processor and outlet *B* is at the mixing location where we connect the mixer. The metal cases of these pieces of equipment, as well as their signal grounds, are now also connected to the safety ground wire of the venue's power distribution wiring, and thus assume the potential difference present across the two outlets' ground contacts. If the processor is now connected to the mixer via a snake, as shown in Figure 13-31, we will have provided a low-impedance connection, via the ground shield of the connecting cable, between the ground points of these two pieces of equipment. This situation creates a ground loop in which current will happily flow all day long due to the potential difference across points *A* and *B*. Because this current is induced from the current flowing in the load wires, it will have the same frequency as the power line, i.e., 60 Hz in North America, 50 Hz in Europe.

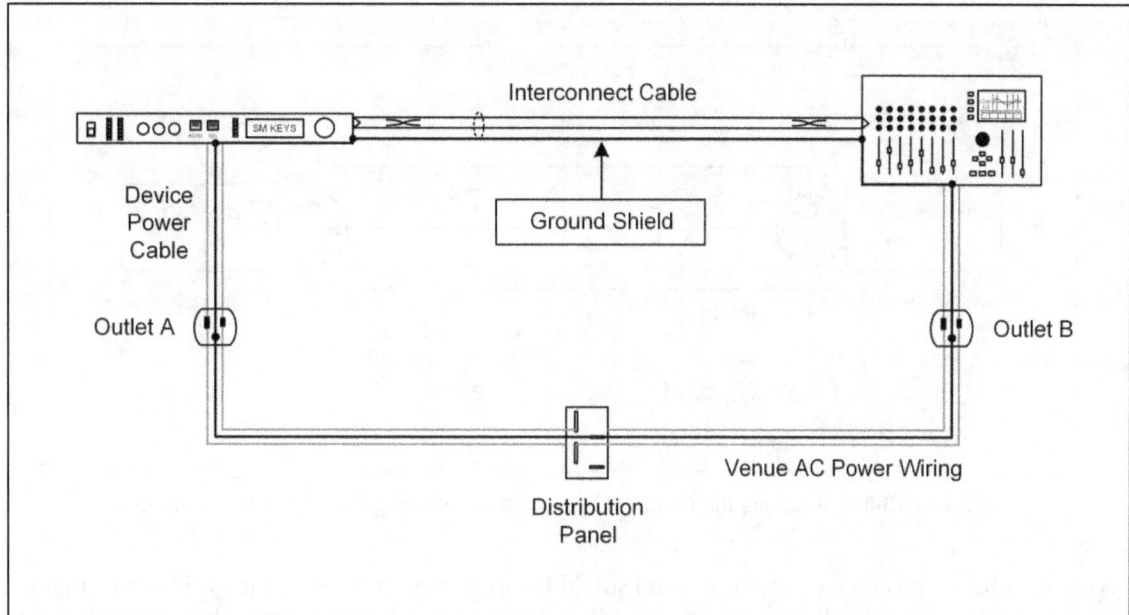

Figure 13-31   Ground Wires of Signal and Power Cables Form a Loop

13.10.2 Noise Infiltration

Now we have an externally induced current flowing in the ground connectors and ground plane of our system components. In poorly designed components, this current can induce unwanted noise into the signal path via internal parasitic capacitive and inductive coupling between the ground and signal traces on their circuit boards.

The current flowing in the ground shield also induces noise in the signal wires of the interconnecting cable. In Figure 13-31, we show the connection as balanced. This is a good approach because if the signal wires are tightly twisted then the noise should be induced equally in both legs of the connection and thus presents a common-mode voltage to the receiver. Recall from Section 5.9.3 that when the receiver sums the two inputs, the common-mode noise voltage will be eliminated—in theory. In practice, though, the interface is not perfectly balanced, and so not all of the noise is eliminated. The common-mode rejection ratio (CMRR) specification of the interface will tell you how well the receiver interface will eliminate the common noise present on both input wires.

13.10.3 Breaking the Loop – Lifting the Ground

The ground shield of the signal cable in a balanced connection is used to protect against electrostatic interference and does not form part of the signal path. We can thus disconnect it and still pass the signal from the output driver to the input receiver. Disconnecting the cable at only one end of the connection breaks the loop, denying the current a path to flow along the run of the cable while still maintaining a connection to ground to drain away electrostatic interference. Although still debated in some circles, most practitioners agree that we should disconnect the cable at the input end, as depicted in Figure 13-32. This is because a long cable shield attached only to the input side acts like a long antenna to pick up noise which can be introduced into the ground plane of the receiver. Some devices feature a built-in ground lift switch, which can be used to disconnect the ground shield at the input connector in this manner.

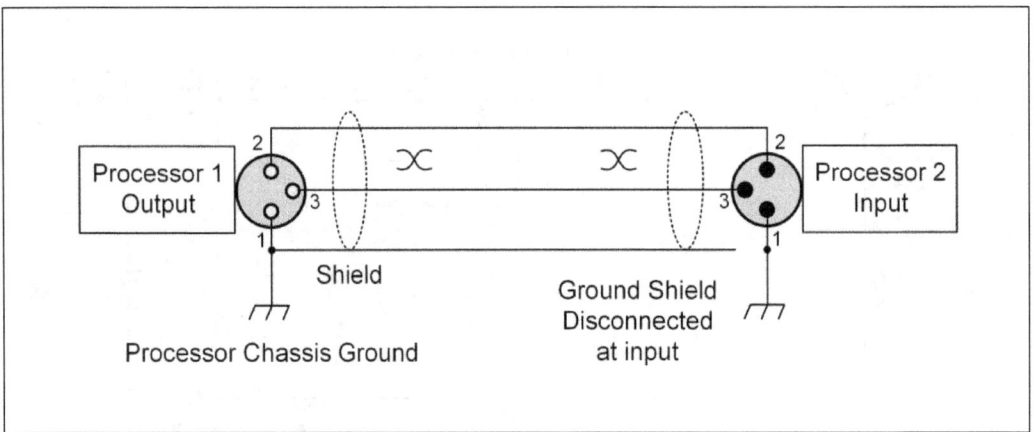

Figure 13-32   Breaking the Ground Loop by Disconnecting the Shield at the Input

With an unbalanced connection, the ground shield forms part of the signal path. This means that the ground shield cannot be disconnected and still maintain a usable signal from the output driver to the input receiver. In a situation like this, any noise induced into the core signal wire is also directly amplified as part of the signal. Therefore, this kind of connection should only be used where the signal level is much higher than any induced noise, and you must be prepared to insert a severe, narrow notch filter at the power line frequency (e.g., via a parametric EQ) later in your signal chain to combat such noise. Alternatively, you can use the approach described below.

13.10.4 Breaking the Loop – Transformer Isolation

Another way to combat noise induced by ground loops is to put an isolation transformer in the signal path, as shown in Figure 13-33. As with all transformers, the isolation transformer breaks the low-impedance path between the connected devices. This means that the low-frequency 60 Hz noise signal that would normally flow in the ground shield between the two devices can no longer take that route, thus eliminating the induced noise current in the signal wires. This approach works for both balanced and unbalanced connections and so is a viable option when the ground connection cannot be lifted.

Besides the primary and secondary windings, true isolation transformers have additional elements to help reduce the transfer of other undesirable noise signals from the primary to the secondary circuits. These consist of two separate (and isolated) shields between the primary and secondary windings, called Faraday shields, that are effective in reducing capacitive coupling between the windings, and hence electrostatically transferred high-frequency noise. When these shields are grounded separately as in Figure 13-33(A), high-frequency noise (e.g., buzz) is shunted to ground and constrained to the circuit in which it occurs.

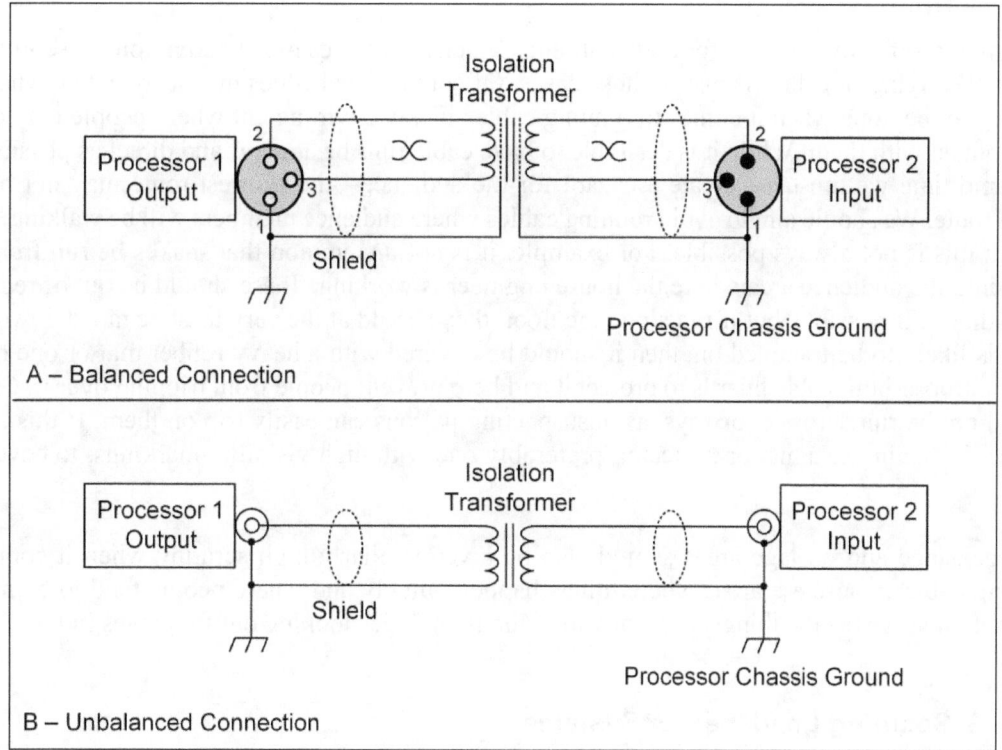

Figure 13-33   Using a Transformer to Break the Ground Loop

Other options for breaking the ground loop exist, such as optical couplers and common-mode chokes, but isolation transformers are most popular. Some manufacturers sell isolation transformers with XLR or RCA connectors already wired and packaged in small units, making this solution easy to deploy.

13.11 Sound System Safety Issues

Safety considerations cannot be overlooked when designing, setting up, or tearing down a sound system. Depending on whether the installation is temporary or permanent, these considerations will be somewhat different. In any case, at some stage of implementation, the following safety issues will need to be addressed: lifting, cabling, structural, electrical, fire, and hearing. We will discuss each of these in the following sections.

13.11.1 Positioning Heavy Equipment

Many items used in the pro-audio industry are very heavy and bulky. Racks full of equipment, big mixers, and loudspeaker bins easily come to mind. Overenthusiastic technicians can hurt themselves and others if not aware of the dangers of improper moving and lifting of such equipment. A Standard Operating Procedure (SOP) should be put in place to cover the emplacement of equipment at all events. This SOP should spell out items such as lifting load limits, personnel requirements for specific bulky or heavy items, lifting techniques, and use of power equipment such as forklifts. All personnel should be aware of the SOP and follow its procedures to prevent injury.

13.11.2 Running Cables

Providing a safe environment for both staff and audience involves mostly common sense in many cases. The laying of cables is one of those cases. Of course, local codes may also dictate what can and cannot be done when it comes to running cables in any environment where people may come into contact with them. While it is desirable to have cable runs be as short and direct as possible, at the same time we must not create a hazard for the audience—the shortest route may not be the safest route. We should aim to avoid running cables where audience members will be walking, even though this is not always possible. For example, it is not uncommon that snakes be run from the stage into the audience area where the house engineer is working. These should be out of reach of the audience if possible, but if run along the floor, they should at the very least be taped down. If a cable is likely to be trampled on, then it should be covered with a heavy rubber mat or one of the many purpose-built cable guards to protect it and help prevent people from tripping over it. Cables should not be run across doorways, as unsuspecting patrons can easily trip on them. If this is unavoidable, again use a mat or protector, preferably one with high visibility markings, to cover the cables.

The backstage and onstage areas should also be given similarly high scrutiny when it comes to running cables. These are areas where things happen quickly and where people tend to be highly focused on very specific things, none of which likely includes looking out for cables in their way.

13.11.3 Securing Loudspeaker Systems

Figure 13-34 shows three common ways of deploying loudspeaker systems, each of which has its inherent dangers. Speaker stands are among the most popular and least expensive ways of setting up a loudspeaker when some elevation off the floor is required. They are probably also the least secure way of mounting loudspeakers. These setups tend to be unstable because we have a large mass high off the floor, and the center of gravity of the structure can easily be moved outside of the base of the stand. This means that if you bump it too hard, the whole arrangement will simply fall over. For this reason, loudspeakers on stands should not be placed in locations where the audience is likely to physically contact the stands.

Stacking of loudspeaker bins and equipment racks is another common practice in the pro-audio industry. Again, use common sense when stacking bins—put the bigger, heavier bin at the bottom. The risk of bins falling over greatly increases when stacked more than two high. In such a case straps, stacking racks, or specialty made locking mechanisms should be used to secure the bins. Since stability decreases rapidly with taller stacks, bins should probably not be stacked more than three high.

Loudspeakers that are suspended from a rigging structure so as to elevate them above the audience level are said to be flown. Flying loudspeakers is a whole different ballgame altogether. It is smart to leave flying of loudspeakers to qualified riggers, who should have liability insurance in case of an accident. If a loudspeaker should fall, it could easily kill whoever it hits, so we really want to make sure that once it is put up, it will stay up. Loudspeakers should not be flown over the audience area and must also be flown in accordance with all applicable local safety regulations. Loudspeakers should only be flown from approved rigging points with approved hardware. Manufacturers normally specify these points and the hardware that should be used.

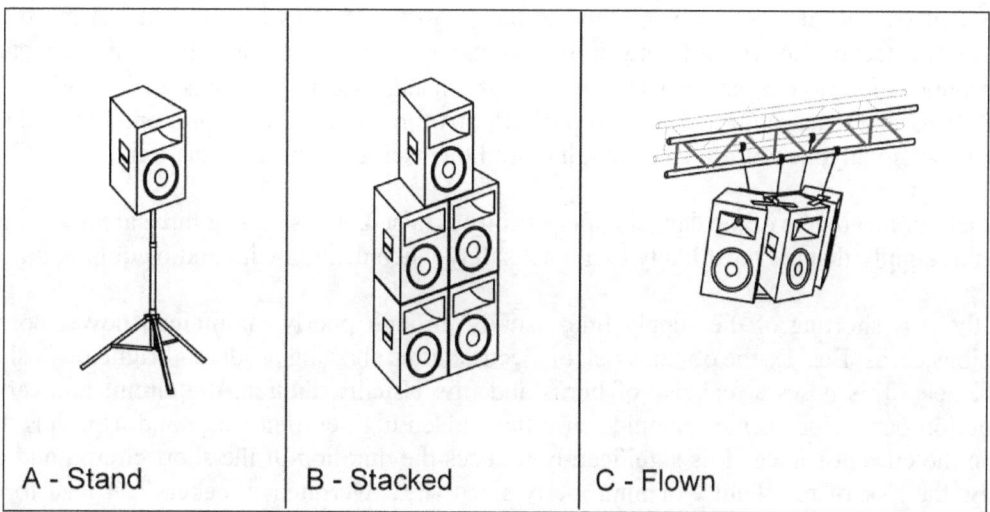

A - Stand   B - Stacked   C - Flown

Figure 13-34 Deploying Loudspeaker Systems

13.11.4 Electrical Safety

Electricity is a wonderful thing. Without it, we could not have a live sound reinforced show as we know it. Yet, without the proper precautions, electricity can be very dangerous for anyone coming into contact with it in the wrong way. As long ago as 1961, Dalziel presented research results showing the effects of electric current on the human body.[8] A few years later, Kouwenhoven also presented data that showed similar results.[9] Table 13-1 is a distilled combination of these results to a level that is pertinent to our discussion. You will notice that some of the ranges of current that cause certain effects overlap considerably. This is because it turns out that women are more sensitive to the effects of shock than men. The really interesting thing to note, though, is that it takes a rather small alternating current of only about 100 mA flowing through the body to inflict serious damage or even death. For comparison, the current available at a typical North American household outlet is usually around 15 A. Notice too, that for a given current level, a shock from 60 Hz AC is more dangerous than a shock from DC. North American utilities provide electricity at 60 Hz, while 50 Hz is typical in Europe.

Effects	Current, mA	
	DC	AC (60 Hz)
Perception threshold – slight tingle	1 – 6	1
Slight shock – not painful	6 – 9	2 – 5
Painful shock – muscular control lost	41 – 76	6 – 16
Severe shock – breathing difficult	60 – 90	15 – 23
Extreme shock – ventricular fibrillation, possible death (3 second exposure)	500	100

Table 13-1 Effects of Electric Shock on Humans

We've often heard the phrase "it's the current that kills you," and now we know that it really does not take that much current to do just that. How much voltage, though, does it take to drive that much current through the body? Well, that depends on the condition of the skin at the point of contact. Dry skin is a fairly poor conductor of electricity. On the other hand, moist, sweaty skin is a fairly good conductor. A tight grip provides better contact than a light touch. Combine these factors and we can get a worse case hand-to-hand resistance of as low as 1 kΩ. A more typical scenario of a technician with a tight grip on a metal tool and a firm hand-hold on the metal case (i.e., ground) of a piece of equipment may give us a hand-to-hand resistance of 3 to 5 kΩ. Ohm's Law tells us that it takes 60 to 100 V to push 20 mA through such a circuit, enough to deliver a severe shock to anyone unfortunate enough to find themselves in such a situation.

Now that our awareness of the dangers of electricity is raised, let us look at three areas surrounding the mains supply that are most likely to impact safety when dealing with audio equipment.

First, there is shorting of the supply line, usually through poorly maintained power cords and extensions cords. Besides the obvious risk of electrocution, shorting produces bright arc flashes and intense heat. This poses a real risk of burns and fire. Usually, though, the circuit breaker at the distribution panel (load center) should sense the sudden high current surge and trip, thus disconnecting the current source. This significantly reduces the duration of the short circuit, and consequently, the risk of fire. Faulty or improperly sized (i.e., overrated) breakers can lead to a very dangerous situation in cases like these.

The second issue—overload—is more common. Overload is like a very mild form of shorting. In this case, we are attempting to draw more current from the circuit than it was designed to safely provide. If the overload is high enough, the breaker should trip and shut down the circuit. Usually, all building wiring will be able to handle the current rating for the circuit, i.e., from the circuit breaker in the distribution panel to the electrical outlet. It is possible, though, to plug an underrated extension cord into an outlet and then overload the extension cord. In this case, the cord will get hot and could become a fire hazard. You should be aware of the current rating of the outlet and all extension cords being used, as well as the current drawn by all devices plugged into the outlet. Make sure that the current handling capacity of the wiring is more than required by all the equipment combined.

Finally, we may have problems with faulty ground connections on our equipment. The typical three-pronged power cord that is used on audio equipment has three conductors in it. One conductor mates to the building's hot (or live) supply conductor, which brings electrical current from the distribution panel to the power outlet. Another one mates to the building's neutral conductor, which completes the circuit back to the distribution panel. Lastly, the third conductor connects to the ground conductor of the building wiring. The metal case of electrical equipment having a power cord terminated with a three-pronged plug is usually connected to the ground conductor of the power cord. In the event that the hot wire touches the case, electricity will flow to ground through the ground conductor, rather than perhaps through a user who happens to be touching the equipment case. If the metal case is somehow disconnected from the ground conductor, or the ground conductor is disconnected from the building ground wiring, the equipment is said to have a faulty ground. Equipment with a faulty ground can be quite dangerous to users. Sometimes, sound technicians will even deliberately disconnect the ground conductor of a piece of equipment to help cure hum problems. This is bad practice and should not be done.

When a user touches the energized metal case of a piece of equipment with a faulty ground, electricity will flow through the user, with potentially deadly results. To guard against the worst-case scenario of electrocution, a ground fault circuit interrupter (GFCI, referred to as residual current device in Europe) should be used on all primary circuits when assembling the sound system. A GFCI may already be present (though unlikely) in the breaker at the head of the circuit in the distribution panel, or more commonly at the electrical outlet, but when a circuit does not already contain a GFCI, a power bar with a built-in GFCI, like the one shown in Figure 13-35, can be used.

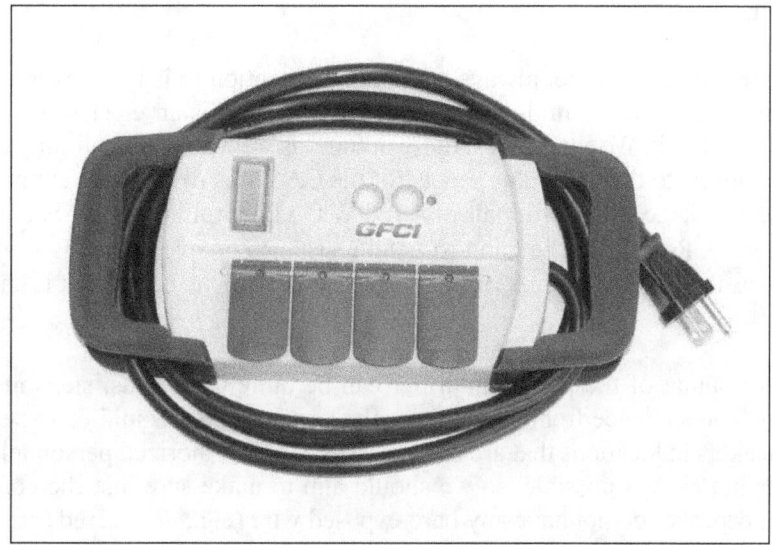

Figure 13-35   Power Bar with Built-in GFCI

A GFCI protects users by constantly monitoring the current flow through itself. If the current flowing out of it (via the hot port) is equal to the current flowing into it (via the neutral port), then there is no current loss through the ground and all is OK. Therefore, we could have a piece of equipment with a faulty ground, and even when the equipment is plugged in and turned on, and its metal case is energized, the circuit may not trip.

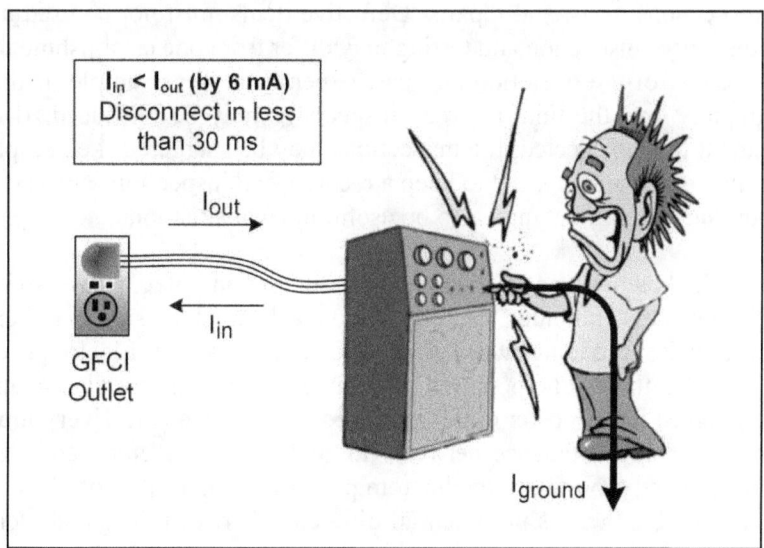

Figure 13-36   A GFCI Can Protect from Electrocution

Suppose now that a user makes contact with the energized case of this equipment, like in Figure 13-36. He will receive a shock, as some of the current will flow through him to the ground. This means that the current flowing back through the GFCI will be less that the current flowing out of the GFCI. When this difference in current flow is bigger than 6 mA, the GFCI determines that a ground fault has occurred and shuts down the circuit in less than 30 ms, thus preventing electrocution of the user.

One area of concern that we do not always pay enough attention to is the power amplifier output. The speaker output from even a mid-sized power amplifier poses an electrical hazard to anyone coming into contact with it. We should treat these outputs as we would any building electrical outlet. For example, according to the Crest data sheets for the CA series of high current power amplifiers, the maximum peak voltage at the terminals of a 700 W CA12 amplifier is 126 V, and the maximum RMS voltage is 90 V. Recall that the hand-to-hand resistance of sweaty palms is about 3 kΩ, and Ohm's Law tells us that at 90 V, 30 mA would be driven through the body. That is enough to deliver a severe shock.

Since the speaker output of the power amplifier can be quite dangerous, steps must be taken to protect both crew and audience from this output. The best approach would be to secure the amplifiers and loudspeakers in locations that are not accessible to unauthorized personnel. Of course, we know that this is not always possible, so we should aim to make sure that the connections at the amplifier and loudspeaker do not have any bare exposed wire (e.g., uncovered screw terminals). If the bare terminals are exposed in an unsecured area, then care should be taken to keep them out of reach of audience (e.g., if a live wire is exposed and the equipment is unsecured, US Department of Labor OSHA requirements (29 CFR Subpart S 1910.303) state that it must be at least 2.44 m above the floor if the voltage carried is more than 50 V[10]). When the loudspeakers are in a non-secure area, it is much better to use locking connectors (e.g., Speakon connectors, or locking 1/4-inch jacks) than non-locking 1/4-inch connectors, which can easily be pulled out from their jacks, and which have exposed metal contacts.

Preventative maintenance is an excellent way to help promote electrical safety. This consists of a set schedule of inspections, tests, and repairs. Defective items must not be used until replaced or repaired. The frequency of inspection and testing may differ from one establishment to the next and may depend on the size of the operation and past experience. For example, if the rate of defect discovery is extremely low, the time between inspections may be increased. If there is a large amount of equipment to be inspected, the inspections may be staggered, i.e., amplifiers one time, processors next time, etc. It is important to keep a record of all inspections and tests. This will help to keep the accounting straight, but may also be useful in legal situations.

Visual inspections should at least cover the mains power cord and plug, power switch, fuse holder, equipment case, patch cords and plugs, and connection jacks to make sure that there are no cracks in insulation, loose ends, shredding, loose screws, or other distress. Electrical tests should at a minimum make sure that there is no potential difference between the metal case and ground of all equipment when plugged into a power outlet, powered on as well as off. Every input on the mixer should be tested for voltage imbalance between hot and cold connectors, and excessive voltage between any connector and ground, when phantom power is on as well as off. Microphones should be tested to make sure that there is no potential difference between their body and ground when phantom power is on. Special testing rigs can make the last two tests very easy to perform, and can save lots of time when there are several units to be tested.

Review Questions

1. What are the high-level considerations that must be taken into account when designing a live sound reinforcement system?

2. How does a system designed primarily for speech differ from one designed primarily for music?

3. What are some of the things that can be done to increase intelligibility in a space where the main problem is high reverberation time?

4. What is the critical frequency range for the transmission of consonant sounds?

5. What is critical distance and how does it factor into the intelligibility considerations of the sound system?

6. Discuss the mixing locations generally available to the FOH engineer and the issues associated with each.

7. What does the dispersion specification of a loudspeaker tell us and how is it useful in designing the sound system?

8. Discuss how the height and aiming angle of a loudspeaker affects its coverage in the audience area.

9. Explain how a cluster can be used to increase the horizontal coverage of the speaker system.

10. What approaches are used to extend the reach of the sound system in a large venue?

11. Explain how feedback develops in the audio signal chain, and discuss one technique that can be used to combat it.

12. What is meant by gain before feedback, and how can it be maximized?

13. Calculate the amplifier power required to achieve a level of 100 dB SPL at the listening location which is 25 m distant from the loudspeaker which has a sensitivity of 98 dB @ 1 W. Assume a requirement of 3 dB headroom, and ignore cable losses in the setup.

14. What is a ground loop? How is it formed, and what are some of the ways to eliminate it?

15. What factors must be considered when designing an indoor sound system compared to an outdoor system?

Chapter 14

Microphone Usage

14.1 Miking – One Approach

In the last several chapters, we have taken a somewhat theoretical look at the operation of various pieces of audio equipment and their interaction with the acoustical environment. We now want to shift gears and look at the more practical aspects of the execution of a live sound show and the application of the knowledge we have learned up to now. From here on out, we will be looking at some practical guidelines for getting through a live sound event. Let us start with how to get the best usage out of the microphone.

When it comes to microphone usage and technique, the first thing we should accept is that there are several ways to approach any given miking application, none of which is necessarily right or wrong. As you gain engineering experience, you will develop a particular style that supports your taste and fulfills the objectives and requirements of the situation. However, we all need a fundamental approach to miking in general—a set of guiding principles, a philosophy, so to speak. Here is one philosophy of miking:

- The first concern is to get the best quality sound from the microphone. The tone and quality of the sound will change depending on where we put the microphone and how it is oriented.

- Our next concern is to pick up only the sound we really want to amplify. All other sounds should be rejected as far as possible.

- Our third concern will be to deal with specific issues that arise in every miking application. These will be different from one application to the next, and we will find hundreds of them. For example, we may have to deal with plosives in a vocal application or with a player's hand blocking the microphone in an acoustic guitar application.

14.2 The Rules of Microphone Usage

With a philosophy like that, can we lay down any real rules for microphone usage? Well, yes and no. It is hard to be strict about how to mike any given application. With that in mind, here are two rules of miking, which, it can be argued, are not really rules at all.

Rule No.1: **There are no hard and fast rules, only guidelines.**
We must be prepared to experiment within certain boundaries to find the best approach to miking any given situation. When we find an approach that works for us, we use it.

Rule No. 2: **If it sounds good, then it is good.**
Given our approach of 'no hard rules,' there are bound to be times when critics will disagree with a particular miking setup. However, this is the one rule that you should remember in the face of your critics. Again, experimenting to find the best setup within the boundaries of our philosophy is good.

14.3 Source-to-Microphone Distance and Sound Quality

Before looking at specific miking applications, let us consider how the source-to-microphone distance may affect the quality of the sound we can get from the microphone. Later on, we will consider specific applications to vocal and instrument miking, but what we discuss here is applicable to both cases. We will break the analysis into three broad areas—close, mid, and distant miking regimes.

14.3.1 Close Miking

Close miking is a technique where the microphone is less than 8 cm from the source, i.e., the instrument or mouth of the performer. This approach provides a full, strong, up front sound. If using cardioids and very short distance (~ 3 cm), the bass end of the spectrum will be emphasized and may need to be rolled off with equalization. Additionally, the SPL at the microphone at this small distance may be quite high in some instances (e.g., at the bell of a trumpet or when miking a bass cabinet), so we must be careful not to push the microphone into a zone where it starts to distort the audio signal, unless of course, that is what we are trying to achieve.

This technique also provides the highest rejection of ambient sounds, providing the maximum gain before feedback of any microphone technique. There will be almost no leakage from other sources. However, because the performer is so close to the microphone, small variations in source-to-microphone distance as the performer moves around will have a dramatic effect on the tonal quality and output level of the signal from the microphone.

14.3.2 Mid Miking

In order to get the best balance of presence and ambience, the source-to-microphone distance can be increased to between 15 and 30 cm. This still gives a good strong level of the performer or instrument at the microphone, but also picks up some of the influences of the environment and ambient sound. The result is a very natural sounding voice, with good body and some ambience. Since the level of performer's voice or instrument is still quite a bit higher that the ambient sound, feedback rejection is still good. As a bonus, as the performer moves around, the output signal level of the microphone remains fairly constant.

14.3.3 Distant Miking

In distant miking, the microphone is quite far from the source—over 75 cm. At this distance the performer or instrument will sound far away because the acoustics of the environment will have a big effect on the sound at the microphone. Ambient sound level at the microphone will be strong compared to the source, and may even be stronger, depending on the exact source-to-microphone distance. Leakage rejection will be low, thus gain before feedback will be the lowest of all miking techniques. However, movement on the part of the performer will have practically no impact on signal level from the microphone.

Unless we are being creative, we must be careful about moving the microphone too far away from the source when miking indoors. This is because, as was discussed in Section 13.3.1, as we move away from the sound source we will reach the critical distance (D_c) where the direct sound from the source and the reflected sound in the reverberant field will be equal in intensity. Beyond D_c, the reflected sound will be higher in intensity than the direct sound. This is an especially important consideration for speech intelligibility, where understanding the performer's words is paramount. The high-level reflected sound can combine with and overwhelm the direct sound, obscuring important cues, and make understanding difficult. Directional microphones can be used to help retain focus on the sound source and extend the range when using the distant miking technique, but should be kept well under D_c for acceptable speech intelligibility. When miking an instrument, we have much more room to experiment with the blend of direct and reverberant sound, as a touch of natural reverb can make an instrument sound more lush and smooth.

14.4 Phasing Issues with Microphone Usage

We just mentioned the issue with speech intelligibility that can occur when the source-to-microphone distance is approaching D_c. However, we can run into phasing issues which may contribute to the degradation of sound quality even when the source-to-microphone distance is quite a bit less than D_c. Let us look at some examples of this now.

14.4.1 Reflected Sound

When direct sound and reflected sound meet at the microphone, the end result will depend on the relative phase of the sound waves. A good example of this is when using a podium microphone. This situation is shown in Figure 14-1(A). In this case, the direct sound and reflected sound will not differ hugely in intensity, and cancellation of certain frequencies will occur due to destructive interference when these sounds combine at the microphone. When the direct and reflected sounds are within 9 dB of each other, this situation becomes problematic,[1, 2] and the resulting frequency response at the microphone has deep nulls. This is the so-called comb filter response and is similar in appearance to that of Figure 14-2. This kind of response may lead to a certain hollowness of the sound which cannot be corrected with EQ.

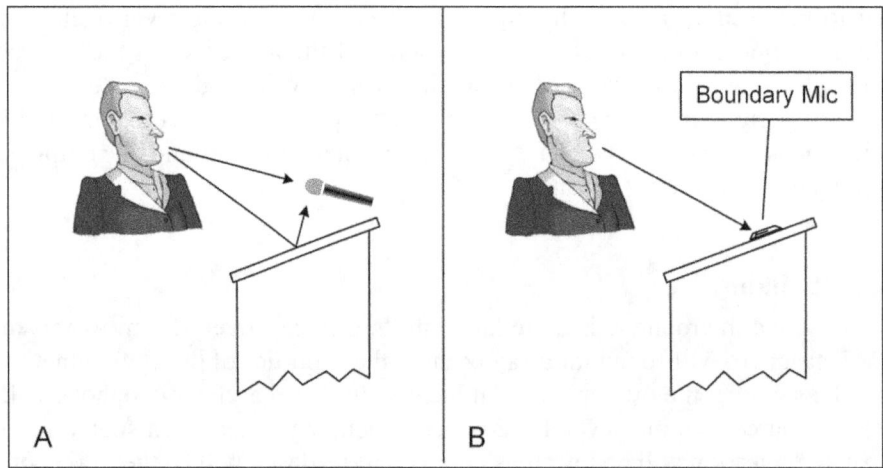

Figure 14-1 Reflected Sound Can Cause Phasing Issues

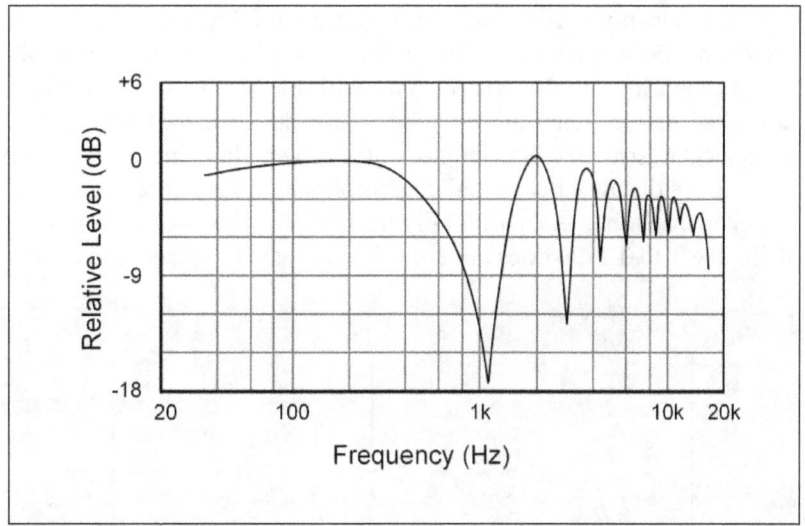

Figure 14-2   Comb Filter Response Resulting from Combined Sound Waves

One solution to this problem is to move the microphone closer to the speaker. The level of the direct sound is then far greater than the reflected sound, and phasing problems are greatly reduced. Additionally, gain-before-feedback is increased. The disadvantage with this approach is that any movement on the part of the speaker can cause the source-to-microphone distance to vary considerably, leading to large changes in audio output. Generally, though, a good approach to reducing problems with reflected sound is to keep the source-to-microphone distance as short as practical and to keep that distance fixed. This is the impetus behind the use of lapel, headset, and earset microphones, where we can keep the source-to-microphone distance fixed even as the speaker moves around. Note we say that the source-to-microphone distance should be as short as practical, not as possible, because when miking some larger instruments, the microphone may need to be backed off a bit in order to capture the true sense of the instrument's tone.

A second approach to dealing with reflected sound is the use of a boundary microphone (also sometimes called a pressure zone microphone, or PZM[a]). This microphone sits flush (or very nearly so) with the surface to which it is attached, like in Figure 14-1(B), and eliminates problems caused by reflection.

14.4.2 Multiple Microphones

Sometimes we may use multiple microphones with a single source. Figure 14-3(A) shows this scenario. In such a case, if the distance from the source to each microphone is different, then the sound arriving at each microphone will be out of phase for some frequencies. This means that the corresponding signal transmitted by each microphone will also be out of phase relative to each other. When these separate signals are then combined in the mixer, we get cancellations at certain frequencies, giving the classic comb filter frequency response. This becomes a real problem when the signals are within 9 dB of each other and, as mentioned earlier, we end up with a hollow sound which cannot be EQed. The best way to solve this phasing problem is to move these two microphones next to each other as shown in Figure 14-3(B). This will drastically reduce the distance between their capsules and ensure that for most frequencies, their separate signals are in phase.

---

[a] PZM is a trademark of Crown Audio, Inc.

There are times, though, when we will want to use multiple microphones and for some reason may not want to move them next to each other. Figure 14-4(A) shows an example of this. In this case, when the microphones are in the same vicinity, we will not be able to avoid the kind of phasing issue we just described above. Each microphone will produce out of phase output for the two sources, with subsequent destructive interference at the mixer. This kind of phasing problem can be minimized by keeping the distance between the microphones at least three times the distance of the farthest source to its microphone, as shown in Figure 14-4(B). This is often referred to as the 3-to-1 rule. Note that using the 3-to-1 ratio does not entirely eliminate the phasing problem, but it reduces it to a large extent.

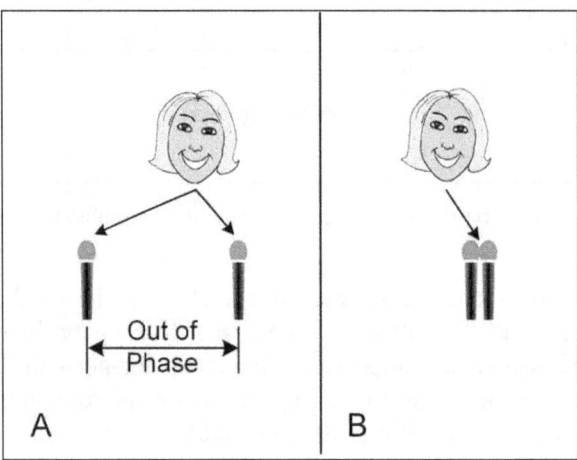

Figure 14-3   Multiple Microphones with Single Source

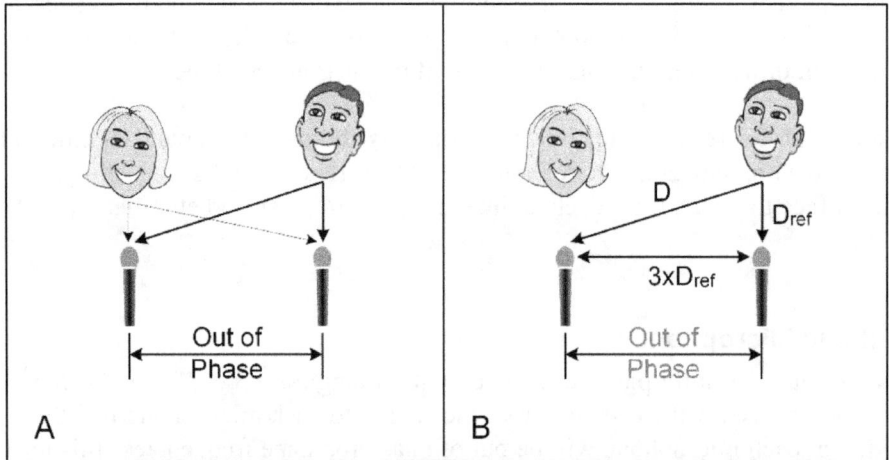

Figure 14-4   Multiple Microphones with Multiple Sources

But why a ratio of 3-to-1? Why not 2-to-1, or 4-to-1? What we are trying to achieve is, for the source on the right in Figure 14-4(B), to make the distance labeled D big enough so that the SPL at the microphone on the right is at least 10 dB more than at the microphone on the left. When the difference in level is 10 dB or more, the comb filter effect caused by the addition of out of phase signals is minimized.[2] Recalling that the SPL drop with distance is given by $20\ log(D/D_{ref})$, we can easily work out that D must be at least 3.16 times the distance D_{ref} in order to give this SPL difference. Now if we recognize that the geometry of the arrangement presents us with a right-angled triangle, we can work out that when D is 3.16 times D_{ref}, the distance between the microphones will be 3 times D_{ref} (see Pythagoras' theorem in Appendix). Just remember, even though this is commonly called the 3-to-1 *rule*, you should use it only as a good starting point for experimentation, and break it when your ears tell you that what you are hearing is what you are trying to achieve.

14.5 Issues Associated with Articulating Certain Sounds

There are certain types of sounds that, when articulated into a microphone, cause some level of grief for the sound engineer and the audience. The results can range from mildly annoying to downright distracting. Since these sounds occur quite frequently in speech, we should make an effort to control their effect on the enjoyment and intelligibility of the program content. In the following section, we will look at the two most common of these sounds—plosives and sibilance—and ways of reducing their effects.

14.5.1 Plosives

A plosive consonant, or stop consonant, is one where high-pressure air is accumulated in the mouth and then suddenly released when these consonants are articulated. A loud thumping sound results from this sudden, explosive release of air from the mouth striking the microphone at high speed. This is especially annoying at the end of some words, but many times at the beginning as well. These consonants include B, D, K, P, and T. To reduce the effects of plosives, we can do the following:

- Since the thump resulting from a plosive consists of low-frequency sound, drastically cut everything below 80 Hz on all voice channels at the mixer.
- Move the microphone out of the direct path of the plosive by moving it slightly to the side, above, or below the mouth, or angling it slightly away from the axis of speech.
- The thump is worst when the air striking the microphone is at its highest speed. When the air first leaves the mouth, it is still accelerating due to the push of the air behind it. Beyond about 8 cm after leaving the mouth the air starts to slow down because that initial push has dissipated. Therefore, make the mouth-to-microphone distance greater than 8 cm.
- Use a pop filter, such as a foam ball or other external windscreen, to reduce the speed of the wind before it strikes the microphone.
- Use an omnidirectional microphone, since plosives are exaggerated by the proximity effect.

14.5.2 Sibilance

Sibilance is that excessive high-frequency hissing that results when consonants like S, TH, and Z are articulated. The sound is caused by forcing air through the narrow space between the teeth and the tongue when the mouth is slightly open. This is called fricative action. To reduce the effects of sibilance, we can do the following:

- Use equalization to cut the high-frequency end of the spectrum.
- Insert an equalizer into the sidechain of a compressor and use it to compress the signal only when sibilance occurs. This is called de-essing.
- Use a dedicated de-esser.

As we just stated, de-essing is the process of reducing or removing the sibilance from a vocal. For live sound reinforcement, a dedicated de-esser is recommended, since the compressor-equalizer setup can take a lot of time to get it just right. An example of a dedicated de-esser that is easy to use is the Drawmer MX50. However, with some practice and a little effort and patience, an effective de-esser can be made using a compressor with an equalizer inserted it its sidechain. Whether using a dedicated de-esser or a compressor, we would put it in the insert path of the vocal channel (or subgroup if sharing it across several vocal channels) on the mixer.

Let us take a look at how we can use a compressor and equalizer to fight sibilance. The setup will be as in Figure 14-5. Sibilance usually occurs in the 5 kHz to 10 kHz range. What we need to do is set the compressor to very quickly compress the audio when excessive energy in this range is detected and very quickly release when the energy level falls back to normal. To get a quick response, we need fast attack and release times.

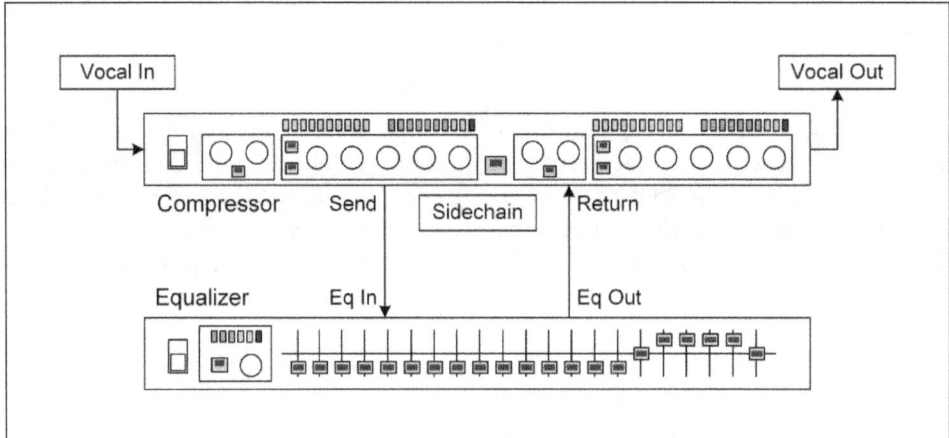

Figure 14-5   De-essing with a Compressor and EQ in the Sidechain

Recall that in a compressor, the level of the input audio signal is used to control the onset and termination of compression as the signal rises above and falls below the compression threshold. The sidechain circuit is used to take that control away from the input signal and give it to another control signal. In de-essing, this control signal is generated by the equalizer.

To generate the control signal, we take the input vocal audio signal and feed it to the equalizer (from the sidechain output or send). On the equalizer, we greatly boost all the frequencies between 3 and 10 kHz while severely cutting all the frequencies below 3 kHz. We feed the output from the equalizer into the sidechain input or return on the compressor. Note that the equalizer is not in the audio signal path—it is merely providing the control voltage, so it need not be of the highest quality.

With the equalizer set to boost the high frequencies and cut low, its output voltage will be high when there is high-frequency content (sibilance) in the input audio signal. We must now fine tune the threshold on the compressor so that when sibilance occurs the control signal will cross the threshold, thus causing the vocal signal to be compressed, and hence reduce the level of the sibilance. In order to sound natural, the compression ratio must be carefully set, and the compressor must be very fast acting in order to just catch the sibilant vocal. As a starting point for experimentation, set the attack around 0.05 ms (50 μs) and release of around 50 ms, and go from there. We don't want to slow down the rise of the control voltage in any way, so use hard knee instead of soft, and peak level detection instead of RMS. Once you find settings that work well, note them carefully so that next time you need to set up for a show, there is already a baseline that can be used as a starting configuration. This will save a lot of time during the next sound check.

14.6 Common Microphone Handling Issues

If you deal at all with inexperienced performers, you will eventually have to deal with problems arising from improper microphone handling. Misuse of handheld microphones is probably the most common transgression committed by performers. A few reassuring words from the audio engineer can go a long way towards helping non-technical, inexperienced performers become comfortable handling the microphone. If we see that a performer is unsure of the proper use of the microphone, we should clearly communicate to that person what is the proper technique. That way, everyone benefits—the performer will be heard clearly and effortlessly, and the engineer does not have to struggle unnecessarily to get them heard.

The most common mishandling issue is probably not holding the microphone close enough to the mouth when speaking softly. This may lead some engineers to turn up the gain on the microphone at the mixer, but this could lead to feedback or leakage from other sources. In this case, the performer must be made comfortable having the microphone 'in their space.' An alternative, if available, would be to use a lapel microphone which keeps the distance to the speaker's mouth fairly constant and is unobtrusive.

The next most common problem is probably pointing the microphone away from the mouth while speaking. This is not the slight off-axis deviation recommended for dealing with plosives, but in some cases the microphone is pointing at a source other than the performer. This can lead to bleed, or worse, feedback if there are stage monitors running and the microphone happens to be pointing in their direction.

When feedback does start to occur, some people instinctively try to shield the microphone with a cupped hand, like in Figure 14-6(A). When the hand is held like this just in front of the microphone, it forms a fairly effective reflector, somewhat like a satellite dish. This can sometimes actually focus the feedback-causing signal into the microphone. If a person insists on shielding the microphone, they should go all the way and cover it up. While this may sometimes create a resonant cavity, it is likely to be more effective at reducing feedback.

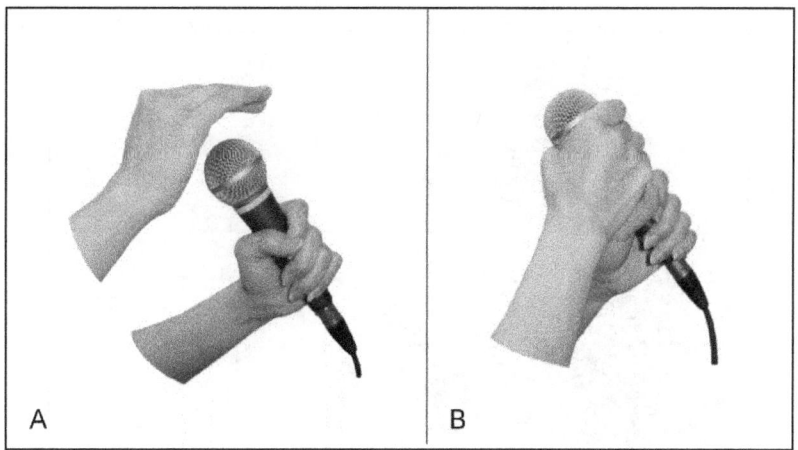

Figure 14-6   Improper Shielding and Holding of Microphone

Some performers try to get creative in holding the microphone. You will recall from Chapter 3 that cardioid microphones get their directional properties because of a set ports designed into the back of the microphone capsule. If a performer covers the area of influence of these ports by holding the microphone as in Figure 14-6(B), the directional characteristics of the microphone will be changed, and the microphone may unintentionally be changed from unidirectional to somewhat omnidirectional. Again, this may manifest itself in unexpected feedback.

While the foregoing discussion focused on issues that may cause grief for the audio engineer, the next problem may also cause grief to the microphone. All of us have seen inexperienced and even some experienced microphone users test the microphone by banging on it, sometimes quite forcefully! This really is bad practice. While a tap on the microphone grill is unlikely to damage a rugged live sound microphone, most microphones can be damaged if hit hard enough. We really want to instill in the performers that the microphone is a sensitive precision instrument and should be treated accordingly.

Communication and education are important factors in ensuring success in proper microphone usage. As engineers, we should instruct inexperienced performers on how to get good sound out of the microphone. If we are using a directional microphone, explain what that means in terms of its pickup pattern, and thus why it is important to point it at their mouth. Explain why you want them to hold it by its body. Then, establish a good working distance from the mouth. After the basics are clear, variations in position, either to achieve various effects or solve some specific issue, can be explored.

14.7 Miking Soloists – Speaking and Singing

Now that we have our approach to miking established, let us look at several miking applications and some guidelines on solving some of the basic issues that arise in these situations.

The handheld vocal microphone is probably the most common usage scenario that we will encounter as an audio engineer. If the microphone can be mounted on a stand, then that is probably the route we should take. Otherwise, the microphone should be properly held by its body, as in Figure 14-7. When handheld, we want to select a microphone that has low handling noise. We should also apply a low-cut filter below 80 Hz on its input channel to help reduce the handling noise as much as possible.

Figure 14-7 Proper Microphone Holding Technique (Source: ©iStockphoto.com/Astakhova)

The distance of the microphone from the mouth is an important consideration. Around 8 to 10 cm is a good place to start for most applications. For good, consistent pickup while speaking, the microphone should be kept at a fairly constant distance from the mouth, and the speaker should direct the speech into the microphone. The microphone may be held at a slight angle to the forward axis of speech to help reduce the effect of plosives. A dynamic microphone will work well in most situations where speech is dominant.

For singing, there are many more variables to contend with. The SPL at the microphone may vary considerably compared to speaking, as the performer may be very expressive. The microphone should be able to handle this wide range of levels, cleanly capturing the quiet passages without distorting the loud ones. The user may also vary the distance of the microphone from their mouth considerable, bringing it very close on soft passages or to take advantage of the proximity effect (assuming a directional microphone), or moving it far away when they are being especially loud. In cases such as above, a compressor may be inserted in the vocal channel to help even out the level variations from the microphone. Good condenser and dynamic microphones both do well in singing applications.

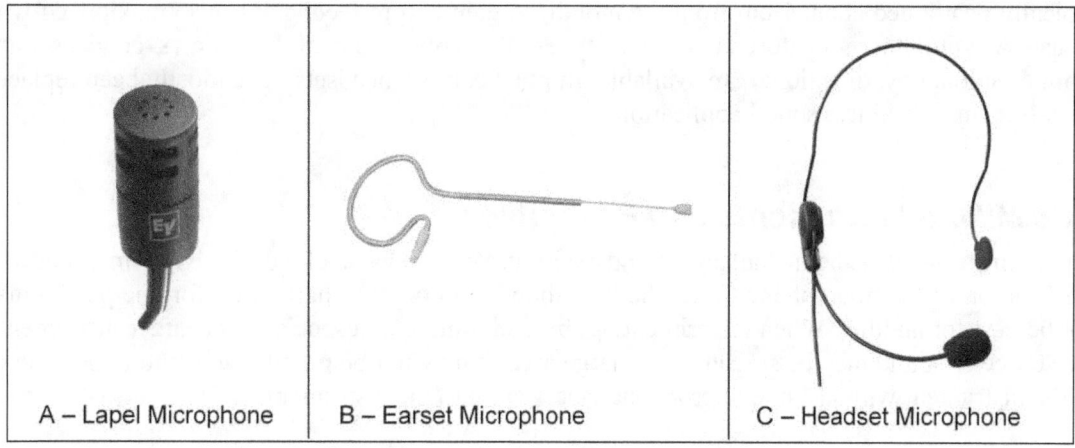

| A – Lapel Microphone | B – Earset Microphone | C – Headset Microphone |

Figure 14-8   Lapel, Earset, and Headset Microphones

When both freedom of movement and freedom of hands are required, miniature microphones in the form of lapel or lavaliere, earset, or headset versions may be used. The lapel microphone, shown in Figure 14-8(A), is often clipped to the lapel of a garment, or the collar, or tie, for example. Because higher frequencies in speech are mostly directed forward (i.e., in the direction of speaking), and these microphones are usually fixed several centimeters below the chin, there is a natural lack of high-frequency energy delivered to the microphone in normal use. Lavaliere microphones may also be buried under clothing or hair, again impairing the delivery of high-frequency sound to its capsule. Manufacturers may compensate for this by tuning the capsule with a high-frequency boost between 4 to 12 kHz. You will have to check the frequency response specification of the microphone to know for sure, and use appropriate equalization if your ears tell you that high-frequency content is somewhat lacking. Additionally, depending on the exact placement of the microphone when attached in the chest area, there may be a boost in the output of the microphone between 700 and 800 Hz due to the resonance created by the chest cavity. Again, you will have to equalize the output of the microphone if your ears suggest that the manufacturer has not compensated for this. As with headset microphones, lapel microphones are giving way to earset microphones for live sound use.

Earset microphones are very small capsules, typically omnidirectional condensers, but also available in directional forms, that sit at one end of a thin flexible boom. The other end of the boom loops around the ear to hold the microphone in place, which is typically around 1 cm from the side of the mouth, away from the direct blast of breath which can cause distracting wind noise. Figure 14-8(B) shows an example of an earset microphone. Manufacturers usually supply a set of small caps that fit over the capsule and which can be used to adapt the frequency response of the microphones in different miking situations. These microphones have a few advantages in use. First, once set, the microphone capsule is fixed at a constant distance from the mouth, and so even as the performer moves around, the output level from the microphone remains constant. Secondly, because the microphone is so close to the mouth, the direct-to-reverberant level is very high, thus helping to maximize gain before feedback. Additionally, because of their very small size (and available skin-tone matched boom color), these microphones are virtually invisible in normal use.

Headset microphones are the predecessors of the earset, but the general concept is the same. As shown in Figure 14-8(C), the capsule is also located at one end of a flexible boom. The other end of the boom is attached to a structure that wraps around the head to secure the microphone in place. Headset microphones are therefore bulkier and heavier than earsets. The microphone capsule is typically positioned about 1 cm from the mouth, so gain before feedback is also maximized with its use. As with other miniature microphone types, all combinations of dynamic vs. condenser and omnidirectional vs. directional are available. In practical use, headsets have mostly been replaced by earsets in typical live sound applications.

14.8 Miking Instruments and Ensembles

Miking instruments is both challenging and exciting. You can have a lot of fun experimenting with the location of the microphone to get the best sound. Of course, what is best for one person may not be best for another. When experimenting, bear in mind that, especially for larger instruments, the source of sound may be spread over a large area. It may not be possible to capture the essence of the instrument with just one microphone that is close to the instrument.

Miniature microphones are great for close miking smaller instruments. They have many advantages, including placement options, excellent isolation, and are almost invisible from a distance. When coupled with a wireless transmitter, they allow the performer almost unlimited freedom of movement.

14.8.1 Acoustic Guitar

Two of the best places to mike an acoustic guitar are over the bridge and over the neck joint (about half way between the sound hole and the twelfth fret). These points yield natural, well-balanced sounds. As a starting point, place the microphones about 30 to 40 cm from the guitar. By moving the microphone towards the sound hole, you can experiment in balancing the warm sounds of the body and bright sounds of the bridge and neck. If you place the microphone directly over the sound hole, you will find that the sound is boomy in comparison to the other locations. The biggest problem with miking the bridge is the interference caused by the performer's hand, and careful positioning and angling of the microphone will be necessary to overcome this.

14.8.2 Grand Piano

Because of its large size (implying a large sound source) and interaction with the room, it can be difficult to close mike a grand piano with a single microphone. However, if you only have a single microphone, aim for a balanced sound by placing the microphone above the middle strings, with the lid fully open. Place the microphone about 30 cm above the strings. The closer you place the microphone to the hammers, the more hammer noise you will get, so bear that in mind if that is an effect you are trying to avoid. Around 20 cm is a good place to start experimenting with the distance from the hammers when trying to balance the attack of the hammers with the mellowness of the strings. The microphone can be moved closer or farther away from the hammers depending on how much of this mechanical noise you want to capture.

With two microphones, a better balance can be achieved by placing one microphone over the bass strings and the other microphone over the treble strings. This will add the brightness of the high end while also adding the depth of the low end. As a starting point for experimentation, use the same distances as for a single microphone, with the lid fully open. Remember to use the 3-to-1 rule when spacing the microphones.

Many other approaches can be used for miking the grand piano, all yielding different results. These include changing the lid position (half open, closed, etc.), aiming at the soundboard beneath the piano, and using surface mount microphones inside the piano. Additionally, we may want to capture some of the ambience of the room by moving the microphones farther from the strings.

14.8.3 Upright Piano

There are just as many ways to mike an upright piano as a grand piano. One way is to place the microphone just at the open top, pointing down into the piano. With only one microphone, experiment with placing the microphone over a specific range of strings to get the desired tonal balance. For more isolation, place the microphone inside the top of the piano. This will also pick up more hammer noise.

As with the grand piano, we can use two microphones, one over the treble and one over the bass strings, for a more natural and balanced sound. Again, remember to use the 3-to-1 rule to minimize phasing problems.

14.8.4 Brass and Woodwinds

The majority of the sound from belled instruments comes from the bell. However, for instruments with finger holes, quite a bit of sound emanates from the holes as well. This is certainly the case with instruments like the saxophone, clarinet, oboe, etc. For a very natural sound, the microphone can be placed in a position where it will pick up sound from both these sources. For example, with the alto and tenor sax, this is just above the bell and aiming at the holes. The closer and more directly the microphone is aimed at the bell, the brighter the sound will be.

For instruments like the trumpet, soprano sax, as well as other straight brass instruments and straight woodwinds, aiming the microphone a few centimeters from the bell gives a bright sound with good isolation. However, whenever we are presented with any instrument in this category with finger holes, we should be thinking of miking these holes as well, as this expands our options for experimentation and creation of more balanced and mellow sounds.

With any instrument in this category, a miniature microphone clipped to the bell will give a very bright, bold, up-front sound with high isolation. Furthermore, if the performer likes to move around, this option affords great freedom of movement.

14.8.5 Flute

The flute is different from other wind instruments mentioned above in that the player blows across a sound hole (called an embouchure) instead of directly into the instrument. This can produce a breathy or windy noise when playing. With the flute, the sound comes from both the embouchure and the open finger holes. Additionally, the flute does not produce a high sound level, so close miking will have to be used to get a good pickup and provide isolation.

If the microphone is placed directly in front of the player, that breath noise we mentioned can become quite dominant in the final sound. This may be an effect you are after, but if not, a large windscreen will have to be used on the microphone to cut down this noise. Large foam windscreens may color the sound in some way that you were not intending, so be on the lookout for this and apply EQ to bring it back to where you want it. Another good technique for overcoming the breath noise is to move the microphone laterally, out of the direct line of the wind. You can then experiment with aiming it at the finger holes and angling back toward the embouchure to balance the combination of sounds from these sources.

Because the sound of the flute is produced so close to the face, an earset microphone worn by the player may be used to capture the sound of the instrument. In this case, an omnidirectional capsule is preferred, as it can capture both the sound from the embouchure as well as the sound holes without discrimination.

14.8.6 Loudspeaker Cabinet

Sometimes, it is necessary to mike a loudspeaker cabinet, e.g., when an electric guitarist is performing and no DI box is available, or the guitarist insists on getting the sound that he gets from his personal amplifier combo. The microphone should be placed quite close to the loudspeaker grille— between 2 and 30 cm. The SPL here can get quite high, so a dynamic microphone is a good choice in this situation. Experiment with the exact location of the microphone to help shape the final sound. As you will expect, when using a cardioid microphone, the sound will become more bassy as the microphone is moved closer to the loudspeaker. If the microphone is moved off the loudspeaker's center, the sound becomes less bright and more mellow.

14.8.7 Pop/Rock Drum Set

The approach taken to miking drums depends on how many microphones are available. We can use as few as one microphone for the entire kit, to as many as one microphone per component. The number of pieces in a typical pop/rock drum set varies considerably too, with 4, 5, 6, 7, and 9 pieces not uncommon. This variety gives us plenty of room for experimentation. The 5-piece set, shown in Figure 14-9 is very popular. The pieces are arranged in a more-or-less semicircular fashion around the drummer, with the bass drum (also called the kick drum) in front and the snare to the left. The high and mid toms (rack toms) are typically mounted on top of the kick drum, while the larger low tom sits on the floor to the right. The hi-hat cymbals are to the left of the snare. The crash cymbal stands between the toms, while the ride cymbal stands between the kick and the snare (for clarity, these cymbals are not shown in Figure 14-9). As you can see, there are a lot of different pieces for a mid-sized set.

When only one microphone is available, place it on a boom about 1 to 1.5 m above the set and aim for a balance of the entire kit. Condenser microphones work well for this because they capture the cymbals very well and so are generally used in the overhead position.

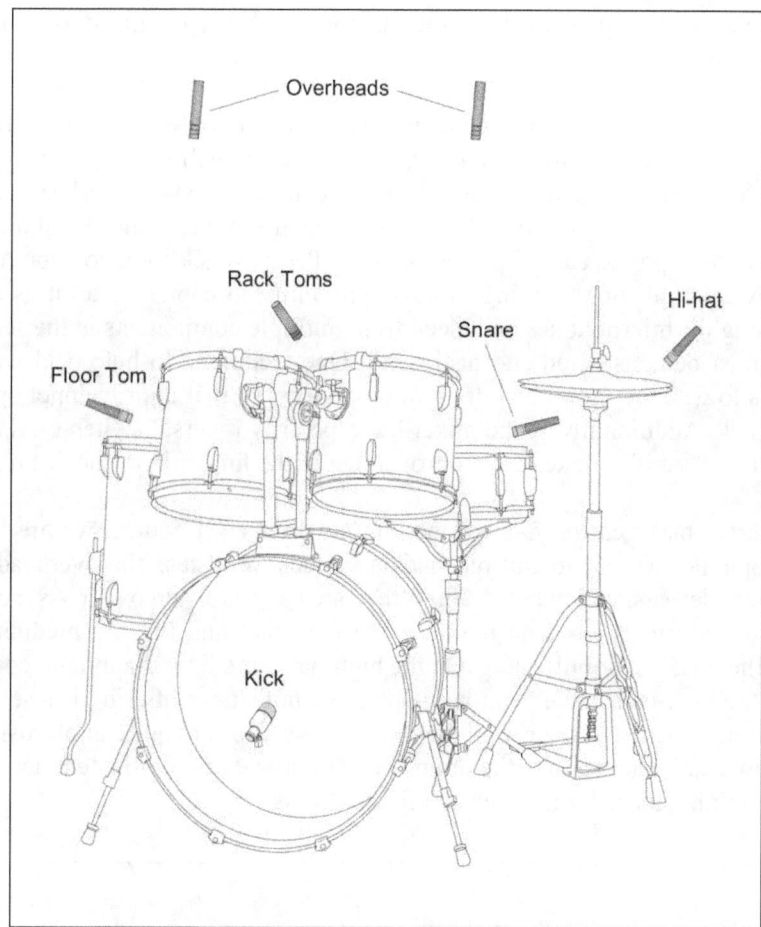

Figure 14-9   Drum Set Miked with Seven Microphones

If we have two similar microphones, both can be placed about 1.5 m above the set, again aiming and mixed for a balance overall sound. If two different style microphones are available, place the condenser about 1 m over the entire kit and the dynamic microphone inside the kick drum (with the front head removed). Experiment with the placement of the microphone inside the kick drum to get the desired tonal quality. It should be just a few centimeters from the skin, and its distance from the beater head strike position can be adjusted to get more or less of the thwack of the strike. Large diaphragm dynamics, such as the Shure Beta 52, are popular for this application because there is high-level low-frequency energy that we are trying to capture. Also experiment with padding the inside of the drum with towels to reduce the boominess and increase the punch of the kick. If the front head is on the bass drum, place the dynamic microphone close to the head, and experiment with moving it from center to edge of the drum to get the desired tone. If we want to emphasize another element of the kit other than the kick, we can experiment with moving the lower microphone closer to that element.

When three microphones are available, we use the techniques as for two microphones, and place the additional microphone on the snare. The snare has that snappy sound that is used for accents in many styles of music. It can sometimes sound too bright and ring quite a bit. If that is the case, it can be dampened with a piece of gaffer tape stuck to the head. Place the additional microphone about 5 cm above the top of the snare, about 5 cm inside the rim, and aiming at the head. Alternatively, if we have two condenser microphones in the set, they can be placed overhead, while the other microphone is placed inside the kick drum.

With four microphones, set up as for three microphones, with two overhead, one on the snare, and one in the kick drum.

With more than four microphones, miking individual components becomes more of an option. The kick, snare, and hi-hat are most important to pick out, so start with those first. The hi-hat microphone can be placed a few centimeters above the rim of the cymbal. The two overheads will capture the rest of the cymbals and toms quite well. The toms can then share any additional microphones. For example, a single microphone can be placed between the two rack toms to capture both of them. With these many microphones working in close proximity to capture the kit as a whole, phase cancellation among the microphones and bleed from multiple components in the separate channels are issues that must be considered and dealt with. One technique to help deal with these issues (mostly bleed) is to gate the individual drum microphones, so that their channel opens only when their drum is struck. Additionally, if the mixer has a polarity reversal switch on the channel strip, we can engage it and see if it makes any improvement to the final mix of the drum set.

Some manufacturers make entire microphone kits for drums. Of course, we are free to mix and match the microphones we use to suit our budgets, needs, or tastes. The overheads are usually a matched pair of condensers, which helps a bit if they are used as a stereo pair. As mentioned earlier, a large diaphragm is usually used on the kick. On the snare and toms, a medium diaphragm is typically used. The SM57 is popular here. On the high-hat, a medium diaphragm condenser may be used to capture the high-end sizzle. Another popular manufacturer offering is a set of microphone clips designed to be attached to the rim of the drum. These clips are quite unobtrusive and hold the microphone a few centimeters above the drum head. Their use can eliminate practically half of the boom stands we might otherwise need when miking drums.

Figure 14-10 Miking a Choir Using Area Miking Technique

14.8.8 Vocal Ensemble

One technique that can be used to mike a choir or other a large ensemble is area miking (stereo techniques can also be used, as discussed later). Condenser microphones work well in this application because they are more sensitive than dynamics and so afford greater reach. The approach is shown in Figure 14-10.

In area miking, the ensemble is logically divided into sections, each of which is covered by a single microphone. The aim is to use as few microphones as possible while following the 3-to-1 rule. In many cases with a choir, each subsequent row stands a little higher that the row in front of it. This helps their voices project farther and makes our job as audio engineers a bit easier. For a choir, the microphones should be placed about 1 m in front of the first row and about 1 m above the highest row. This means that each microphone should cover a section that is around 3 m wide. Don't forget, because of the inverse square law, the front rows will sound much louder than the rear rows at the microphone's position. Aiming each microphone at the mid to rear rows will help balance the levels (assuming the microphones are directional).

14.8.9 Instrument Ensemble

A mid-sized ensemble playing in a well-designed concert hall may not need any reinforcement. However, in many cases, the performance venue will be less than ideal and some reinforcement will be necessary. The area miking technique is also used with instrumental ensembles. However, additional microphones will also likely be used to enhance certain effects or to highlight certain sections or specific instruments. Some ensembles also have singers, in which case vocal microphones will be required.

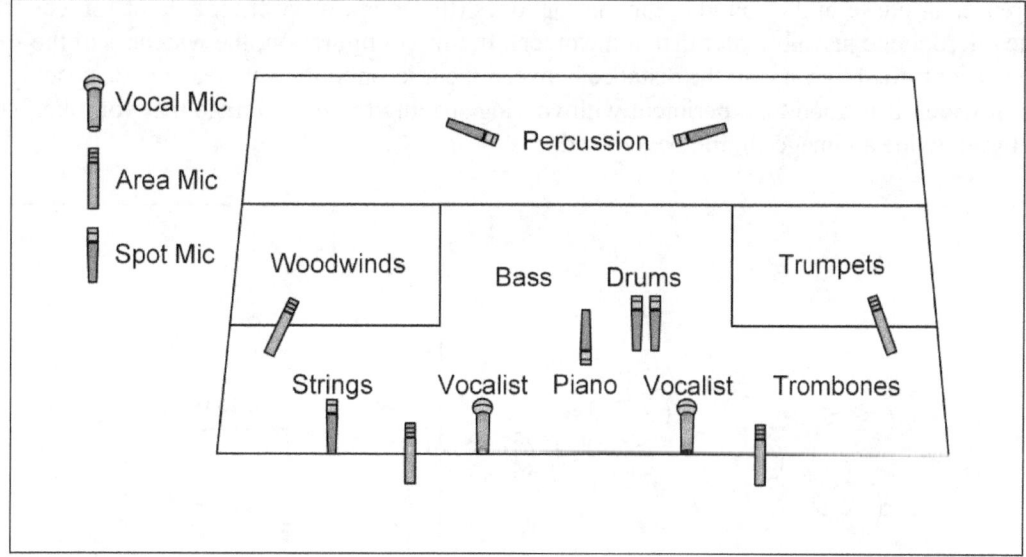

Figure 14-11   Ensemble Miking with Area and Spot Microphones

Figure 14-11 shows an example arrangement of a medium sized concert ensemble. Area miking is used to capture the sense of the band and the reverberation of the venue. We have already discussed miking techniques for most of the individual elements of the band. We can balance weaker elements of the band with louder ones by using spot microphones to capture more of their sound, or highlight a particular element by capturing more of its detail. For example, we may have relatively weak congas or a xylophone in the percussion section, which would not be able to compete with the main drum set in the rhythm section; perhaps we may want to bring out the piano or clarinet in certain sections of a piece. Knowing what the ensemble needs to achieve sonically is critical information that should be used in placing the microphones.

14.9 Stereo Miking Techniques

Live sound reinforcement scenarios seldom use stereo miking techniques to present a stereo image to the audience. Except for a few special or theatrical effects (and many of those are done with panning automation), the audience usually hears the performance in mono. However, stereo techniques are useful when we would otherwise use two microphones to mike a group of performers, or even a large instrument, e.g., a choir or musical ensemble, drum applications with two overhead microphones, etc. All of these applications can make use of one of the stereo techniques presented, and the microphone signals can be appropriately panned to present a slightly wider sonic image to the audience. Furthermore, if you are making a live recording or providing a feed for broadcast, then you will be called upon to provide a number of stereo feeds that can be manipulated separately from the FOH mix. Additionally, as we mentioned in Section 13.8.4, stereo feeds are frequently used to provide in-ear monitor users with a sense of space and presence in an attempt to eliminate the isolation caused by having their ears plugged up with earphones.

14.9.1 AB Stereo

The microphone setup for the AB stereo technique is shown in Figure 14-12. In AB stereo, two omnidirectional microphones are placed parallel to each other, about 18 to 60 cm apart, and pointing towards the audio source. Since the microphones are omnidirectional and fairly close together, the intensity level at each microphone is almost the same, and the stereo image is produced by small differences in phase of the sound at the microphones (the sound wave from a point source arrives at one microphone just a bit later that at the other). In this configuration, the wideness of the stereo image is controlled by varying the distance between the microphones—increasing the spacing will make it wider. It is good to experiment with varying spacing to get an image that you like, but an unnaturally wide an image should be avoided.

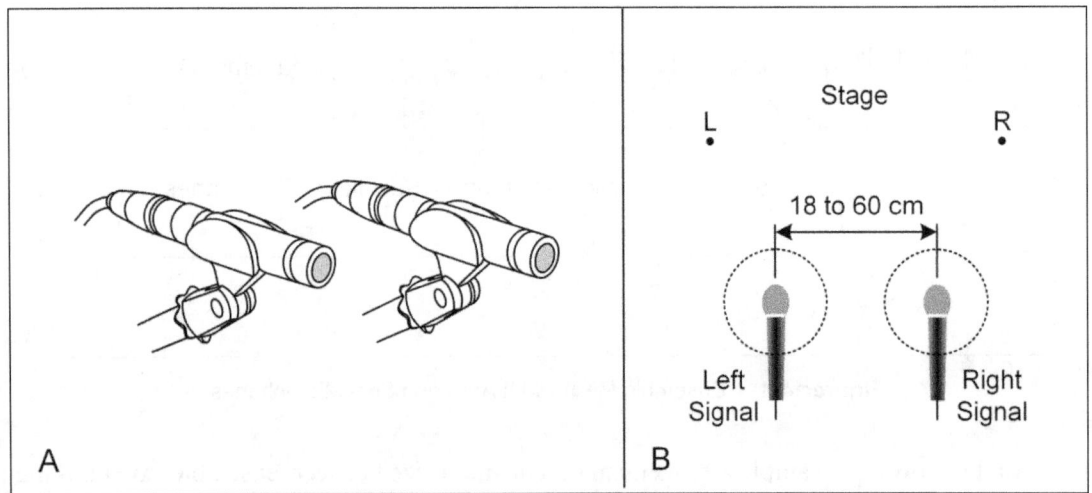

Figure 14-12 AB Stereo Miking Setup

If you are going to be recoding these signals, it is also a good idea to check for mono compatibility. Simply mix the microphone signals and sum them to a mono bus. Then listen to the resulting output and see if any phase cancellations adversely color or destroy the tone of the signal. You may be surprised how hollow or otherwise skewed the combined signals can sound. If any of that happens, just slightly adjust the spacing between the microphones until an acceptable mono mix is obtained.

14.9.2 XY Stereo

The XY stereo technique uses two cardioid microphones angled 90° apart. The idea is to have the microphone capsules occupy the same point in space, so this technique is sometimes called coincident miking. Since actual coincidence is physically impossible with real microphones, the microphones are arranged so that the capsules are stacked one above the other (assuming the microphones are in the horizontal plane), as shown in Figure 14-13. In this arrangement, there is no phase difference (at least in the horizontal plane) in sound waves arriving from a point source, and the stereo image is produced by intensity differences of the sound at each capsule. That is why we use cardioid microphones for this technique—output level is dependent on the direction of the sound source from the microphone.

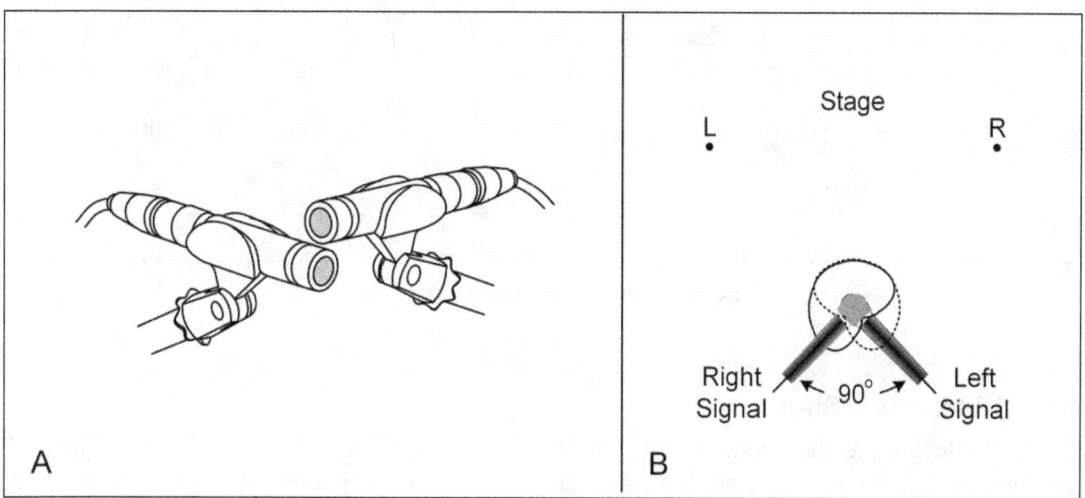

Figure 14-13   XY Stereo Miking Setup

This technique may not provide as wide a stereo image as the AB method, since it depends on the reduction in sensitivity of the microphone at a maximum of 90° off axis. However, this method has good mono compatibility, since there is no difference in phase of the sound wave at the two capsules. There is still the issue of coloration of sound caused by off-axis response of the microphones to deal with, but unlike with the AB technique, that is something that can readily be fixed with EQ (the hollowness of mono sound derived from the AB technique is caused by destructive interference, typically eliminating an entire frequency component from the audio material, which no amount of EQ can restore).

14.9.3 ORTF Stereo and NOS Stereo

ORTF stands for *Office de Radiodiffusion-Télévision Française*. Up until its breakup in 1975, it was the national agency responsible for radio and television broadcasts in France. The setup for this technique is shown in Figure 14-14. In this configuration, two cardioid microphones are placed with their capsules 17 cm apart. The microphones are also angled away from each other by 110°. Because the cardioid microphones are both angled and separated, this technique uses a combination of both phase and intensity differences to produce its stereo image.

This technique produces a wider stereo image than the XY method, but with good phasing characteristics. Thus, its mono compatibility is pretty good, and so this method may be a good compromise between the XY and AB techniques.

A very similar technique to ORTF miking is the NOS technique. NOS stands for *Nederlandse Omroep Stichting*, the Netherlands Broadcasting Foundation. In this technique, the two cardioid microphones are arranged as for OFRT, but the capsules are placed 30 cm apart, while the angle between the microphones is reduced to 90°.

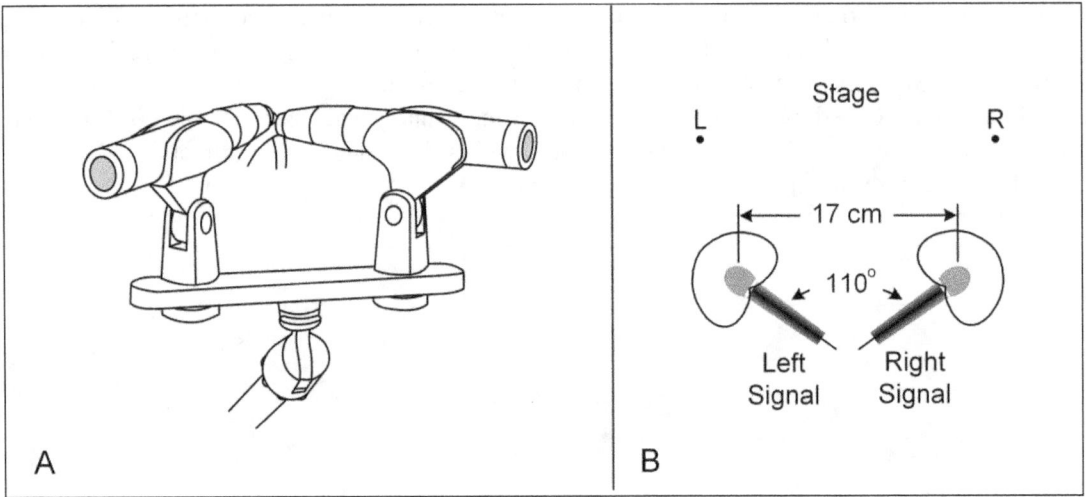

Figure 14-14 ORTF Stereo Miking Setup

14.9.4 Middle-Side Stereo

Like the XY technique, the middle-side (sometimes called mid-side or MS) stereo technique is also considered a coincident technique, in that we try to get the microphone capsules as close to each other as possible. This means that phasing issues between the microphone signals are minimized and, like the XY technique, this method gives good mono compatibility. It is thus one of the methods likely to be used if you are providing a feed for recording or broadcast.

The MS technique uses two different types of microphones—one with a cardioid pickup pattern (actually, an omni microphone would work just as well) and one with a figure-8 pattern. Figure 14-15 shows how the microphones are set up for use. The cardioid microphone is set to point towards the middle of the stage. This is the middle microphone and provides the middle signal. The figure-8 microphone is set up so that its faces are perpendicular to the axis of the cardioid microphone. This microphone provides the side signal.

When we look at the MS technique, it is not initially obvious how we get a stereo signal from this setup, so let us take a closer look at the way it is done. Simonton[3] describes the decoding process by representing the influence of the sound waves on the microphone as scalars R and L, and we follow his analysis here. Recall in Section 3.7.4 we said that the figure-8 pattern is typically obtained from a dual diaphragm capsule wired such that the signals from the two diaphragms are of opposite polarity. That means that a sound wave in its positive pressure region (i.e., the part of the wave that will cause the diaphragm to move inwards) impinging on the front face of the microphone will drive its output signal positive, while the identical scenario acting on the rear face (i.e., pushing the rear diaphragm inwards) will drive the output signal negative. For argument sake, let's say that when we set up our microphones for MS stereo, the front face of the figure-8 microphone is pointing to the right of the stage and the rear face is pointing to the left. We could represent the influence of a wave's positive pressure coming from the right on the microphone's output as $+R$ (since it will drive output voltage positive) and represent the influence of a positive pressure from the left as $-L$ (since it will drive the output voltage negative). Thus, at any instant when the microphone is being influenced by positive pressure from both the right and the left of the stage, we could represent the output signal as $+R-L$. This is depicted in the top left side of Figure 14-16.

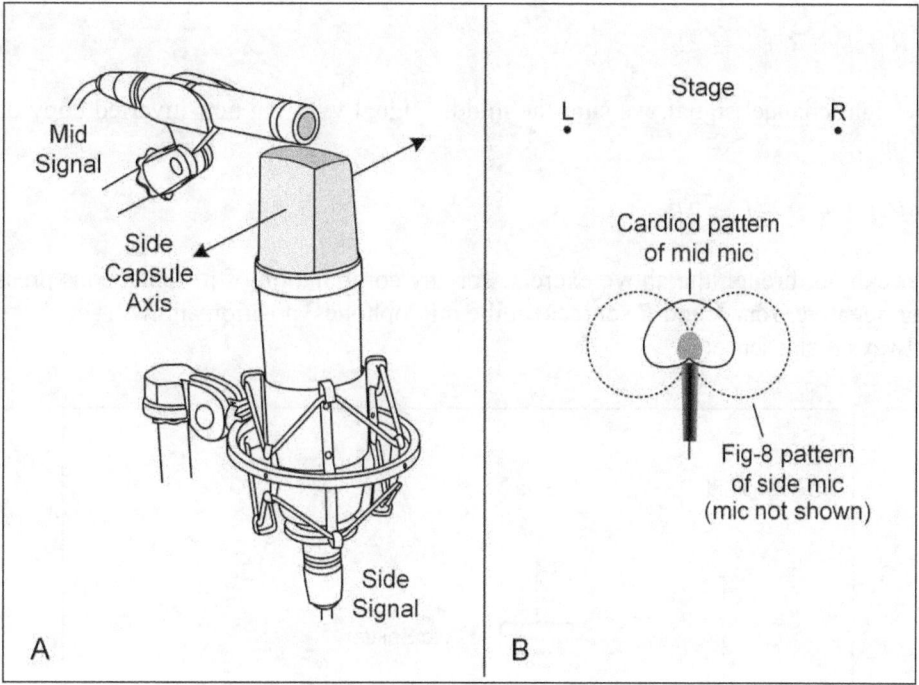

Figure 14-15   Middle-Side Stereo Miking Setup

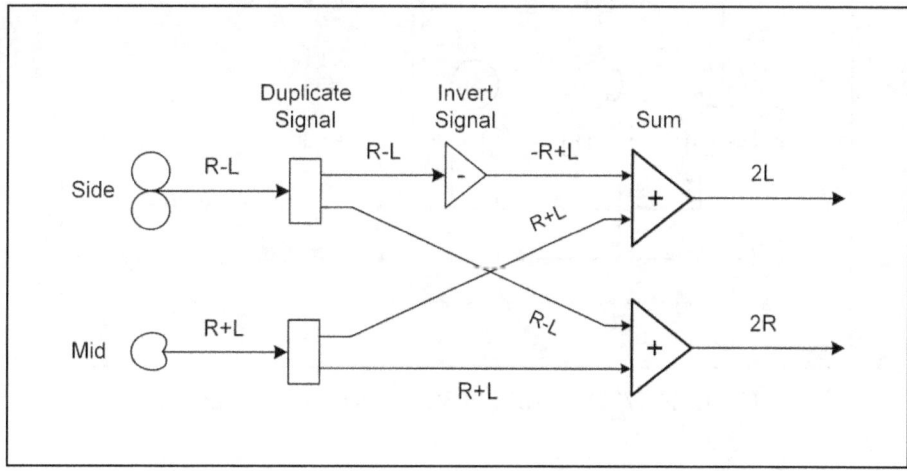

Figure 14-16   Decoding the Middle-Side Stereo Signals (after Simonton[3])

Let us assume now that these waves affect the middle microphone (cardioid) to the same extent as the side (figure-8) microphone. We can see from its orientation that the positive pressure forces from the right and left of the stage will each act to drive the output of the cardioid microphone positive. If we consider the instantaneous output signal for the same point in time as we did for the figure-8 microphone above, we could represent the output of the cardioid microphone as **+R+L**.

Now that we have the separate microphone signals, Figure 14-16 shows us how we manipulate them to create the stereo signal. The first thing we do is duplicate the signals so we have two copies of each. We then take one of the side signals and invert it (the positive parts of the signal become negative, and vice versa). Thus, the signal that was **+R−L** becomes **−R+L**. We then sum this signal with a copy of the middle signal, **+R+L**, to give us the left channel signal as follows:

$$R + L - R + L = 2L$$

To get the right channel signal, we sum the middle signal with the non-inverted copy of the side signal, giving:

$$R + L + R - L = 2R$$

The reader can go through the above exercise for any combination of instantaneous pressure (i.e., positive or negative from **L** and **R** sources) at the microphones' position and will see that the stereo outputs always make sense.

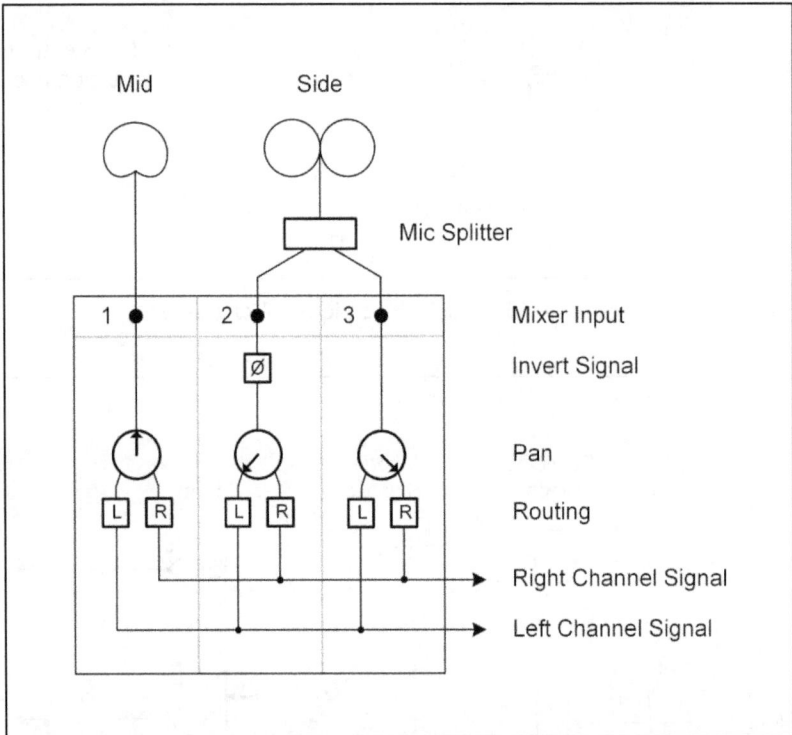

Figure 14-17   Decoding MS Stereo Using Mixer Channels

Sometimes, the signals from the two microphones are fed to a dedicated MS matrix to decode this stereo signal, but any mixer with a polarity invert switch on its channel strip can be used to do the same thing, as shown in Figure 14-17. As before, we duplicate the side signal (use a microphone splitter) so that we have two copies of it and send the three signals (one middle and two sides) to three separate mixer inputs. We invert the polarity on *one* of the side channels. Now simply choose two busses on the mixer (say **L** and **R**) and route the three channels to these busses as follows. Send the middle signal equally to both busses (i.e., pan center). In our configuration, we need to add the inverted side signal to get the left channel, so pan the polarity inverted channel all the way left. That sends it only to the *L* bus. Similarly, pan the non-inverted side channel all the way right so that it goes only to the *R* bus. This now gives us the stereo image that we are after on the *L* and *R* busses.

Note that in the MS stereo technique, unlike other stereo techniques, we do not use the pan pots to adjust the width of the stereo image. The width of the stereo image is adjusted by changing the ratio of the middle signal to the side signal. If the level middle signal is increased, the width of the stereo image will decrease, and conversely, decreasing it will cause the image to get wider.

Review Questions

1. Compare the close, mid, and distant miking techniques and how they affect the quality of audio captured from the microphone.

2. How can reflected sound cause problems with phase cancellation at the microphone? How can this be mitigated?

3. Explain how phasing problems can develop when using multiple microphones to capture sound from a single source.

4. What is the 3-to-1 rule in miking? Why is it effective?

5. What is a plosive, and how is it reduced in microphone usage?

6. What is sibilance?

7. Explain how a compressor can be used with an equalizer to make a de-esser.

8. What happens to its pickup pattern when a user holds a cardioid microphone so as to block the rear port?

9. What are some of the issues to be considered when miking a grand piano?

10. How would you go about miking a pop drum set when two condenser and two dynamic microphones are available?

11. Explain how to effectively mike a choir.

12. Compare and contrast the AB stereo miking technique with the XY miking technique.

13. Explain the setup for the middle-side stereo technique.

14. How can we use a mixer to derive the left and right stereo signals from a middle-side microphone feed?

15. When providing a stereo feed for broadcast or recording, why is it important to check mono compatibility?

Chapter 15

Pre-Performance Tasks

15.1 Preparation – The Key to a Successful Performance

In this chapter, we will continue our look at the practical aspects of executing a live show. From the engineering point of view, clearly the work starts long before a single performer sets foot on the stage, and it should go without saying that there is a huge amount of preparation that goes into making any show a success. The performance of the sound engineering team, even though typically out of sight of the audience when things are going smoothly, is every bit as important as the performance on stage. Get it wrong, and you will find that the sound engineering team quickly becomes the focus of the spotlight. While every eventuality cannot be foreseen and planned for, as you gain experience, you will learn the most common issues that are likely to pop up.

There are a number of areas that require special attention in preparation for a performance. We will examine these areas in some detail below, since careful preparation in these areas is key to a successful show.

15.2 Technical Briefing

The technical briefing is information given to the technical team to help ensure that the show goes smoothly. This is therefore a key element in the preparation for any non-trivial show. In this briefing, the team is identified, and the command chain is clearly established, with people in charge authorized to make appropriate decisions. Everyone should know their roles and responsibilities and understand that they are part of a team that needs to work together to get the job done in the most efficient and effective manner. Certain key members of the team should be aware of emergency backup plans and when and how to activate these plans for their area of responsibility. This briefing is meant to cover information pertinent to the critical elements of the performance, such as the layout of the equipment, the timing of major elements of the performance (e.g., setup, sound check, set changes), interaction with other teams (e.g., stage management, lighting, facilities, security), and so on.

15.3 Equipment Emplacement, Hookup, and Power-up

Many items used in the pro-audio industry are very bulky and heavy. We should therefore try to avoid doing unnecessary work and not move equipment around more than we have to. In other words, know where everything is supposed to be placed at the show site and place it in its correct location the first time. This is facilitated by having a layout plan.

The layout plan does not have to be highly sophisticated. The pertinent information at this point for the road crew is the equipment locations with respect to the stage and technical areas. We discussed the creation of the layout plan in Section 13.4, which detailed the other important information that goes into this plan.

When it comes to interconnecting the various pieces of equipment, the benefits of using a checklist and proper labeling system cannot be overemphasized. The savings in time and reduction in hook-up errors can be substantial. For analog connections, the use of snakes, sub-snakes, and cable bundles where possible will greatly aid in simplifying the interconnection of various components, but only if the cable connectors are accurately labeled. For digital systems, interconnections should be somewhat simplified, since we can usually plug the data cable into any free port on the switch, but we must still be careful not to create unintentional switching loops in the network.

Once all the system components have been connected, they should be inspected by a peer team member, i.e., by someone other than the person who connected them. Based on past experience, we know that a hookup error is more likely to be spotted by a peer than by the one who did the initial connections. Any hookup errors should be corrected and recorded. Hookup errors should be tracked so that trends can be identified and corrective action can be taken to improve performance.

For large systems, a power-up checklist should be used. Such a checklist will be somewhat unique to each system, but all will include important aspects such as any safety precautions to be completed; gain settings for power amplifiers, effects, and mixer channels; channel mutes on mixers; power-on sequence for all equipment; etc. In any size system, proper power-up sequence should be followed—outboard effects, followed by the mixer, and lastly the power amplifiers, one at a time. This will prevent possibly damaging, high-level transients from traveling all the way to the loudspeakers (that loud pop we hear when we turn on the mixer after the amplifier), as well as possibly tripping the electrical supply breaker due to sudden high-level current rush when several amplifiers are simultaneously switched on. After power-up, a quick visual inspection of the system should be performed to make sure that everything still looks ok (no smoking amplifiers, arcing cables, etc.).

15.4 System Testing – The Equipment Check

At this stage, the core elements of the system have been interconnected, inspected, and powered up. The next step is to test every channel that will be used (both actively and standby) in the performance before proceeding to tune the system.

The goal of the equipment check is to verify that the audio sources, i.e., the microphones, DI boxes, keyboards, electric guitars, etc., and their signal paths through the mixer to the loudspeakers are functional, and that the system as a whole is able to deliver the audio at the SPL and quality required for the performance. This testing is typically performed by the technical team, not the performers. With larger or more established bands, their backline technicians will be responsible for placing their instruments on stage. They will help to troubleshoot any connection issues with their equipment and ensure that their signal levels are appropriate to drive the system. During this part of the system test, we are trying to discover and fix any hidden technical problems with the system before the performers come in for the sound check, or at the very least, before the show starts. Most commonly, these will be hookup, routing, noise, and electrical issues.

Start with each channel of the FOH system and test each amplifier separately. Some mixers have a built-in tone generator that can simplify this kind of testing in case the actual on-stage source (e.g., a keyboard) is not yet in place, but an external signal generator is more flexible in troubleshooting the entire audio path from source to speaker. Listen to each amplifier-loudspeaker combination to make sure it sounds good, not distorted, and can deliver the sound level that will be required during the performance. Test each loudspeaker unit one at a time, making sure that all the drivers are playing without distortion. Then listen to all the units together to get a sense of the overall sound, listening for any harshness or other undesirable changes in the response due to the combined sound from the individual units (you may be able to correct these with EQ or other processing, or with a slight adjustment to the alignment of the speaker units). Do the same for the subs—separately and then together. Next, move on to testing each stage monitor channel in the same way. Finally, run the whole system with all amplifier channels driven at a level that will be used in the show. We want to run the amplifiers at a level such that they will be drawing a current load similar to show levels. That way, we can make sure that the electrical supply circuits are adequate and will not trip later during the middle of the performance.

15.5 System Tuning – Ringing Out

Once the signal paths have been shown to be complete, we can move forward with tuning the system. While performing this activity, we will determine the hot spots that are prone to feedback. With wireless microphones, we will try to discover where all the dead spots and dropout zones are as well. Since we know that vocalists are likely to move around while performing, this information will be communicated to them as part of the sound check briefing.

Ringing out is the term given to the activity of removing as much feedback from the sound system as possible. The aim is to make the system as acoustically stable as practical while maximizing gain before feedback. Ringing out can be a strictly manual exercise. However, some devices exist that assist users in this activity by automating parts or all of the process. We will look at this in detail shortly, but first, let us review what we know about feedback.

15.5.1 Feedback

Recall from Section 13.8.1 that feedback occurs when the sound emitted from the loudspeaker enters the microphone, passes through the electronic processors, and finally out the loudspeaker again. If the gain around this loop is less than 1, the feedback signal will die out on its own, otherwise the system will start to ring and the sound level may increase until we have a full-blown howl. Figure 13-19 is a depiction of the path the energy takes as it flows around this loop.

As the volume level of a system is increased, the frequency response at any given microphone location may change. That is because the amount of energy reflected from surfaces in the area increases, causing the results of interaction with the direct sound to change. In a system close to feedback (i.e., some resonance frequency of the system is being excited), large peaks can appear in the frequency response when this interaction reaches a critical level, as shown in Figure 13-21. Any increase in level at this point which pushes the loop gain to 1 or more will cause the system to ring at the frequency of the highest peak.

15.5.2 Equalizing Feedback

We see then, that when feedback occurs, it initially starts at a single frequency. The frequency response of the system at that time will show a spike at the frequency where feedback is occurring. In Section 13.8.2, we discussed how an equalizer can be used to apply a notch filter centered at the problem frequency to eliminate that spike in the frequency response.

While a parametric equalizer gives us control over the key parameters of center frequency and bandwidth, and would conceptually be the perfect tool for this application, the center spacing of a

31-band equalizer is close enough to allow its use here as well. Moreover, since a 31-band equalizer is a bit easier to use than a parametric equalizer, it is more popular in this application. With a bit of practice, it becomes fairly easy to guess the frequency range of the ringing and select the right slider on the first or second try.

There will likely be a number of frequencies at which the system is prone to ringing. Each one will manifest itself at some gain setting. Thus, the process of finding a problematic frequency and correcting it will have to be repeated many times. Our goal is to flatten the frequency response in the region of the ringing. This stabilizes the system and allows us to operate it at higher gain without ringing at these frequencies.

15.5.3 Manual Ringing Out

With that understanding of what we are trying to do with an equalizer when fighting feedback, a manual procedure becomes fairly straightforward—basically find each problem frequency and apply a cut with the EQ at that frequency. Since we need to process all of the audio signal when fighting feedback, the entire signal must go through the equalizer. Equalizers should be inserted into both the FOH and monitor output paths of the mixer, using either the insert jack, or an equivalent break in the audio path, as shown in Figure 15-1.

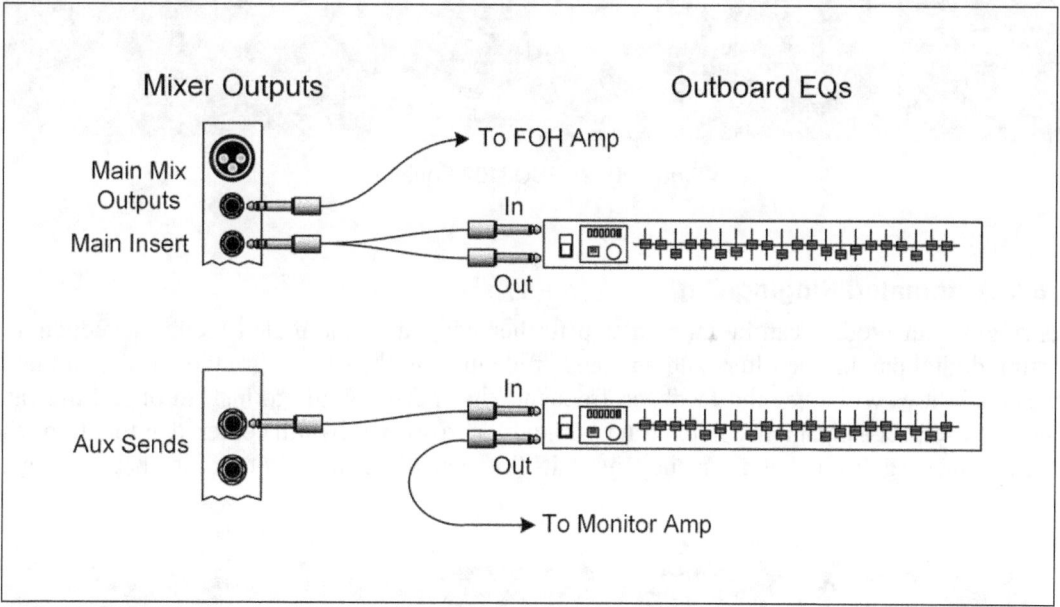

Figure 15-1   Inserting an Equalizer in Audio Path to Control Feedback

The step-by-step manual procedure for ringing out is as follows:

1. Determine the maximum number of microphones that will be in use at any one time. Put them at their intended locations and open their channels.
2. Start with all EQ sliders in the flat position and increase the system gain until ringing starts.
3. Find the EQ slider closest to the ringing frequency and cut the level until ringing stops. Initially this will be a trial and error process, but with experience, will become easy. You may have to quickly boost (and then reduce) a slider to help locate the offending frequency.

4. Increase the system gain again until ringing starts. Then repeat the process of finding and cutting the feedback frequencies on the equalizer. Stop the process when no more feedback can be taken out of the system.

5. Finally, reduce the overall system gain to a level below the last step.

15.5.4 Semi-automated Ringing Out

Some equalizers assist the user in ringing out the system by identifying the frequencies which are ringing. They use the fact that when ringing starts, sharp peaks appear in the frequency response of the system. The equalizer uses its spectrum analyzer circuitry to examine the frequency response for large peaks. In manual equalizers with sliders, identification is usually done by lighting up the sliders. Several sliders on the equalizer may be lit, but the one closest to the feedback-causing frequency will be lit most brightly. The equalizer shown in Figure 15-2, a Behringer FBQ3102, presents its information like that. Once identified, the user simply cuts the gain of the offending frequency until ringing stops. The ringing out procedure with this type of equalizer is then similar to that of the manual process.

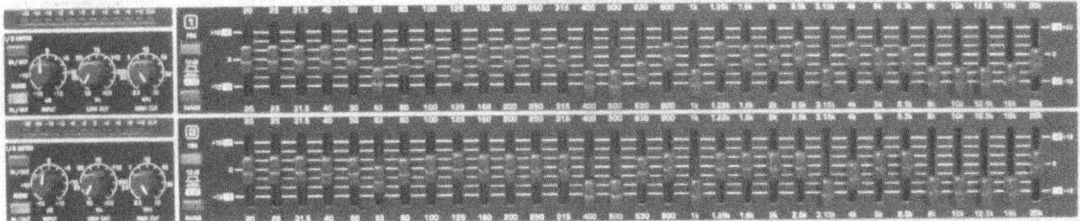

Figure 15-2   FBQ3102 Equalizer

15.5.5 Automated Ringing Out

This ringing out process can be taken a step further with fully automated feedback reduction by utilizing digital parametric filters and a means of identifying the ringing frequency. A unit that has these capabilities will, once the feedback frequency has been identified, instead of just indicating this to the user, set a parametric notch filter with very narrow bandwidth centered at that frequency. It then reduces gain at that frequency until its internal algorithms determine that ringing has stopped.

Figure 15-3   FBQ2496 Feedback Destroyer Pro

Figure 15-3 shows a Behringer FBQ2496 Feedback Destroyer Pro, which has 20 digital parametric filters per channel. It can be used as an automated feedback suppressor when ringing out the sound system.

15.6 The Sound Check

The next large task that we must get through in preparation for a show is the sound check. This is a very important part of the pre-show process where performers and engineers gather and exchange the information that will allow the show to progress smoothly, FOH and monitor levels are set, and initial mixes are crafted for each act.

15.6.1 Who is the Boss?

One of the key things to understand about the sound check is who and what it is for. The sound check is primarily for the audio engineer. The audio engineer must therefore be in charge of the sound check, not the performers or their producer/manager. Sound checks must usually be done in a short period of time. However, some performers will not demonstrate the kind of discipline needed to optimize the use of this time. The engineer must therefore display strong leadership in order to accomplish the goals of the sound check within the available time.

15.6.2 Engineer-Performer Communication

Information exchange between the engineer and the performers is crucial to having a good show. The sound check should provide an opportunity to exchange information that is important to the success of the show. The engineer-performer briefing is where this information is exchanged. Remember, communication is a two-way street!

The engineer should get certain basic information from each performing act. Each group will like to do things differently, so it helps to get a clear picture of their playing positions and any unusual gimmicks performed during the show (like running in front of main house loudspeakers with the live microphone).

A set-list with the songs planned, giving an indication of things like instrument changes in the middle of sets, additional effects requests, and so on, is very useful. This list will be annotated during the sound check by the engineer, with mixer and effects settings. If the band prefers a particular monitor mix, or like to have a peculiar tone to their music, this should be communicated as well. Finally, the sound check is a good place to work out some method of letting the engineer know that the monitor level needs to be raised.

In the other direction, there is certain information the engineer should give to the performers. General guidelines of what not to do on stage during the performance would be useful, e.g., don't unplug equipment without letting the engineer mute the channel; don't turn up the EQ on equipment feeding the mixer. Also, let the performers know where the hot spots for feedback are and, if using wireless systems, where dropout zones are. You will also need to establish how you will communicate with the performers on stage, e.g., hand signals or talkback.

15.6.3 Key Goals of the Sound Check

At this stage, any information that is necessary for a good show should have been exchanged between the engineer and the performers. Furthermore, it must be clearly understood that the sound check is not a rehearsal for the performers. Do not allow the performers to waste time practicing what should have been done long before this time. The performers should be asked to perform only as much of a particular song to give the engineer the information needed to make necessary adjustments to the sound system. For example, performers may be asked to perform the loudest part of a piece, or the softest part. Solos may be requested to check clarity and tone.

This is the time to establish the initial house mix. To set tonality, try initially to adjust the EQ from the mixer channel strips. The effects processors will need to be adjusted for balance and clarity. On the FOH equalizer, only change the EQ settings that were established earlier to control feedback if really necessary. A compromise may have to be made between tone, balance, and level to keep feedback under control. Don't try to get everything perfect at this stage. Remember, when the audience comes in, absorption characteristics of the venue will change and the sound will be different, so the system will have to be readjusted. During the sound check, you need to get the mix good enough to start the show.

What are we aiming for in the house mix? A full, distortion-free sound that is well balanced, where individual instruments can be heard and lead instruments and vocals cut through. Note that 'cut through' does not mean blasting over the levels of all other instruments, but that the level in question should be high enough to be heard above the background instruments. EQ should be well-adjusted to make the mix non-fatiguing. A fatiguing mix will have an excessive amount of upper-mid to high frequencies.

The initial stage monitor mix for the performers is also set up at this point. The performers will need a certain volume and clarity for the mix to be effective. Therefore, the need for volume and clarity will have to be balanced with the need to control feedback. The settings that were established earlier on the stage monitor equalizers to deal with feedback should not to be changed unless necessary.

What are we aiming for in the monitor mix? Just enough that the performers can stay in time and in key with the rest of the band, receive any cues that they need to maintain situational awareness, and not overstress their voices or overblow their instruments—in other words, as little as practical. The less we put in the monitor mix, the better our chances of having a quiet stage environment. Keeping stage levels low will help us get a cleaner house sound, as the microphones will pick up less spill from the stage monitors and on-stage cabs (like bass and guitar amplifiers). Similarly, we stand a better chance of controlling feedback. This is where in-ear monitors can really shine and prove to be an engineer's friend. One approach to setting initial monitor levels is find the weakest source of unamplified sound on stage, usually an instrument like an acoustic guitar or the vocals, and set the other instrument levels around that. Try and get the bass and drums (usually the loudest musicians on stage who create the most spill problems) to play down. This will help you to manage the levels of the weak instruments. Getting the band to play down is not always easy and is probably not likely to happen if it is not used to trying. Well, no one said the monitor engineer's job was easy!

15.6.4 Sound Checking the Band

There are all kinds of ways to run through the sound check. Here is one suggested sequence that you can use as a start. As you gain experience, you will no doubt make up a sequence that works best for you.

First, get an initial house sound. Start with the drum kit. This will take the longest time because there are so many pieces. Within the kit, start with the kick, then move to the snare, then the hi-hats, toms, and cymbals. Get the drummer to play a variety of material, and aim to get the kit balanced within itself. It is a good idea to route the entire kit to a subgroup, so that the level of the entire kit can be controlled via a single fader later on.

Next, move to the bass guitar. EQ as necessary to get the mood the player wants while still controlling the level to achieve a good distortion-free tone.

When the bass is done, move on to the rhythm section (keyboards and rhythm guitar) and then lead guitar. The lead guitar may require more presence to cut through the mix later. Presence is a careful blend of level, EQ, and reverb.

Next, check the vocals, starting with the backing singers. Aim for clarity and good tone. For the lead vocals, clarity is extremely important. There is nothing worse than being at a show and not being able to understand what the singer is saying, so the lead singer will have to cut through the mix. Some things that can adversely affect clarity are too much low-frequency boost and too much reverb.

When you are happy with the house sound, you can move on to setting up the stage monitor levels. Start with the lead vocals, then the backing vocals. Then set the other instrument levels around the vocal levels. Remember that the vocals can be overpowered by loud stage amplifiers and the drums which will spill into the microphones, so their levels must be set appropriately. If band members are having problems hearing, the first line of attack should be to get the band to play softer.

When all the individuals have been checked, get the entire band to play a variety of pieces, i.e., parts of their material that will give a feel for the range of sounds that will have to be handled. Try to balance all of the separate instrument and voices to form a cohesive mix. You will want to hear the most tonally different pieces, the loudest and softest, and so on. Run the sound check at the levels that will be used during the actual show. This is important because feedback may not occur at lower levels, and the tonality of the music will be different at high levels when compared to low levels.

15.6.5 Multi-Act Shows

With multi-act shows, sound-check performers in reverse of the order they play. That way, at the end of the sound check, the system is already set up for the first band that will perform. Not only that, the headline act which carries the show and usually does not open, has more time up front to sound check properly.

One important aspect of the sound check is to keep a record of mixer and effects setting for each new performer in multi-act shows. This is critical to having the show run smoothly and will help you appear to be professional. Use any automation to your advantage. For example, on mixers that can store settings, save each act's setting and recall them when they go on stage. This is where digital mixers really stand out. Do the same for digital effects processors. Just remember to keep track of what settings correspond to which act.

Review Questions

1. Discuss the value of a layout plan in the early stage of preparation for a performance.
2. What is a technical briefing, and why is it important in preparation for a live performance?
3. What is a set-list? How is it useful in preparing for a complex performance?
4. What is the proper power-up sequence of equipment typically found in a live sound reinforcement deployment?
5. What is the purpose of testing the entire sound system after it is powered up? Describe a procedure for such testing.
6. Discuss in detail how an equalizer can be used to ring out the sound system.

7. Compare and contrast the use of various types of equalizers (graphic, parametric, digital) in combating feedback in the sound system.

8. What are the key differences between the line check and the sound check?

9. What are the main goals of the sound check?

10. What kind of information is important to exchange between the engineer and performers during the sound check?

11. Discuss an approach of establishing an initial FOH mix for a show.

12. Discuss one approach to creating the initial monitor mix.

13. Describe how a typical pop/rock band may be sound checked.

14. Discuss the rationale for sound-checking multi-act shows in reverse order of performance.

15. How can a digital mixer make engineering a multi-act show easier?

Chapter 16

Performance Tasks

16.1 State of Mind – Alertness, Focus, and Agility

The audience has come in. The MC gets out his last few syllables. The band is primed and ready. It's time. All the hours of preparation have come down to this moment. For the remainder of the show, you will do your best to make the performers on stage sound their best. If you do a good job, you will remain in the background. That is the way you should aim for it to be. If you mess up, the spotlight will suddenly be on you rather than the performers, and you really do not want to be in the spotlight in those kinds of circumstances.

It should go without saying that during the performance, the house and monitor engineers need to be constantly alert and focused so as to be able to quickly respond to the myriad issues that occur in such a dynamic environment. One of the challenges will be to maintain the right state of mind for the duration of the show. Despite having worked for several hours already in the pre-performance phase of the show, you will now have to find the energy to get through the next few hours of actual performance, remaining alert and focused. Fatigue and distractions can sometimes lead an engineer to be asleep at the switch. One approach which helps to get through a performance is to work as a team. There should be a chief engineer and an associate. While the chief is at the top of the command chain, both engineers are nonetheless always in sync, both understand what is happening on stage, both know the state of the board and effects, both know the flow of the show, and each can take over the mixing task from the other at any time.

16.2 Performance Tasks

During the performance, there are many things that you will have to do while manning your post at the mixer. Again, it really helps to have an assistant to share the load when things start to heat up. Here is a list of important tasks that you must be prepared to do for the duration of the performance.

Generally manage the mix: This will be your main task, and we will spend some time elaborating on this in the next few pages. With your settings recalled from the sound check as a starting point, you will find that you will want to tweak a level here, adjust EQ there, add a little more effect here, etc.

Adjust solo levels: On solo sections, the level and tone of the solo may need to be adjusted so that it cuts through the mix and stands out. Set sheets are a great help with this. They identify the solo, giving the location and instrument in each song.

Adjust monitor levels: A performer may signal that his monitor level is too low or that he wants more of something, usually 'me.'

Mute unused microphones: When certain microphones are no longer used, they should be muted to reduce spill and feedback. Conversely, be alert to when a performer is about to start using a muted microphone—it will need to be brought to life. If you need to constantly manipulate sets of microphones like this, e.g., choirs mixing with bands, or theatrical work with short scenes, then a mixer with mute groups or other channel automation will be very handy for managing these kinds of scenarios.

Handle effects changes: Sometimes, the effects will need to be changed in the middle of a song, as requested by the performer. There are other times when effects will need to be changed too, and we will discuss these later.

Know where things are: Vocalist have a habit of moving microphones around. Keep an eye on where each vocal microphone ends up so that you can relate it to its correct channel on the mixer. Color-coding the microphones can help with this. In the performer briefing during the sound check, it may help to strongly encourage the vocalist to place microphones back in the same stands from which they were taken.

Handle distractions: The engineering area should be secure from audience intrusion. In the simplest case, it may just be taped off. In the worst case (without security personnel to keep the audience at bay), you may have to contend with distractions from the audience members who get too close to the equipment and the mixing area, ask distracting questions, tell you how to do your job, and otherwise take attention away from the task at hand.

Damage control: All kinds of thing can go wrong during a performance—and they will. Mistakes will happen. Sooner or later, one of the musicians on stage will suddenly and dramatically do something that will put you in a bind (e.g., change the level or EQ of his instrument, or stand in front of a house loudspeaker while singing). Missed cues happen all the time, and in the heat of the show, communications can get garbled. You need to be alert enough to catch the problem early, before it becomes a real disaster. Whatever the course of recovery, do it gracefully; make your changes on the board smoothly enough so that the audience barely notices the mistake.

16.3 Mixing

The largest single task of the audio engineer during the performance is mixing. This consists of combining all the separate sounds produced by the various performers in a well-balanced way, polishing that combination with the use of various effects, and delivering the final product to the audience, or in the case of the monitor mix, to the performers themselves. Now, if you ask ten experienced engineers how to create a good mix, you will get ten different answers, which suggests that there is no right or wrong way to go about producing a mix. What we will do here is give some broad guidelines to help ensure that the fundamentals are in place and some suggestions on how to refine the mix after that. Once we get past the fundamentals, who can really say what is right and what is wrong? Make sure that the fundamentals are firmly in place and then build on that to develop your own style. Who knows what unique trademark sound you may develop as you work to perfect your mixes!

16.3.1 Lay the Fundamentals First

It is important to establish the proper understanding of the foundation of a good mix. We will build this foundation on vocal clarity, the body of the time keepers (drum and bass), and the cohesiveness of the overall mix.

Firstly, the audience will not be happy if they have to strain to hear what is being said or sung. That means that the vocals need to be clear, intelligible, and not crowded or masked by other instruments. Particularly, the vocal level must be high enough that it can be heard above the instruments. If we cannot hear the vocal to begin with, we won't have anything to work with. To help the vocals cut through the mix, we may have to use EQ to adjust the spectral balance of various elements competing with voice in the range that is critical for intelligibility. You will recall from Section 13.2.1 that we identified that range to be about 2 kHz to 9 kHz, where the consonant sounds convey intelligibility information. Figure 16-1(A) shows the situation where the vocals are being drowned by background elements in the mix. By slightly boosting the EQ of the vocals in the critical region and slightly cutting the competing elements as shown in Figure 16-1(B), we can give the vocals some breathing room so that they can cut through the mix and present an intelligible source to the audience.

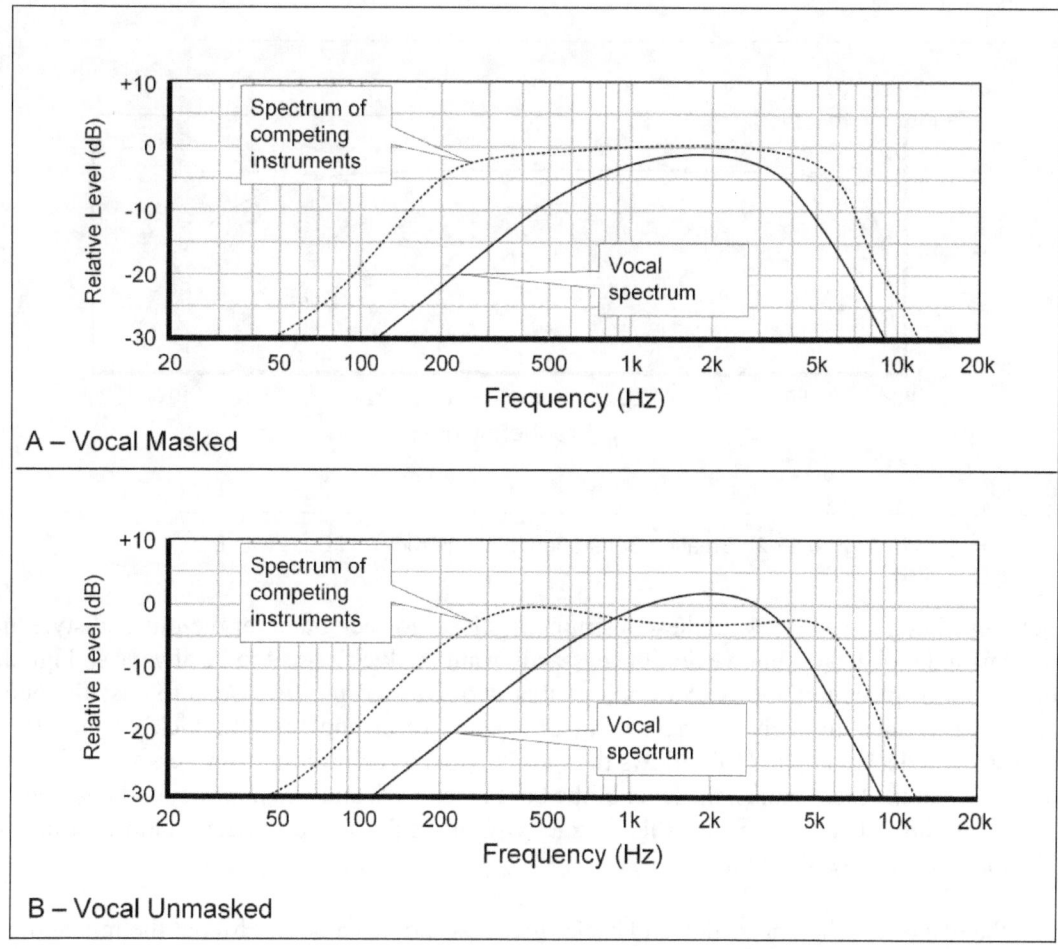

Figure 16-1   Unmasking the Vocals in the Mix

We also need to be alert to the acoustical properties of the venue, especially its reverberation time and how that impacts vocal clarity. Recall again from Section 13.2.1 that long reverberation times in rooms can muddle the consonant sounds and make understanding difficult. Therefore, we must be careful to adjust the amount of effects processing that we apply to the vocal, e.g., we do not want to make a bad situation worse by adding reverb to a vocal channel when the venue already has a long reverberation time.

Next, the kick drum and the bass guitar, which set the pace and keep the timing of the piece, need to complement each other in the mix. The kick drum delivers lots of energy in its fundamental range between 40 Hz to 100 Hz. That is where we get that deep thump sound. The kick also has lots of high-frequency energy. The thwack sound that comes from the sharp attack of the beater head striking the drum has energy well up into the 10 kHz range. Meanwhile, the bass guitar's fundamental frequency range is about 31 Hz (low B) to 392 Hz (highest fretted G), with harmonics well above 5 kHz and even higher, depending on the playing style. However, it is in the range from 300 Hz to 2 kHz that the character of the bass is really established and most easily manipulated. That is good news for us because we can make judicious use of EQ to emphasize slightly different regions of the spectrum and create a slight separation between these instruments, as shown in Figure 16-2. We do not want to cut the higher frequency content of either of these instruments, otherwise they will sound dull and unnatural. We then balance the two in the mix so that they coexist without overwhelming each other.

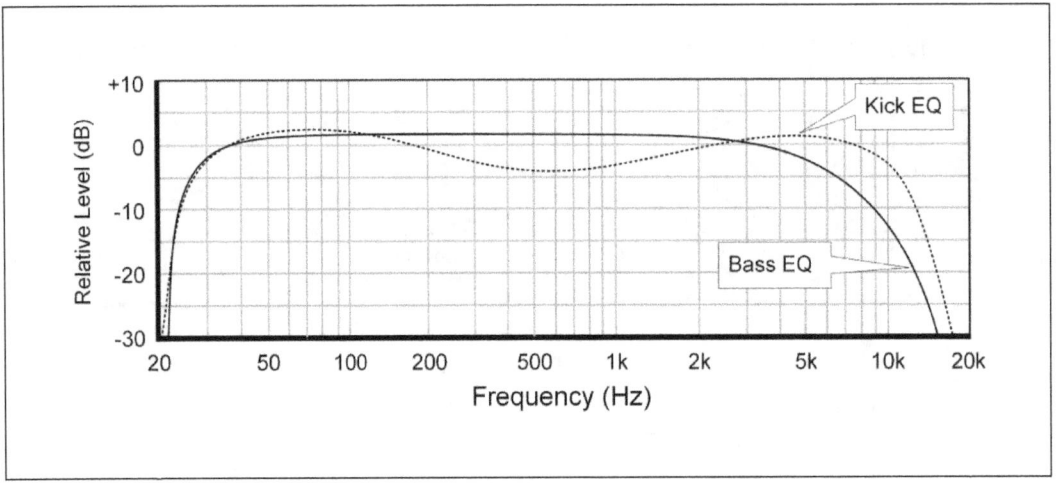

Figure 16-2 Creating Separation between Kick and Bass with EQ

Sometimes, though, we may want these instruments closer to each other, or the musical style may be punchy and kick heavy, but we don't necessarily want the kick's level to be always so high that it sounds out of place in the mix. A technique that is often used in this case is to pass the output from the bass guitar through the main input of a compressor and feed the kick's signal to the compressor's sidechain input. This allows the kick's signal to control the compressor's VCA gain. That way, when the kick sounds, the level of the bass is reduced, and when they sound together, the kick will cut through and dominate. This is exactly what we discussed in Section 8.6.7, which has the details of how to set this up.

Lastly, the rest of the rhythm section and backing vocals should smoothly fill out the mix and give it body. If we did a good job during the sound check, we should already have workable levels for the drum set as a whole, keyboards, rhythm guitar, backing vocals, etc. Now we concentrate on getting that balance just right. Think of it as if you are painting a picture, with the lead vocals, bass, and kick as the key elements of a theme and the other instruments as the background, supporting the theme. The point here is that, for the most part, they are subordinate, building harmony and filling the sonic holes in the picture. Therefore, their levels and tone, except when performing as a solo instrument, need to be supportive of that role. As with the bass and kick, we may use equalization to 'move' the instruments around to occupy different regions of this sonic landscape, and help us achieve the basic overall balance of the mix.

16.4 Refining the Mix

Once the fundamentals are in place, we can move on to polishing and refining the mix—the artistic side of mixing. There are three general areas that are manipulated to achieve this, *viz.* the EQ, the effects, and the dynamic range of the mix over time. Let's look at these in more detail.

16.4.1 Equalization and Overall Tone of the Mix

Equalization plays a huge role in the success of the mix. It affects several areas, from overall impression, to separation and spaciousness, to clarity and fatigue. The use of EQ to control the overall tone and feeling of the mix is different from all of the uses of EQ we have looked at up to now. Here we are talking about EQing the final product, as with an equalizer inserted between the mixer and the power amplifier. We have discussed the use of an equalizer in this position before to combat feedback. Now, though, equalization is used not so much for problem rectification as for style, to set the mood, and for personal taste. The key to success in using the equalizer in this way is to control the amount of energy distributed across the audio spectrum and to understand how that energy distribution affects the character of the final mix.

Between 60 Hz and 300 Hz, too much energy will give the mix a boomy character, while too little will leave the mix sounding thin. In the middle frequencies, up to 7 kHz or so, too much energy will impart a harsh feeling to the mix and lead to listening fatigue, while too little energy will give the impression that mix lacks body or fullness. On the high end, too much energy will give the mix an edgy sound and quickly tire the ears, while too little energy will have your mix sounding dull and uninteresting. So, if you are a beginning sound engineer, it will be useful to get feedback from others regarding the overall tone and feel of your mix—whether it is thin, dull, edgy, bright, full, etc. Use this information to fine tune and perfect your style.

16.4.2 Equalization and Spectral Mixing

When we were discussing the fundamentals of mixing, we used EQ to pull the lead vocals out of the background clutter and to separate the kick from the bass, giving each their own breathing room. This is an example of spectral mixing. However, this technique is not at all limited to vocals, kick, and bass. You can make the mix more interesting and at the same time increase the clarity and spaciousness of the mix by using EQ to put competing instruments in their own space. To do this, just use EQ to boost (or cut as appropriate) each instrument in a different region of the spectrum. That way, the instruments will be emphasized in different frequency ranges and achieve a measure of separation from each other. This will help to untangle the mix and add a bit more clarity and spaciousness.

16.4.3 Equalization and Vocal Clarity

Equalization plays a strong role in the clarity of vocals and speech intelligibility. In Section 13.2.1, we mentioned that the frequency range that is especially important in the articulation of consonants is roughly 2 kHz to 9 kHz. On vocal channels, too much cut in this range will cause serious intelligibility problems as the vocal disappears in the mix. Conversely, on other channels or the final mix, too much boost will also lead to a reduction of clarity and increase listening fatigue. Furthermore, we know that low-frequency sound is very effective at masking speech, and with directional microphones, the possibility of low-frequency boost due to the proximity effect is always present. Therefore, for vocal channels, everything below 80 Hz should be cut to minimize masking from low-frequency noise.

16.4.4 Effects Usage

The way you use effects will also have a large impact on your mix. Poor or wrong use of effects can ruin an otherwise good mix. The key is to know what you are trying to achieve and how each effect will help you achieve your goal. Of course, to attain that level of confident expertise requires a good deal of practice. Experiment with all your effects to learn how they affect the sound of the source material before using them in a live show. Learn what instruments each effect sounds good or bad on, and develop your own rules of where and when it makes sense to apply an effect. For example, you would not want to add reverb and chorus on a bass guitar, but it can make a singer sound smooth. Similarly, you may not want to add a lot of delay to drums, but a similar amount added to the backing vocals can make them sound fuller.

Effects settings are not static for the duration of the performance. Sometimes, as the dynamics of the songs change, you may want to adjust the effects to match the mood (e.g., add a little more reverb on a ballad, but reduce it on the punchy rock number; delay/echo will need to be changed when the tempo changes, etc.). Additionally, effects like reverb should be reduced or removed for speech (e.g., announcements, monologue, etc.), but put back when the vocalist starts to sing.

16.4.5 Dynamic Range

In Chapter 8, where we discussed dynamic range, the reference was directed toward program material that lasted relatively short periods of time, i.e., individual songs. In the current context, though, dynamic range refers to the entire show, not just a single song. If the level of the performance is more or less constant for the entire show, it soon becomes monotonous and much less interesting than if there were some life and movement in the level of the presented material. Of course, achieving this does not rest solely on the sound engineer—the producers and performers have a big role to play in the realization of this in their choice of material and delivery on stage. Besides making sure that solos stand out and vocals cut through the mix, the levels of other instruments and entire sections of the band may need to be adjusted to set the right tone and mood of each piece. For example, you would not expect the backline to have the same balance in a ballad as in a pumping rock anthem. You would not expect the overall level of the songs to be the same either. You have to manage both the balance of the sections and the overall level during the performance to keep it interesting.

16.5 The Monitor Mix

The actual job of providing effective and clean monitor mixes to several musicians is quite difficult—at least as difficult, if not more so, as providing a good FOH mix. It is a combination of art, science, and communication that takes a lot of practice to get right. The audio distribution to the various performers is highly selective. For example, the drummer and the bass guitarist need to hear each other because they keep the timing for the performance and must be in sync. However, the bass guitarist typically has his own bass cab on stage and may be controlling its output level independent of any mix engineer. Between the two of them they create the highest sound levels on stage and can easily drown out all other musicians. Lead vocalists, meanwhile, rely on the keyboards or guitar to help them stay in key, but must also hear some drums (kick/high-hats) to stay in time, and must hear all other cues (e.g., when the lead guitarist finishes his solo) to know where they are in the performance. Meanwhile, some keyboardists sub-mix their own monitors on stage, but they also need to hear the singers and the bass. It is a seemingly never-ending struggle, with various musicians asking to hear more of this and less of that feed. This is one area where communication and the right combination of mixer and monitors will keep everyone involved sane.

16.6 Taking Notes

Does anyone really have time to take notes in the heat of the performance? Not at all likely! However, once a piece has settled down, it is good to take thirty seconds to jot down information that can later be used to enhance the learning experience associated with the show. Information such as specific equipment used and their settings when something sounds really good, or complaints about something sounding bad when you thought it was good, is especially useful for debrief. This information can reinforce the learning experience and give you a head start for your next show as you recall experiences from one show to the next. Over time, you will build up a wealth of material that you can draw on if you ever need to make a presentation, train other engineers, or just review periodically as a personal resource to help reinforce your own experience.

Review Questions

1. Discuss, at a high level, the tasks that the FOH engineer must perform during the performance phase of a show.
2. Why is it important that unused microphones be muted during the performance?
3. How can a mixer with mute groups be used to help manage the use of several sets of microphones?
4. What is the frequency range that is critical to speech intelligibility?
5. Explain how we can use EQ to unmask the vocal in a mix and help it 'cut through.'
6. How can we use a compressor to separate the kick from the bass in a mix?
7. Why is it important that the bass and drum players be able to hear each other in the mix?
8. How does the amount of energy in the midrange (1 kHz to 7 kHz) affect the mix?
9. What is spectral mixing?
10. Discuss the use and management of effects during the course of the performance.
11. What does it mean to manage the dynamic range of a show?
12. Discuss the difference between using EQ to combat feedback vis-à-vis setting the mood of a performance.
13. Explain how inappropriate use of EQ and effects can impact speech intelligibility.
14. Explain the key components to laying the foundation of a good mix.
15. Discuss how a digital mixer with scene recall can be used to help manage the flow of performances in a show.

Chapter 17

Post-Performance Tasks

17.1 We're Almost There

We made it through the show. Fantastic! We can now take a few moments to breathe a little easier and congratulate ourselves on a job well done. However, while the end of the road may be in sight, it is still not time to relax just yet. There is still a lot of work to be done, as now we move on to the post-performance tasks. Once again, clear roles and teamwork will help us get through this phase in the most efficient manner. There are three broad areas that we will look at below: power down, tear down, and technical debriefs.

17.2 Power Down

The performers have all left the stage and it is time to power down the equipment. Again, if there is a lot of distributed equipment, it is good practice to have a short checklist. While shutdown is a simpler task than power up, it must still be approached with some discipline. Shutdown order is basically the reverse of power up order. Start with the power amplifiers and work backwards through the audio chain to the effects processors. This will prevent high-level spikes from propagating through the system and damaging the loudspeakers. It is also a good practice to zero out the mixer (i.e., EQ and Pan to 0; Gain, Trim, Sends, and Faders to $-\infty$; Channels muted with no bus assignments) so that next time you use the board, you start from the same default state. Digital mixers are a blessing for this, as this would essentially be a simple recall operation on that type of mixer.

17.3 System Teardown

Equipment teardown is not simply a matter of unplugging everything and throwing them into boxes. Teardown should be carefully planned so that it occurs in the most efficient manner. For example, the cable tech may disconnect the microphone end of all wired microphone cables in one pass, allowing the microphone tech to gather up all microphones in parallel with the roll-up of mic cables, or the batteries in all wireless microphones may be removed as a single-pass checklist item, eliminating the *did-we-do-that* uncertainty of this teardown task. Clearly, you will have to come up with a strategy that makes sense for your size of team and deployed equipment.

All deployed equipment should be clearly marked for easy identification and checked off when packaged. In multi-provider productions, you want to make sure that you take only your equipment. Just as important, you want to make sure that no other provider mistakenly takes your equipment and that none of your equipment gets left behind.

Once again, each team member should know his primary responsibility, and the group should work as a coordinated team. At the end of the show, many people tend to be less careful than during setup because they are already tired from the show activities and just want to get their tasks over and done with. Therefore, we may need to force ourselves to exercise extra diligence during teardown.

17.4 Equipment Security

It is important to secure the equipment areas before teardown begins. Things happen fast during teardown, and unauthorized persons milling about the stage and technical areas can lead to distractions and increase the risk of injury or loss of equipment.

Besides reducing the risk of injury, security helps to keep small, expensive items like microphones and body packs from disappearing during teardown. Only team members should be handling the teardown, with clear responsibilities for particular items. Coupled with a system of equipment identification and inventory, this should help reduce the loss of small pieces of loose equipment.

17.5 Post-performance Debriefings

A debriefing after a show is an important part of the learning process. This is where we take the opportunity to reinforce good behaviors and come up with strategies of how to correct bad behaviors. The whole aim is to better the team's performance, even if the team consists of only yourself. There are two reviews that are particularly valuable in this regard. These are the review with the show's producers and performers, and the review with the technical team.

17.5.1 Performer Technical Debriefing

It is not likely that the technical team leader will usually get to have a post-performance review with any performers. More common is a debriefing with the producers. However, if the opportunity is available to meet with some of the performers, it should be taken. Note that this is a technical review, not a musical or artistic review. The aim is not to tell anyone how to run a production, how to sing, or play their guitar. This is shaky ground and you must be careful not to alienate a source of information that could help you improve your own performance. The intention is to have the bulk of the information flow from the performer to you, the sound engineer.

The first goal of the review is to get feedback from the performers or producers on what they feel the technical team could have done better to improve both their own performance and the performance of the artistes. This may include comments on areas like preparedness, briefings (pre-performance), communication methods, monitor levels, information flow, effects handling, mixing effectiveness, and sound quality.

Next, if the performers are inexperienced or if they ask for feedback on their performance, highlight examples of good *technical* execution by the performers. Encourage this behavior. Finally, point out areas where their *technical* performance was weak and offer tips for improvement. This will most likely involve things like microphone usage, interfering with on stage equipment settings, etc.

17.5.2 Team Technical Debriefing

After the show, and after the debriefing with the show's producers and performers (if you had one), it is important for the technical team to meet and review their own performance. The goal of this review is not to single out anyone for ridicule, but to identify areas of strength and weakness of the team so that future performances can be better executed from a technical perspective.

The review should be held shortly after the show, when it is still fresh on everyone's mind. Notes taken from the pre-performance and performance phases of the show should be reviewed. It is especially important to identify trends. Good performance and trends should be highlighted and encouraged. For bad trends, causes should be identified, and a consensus should be reached on ways to improve or correct them.

Review Questions

1. How can having a set of clearly defined roles for team members help in efficiently managing the post-performance phase of a show?

2. What happens if an outboard effects processor or the mixer is powered off when the power amplifier is still powered on?

3. What is the recommended power-down sequence for live sound system?

4. Why is it good practice to zero-out the mixer at the end of a show?

5. How does using a digital mixer help when it comes to zeroing-out the mixer?

6. Why is it important that the equipment teardown be planned rather than just a rushed, haphazard exercise to get everything packed up as fast as possible?

7. What are some of the things that can be done to improve safety and security during the post-performance equipment teardown?

8. What is the role of equipment identification and inventory systems in the teardown phase?

9. What are the objectives of a performer debriefing?

10. What are the objectives of the team technical debriefing?

Appendix: Mathematics Review

In this section we will review the basic mathematical notation and concepts that are encountered throughout the book and which are not explained in detail where encountered. Hopefully, this will allow those who are a bit rusty or unfamiliar with some of these concepts to make full use of the material in these pages.

Abstraction

Sometimes when we speak generally about numbers, we represent them using letters. For example, we may use the letter *n* to represent the number 10, or any number in the range 1 to 1000, or any number imaginable.

Zero Reference

A number gets its value by reference to where zero is. For example, zero on the Celsius temperature scale is chosen to be the freezing point of water. The zero reference point, however, could have been chosen to be the temperature associated with any physical phenomenon. Once chosen, temperatures (numbers) bigger than the zero reference are positive. Temperatures (numbers) below the zero reference are negative.

Sign of a Number

A number may be bigger than zero, in which case it is called a positive number, or may be less than zero, in which case it is called a negative number. A positive number is shown either with a + sign in front of the number, or with no sign. Thus, 10 and +10 are both the same, *viz.* positive 10.

A negative number is shown with a minus sign in front of the number, such as -10, (minus 10).

Note the following:

- When multiplying or dividing two positive numbers, the result is always positive.
- When multiplying or dividing a positive number by a negative number, the result is always negative.
- When multiplying or dividing two negative numbers, the result is always positive.

Large vs. Small – the Relative Value of a Number

The notion of largeness is relative, i.e., a number that is considered large in one domain may not be considered large in another domain. A number is small if it is 'close' to zero. A number is large if it is 'far' from the zero reference. Thus, for example, 0.01 is a small number, while 10000 and -100120 are both large numbers. Note that a number can be negative and still be considered large. The notion of 'largeness' has nothing to do with the sign of the number, only how near it is to zero.

Lists and Ranges

When a set of numbers are somehow associated, we may represent them as being in a list. Say for example that the numbers 10, 23, and 16 were somehow associated with each other. We may show them in a list like:

$$(10, 23, 16)$$

If we were using letters to represent the numbers in the list, it may look like this:

$$(a, b, c)$$

Here a would represent 10, b would represent 23, and c 16. We could use the same letter to represent the numbers in the list and differentiate the list members by using a subscript, as follows:

$$(a_1, a_2, a_3)$$

In this case, a_1 would represent 10, a_2 would represent 23, and a_3 would be 16.

Sometimes, we may want to describe a large list of numbers, but it may be inconvenient to write out the list in long form. In this case, we can use the shorthand notation:

$$a_i, \ i = 1 \ to \ 10$$

This is the same as writing out the list in its entirety:

$$(a_1, a_2, a_3, a_4, a_5, a_6, a_7, a_8, a_9, a_{10})$$

As can be seen, the shorthand notation saves us a lot of writing.

Powers

A number can be multiplied by itself a number of times, and we use a shorthand notation for that as well. For example, if we multiply a number by itself 2 times, we have:

$$a \times a = a^2$$

This is called 'a squared.' For example:

$$10 \times 10 = 10^2 = 100$$

Similarly:

$$a \times a \times a = a^3$$

is called a cubed. Generalizing further, a^n is called 'a to the power n' and is equivalent to:

$$a \times a \times \cdots \times a, \quad a \ multiplied \ by \ itself \ n \ times$$

When written like this, a is called the *base*, and n is called the *exponent* or the *index*.

Logarithms

Logarithms allow us to rescale a range of numbers, where the extent of the range may be very large, to make it easier for us to work with those numbers. In the end, we end up with a new range of numbers whose extent is smaller than the original set, but still perfectly represent the original. Suppose:

$$a^n = y$$

Recall, *a* is called the base and *n* is the exponent. We could write this out explicitly:

$$base^{exponent} = y$$

Given that the above equation holds, then we can define the logarithm as follows:

$$\log_a(y) = n$$

which says that the logarithm to base *a* of *y* is *n*. Using *base* and *exponent* notation, the definition is:

$$\log_{base}(y) = exponent$$

The case where the base is equal to 10 is the *common logarithm* that we use in audio engineering.

Now, if we know the logarithm of a number, say:

$$\log_{10}(y) = 2$$

where we can easily see that the base is 10 and the exponent is 2, then we can find the original number, *y*, by referring back to the definition:

$$base^{exponent} = y$$

i.e.

$$10^2 = y = 100$$

This is sometimes referred to as finding the antilog of the number. Logarithms were given full treatment in Section 1.8.2.

Roots

The *square root* of a number *n* is written as $\sqrt{n}$. It is equal to the number that you would have to multiply by itself (2 times) to get *n*. That sounds confusing, so consider this. We know that:

$$n \times n = n^2$$

Therefore:

$$\sqrt{n^2} = n$$

Or, as another example, if:

$$\sqrt{x} = a$$

then:

$$a \times a = x$$

Summation

Summation is nothing more than adding up all the members of a list. Therefore, if our list was:

$$(a, b, c, d, e)$$

then summation would give us the result:

$$a + b + c + d + e$$

We can represent the above summation by the following notation:

$$\sum (a, b, c, d, e)$$

When we use shorthand to generalize the list, the notation looks like this:

$$\sum_{i=1}^{n} (a_i)$$

which is simply equivalent to:

$$a_1 + a_2 + a_3 + \cdots + a_n$$

Arithmetic Mean

The *arithmetic mean*, also known as *arithmetic average*, or just *average*, is a number that describes the 'more or less' middle of a set of numbers. Suppose we have only two numbers in our set, then the average would be exactly midway between the two of them. For example, the average of 10 and 20 is 15. If we have more than two numbers in the set, we would add up all the numbers and divide that result by the number of numbers in the set to get the average. Suppose we have the following list of 5 numbers:

$$(3, -7, 0, 7, -3)$$

then the average would be:

$$\frac{3 + (-7) + 0 + 7 + (-3)}{5} = 0$$

If we use general notation and represent our list by shorthand using letters, i.e., a_i, $i = 1$ to 5, we could represent the average as:

$$\frac{\sum_{i=1}^{n}(a_i)}{n}$$

RMS Mean

RMS, or *root mean square*, is another method of finding the average of a set of numbers. We use this method when we may have negative numbers in the list and we want to consider the 'largeness' of each number and not its absolute value, i.e., remove the effect of some values being negative. So, for example, if we were sampling the change in air pressure at a point by which a sound wave was moving, and our zero reference was the air pressure when there was no sound, we would end up with a set of pressure values that are both negative and positive. To find the average of our set of numbers in this case, we would use the RMS mean.

Suppose we did such an experiment and got the following set of numbers:

$$(3, -7, 0, 7, -3)$$

To find the RMS average, we would first square all the numbers in the set. Remember, to square a number simply means to multiply it by itself. Remember, too, that when multiplying two negative numbers, the result is always positive. Thus, we get:

$$(3^2, -7^2, 0^2, 7^2, -3^2) = (9, 49, 0, 49, 9)$$

Next, we find the regular arithmetic mean of the set, thus:

$$\frac{9 + 49 + 0 + 49 + 9}{5} = 23.2$$

Finally, we find the square root of the mean:

$$\sqrt{23.2} = 4.8$$

We can see that the RMS average of this set of numbers is quite different from the straight arithmetic mean, which is 0.

If we use general notation to express the RMS mean of a set of numbers, we get the following expression:

$$Mean_{RMS} = \sqrt{\frac{\sum_{i=1}^{n}(a_i^2)}{n}}$$

You will see the RMS mean used to find the average of many physical quantities which can have both positive and negative values over time, such as signals with alternating voltage and current.

Pythagoras' Theorem

Suppose we have a right-angled triangle, *ABC*, like the one shown below.

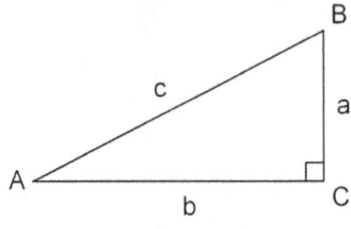

A Right-Angled Triangle

The lengths of the sides are shown as *a*, *b*, and *c*. The angle at *C* is 90°. Pythagoras' Theorem tells us the relationship between the sides of the triangle. It is:

$$a^2 + b^2 = c^2$$

This allows us to find the length of any side of the triangle if we know the length of the other two. For example, suppose we know the length of sides *a* and *c* to be 2 m and 4 m, respectively, and we want to find the length of side *b*. We can write this down as:

$$2^2 + b^2 = 4^2$$

We can rearrange this as:

$$b^2 = 4^2 - 2^2$$
$$= 16 - 4$$
$$= 12$$

Thus:

$$b = \sqrt{12} = 3.46 \text{ m}$$

We used this theorem to find the distance between microphones when discussing the 3:1 rule in Section 14.4.2.

Acronyms and Abbreviations

Collected here for easy reference are all the abbreviations used throughout the book.

AAF *Anti-aliasing filter:* Low-pass filter at the input of an ADC. It is used to attenuate signals above the Nyquist frequency in order to prevent aliasing.

AC *Alternating current:* Current characterized by the periodic reversal of direction of electron flow in the conductor.

ADC *Analog to digital converter:* Device that converts an analog signal to digital format.

AES *Audio Engineering Society:* International organization of professional audio engineers, scientists, students, and artists heavily involved in the promotion of audio technology and creation and maintenance of international audio standards.

AFL *After-fader listen:* Monitoring mode on mixer where signal is taken after the channel fader.

AIF *Anti-imaging filter:* Low-pass filter at the output of a DAC and used to attenuate signals above the Nyquist frequency to eliminate jaggedness of the output signal.

ALCONS *Articulation loss of consonants:* A machine measurement of predicted speech intelligibility based on the reverberation characteristics of the acoustical space.

ANSI *American National Standards Institute:* Private organization in the United States that creates and maintains wide ranging guidelines and standards, and that accredits other programs that assess conformance to international standards.

ASA *American Standards Association:* Standards organization in the USA that later became ANSI.

AVB *Audio Video Bridging:* A set of protocols developed by the IEEE that allow low latency time-synchronized audio and video to be transported across an Ethernet network. These standard protocols include IEEE 802.1AS, 802.1BA, 802.1Qat, and 802.1Qav.

AWG *American wire gauge:* Old standard used primarily in North America for measuring the diameter of nonferrous (e.g., copper, aluminum) electrical wire.

BJT *Bipolar junction transistor:* Three-terminal (base, collector, emitter) semiconductor device in which a small current flowing between the base terminal and one other terminal controls the amount of current flowing between the collector and emitter.

CD *Compact disc:* Digital storage medium utilizing and optical disc. A standard size CD is 12 cm in diameter and can store 80 minutes of audio, while a mini-CD ranges in size from 6 to 8 cm.

CFR *Code of Federal Regulations:* The codification of the general and permanent rules published in the Federal Register by the executive departments and agencies of the US Federal Government.

CMRR *Common-mode rejection ratio:* In a differential amplifier, measures the ability of the amplifier of reject a signal that is present at both inputs simultaneously.

COTS *Commercial off the shelf:* Refers to products that are commercially available and that can be used as-is to easily integrate with other products to create a functional system.

CPU *Central processing unit:* In a computer system, this is the main electronic unit that executes the program instructions and allows the computer to perform useful functions.

CRC *Cyclic redundancy check:* A mathematically derived code transmitted with data and used by the receiver to detect erroneous changes in the data after it was transmitted.

DAC *Digital to analog converter:* Device that converts a digital signal to analog form.

DC *Direct current:* Current flow characterized by one-way, non-reversing flow of electrons.

DCA *Digitally controlled amplifier:* Digital version of VCA found on digital mixers. The DCA's gain is controlled by a digital signal.

DHCP *Dynamic host configuration protocol:* Protocol used by device connecting to an IP-based network to obtain its IP address (and other important addresses and information) from a server in order to allow communications with other network devices.

DI *Directivity index:* A measure of how much a loudspeaker concentrates its output into a beam as compared to an omnidirectional radiator.
Direct input or *direct injection:* A method of transforming speaker and line-level signals into microphone-level signals before sending to mixer input.

DIN *Deutsches Institut für Normung:* German Institute for Standardization.

DIR *Digital interface receiver:* A receiver on a digital processor that acts as an input device for already digitized signals before passing them on to the internal processor.

DIT *Digital interface transmitter:* A unit in a digital processor that takes digital signals from the internal processor and transmits them in digital form outside the processor.

DSCP *Differentiated services code point:* A field in the IP header of transmitted data, used to indicate the quality of service level (e.g., best effort, expedited, etc.) that the sender wants to be applied to the data.

DSP *Digital signal processor:* Any device that manipulates digital signals.

EIA *Electronic Industries Alliance:* Now disbanded alliance of trade associations for electronics manufacturers in the United States, formerly involved in the creation and maintenance of several standards in the manufacture, testing, and deployment of electronic equipment. EIA standards are now managed by the Electronics Components Association (ECA).

EIN *Equivalent input noise:* A method of calculating the input noise of a microphone preamp by setting its gain very high and then figuring out what the noise would be at a lower gain setting.

EMF *Electromotive force:* A loosely used term describing the potential difference, measured in volts, across the terminals of a conductor that is acted upon by a varying magnetic field, or across the terminals of a battery.

EMI *Electromagnetic interference:* Unwanted current induced in a conductor via inductive coupling with a nearby conductor because a varying current in the nearby conductor creates a changing magnetic field.

EQ *Equalizer:* Device that shapes the frequency response of an audio system.

ESI *Electrostatic interference:* Unwanted current induced in a conductor via capacitive coupling with a nearby conductor because a varying current in the nearby conductor creates a changing electric field.

FCS *Frame check sequence:* A mathematically derived code transmitted with a layer 2 data frame and used by the receiver to detect erroneous changes in the data after it was transmitted.

FET *Field-effect transistor:* Three-terminal (gate, source, drain) semiconductor device in which a voltage applied to the gate terminal creates an electric field that controls the amount of current flowing between the source and drain. A MOSFET is a type of FET.

FFT *Fast Fourier transform:* An efficient mathematical method of decomposing a complex signal into its individual frequency components.

FM *Frequency modulation:* A method of wireless transmission where information is encoded by modulating the frequency of the carrier.

FOH   *Front of house:* Designation for the main audience area, and the mixer, amplifiers, and loudspeakers that service that area.

FQTTS   *Forwarding and Queuing for Time Sensitive Streams:* Protocol that is part of the AVB suite and used within network switches to give priority to registered data streams to reduce latency and jitter of the streams.

FTC   *Federal Trade Commission:* An agency of the United States government primarily focused on consumer protection.

GFCI   *Ground fault circuit interrupter:* Device that disconnects the main AC flow when input and output currents are different by more than 6 mA.

HPF   *High-pass filter:* Filter that passes signals above its cutoff frequency and attenuates signals below the cutoff frequency.

HVAC   *Heating, ventilation, and air conditioning:* Equipment and interconnected systems that provide the heating, cooling, humidity control, and fresh air circulation of an enclosed space.

IC   *Integrated circuit:* Solid state device consisting of many components (mostly transistors and resistors) etched from a single piece of semiconductor material.

IEC   *International Electrotechnical Commission:* Organization that creates and maintains international standards relating to electrical and electronic technologies.

IEEE   *Institute of Electrical and Electronics Engineers:* A professional association that advances technical standards in the electrical, electronic engineering, telecommunications, and computer engineering fields.

IEM   *In-ear monitor:* Small earbud earphones that fit in the ear canal of the wearer and used to provide monitor or foldback signal to performers.

IGMP   *Internet group management protocol:* Protocol used at layer 3 of the Internet model to establish multicast group membership. Some switches can snoop the IGMP messages to determine if there are any attached devices belonging to a particular group.

IHF   *Institute of High Fidelity:* Defunct organization of North American Hi-Fi manufacturers.

IMD   *Intermodulation distortion:* Distortion of the input signal by a processor when two frequency components within the input signal interact with each other to produce other frequencies that were not present in the input signal, and whose frequencies are not harmonically related to either original signal component.

I/O   *Input/Output:* Term used to refer to ports allowing signals to ingress or egress a device.

IP   *Internet protocol:* Protocol used at layer 3 of the Internet model for communicating devices and which determines the path through the connecting network.

ISO   *International Organization for Standardization:* Multinational body responsible for developing standards that promote exchange of technology, goods, and services.

ITU-R   *International Telecommunication Union – Radiocommunication Sector:* International organization that coordinates telecommunication standards. The Radio Communication unit deals with radio-frequency spectrum management.

LAG   *Link aggregation group:* A grouping of multiple physical Ethernet links between two devices so that they can be used as a single logical link with the combined bandwidth of the separate physical links.

LAN   *Local area network:* A digital communications network that spans a small geographic area and is under the administrative control of a single entity.

LFO   *Low-frequency oscillator:* Electronic device used to create a simple low-frequency (typically below 50 Hz) signal (typically sine, triangle, saw-tooth) used to modulate the main audio signal in an effects processor.

LPF   *Low-pass filter:* Filter that passes signals below its cutoff frequency and attenuates signals above the cutoff frequency.

LSB   *Least significant bit:* In a binary representation of a number, this is the bit that gives the units value, carrying the lowest weight, and is usually the rightmost digit when the number is written left to right.

MAC   *Media access control:* A function at the data link layer of communicating entities that controls when the device takes control of the communications medium to transmit its data.

MAN   *Metropolitan area network:* A digital communications network that spans city-wide geographic area.

MOSFET   *Metal oxide semiconductor field effect transistor:* A rugged transistor device used in the output stages of high current amplifiers.

MS   *Middle-side:* Stereo miking technique using a figure-8 microphone with its axis at 90° to the sound source (side signal) and a cardioid microphone pointing at the source (middle signal).

MTU   *Maximum transmission unit:* The size (in bytes) of the biggest PDU that can be accepted by a lower layer for transmission.

NOS   *Nederlandse Omroep Stichting:* (With reference to miking) Stereo miking technique using two cardioid microphones angled 90° and spaced 30 cm apart.

OCA   *Open Control Architecture:* A protocol architecture describing the framework for the control, monitoring, and connection management of devices in a media network.

ORTF   *Office de Radiodiffusion-Télévision Française:* (With reference to miking) Stereo miking technique using two cardioid microphones angled 110° and spaced 17 cm apart.

OSHA   *Occupational Safety and Health Administration:* Arm of the US Department of Labor that sets and enforces standards to assure workforce safety.

OSI   *Open Systems Interconnection:* A standard model of communicating systems that abstracts the communication functions as a set of 7 layers, where a given layer provides a communication service to the layer above it without exposing the internal implementation of the function.

PAM   *Pulse-amplitude modulation:* A form of modulation used in data transmission where the encoded information is determined by the amplitude (level) of a series of pulses.

PC   *Personal computer:* A small, low-cost, self-contained computer typically used for personal and individual computing needs.

PCM   *Pulse-code modulation:* A process of converting an analog signal to digital form by first sampling the amplitude of the signal, quantizing the measured amplitude against a fixed set of values, and finally encoding the set of measure values in binary form.

PDU   *Protocol data unit:* A single unit of information that is passed from an entity in one layer to an entity in the next adjacent layer and carries the information that the entity wants to send to, or received from, its remote peer.

PFL   *Pre-fader listen:* Monitoring mode on mixer where signal is taken before the channel fader.

PTP   *Precision time protocol:* Protocol used to synchronize clocks of participating network nodes within sub-microsecond accuracy.

PWM   *Pulse-width modulation:* Process in which the width of the pulses of a rectangular pulse wave is changed (i.e., modulated) by some controlling signal (the modulated wave is then used to control the power delivered to a load).

Q   *Directivity factor:* A measure of how much a loudspeaker concentrates its output into a beam as compared to an omnidirectional radiator.
Quality factor: The center frequency of a band-pass filter divided by its bandwidth.

QoS   *Quality of service:* A description of the level of service applied to a data stream in terms of its priority treatment within a network device, directly affecting such things as packet delay, jitter, and guaranteed delivery of the data.

RMS   *Root-mean-square:* The square root of the average of the squares of a set of numbers.

RSTP *Rapid spanning tree protocol:* Improved version of the spanning tree protocol that detects loops in a switched network and deactivates switch ports to ensure that the active data paths between all connected endpoints are loop-free.

RTCP *RTP control protocol:* Protocol used to provide statistics about an RTP session. Such statistics may be used to infer some QoS information about the session by a control application.

RTP *Real-time transport protocol:* Protocol used to transmit audio and video data streams over an IP network.

SFP *Small form-factor pluggable:* A small, hot-swappable optical transceiver used to allow fiber optic cable to interconnect communicating network devices.

SIL *Sound intensity level:* A logarithmic measure of the strength of sound where the reference intensity is taken to be 10^{-12} W/m^2 (in air).

SMPTE *Society of Motion Picture and Television Engineers:* Organization of engineers, technicians, and other disciplines that creates and maintains standards, recommended practices, and guidelines for the motion imaging industry.

SNMP *Simple network management protocol:* Protocol used to collect and transport information about managed devices in an IP network to a control application, and to allow altering certain parameters of these devices by the control application.

SNR *Signal to noise ratio:* Ratio of pure signal power to the noise power for a given signal. You will sometimes see this written as S/R.

SOP *Standard operating procedure:* Set of guidelines or rules for executing certain procedures as established by an entity.

SPB *Shortest path bridging:* Ethernet protocol that allows load sharing of data over multiple active paths which would otherwise cause the network to be wired in a loop.

SPL *Sound pressure level:* A logarithmic measure of the strength of sound using 20 μPa as the 0 dB reference (in air).

SRP *Stream reservation protocol:* Protocol that is part of the AVB suite and used to register a data stream and reserve resources on the network switches along the entire path from talker to listener.

STI *Speech transmission index:* An indirect measure of the quality of an audio transmission channel as it relates to the intelligibility of speech.

TCP *Transmission control protocol:* Protocol used at layer 4 of the Internet model for communicating devices and allows for reliable delivery of data to a remote destination.

THD *Total harmonic distortion:* A measurement of the distortion created by a device when it introduces harmonics (i.e., not originally present) to the input signal.

TRS *Tip-ring-sleeve:* Designation of an audio plug where the shaft is divided into three conductive bands: the tip, a ring section, and the sleeve (ground).

TS *Tip-sleeve:* Designation of an audio plug where the shaft is divided into two conductive bands: the tip, and the sleeve (ground).

TSN *Time Sensitive Networking:* A newer name for Audio Video Bridging network. See the entry for AVB.

TTL *Time to live:* A device-to-device hop limit applied to network data to ensure that such data is not endlessly forwarded around a network. When the TTL expires, the data is discarded.

UDP *User datagram protocol:* Protocol used at layer 4 of the Internet model for communicating devices. Delivery service is unreliable, but it is used when fast, low-overhead, delivery service is needed.

UHF *Ultra high frequency:* Radio frequency range from 300 MHz to 3 GHz.

USB *Universal serial bus:* A widely used specification that defines a high-speed serial communication method between devices and host controller (e.g., a computer).

VCA   *Voltage controlled amplifier:* An amplifier whose gain is determined by a control signal of varying voltage.

VHF   *Very high frequency:* Radio frequency range from 30 MHz to 300 MHz.

VLAN   *Virtual LAN:* A subdivision of a LAN that is partitioned and isolated at the Ethernet layer based on an ID that is assigned for all devices in that subdivision. Devices in different VLANs cannot directly communicate with each other.

WAN   *Wide area network:* A digital communications network that has continental or international geographic span and whose administrative control and ownership is typically shared my multiple entities.

WAP   *Wireless access point:* A device that is used to allow a computer or tablet to connect wirelessly to a (usually wired) network, typically using Wi-Fi standards.

Wi-Fi   Wireless technology that realizes the implementation of a wireless LAN based on IEEE 802.11 standard.

WPA   *Wi-Fi protected access:* A set of protocols that are used for authentication of users and to encrypt their communications on wireless Wi-Fi networks.

References

Chapter 1 – Fundamentals

1. Howard, D.; Angus, J. *Acoustics and Psychoacoustics*, Fourth Edition; Focal Press, 2009. See discussion in Chapter 1, "Introduction to Sound", pg. 25.
2. Everest, F.A. *Master Handbook of Acoustics*, Fourth Edition; McGraw-Hill, 2001. Chapter 2, "Sound Levels and the Decibel", pg. 39.

Chapter 2 – The Environment

1. Cramer, O. "The Variation of the Specific Heat Ratio and Speed of Sound in Air with Temperature, Pressure, Humidity and CO_2 Concentration", *Journal of the Acoustical Society of America*, 1993(5), pp 2510 – 2616.
2. Bohn, D. "Environmental Effects on the Speed of Sound", *Journal of the Audio Engineering Society*, Apr 1988, Vol. 36, No. 4.

Chapter 3 – Input Devices

1. Eargle, J. *The Microphone Book*, Second Edition; Focal Press, 2004. Chapter 4, "The Pressure Gradient Microphone". pg. 54.
2. Eargle, J. *The Microphone Book*, Second Edition; Focal Press, 2004. Chapter 5, "First-Order Directional Microphones". pg. 84.
3. Boré, G.; Peus, S. "Microphones for Studio and Home Recording Applications", pg. 33, http://www.neumann.com/download.php?download=docu0002.PDF (accessed Nov 2013).
4. Horn, D. *Basic Electronics Theory*, Fourth Edition; TAB Books, 1994. Chapter 10, "Impedance and resonance circuits", pg. 140.
5. *Consultation on the Technical, Policy and Licensing Framework for Wireless Microphones*, SMSE-019-17; Innovation, Science and Economic Development Canada, Nov 2017.

Chapter 4 – Output Devices

1. Mitchell, J. "Loudspeakers". In *Handbook for Sound Engineers*; Fourth Edition, Ballou, G., Ed.; Focal Press, 2008; pg. 600.
2. Elliott, R. "Benefits of Bi-Amping (Not Quite Magic, But Close) – Part 1", http://sound.westhost.com/bi-amp.htm (accessed May 2012).
3. Meyer, P. "DSP Beam Steering with Modern Line Arrays", Meyer Sound Technical Report, Dec 2002.
4. Stage Accompany "TechDoc: SA AH36", http://www.stageaccompany.no/wp-content/support/ah36_techdoc.pdf (accessed May 2019).

Chapter 5 – Amplifiers

1. "The Class-I Amplifier", http://www.crownaudio.com/media/pdf/amps/137234.pdf (accessed 6 Jun 2016).

2. Tripathi, A.; Delano, C. U.S. Patent No. 5,777,512, Method and apparatus for oversampled, noise-shaping, mixed-signal processing, issued Jul 7, 1998.
3. "Introduction to the Class TD Technology", Lab.gruppen Technical Note, Number 9, Jan 2002.
4. CEA-490-A, Standard Test Methods of Measurement for Audio Amplifiers, Dec 2004.
5. IEC 60268-3:2001, International Standard, Sound System Equipment – Part 3: Amplifiers, 3rd edition.

Chapter 6 – Interconnecting Devices
1. Whitlock, B.; Floru, F. "New Balanced-Input Integrated Circuit Achieves Very High Dynamic Range In Real-World Systems", *AES 117th Convention Paper 6261*, Oct 2004.

Chapter 8 – Audio Effects Processors
1. Howard, D.; Angus, J. *Acoustics and Psychoacoustics*, Fourth Edition; Focal Press, 2009, pp 437 – 446.
2. Smith, S.W. "Recursive Filters" in *The Scientist and Engineer's Guide to Digital Signal Processing,* http://www.dspguide.com/ch19/4.htm, online book, (accessed Sep 2011).
3. Smith, J. O. "Allpass Filter Sections" in *Introduction to Digital Filters with Audio Applications,* http://ccrma.stanford.edu/~jos/filters/Allpass_Filter_Sections.html, online book, (accessed Sep 2011).
4. Hidaka, T. et al, "A New Definition of Boundary Point Between Early Reflections and Late Reverberation in Room Impulse Responses", *Journal of the Acoustical Society of America,* Jul 2007, Vol. 122, No. 1, pp 326 – 332.

Chapter 9 – Analog Mixers
1. Buick, P. *Live Sound: PA for performing musicians*; PC Publishing, 1996. Chapter 5, "The mixer". pg. 77.

Chapter 11 – Digital Data Networks
1. IEEE 802.3-2002, IEEE Standard for IT - Telecommunications and Information Exchange Between Systems – LAN/MAN - Specific Requirements – Part 3: CSMA/CD Access Method and Physical Layer Specifications.
2. IEEE 802.11-2016, IEEE Standard for Information technology – Telecommunications and information exchange between systems Local and metropolitan area networks – Specific requirements – Part 11: Wireless LAN Medium Access Control (MAC) and Physical Layer (PHY) Specifications.
3. ISO/IEC 7498-1:1994, Information technology – Open Systems Interconnection – Basic Reference Model: The Basic Model – Part 1, Second Edition, 1994.
4. FitzGerald, J.; Dennis, A.; Durcikova, A. *Business Data Communications and Networking*, Twelfth Edition; Wiley, 2015. Chapter 1, "Introduction to Data Communications", pg. 9.
5. Schulzrinne, H. et al, *RTP: A Transport Protocol for Real-Time Applications*, RFC 3550, Jul 2003.

Chapter 12 – Digital Audio Distribution Networks
1. Schulzrinne, H., et al, *RTP: A Transport Protocol for Real-Time Applications*, RFC 3550, July 2003. Section 6.2, "RTCP Transmission Interval".
2. Lester, M.; Boley, J. "The Effects of Latency on Live Sound Monitoring", *AES 123rd Convention Paper*, Oct 2007.

3. Dreher, A.; Mohl, D. "Precision Clock Synchronization: The Standard IEEE 1588", Rev 1.2. Hirschmann Automation and Control, White Paper, retrieved Aug 2018.
4. Audinate Pty Ltd., "Networks and Switches", https://www.audinate.com/resources/networks-switches (accessed Nov 2017).
5. IEEE 1588-2008, IEEE Standard for a Precision Clock Synchronization Protocol for Networked Measurement and Control Systems.
6. IEEE 802.1AS, Standard for Local and Metropolitan Area Networks - Timing and Synchronization for Time-Sensitive Applications in Bridged Local Area Networks.
7. Case, J., et al., *A simple Network Management Protocol*, RFC 1157 https://tools.ietf.org/html/rfc1157, May 1990.
8. IEC 62439-3:2016, Industrial communication networks – High availability automation networks – Part 3: Parallel Redundancy Protocol (PRP) and High-availability Seamless Redundancy (HSR).
9. IEEE 802.1D-2004, IEEE Standard for Local and metropolitan area networks: Media Access Control (MAC) Bridges.
10. IEEE 802.1aq-2012, IEEE Standard for Local and metropolitan area networks – Media Access Control (MAC) Bridges and Virtual Bridged Local Area Networks – Amendment 20: Shortest Path Bridging.
11. IEEE 802.1AX-2014, IEEE Standard for Local and metropolitan area networks – Link Aggregation.
12. AES67-2018, AES standard for audio applications of networks - High-performance streaming audio-over-IP interoperability.
13. Mann, E., et al, *AVB Software Interfaces and Endpoint Architecture Guidelines – AVnu Alliance Best Practices*, Revision 1.0, Dec 2013.
14. Pearson, L. *Stream Reservation Protocol – AVnu Alliance Best Practices*, Revision 1.0, Nov 2014.
15. Perales, V., et al, "Milan Whitepaper", https://avnu.org/milan-whitepaper/, (accessed Sept 2019).
16. The Death of Analogue and the Rise of Audio Networking, RH Consulting, 2015, http://www.audinate.com/rise-of-audio-networking (accessed Apr 2018).
17. So you're adding Dante to your network? Here is all you need to know, https://www.audinate.com/sites/default/files/PDF/adding-dante-to-your-network-audinate_0.pdf (accessed Apr 2018).
18. ST 2110-30:2017 - SMPTE Standard - Professional Media Over Managed IP Networks: PCM Digital Audio.
19. Berryman, J. "The Open Control Architecture", *Journal of the Audio Engineering Society*, Apr 2013, Vol. 61, No. 4.
20. AES70-1-2018, AES standard for audio applications of networks - Open Control Architecture - Part 1: Framework.
21. AES70-2-2018, AES standard for audio applications of Networks - Open Control Architecture - Part 2: Class structure.
22. AES70-3-2018, AES standard for audio applications of Networks - Open Control Architecture - Part 3: Protocol for TCP/IP Networks.

Chapter 13 – Designing the Sound System

1. Jones, R. "Speech Intelligibility Papers", http://www.meyersound.com/support/papers/speech/ (accessed Jul 2012).
2. "Behavior of Sound System Indoors" Chapter 6, in *Sound System Design Reference Manual*, JBL Professional, 1999.
3. Mijic, M.; Masovic, D. "Reverberation Radius in Real Rooms", *Telfor Journal*, 2010, Vol. 2, No. 2, pp 86 – 91.

4. Mason, B. et al, "Horizontal Directivity of Low- and High-Frequency Energy in Speech and Singing", *Journal of the Acoustical Society of America,* Jul 2012, Vol. 132, No. 1, pp 433 – 441.
5. Peutz, V. M. A. "Articulation Loss of Consonants as a Criterion for Speech Transmission in a Room", *Journal of the Audio Engineering Society*, Dec 1971, Vol. 19, pp 915 – 919.
6. Faiget, L.; Ruiz, R. "Speech Intelligibility Model Including Room and Loudspeaker Influences", *Journal of the Acoustical Society of America,* Jun 1999, Vol. 105, No. 6, pp 3345 – 441.
7. Mathews, P. "Unwinding Distribution Transformers", *RaneNote 159,* Rane Corporation 2005.
8. Dalziel, C. "Deleterious Effects of Electric Shock", as cited in *Safety Basics, Handbook for Electrical Safety,* Cooper Bussman, 2001, www.bussmann.com/library/docs/Safety_Basics_Book.pdf (accessed Feb 2010).
9. Kouwenhoven, W. "Human Safety and Electric Shock", *Electrical Safety Practices, Monograph 112,* Instrument Society of America, 1968.
10. US Department of Labor, OSHA Regulation 29 CFR Subpart S 1910.303, https://www.osha.gov/pls/oshaweb/owadisp.show_document?p_table=STANDARDS&p_id=9880 (accessed Mar 2013).

Chapter 14 – Microphone Usage

1. Eargle, J. *The Microphone Book*, Second Edition; Focal Press, 2004. See discussion in Chapter 18, pg. 307.
2. Everest, F.A. *Master Handbook of Acoustics*, Fourth Edition; McGraw-Hill, 2001. See discussion in Chapter 17, "Comb-Filter Effects", pg. 379.
3. Simonton, J. "How MS Stereo Works", http://www.paia.com/ProdArticles/msmicwrk.htm (accessed Jul 2013).

Index

relationship to decibel, 12
Logarithmic scale, vs. linear, 166
Longitudinal wave, 7
Loudspeaker
  active system, 75
  area coverage, 273–83
  array, 73, 280
  beamwidth, 78
  cables for connecting, 119–21
  choosing, 84
  design, 58
  directivity index, 79
  dispersion, 73, 76, 273
  distributed system, 282–83
  enclosure types, 70–74
  flying, 302
  frequency response, 66, 69, 70–74, 80
  front of house, 84, 273
  impedance, 69, 76
  management systems, 283
  monitor. See Stage monitor
  overload protection, 75
  power handling tests, 82
  power specification, 83
  resonance, 72
  sensitivity, 80, 294
  stacking, 302
  three-way system, 59
  transient response, 71
  two-way system, 59
Low frequency oscillator (LFO), 162, 166, 168, 169, 357
Low-pass filter. See Filter, low-pass

M

MAC address. See Media access control
Magnetic flux, 37, 59
MAN. See Metropolitan area network
Media access control
  boadcast address, 228
  defined, 221, 358
  host address, 217, 221
  multicast address, 229
Metropolitan area network, 218, 358
Microphone
  capsule, 37, 40
  choosing a, 51, 55
  circuit, 39
  condenser, 39–41, 51
  design, 36
  dual diaphragm, 48
  dynamic, 37, 51
  earset, 318, 320
  feedback control, 288
  handheld usage, 315–16
  handling noise, 42
  headset, 318
  high impedance, 108
  lapel, 52, 317
  lavaliere, 52, 317
  maximum SPL, 51
  output impedance, 50
  pickup patterns, 44–48
  pressure zone (PZM), 311
  ribbon, 48, 51
  self-noise, 42
  sensitivity, 51
  splitter, 202–4
  wireless, 52–55, 155
Microphone level, 15, 86
Middle-side (MS) stereo, 326–29, 358
Miking
  3-to-1 rule, 312
  area miking, 323–24
  of choir, 323
  of drum set, 320–22
  of ensemble, 323
  of musical instruments, 318–20
  of speaker cabinet, 320
  of vocals, 316–18
  opposed, 182
  podium, 310
  source to microphone distance, effects of, 309–10
  stereo techniques, 324–29
Milan protocol, 259
Mix matrix, 190, 200
Mixer, 2
  aux return, 190, 194
  aux send, 178, 180, 183, 187, 194
  bus, 180
  channel fader, 184, 185, 210
  channel insert, 180, 181
  choosing a, 206
  functions of, 177
  impedance matching, 50
  input impedance, 50
  input signal flow, 179
  insert point, 193
  layout, 177
  main mix, 192
  metering, 192
  monitoring, 181, 183, 191
  mute groups, 188
  network address configuration, 225
  output signal flow, 180
  pan, 180, 183
  powered, 177
  pre-fader listen, 180, 184, 191, 199, 202, 358
  routing, 180, 183
  solo, 184, 191, 199, 202
  subgroup, 178, 180, 181, 184
  VCA groups, 186–89, 209
Mixing
  bass and drums, 342

Psychoacoustic effect, 19
Psychoacoustic interface, 2
PTP. *See* Precision time protocol
Pulse-amplitude modulation, 231, 358
Pulse-code modulation, 132, 358
Pulse-width modulation, 90
Pumping (compressor effect), 150
Pythagoras' theorem, 354

Q

Q. *See* Filter, Q
QoS. *See* Quality of service
Quadrature phase, 163, 167
Quality Factor. *See* Filter, Q
Quality of service, 227, 239, 247, 258, 260, 358
Quantization, 131
Quantization noise, 131, 136

R

Rapid spanning tree protocol, 254, 359
Rarefaction of air. *See* Sound wave creation
Reactance, 50
Real-time transport protocol, 242, 245, 252, 359
Reconstruction filter. *See* anti-imaging filter
Rectangular coordinate system, 6
Reflection of sound. *See* Sound, reflection
Relative phase. *See* Sound wave, Relative phase
Resistance, to current flow, 49
Resonance, of room, 287
Reverb. *See* Reverberation effect
Reverberant sound field, 268
Reverberation effect
    damping, 173
    decay time, 173, 175
    design, 173
    gated reverb, 176
    pre-delay, 172
    reverse reverb, 176
    use in mix, 176
Reverberation time. *See* Reverberation, decay time
Reverberation, of enclosed space
    decay time (RT60), 31, 172, 270
    direct sound, 31
    early reflections, 31, 170
    effect on intelligibility, 266
    late reflections, 31, 170
Reverse stereo sross stage mix. *See* Side fill monitor
Ribbon driver. *See* Driver, ribbon
Ringing out, 333–34
RJ45 plug. *See* 8P8C modular connector
RMS. *See* Root mean square
Robinson-Dadson curves, 20
Roll-off. *See* Filter, roll-off
Root mean square (RMS)
    defined, 81, 358
    explained, 353

power, 83, 94
    signal level, 81, 93, 97, 148, 150
Router. *See* Network device, router
RT60. *See* Reverberation, decay time
RTP. *See* Real-time transport protocol

S

S/N ratio. *See* Signal-to-noise ratio
Sabin, 32
Sampling. *See* Digital signal, sampling
Sampling rate. *See* Digital audio processor, sampling rate
SDP. *See* Session description protocol
Sealed box speaker, 71
Servo-balanced output. *See* Balanced output stage
Session description protocol, 242
SFP. *See* Small form-factor pluggable
Shadow zone, 34
Sibilance, 313
Side fill monitor, 291
Sidechaining. *See* Compressor, sidechain
Signal dropout, wireless systems, 53
Signal levels
    defined, 14
Signal processor, 2
Signal splitter, 202–4
Signal-to-noise ratio, 54, 101, 136, 196, 202, 359
SIL. *See* Sound Intensity Level
Simple network management protocol, 252, 359
Simplex transmission, 219
Sine wave, 23, 158, 168
Small form-factor pluggable, 219, 359
Snake, 55
SNMP. *See* Simple network management protocol
SNR. *See* Signal-to-noise ratio
Solo in place, 199
    See also Mixer, solo.
Sound
    absorption of, 32, 174
    absorption variance with temperature and humidity, 34
    definition, 3
    diffraction of, 34
    diffusuion of, 31
    energy transfer, 3, 8
    intensity increase with reflective surface, 9
    intensity of, 7–9
    loudness of, 19
    reflection of, 30
    sound power, 8, 294
Sound check, 33, 198, 212, 335–37
Sound intensity level (SIL), 16–17, 359
Sound power. *See* Sound, sound power
Sound pressure level (SPL), 61, 70, 81, 84, 359
    change with distance from source, 268
    definition, 17
    measurement, meter, 17, 269